Topics in Nanoscience

Part I
Basic Views, Complex Nanosystems: Typical Results and Future

Series on the Foundations of Natural Science and Technology

ISSN: 2010-1961

Published:

Vol. 15 *Topics in Nanoscience*
In 2 Parts
Part I: Basic Views, Complex Nanosystems: Typical Results and Future
Part II: Quantized Structures, Nanoelectronics, Thin Films
edited by W. Schommers

Vol. 14 *Elastic Constants in Heavily Doped Low Dimensional Materials*
by K. P. Ghatak and M. Mitra

Vol. 13 *Scalar Field Cosmology*
by S. Chervon, I. Fomin, V. Yurov and A. Yurov

Vol. 12 *The Origin of Natural Order: An Axiomatic Theory of Biology*
by Q. Zhao

Vol. 9 *The Timeless Approach: Frontier Perspectives in 21st Century Physics*
by D. Fiscaletti

Vol. 8 *Quantum Effects, Heavy Doping, and the Effective Mass*
by K. P. Ghatak

Vol. 7 *Magneto Thermoelectric Power in Heavily Doped Quantized Structures*
by K. P. Ghatak

Vol. 6 *Nano-Engineering in Science and Technology: An Introduction to the World of Nano-Design*
by M. Rieth

*For further details, please visit: http://www.worldscientific.com/series/sfnst

(Continued at end of book)

Series on the Foundations of Natural Science and Technology — Vol. 15

Topics in Nanoscience

Part I

Basic Views, Complex Nanosystems: Typical Results and Future

Editor

Wolfram Schommers

Universty of Texas at Arlington, USA &
Karlsruher Institut für Technologie (KIT), Germany

World Scientific

NEW JERSEY • LONDON • SINGAPORE • BEIJING • SHANGHAI • HONG KONG • TAIPEI • CHENNAI • TOKYO

Published by

World Scientific Publishing Co. Pte. Ltd.
5 Toh Tuck Link, Singapore 596224
USA office: 27 Warren Street, Suite 401-402, Hackensack, NJ 07601
UK office: 57 Shelton Street, Covent Garden, London WC2H 9HE

British Library Cataloguing-in-Publication Data
A catalogue record for this book is available from the British Library.

Series on the Foundations of Natural Science and Technology — Vol. 15
TOPICS IN NANOSCIENCE
In 2 Parts
Part I: Basic Views, Complex Nanosystems: Typical Results and Future
Part II: Quantized Structures, Nanoelectronics, Thin Films

ISBN 978-981-124-236-6 (Set_hardcover)

ISBN 978-981-124-267-0 (Part I_hardcover)
ISBN 978-981-124-268-7 (Part I_ebook for institutions)
ISBN 978-981-124-387-5 (Part I_ebook for individuals)

ISBN 978-981-124-269-4 (Part II_hardcover)
ISBN 978-981-124-270-0 (Part II_ebook for institutions)
ISBN 978-981-124-388-2 (Part II_ebook for individuals)

For any available supplementary material, please visit
https://www.worldscientific.com/worldscibooks/10.1142/12419#t=suppl

Desk Editor: Nur Syarfeena Binte Mohd Fauzi

Typeset by Stallion Press
Email: enquiries@stallionpress.com

Printed in Singapore

Preface to Part I

The book, *Topics in Current Nanoscience*, consists of two parts and 16 chapters written by an international group of experts. Both volumes deal with basic questions of nanoscience, and most of the chapters cover the background, typical applications and future developments

The eight chapters of part 1 involve closed blocks and deal with the following topics:

- Nanoassemblies of Porphyrin Derivatives
- Aspects of Time: On the Realistic Treatment of Self-organization Processes
- Multi-Spinor Nonequilibrium Quantum Transport: Entanglement-Induced Delocalization and Topological Insulators
- Layered Nanodiamond Hybrid Materials
- Heat Transport and Thermal Characterization of Nanoscale Materials and Devices
- Thermoelectric Phenomena at Nanoscale Level
- Application of Green's Functions and Difference Equations in Theoretical Analysis of Nanostructures
- Heisenberg's Uncertainty Principle and the Quantum Capacitance in Quantum Wire Field Effect Transistors

Both volumes of the book deal not only with the basic questions of nanotechnology and nanoscience but also with those topics of physics and chemistry that are essential for future developments in this field. The essential features should also be accessible to advanced students.

Each chapter gives the state of the art, and new aspects are indicated and proposed. This is important because in the relatively new discipline "nanoscience" theoretical approaches and models have to be modified and replaced, respectively. Then, the experimental devices and methods have to be adapted to the new theoretical findings. The book explicitly points to such situations in some chapters.

For the development of systems at the nanolevel, the following step was essential: With the construction of the scanning tunneling microscope, nanoscience became an important discipline. Single atoms could be manipulated in a controlled manner, and it became possible to change matter at its "ultimate" level; it is the level on which the properties of matter emerge.

This possibility enables one to construct and to produce devices, materials, etc., with very small sizes and completely new properties. This opens new perspectives for technology and is in particular relevant in connection with nanoengineering.

Nanosystems are unimaginably small and very fast. No doubt, this is an important characteristic. But there is another feature, possibly more relevant, in connection with nanoscience and nanotechnology. The essential point here is that we work at the "ultimate level". This is the smallest level at which the properties of our world emerge and functional matter can exist. In particular, at this level, biological individuality comes into existence. This situation can be expressed in absolute terms: *This is not only the strongest material ever made, this is the strongest material it will ever be possible to make* (D. Ratner and M. Ratner, *Nanotechnology and Homeland Security*).

This is a very general statement. All aspects of matter at the nanolevel are concerned here. Through the variation of the composition, various forms of matter emerge with different items. The chapters of this book demonstrate this.

Nanosystems are usually small, but they offer nevertheless the possibility to vary the structure of atomic (molecular) ensembles, creating a diversity of new material-specific properties. A large variety of experimental possibilities come into play and flexible theoretical tools are needed at the basic level.

This is reflected in the different disciplines: In nanoscience and nanotechnology, we have various directions: materials science, functional nanomaterials, nanoparticles, food chemistry, medicine with brain research, quantum and molecular computing, bioinformatics, magnetic nanostructures, nanooptics, nanoelectronics, etc.

The properties of matter, which are involved within nanoscience and nanotechnology, are ultimate in character, i.e., their characteristic properties come into existence at the basic level. The book underlines this fact. Clearly, only for a part of possibilities could be considered.

All that will have consequences for the treatment of nanosystems, in particular in connection with the situation at the ultimate level. Concerning this point, let us remark the following: What does "ultimate" mean, just with respect to the theoretical treatment of nanosystems? This is an essential point because the experimental possibilities are dictated by the theoretical frame. Also, the constructive feasibilities of technological devices are delimited through the theoretical background, that is, the prediction of new effects and the construction of novel devices is basically influenced by the theoretical assessment.

At the ultimate level, the systems are exclusively described by the basic theoretical laws and have to be applied consequently. Here, the basic conceptions of theoretical physics are in the centre.

Although we have here a lot of directions and disciplines, respectively, which are partly very different in character, we are working in nanotechnology at the same theoretical footing, in contrast to the traditional technologies. In nanotechnology, we have "one" theory for all these disciplines and phenomena, and this theory is given by the basic "laws of theoretical physics".

Clearly, not all basic theoretical laws and methods are of relevance, but we have to select these laws, which point on typical nanoeffects. These selected tools determine the strategy for predicting experimental effects and also the construction rules for devices.

On the other hand, new theoretical laws and methods may be created on the basis of novel experimental nanoeffects and this would extend the range of experimental possibilities as well as the range of nanoengineering. This situation will be the source for new effects and devices. Various chapters of this book point to this situation, i.e., on the creative interplay between theory and experiment is underlined in some cases.

Microtechnique and macrotechnique belong to the realm of traditional technologies. Here, engineers do not work at the ultimate level. Here, more or less phenomenological approaches are applied which, in general, cannot be deduced from the basic physical laws, that is, each discipline has its own description.

The future nanoengineer will work on the basis of "one" theory, i.e., on the basis of theoretical physics. This means that technology has reached the level of science. The consequences are obvious. In this connection, let us quote two points:

1. Since we are working in nanoscience at the ultimate level, the properties of matter and functional matter are defined in absolute terms. What does "absolute terms" mean? At this level, the principal properties cannot be improved further. We have reached the highest level. It is the level at

which the properties emerge. At this level, biological individuality also comes into existence, that is, life has its origin at this level.

2. Working at the level of theoretical physics also means that any change in the basic physical laws will directly influence nanoscience. No intermediate step is necessary. This can lead to completely new perspectives in connection with applications. Therefore, working on nanoscience and nanotechnology also means development of the basic laws of theoretical physics further if there is a necessity for the same. This could be important just in connection with quantum theory. New experimental results indicate that there is in fact a need for that.

Within the frame of brain research, we must even go a step further because here the "world view" is of particular relevance. The theoretical world view must be in accordance with the world view, which the mind (brain) develops of the world outside. Also, this topic will be discussed in this book, that is, the chapters of this book, just in connection with those cases where we touch medicine and biological systems.

In nanoscience, interacting atoms and molecules are in the centre. However, is such an ensemble of interacting atoms (molecules) in all cases appropriate for solving sophisticated questions? Just in connection with brain functions, a holistic view is probably more appropriate. Then, the ensemble has to be treated as a unified whole. Do future treatments require such a view? This point is also a topic for which nanoscience is fundamental.

Are the present basic theoretical methods sophisticated enough to cover the effects in nanoscience sufficiently? We have to develop them further and this is because nanoscience reveals principal new effects that are not known and considered so far. However, new effects often need novel concepts, not only with respect to the experimental situation but also in connection with theoretical descriptions. The book particularly accentuates this line, for example, in connection with self-organizing processes, which belong to the heart of nanoscience, a "quantum aspect of time" is needed, which is however not developed so far. This point is addressed in the book.

Wolfram Schommers
University of Texas at Arlington, USA
and
Karlsruhe Institute of Technology, Germany

Karlsruhe, January, 2021

Contents

Preface to Part I v

1. Nanoassemblies of Porphyrin Derivatives 1

Osman Barış Malcıoğlu and Şakir Erkoç

1.1 Introduction 1
1.2 Structure 3
1.2.1 General structural features of porphyrin and porphyrin derivatives 3
1.2.2 Self assembly 6
1.3 Electronic and Optical 10
1.3.1 UV–VIS features and the Gouterman model . . . 10
1.3.2 Beyond Gouterman 12
1.4 Some Effects Observed in Assemblies 15
1.4.1 Phototautomerization reaction in assemblies, holography, molecular switches 15
1.4.2 Host–guest devices and local field effects 18
1.4.3 Resonant energy transfer, exciton delocalization, concentration quenching 24
1.4.4 Renormalization of energies and band formation in thick films and crystals 26
1.4.5 Polaritons, all organic lasing chambers 31
References 32

2. Aspects of Time: On the Realistic Treatment of Self-organization Processes 39

Wolfram Schommers

2.1 Origin of Time: Empirical Facts 39
2.1.1 Quantum treatment of time 39

2.1.2 Space impressions 40
2.1.2.1 The elements of space and time are not physically real 41
2.1.2.2 Space feeling, time feeling and the arrow of time 42
2.1.3 One-to-one correspondence 42
2.1.4 Basic statements 44
2.1.5 The features of the time-machine 45
2.1.5.1 Couplings 45
2.1.5.2 The San Francisco cable car 45
2.1.6 Switching off the time-machine 46
2.1.7 The system-specific time 47
2.1.7.1 Physically real interactions create the system-specific time 48
2.1.7.2 Summary and concluding remarks . . . 48
2.2 The Container Model of Traditional Physics 50
2.2.1 Confrontation 50
2.2.2 Space structures exist twice 51
2.2.3 The container model without time-machine 53
2.3 The Realistic Case 53
2.3.1 Summary of the main facts 54
2.3.2 Autonomous entities 55
2.4 Basics 55
2.4.1 Coupling phenomena 55
2.4.2 The world outside and its picture 56
2.4.2.1 What can we observe? 56
2.4.2.2 The complete information 56
2.4.2.3 The time structure of material objects . 57
2.4.3 The general case 58
2.4.3.1 Functional connections 59
2.4.3.2 Selections with respect to time 59
2.4.4 Fictitious realities 60
2.4.4.1 The quantum-theoretical treatment . . . 61
2.4.4.2 Connections 62
2.4.5 Specific properties of the wave functions 64
2.4.6 Physically real objects represented in space and time 66
2.4.7 The traditional view 67
2.4.7.1 No quantum aspect of time 68
2.4.7.2 Point-like particle in space 68

2.5 Projection Principle 69
2.5.1 The basic equations 70
2.5.2 Probabilities 71
2.5.3 Interactions 72
2.5.3.1 Constant momentums and energies . . . 72
2.5.3.2 Momentum-energy fluctuations 73
2.5.3.3 Object-definition in space and time . . . 74
2.5.3.4 No separated objects 78
2.5.4 Analysis of the interaction terms 78
2.5.4.1 The meaning of $V(i\hbar\partial/\partial p_x, i\hbar\partial/\partial p_y, i\hbar\partial/\partial p_z, -i\hbar\partial/\partial E)$ 79
2.5.4.2 The Meaning of $V(x, y, z, t)$ 80
2.5.4.3 Summary 80
2.5.5 Uncertainty relations 81
2.5.6 Energy and time representation 83
2.5.7 Self-organizing processes 85
2.5.7.1 General definition 85
2.5.7.2 Traditional treatment 85
2.5.7.3 Projcction thcory 87
2.6 Summary and Final Remarks 90
References 93

3. Multi-Spinor Nonequilibrium Quantum Transport: Entanglement-Induced Delocalization and Topological Insulators 95

Felix A. Buot

3.1 Introduction 95
3.2 Superfield Quantum Transport Equations 100
3.3 Transformation to the Eight Tensor Variables 101
3.3.1 Products of cube matrices in quantum transport equations 101
3.4 Entanglement-Induced Localization and Delocalization . . 102
3.4.1 Hierarchy of entanglements: Alternating sequence of commutators and anticommutators as function of N_s 105
3.4.2 Local and nonlocal terms 106
3.4.3 Integer quantum Hall effect topological insulator . 107
3.4.4 Metallic edge states and bilayer superconductivity 108

3.4.5 Localization and delocalization 109
3.4.5.1 Crossover from weak localization to weak antilocalization 109
3.4.6 Quantum spin Hall effect topological insulator . . 110
3.4.7 Topological Kondo insulator 111
3.5 Concluding Remarks 111
Acknowledgements 112
Appendix 112
A. Binary Product of 'Cube Matrices' 112
A.1 Total scalars or tensors of rank zero 112
A.2 Dirac spin vectors and torques 112
A.3 Isospin vectors and torques 113
A.4 Pseudospin vectors and torques 113
A.5 Dirac spin–isospin dyadics, torques and entangled torques 113
A.6 Dirac spin–pseudospin dyadics, torques and entangled torques 114
A.7 Isospin–pseudospin dyadics, torques and entangled torques 115
A.8 Dirac spin–isospin–pseudospin tensors, torques, and entangled 2 and 3 torques 115
B. Quantum Transport Equations for $N_s = 3$ 116
C. Transport Equations in Phase Space 132
C.1 "Lattice" Weyl transform of differential operators 132
C.2 Lattice Weyl transform of product of arbitrary operators 133
C.2.1 Poisson bracket expansion 133
C.2.2 Integral representation 133
C.2.3 Local and nonlocal integral kernels . . . 134
References 134

4. Layered Nanodiamond Hybrid Materials 139

Markus Mohr and Hans J. Fecht

4.1 Introduction 139
4.2 CVD Techniques for Diamond Growth 142
4.2.1 Basic principles of CVD 142
4.2.2 Hot filament CVD 144
4.2.2.1 Seeding and substrate pretreatment . . 144

4.2.2.2 Microcrystalline vs. nanocrystalline diamond . . . 146
4.2.2.3 Optimization of process conditions . . . 148
4.3 Physical and Mechanical Properties . . . 152
4.3.1 Fracture strength . . . 154
4.3.2 Young's modulus . . . 155
4.3.3 Poisson's ratio . . . 156
4.3.4 Wear behavior . . . 157
4.4 Microfabrication Process Technology . . . 159
4.4.1 General aspects . . . 160
4.4.2 Diamond microtechnology and RIE micromachining . . . 161
4.5 Microengineering Applications . . . 164
4.5.1 Nanosharp cutting tools and scalpels . . . 164
4.5.2 Micromechanical sensors . . . 166
4.5.3 Bioapplications and future developments . . . 170
Acknowledgements . . . 174
References . . . 174

5. Heat Transport and Thermal Characterization of Nanoscale Materials and Devices 177

Xiaomeng Wang and Jin He

5.1 Introduction . . . 177
5.2 Background . . . 178
5.2.1 Thermal transport property at nanoscale . . . 180
5.2.1.1 One-dimensional nanostructure . . . 181
5.2.1.2 Two-dimensional nanostructure . . . 183
5.2.1.3 Nanocomposite . . . 186
5.2.1.4 Carbon allotropes . . . 187
5.2.2 Characterization methods of nanoscale materials . 190
5.2.2.1 Three omega method . . . 192
5.2.2.2 Transient thermoreflectance method . . 196
5.2.2.3 Suspended thermal bridge method . . . 198
5.3 Application . . . 203
5.3.1 Thermoelectrics . . . 204
5.3.1.1 Background of thermoelectric . . . 204
5.3.1.2 Thermoelectric efficiency . . . 205
5.3.1.3 Thermoelectric applications . . . 207

5.3.2 Nanoelectronics 208
5.3.2.1 Heat pipe 209
5.3.2.2 Thermal interface materials (TIMs) . . 210
5.4 Future Development 214
5.4.1 Materials informatics 215
5.4.1.1 Materials selection 215
5.4.1.2 Structure designing 217
5.4.1.3 Challenges 220
5.4.2 Application development 220
5.4.2.1 Thermal insulation 222
5.4.2.2 Thermoelectrics 223
5.4.2.3 TIM 224
5.5 Conclusions 225
Acknowledgements 225
References 226

6. Thermoelectric Phenomena at Nanoscale Level 241

Moumita Dey and Santanu K. Maiti

6.1 Introduction 241
6.2 Theoretical Framework 246
6.2.1 Figure of merit ZT 247
6.2.2 Calculation of Seebeck coefficient S 247
6.2.3 Calculation of electrical conductance G 249
6.2.4 Calculation of thermal conductance κ_{el} 249
6.2.5 Evaluation of transmission probability 250
6.2.5.1 Introduction to S matrix 251
6.2.5.2 Green's function 256
6.2.5.3 Relation between S matrix and Green's function 264
6.2.5.4 Partitioning the Hamiltonian matrix . . 266
6.2.5.5 Alternate derivation of Green's function for a multi-channel wire 268
6.2.5.6 Derivation of self-energy expression for a semi-infinite multi-channel lead 272
6.2.5.7 Expression of self-energy for single- and multi-channel leads in terms of TB parameters 275

6.2.5.8 Coupling matrix 278
6.2.5.9 Evaluation of transmission function . . . 279
6.2.5.10 Spectral function 280
6.2.5.11 Density of states 283
6.3 Molecular Systems as Functional Elements 284
6.3.1 Molecular wire with multiple loops 284
6.3.1.1 Molecular system and the TB Hamiltonian 285
6.3.1.2 Numerical results and discussion 287
6.3.2 Molecular wire with helical geometry 292
6.3.2.1 Helical geometry and the TB Hamiltonian 292
6.3.2.2 Numerical results and discussion 294
6.4 Closing Remarks and Outlook 306
Acknowledgements 307
References 307

7. Application of Green's Functions and Difference Equations in Theoretical Analyses of Nanostructures 311

Vjekoslav Sajfert and Jovan P. Šetrajčić

7.1 Introduction 311
7.1.1 Broken symmetry problem 312
7.1.2 Progress of nanotechnology 313
7.1.3 Some relevant experimental data 315
7.1.4 Expectations of nanoscience research 317
7.1.5 Necessity of theoretical research 319
7.2 Solving of Difference Equations of Second Order By Translation Operators 320
7.2.1 Difference equations with constant and valuable coefficients 321
7.2.2 A more general nonlinear problem 325
7.2.3 Appropriate example 327
7.3 Green's Functions in Solid State Physics 328
7.3.1 Quantum two-time temperature Green's functions 331
7.3.2 Applying the GF method — Examples of bulk cases 346

7.4 Application of Green's Functions and Operator Solving of Difference Equations 357
7.4.1 Phonons in ultrathin films 359
7.4.2 Model of phonons in thin films 361
7.4.3 Excitons in nanolayered structures 373
7.4.4 Electrons in quantum dots 378
7.5 Concluding Remarks 382
Appendix: Statistical and Dynamical Equivalence of Different Elementary Cells 384
Acknowledgements 398
References 398

8. Heisenberg's Uncertainty Principle and the Quantum Capacitance in Quantum Wire Field Effect Transistors 413

J. Pal and K. P. Ghatak

8.1 Introduction 413
8.2 Theoretical Background 417
8.2.1 The quantum capacitance QWFET of nonlinear optical materials 417
8.2.2 The quantum capacitance in QWFET of III–V, ternary and quaternary materials 418
8.2.3 The quantum capacitance in QWFET of II–VI . . 419
8.2.4 The quantum capacitance in QWFET of IV–VI . 420
8.2.5 The quantum capacitance in QWFET of stressed Kane-type materials 420
8.2.6 The quantum capacitance in QWFET of stressed Kane-type materials 420
8.2.7 The quantum capacitance in QWFET of GaP . . 421
8.2.8 The quantum capacitance in QWFET of $PtSb_2$. 421
8.2.9 The quantum capacitance in QWFET of Bi_2Te_3 . 422
8.2.10 The quantum capacitance in QWFET of Ge . . . 422
8.2.11 The quantum capacitance in QWFET of $GaSb$ . . 423
8.2.12 The quantum capacitance in QWFET of II–V . . 423
8.3 Results and Discussion 424
References 440

Index 445

Chapter 1

Nanoassemblies of Porphyrin Derivatives

Osman Barış Malcıoğlu* and Şakir Erkoç†

Department of Physics, Middle East Technical University, 06800 Ankara, Turkey

† *erkoc@metu.edu.tr*

1.1. Introduction

Nanoassemblies of the porphyria are attracting renewed attention in the organic nanophotonics community due to their prowess in controlling and manipulating light at subwavelength dimensions [1–8]. Porphyrin assemblies can be manufactured to act both as the passive and active waveguide simultaneously [4]. The layered Van der Waals bound material confines light in 1D and propagates it in 2D, promising, among other applications, highly efficient yet thin solar energy harvesters, due to several orders of magnitude increase in the optical path [9].

The subject of porphyrins spans many disciplines, including physics, chemistry, biology, medicine, geology, chemical engineering, paleobiology, solar energy, microelectronics, lasers, quantum computers [10, 11]. Of course, it is not possible to compress all the vast volumes written on the matter in a chapter. Instead, a short crash course will be introduced, with reference suggestions for further details, followed by a discussion on recent developments, both on theoretical and experimental fronts. We shall base discussions on the source of various effects using two prototypical porphyria: the unmetalated Tetraphenylporphyrin[1] (H2TPP) and the metalated Magnesium tetraphenylporphyrin[2] (MgTPP).

[1] 5,10,15,20-tetraphenyl-21,23-dihydroporphyrin.

[2] magnesium; 5,10,15,20-tetraphenylporphyrin.

H2TPP resembles naturally occurring porphyrins, but unlike the natural counterparts which are difficult to study due to their low symmetry and the presence of polar substituents, H2TPP is hydrophobic, symmetrically substituted, and easily synthesized. Furthermore, H2TPP by itself has quite a number of novel applications, being a photosensitizer [12] and potential applications in single-molecule electronics [13], to name a few. H2TPP is prone to metalation on suitable surfaces [14–16]. When discussing porphyrin assemblies, it is important to note that the magnitude of the π–π interaction between the macrocycles in the crystalline phase gets enhanced if metalated, but the crystal structure remains the same [17].

The extreme stability of metalloporphyrins is evidenced by their presence in sedimentary rocks and in crude petroleum oils [18–21]. Magnesium is a metal core which is hosted by the porphyrin in a relatively straightforward manner, which makes it easier to approach theoretically. MgTPP is suggested as a promising solar energy harvester [22] and an opto-electronic nose [23]. Metalated porphyrins have spectacularly similar quasi-particle spectra [14].

Metal magnesium is present in chlorophyll molecules, similar to iron in HEME, the other broadly occurring natural tetrapyrrole [24]. Nature favors the metal magnesium for chlorophyll-type light-harvesting molecules, as the non-redox metal magnesium helps in the light-harnessing process while not causing cellular damage. Magnesium porphyrins are exciting molecules owing to their relevance with chlorophylls of photosynthesis. A large number of biologically relevant magnesium porphyrins have been synthesized, including magnesium bicarbonate complexes, with the reaction of synthetic magnesium porphyrins with carbon dioxide (CO_2). This has resulted in a new role of carbon dioxide in photosynthesis.

A variety of biological activities exhibited by porphyrins exist due to the fact that natural and synthetic porphyrins have relatively low toxicity *in vitro* and *in vivo,* and they possess antitumor and antioxidant effects [25]. On the other hand, MgTPP-based coordination complexes show antibacterial activities. These features have received considerable attention on MgTPP.

The topic porphyrins is one of the most studied subjects both experimentally and theoretically attracting the attention of the researchers. As this is an active research area, we have concentrated on the studies performed in the last decade in this review.

1.2. Structure

1.2.1. *General structural features of porphyrin and porphyrin derivatives*

The common literature definition of porphyrins may be expressed as follows:

> *"A large class of deeply colored red or purple, fluorescent crystalline pigments, of natural or synthetic origin, having in common a substituted aromatic macrocyclic ring consisting of four pyrrole-type residues, linked together by four methane bridging groups [26]."*

The smallest unit of a porphyrin is a five-membered heterocycle called as pyrrole (Fig. 1.1).

A porphyrin contains four pyrrole units together with four methine bridges (Fig. 1.2).

A macrocycle molecule, porphyrin, may be constructed by linking together four pyrrole units via methine bridges ($C_{20}N_4H_{14}$). This macrocycle forms an aromatic ring (Fig. 1.3); it is also called as porphin (2HP).

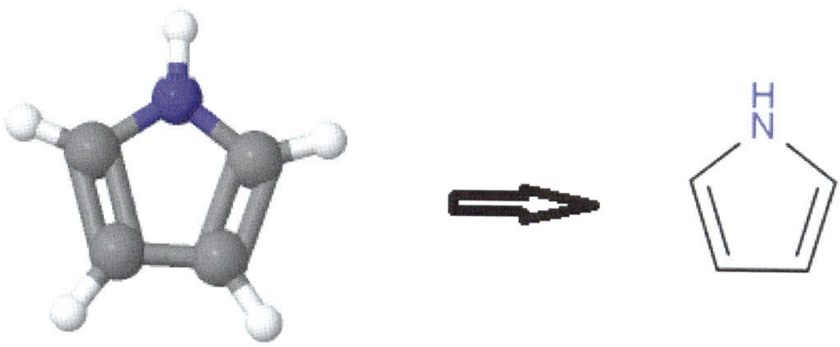

Fig. 1.1: The structure of pyrrole (C4H5N).

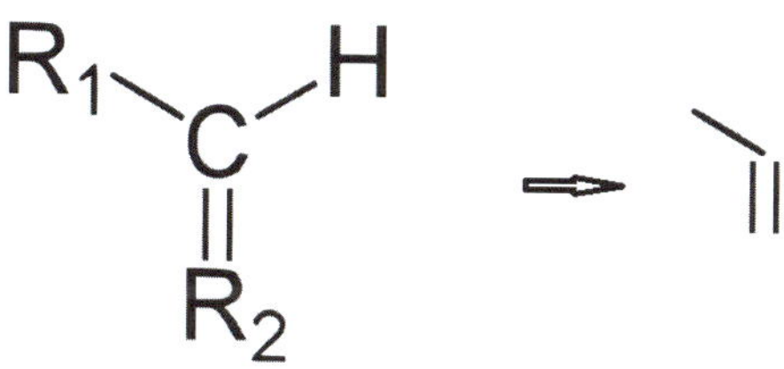

Fig. 1.2: The methine bridge.

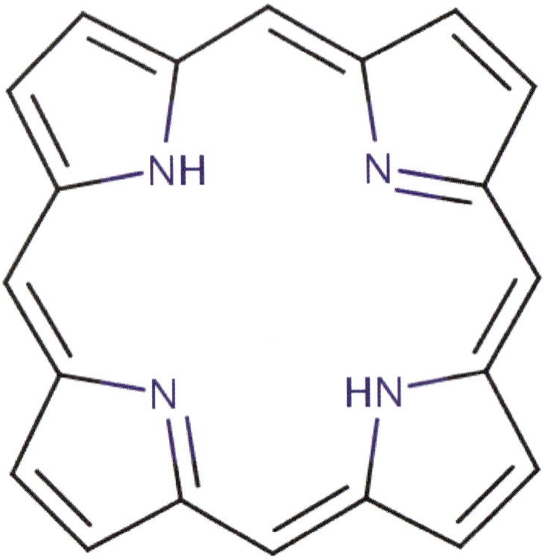

Fig. 1.3: The structure of unsubstituted porphyrin macrocycle, porphin (2HP), an aromatic ring (C20N4H14).

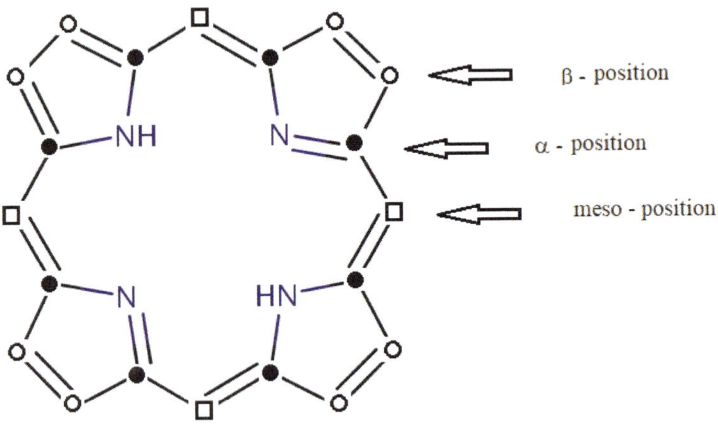

Fig. 1.4: The substitution positions on a porphyrin macrocycle. There are eight α-carbons, eight β-carbons, and four meso-carbons.

Porphyrin derivatives are obtained by substituted porphyrin macrocycles. There are certain substitution positions on a porphyrin macrocycle (Fig. 1.4).

According to IUPAC nomenclature, all atoms (except hydrogens) are numbered as follows (Fig. 1.5):

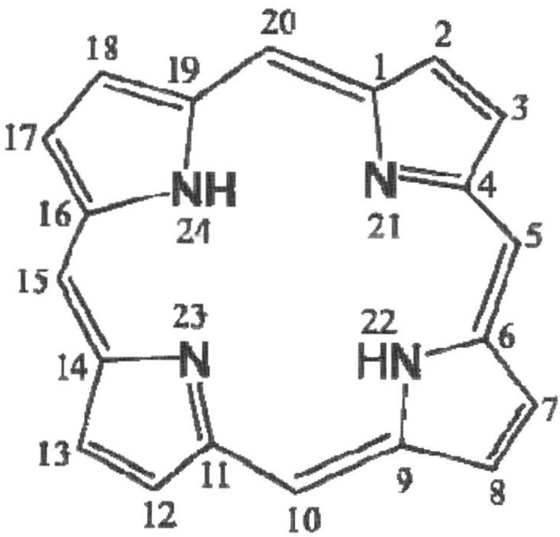

Fig. 1.5: The IUPAC nomenclature on a porphyrin macrocycle.

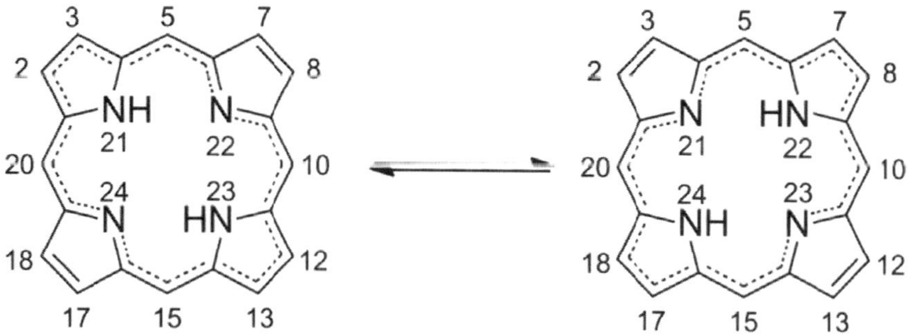

Fig. 1.6: Porphin (2HP), the parent macrocycle of porphyrins. The numbering scheme is also shown according to the current IUPAC nomenclature.

The parent compound of all porphyrins is porphin (Fig. 1.6), a planar hetero-macrocycle with D_{2h} symmetry. It consists of four pyrrole moieties, which are linked, at their α-positions, by methin groups (= CH−). The conjugated π-system of porphin contains 18 electrons along the shortest cyclic path thus obeys the 4n + 2 Hückel rule for aromatic systems. As can be seen in Fig. 1.6, two of the peripheral double bonds in opposite pyrrolic rings are cross-conjugated, and thus, their presence is not necessary for aromaticity (positions 2–3 and 12–13 on the right side of Fig. 1.6) [27].

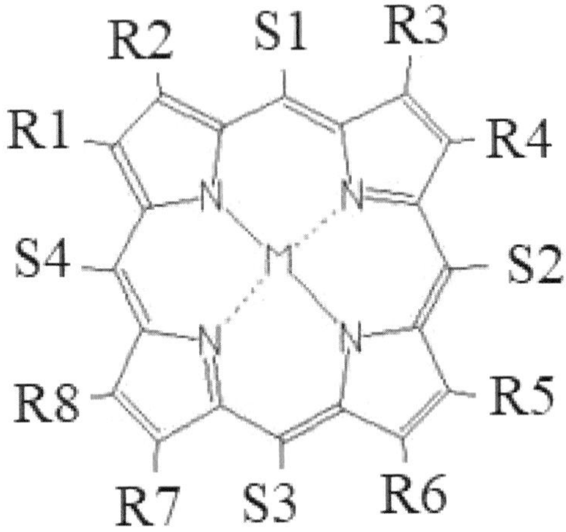

Fig. 1.7: Functionalized porphyrin. M is a metal atom. Ri (i = 1–8) and Si (i = 1–4) are various substituents, such as methyl groups, and ethyl groups.

More detailed information and examples about the numbering systems on porphyrin macrocycles can be found in Milgrom's excellent book, *The Colours of Life* [10].

A small variation on the tetrapyrrolic macrocycle leads to a wide diversity of biochemical functions to porphyrin. Substituents at different positions and combinations generate many possible different derivatives. On the other hand, the main function of porphyrins in nature is to bind metal atoms at the center, which plays an important role in biochemical events. Depending on the central metal atom type with various substitutions at the substitution positions, the functionalized porphyrin represents a different biochemical material (Fig. 1.7). For example, if the central metal atom is iron (Fe), the structure is called *heme*; if the central metal atom is magnesium (Mg), the structure is called *chlorophyll*; if the central metal atom is cobalt (Co), the structure becomes a part of *vitamin* B_{12}.

Porphyrins can also be classified into natural and synthetic porphyrins. The spectral properties, macrocyclic aromaticities and properties of natural porphyrins (1 and 2 of Fig. 1.8) are comparable to those of synthetic porphyrins (3 and 4 of Fig. 1.8) [28].

1.2.2. *Self-assembly*

Self-assembly concerns the spontaneous association of either a few or many components, resulting in the generation of either discrete oligomolecular

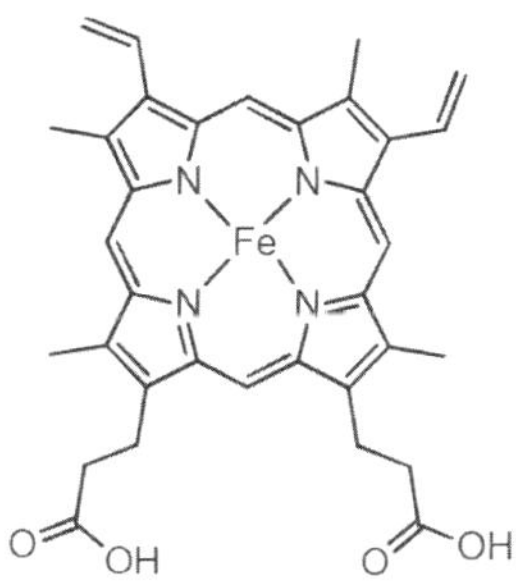

Protoporphyrin IX (1)

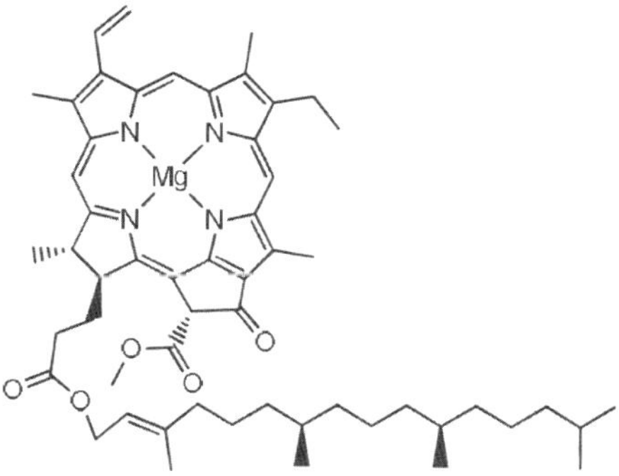

Chlorophyll a (2)

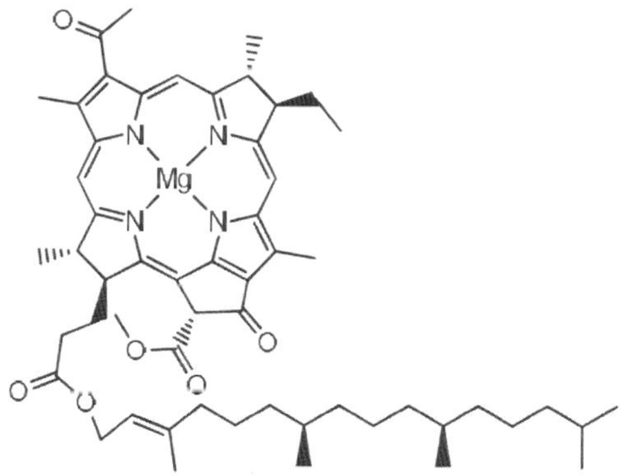

Bacteriochlorophyll (3)

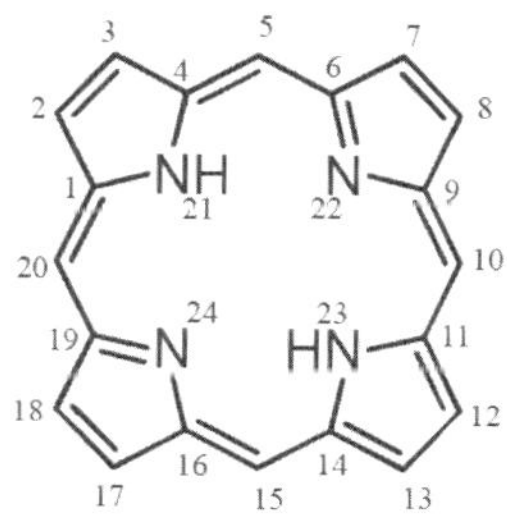

Porphine (4)

Fig. 1.8: Examples of natural (1 and 2) and synthetic (3 and 4) porphyrins.

supramolecules or of extended polymolecular assemblies, such as molecular layers, films, and membranes [29]. The formation of supramolecules results from the recognition-directed spontaneous association of a well-defined and limited number of molecular components under the intermolecular control of the non-covalent interactions that hold them together.

Self-assembly may result in a range of collective properties, such as directed light propagation, electronic conduction, magnetism, and correlated protonic motions. Numerous biological supramolecular structures result from self-assembly and living systems offer the ultimate in self-organization.

Self-assembly of porphyrins, particularly tetraphenylporphyrins (TPPs), usually take place in two-dimensional form on metal surfaces [27]. For metalated TPPs (MTPPs), the attractive interactions prevail in most cases, resulting in the formation of close-packed islands already at submonolayer coverage. On weakly interacting substrates, such as Ag and Au surfaces, the island structure is mainly determined by the molecular symmetry, frequently resulting in near-quadratic lattice geometries. More strongly interacting surfaces such as Cu(110) can induce the formation of adsorbate structures with increased complexity, but the intermolecular interactions are still attractive. In contrast, intermolecular repulsion has been observed for the unmetalated H2TPP on strongly interacting surfaces, whereas attractive interactions prevail on weakly interacting surfaces.

Combined DFT and MD simulations of H2TPP/Au(111) confirmed that the molecular self-assembly is dominated by the T-type interactions between the phenyl substituents rather than by the porphin core. Adsorbed metallo-tetraporphyrins generally form densely packed ordered structures, even on usually strongly interacting surfaces.

Porphyrins have many structural and functional properties and are considered as one of the important materials for the preparation of self-assembled nanomaterials. Although porphyrins and related molecules have long been known to self-assemble and form large-scale aggregates in solution, these structures are typically irregular and not suitable for many technological applications. However, the self-assembly of porphyrins in solution can be exploited to produce well-defined discrete (isolable) porphyrin nanostructures [30]. The facile synthesis and robust nature of these porphyrin-based materials allow access to a novel class of nanomaterials with potential applications in important areas, such as catalysis, sensors, and solar power.

Discrete self-assembled porphyrin nanostructures with well-defined shapes and sizes can be synthesized using a range of methods, including ionic self-assembly, reprecipitation, and coordination polymerization. Structures that have been prepared to date include hollow tubes, rods, fiber bundles, hollow hexagonal nanoprisms, sheets, solid spheres, hollow spheres, and a multitude of shaped nanocrystals. Micron-sized biomorphs, such as four-leaf clover-like structures, with nanoscale features have also been obtained. The porphyrin nanostructures have obvious potential applications in areas, such as light-harvesting, artificial photosynthesis, dye-sensitized solar cells, catalysis, electronics, nonlinear optics, and sensors. Many of the porphyrin-based nanostructures are suitable for such technological applications because they are monodisperse and retain their structures when deposited onto surfaces. Tunable binary porphyrin-based nanostructures and biomorphs could potentially play a significant role in the development of solar devices where it is important to gain greater control over light-harvesting, exciton migration, free-carrier formation and transport, and catalytic processes.

The self-assembly of porphyrins is mainly dependent on various intermolecular non-covalent interactions (hydrogen bonding, π–π stacking, hydrophobic, electrostatic interactions and van der Waals forces), porphyrin nanostructures with a certain size, shape and function can be provided through careful molecular and supramolecular design. Porphyrin-based nanostructures have been extensively applied for visible-light photocatalysis. However, the photocatalytic efficiency of the as-obtained porphyrin nanostructures is still limited by the fast recombination of photoinduced electron–hole pairs. Towards this issue, a combination of porphyrin and other optoelectronic functional materials provides routes to design new nanohybrid systems and thereby seems ideal for fulfilling enhanced light-harvesting and charge-transfer functions, which leads to a final improved solar energy conversion efficiency [31].

Due to the uncontrollability of intermolecular interactions and complexities of the self-assembly process, morphology control of organic micro/nanostructures is one of the major problems. There are two basic strategies to grow and assemble ordered nanostructures of various organic functional materials. One strategy is a "top-down" method, such as lithography, for the fabrication of arrays of organic materials. This method is difficult to apply to fragile organic materials, especially for single crystals of small molecules. The other strategy employs the "bottom-up" process, typically self-assembly methods, to directly grow and align nanostructures on substrates at low cost. The self-assembly of nanostructure aggregates from organic building blocks via a "bottom-up" strategy has attracted considerable attention due to their wide applications.

Competition of intermolecular metal–ligand coordination bonding with the π–π interaction between tetrapyrrole rings in tetrapyrrole molecules leads to a different molecular packing conformation and in turn a different self-assembled nanostructure morphology. A long-range π–π interaction tended to gradually become dominating in the crystals, and this tendency was the reason that led to an assembly morphology transformation [32]. Figure 1.9 shows schematically the formation of self-assembled nanostructures of MTPPs under different conditions.

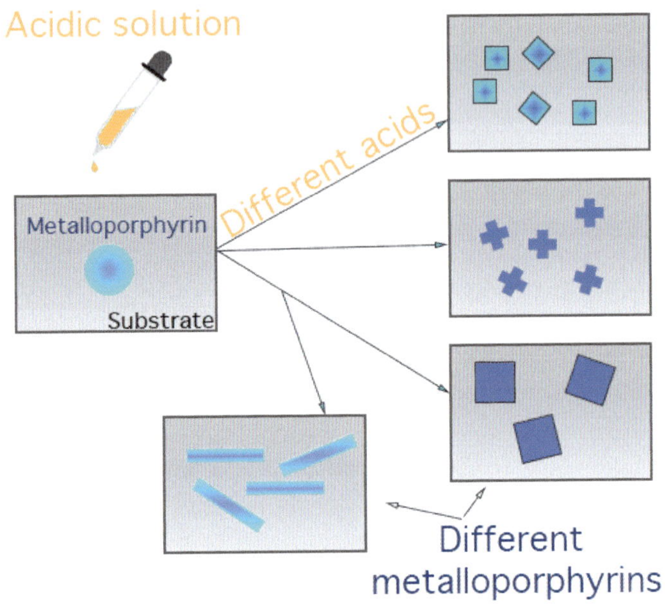

Fig. 1.9: Schematic illustration of controlled assembly of MeTPP under different acidic conditions.

Due to their distinctive π-conjugated architectures, excellent photosensitivity and biochemical functionality, porphyrins have been widely used as building blocks in supramolecular self-assemblies. Functional groups can be easily introduced into the porphyrin moieties, endowing the obtained supramolecular systems with interesting properties [33]. α,α,α,α-meso-tetrakis(2-aminophenyl)porphyrin (α,α,α,α-H2TamPP) and its derivatives possess unique "picket fence" topological structures and attractive features. They have been previously used as binucleating ligands, building blocks for cages, anion receptors, synthetic models for oxygen-binding hemoproteins and cytochrome oxidase as well as chloroperoxidase analogs. The unique structure of α,α,α,α-H2TamPP makes it very suitable as a building block to prepare supra-amphiphiles for the controllable self-assembly.

1.3. Electronic and Optical

1.3.1. *UV–VIS features and the Gouterman model*

The UV–VIS spectra of porphyrin compounds share a distinct signature. The spectra has three basic regions. A typical UV–VIS spectra of

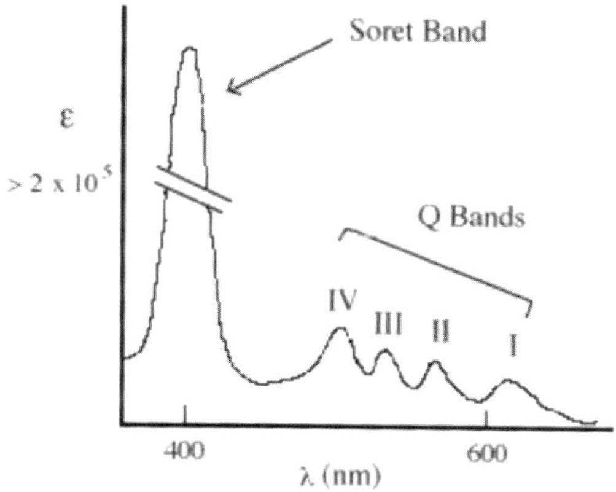

Fig. 1.10: Typical UV–VIS spectra of the porphyrin.

porphyrins is shown in Fig. 1.10. The Q band is relatively weak and occurs in the visible region. The Q band consists of a degenerate electronic transition for divalent metalloporphyrins and two separate electronic transitions for free-base porphyrins. The intense Soret or B band occurs in the near-UV and is often accompanied by a closely related N band of lower intensity. The B band for divalent metalloporphyrins is assumed to be a degenerate pair. The higher UV bands in the third region are broad and are often of near-uniform intensity. The intensity and color of porphyrins are attributed to the highly conjugated 18 π-electron ring.

The essence of this spectrum is captured by a highly successful symmetry cascade model proposed by Gouterman, the so-called "four-orbital model" [34–36]. The four orbitals mentioned are the two highest occupied π orbitals (HOMOs) and the two lowest unoccupied π^* orbitals (LUMOs). In this model, the Q and B bands arise from a linear combination of one-electron transitions between these four orbitals. For the lower Q band, the transition dipoles nearly cancel giving rise to the relatively weak absorption in the visible region. For the higher-energy B band, the transition dipoles reinforce, resulting in the very intense absorption in the UV region. The Gouterman model is illustrated in Fig. 1.11.

Gouterman model relies on symmetry cascade principle. The HOMOs of the planar D_{4h} porphyrin skeleton (i.e. without the pyrrolic Nitrogens) are a_{1u} and a_{2u} , while the LUMOs are degenerate e_g orbitals. Transitions between these orbitals gave rise to two excited states, both of ^1E_u character. Orbital mixing splits these two states in energy, creating a higher-energy

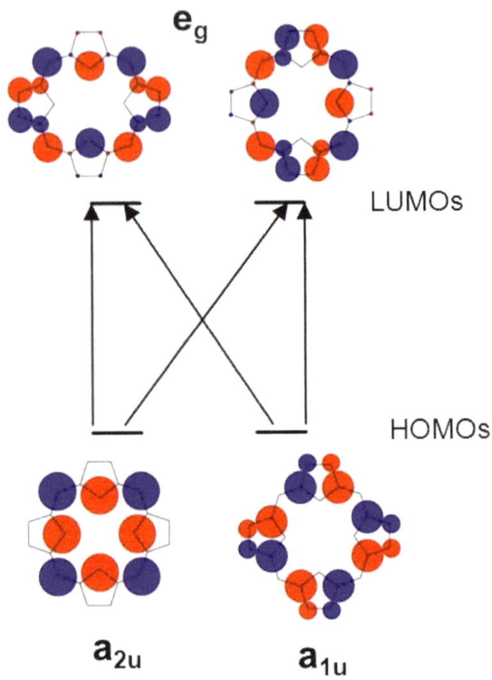

Fig. 1.11: The four-orbital model of Gouterman.

^1E_u state with greater oscillator strength, giving rise to the Soret band, and a lower-energy ^1E_u state with less oscillator strength, giving rise to the Q band.

In order to demonstrate the effectiveness of this symmetry cascade approach in capturing the essence of porphyrin UV–VIS spectra, let's apply it on a case where the porphyrin skeleton assumes a C_{2v} symmetry. The a_{1u} and a_{2u} HOMOs become a_2 and b_1, respectively, and the LUMOs split into a_2 and b_1 orbitals. Transitions between these give rise to two $^1A_1$ and two $^1B_2$ excited states. The relative intensities of the bands remain unchanged, the Q bands are still much weaker than the Soret; however, the Soret band would now be expected to split, unless the transitions to these states are accidentally degenerate.

1.3.2. *Beyond Gouterman*

Although the Gouterman picture captures an important aspect of the spectral features, it falls significantly short on capturing the exact peak positions. There are numerous interesting reasons behind this. First of all, a variety of factors affect this point group symmetry. For example, the pyrrolic hydrogen in free-base porphyrin lower the symmetry group to D_{2h}. The pyrrolic hydrogen are actually mobile under the influence of external Electric fields (phototautomerization). One well-known effect that arise from this mobility is called the "persistent spectral hole burning" effect, where a particular absorption peak disappears after a certain amount of exposure [37]. The pyrrolic hydrogen rearrange in the macrocycle such that the absorption cross-section is minimized. When the macrocycle is physically constrained in a particular orientation (i.e. in a polymer matrix for OLED devices), the pyrrolic hydrogen can even occupy neighboring sites if the net external energy supplied can make the initially energetically unfavored configuration more favorable than the degenerate ground state. This further lowers the symmetry. In metalloporphyrins, if the cationic radius of the metal is in the range of 55–80 pm, which corresponds to the sphere in the porphyrin core surrounded by the four pyrrolic Nitrogen, the core assumes a near-planar D_{4h}, otherwise some ruffling effects appear, such as the S_4 in Nickel(II) Tetraphenylporphyrin [38, 39]. Each and every substituted derivative of porphyrin and their complexes lead to different symmetry point groups.

Another factor is the inclination of the porphyria to non-radiative transfer of energy evidenced by considerable quenching. The quenching of luminescence is described by Stern and Volmer [40]. An initial photoexcitation places a luminophor X (a molecule or phosphor that gives out light) in an excited state X^*, with some finite lifetime:

$$X + hv_{ex} \rightarrow X^* + \text{heat}$$

The X^* can emit a photon and return to the ground state: $X^* \rightarrow X + h_{v_{em}}$. or return to X non-radiatively: $X^* \rightarrow X + \text{heat}$.

The Jablonsky diagram for the internal processes are depicted in Fig. 1.12.

The ultraflexible nature of the Porphyrin skeleton gives rise to ultrafast relaxation dynamics [41, 42]. This effect can easily be observed when the emission and extinction spectra of the Q bands is compared, where the ultrafast dynamics wins over the pure electronic pathways, and strong peaks of vibrionic nature are observed (Fig. 1.13). For example, in the absorption and emission spectra free base Tetraphenylporphyrin (H2TPP) [34, 36, 43–45], the measured timescales indicate 100–200 fs for intramolecular vibrational energy redistribution of the captured photon,

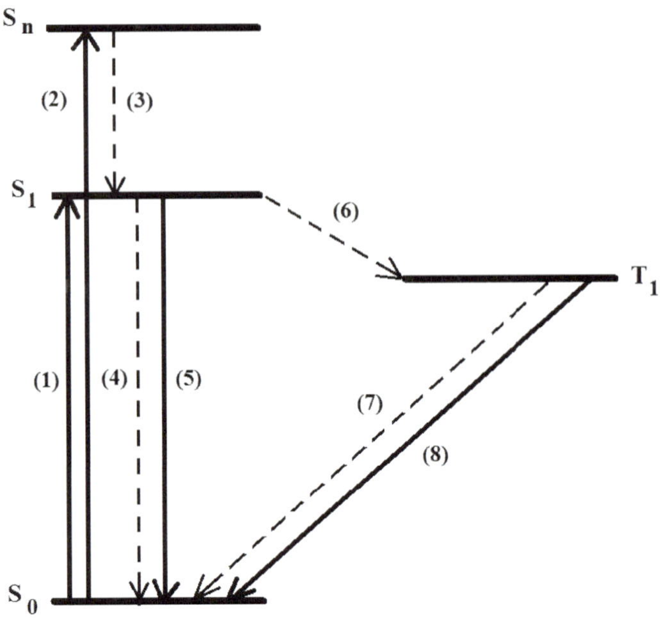

Fig. 1.12: Jablonsky diagram of a molecule: (1,2) excitation (absorption); (3,4) internal conversion; (5) fluorescence; (6,7) intersystem crossing; (8) phosphorescence.

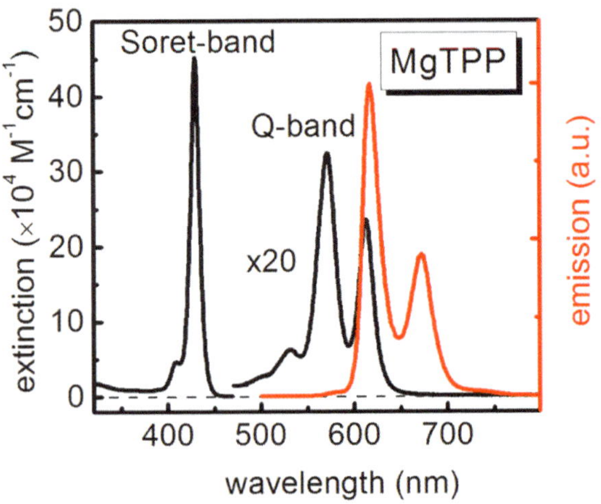

Fig. 1.13: The vibronic nature of the UV–VIS spectra.

while decay of the equilibrated Q_x population occurs on the nanosecond timescale by intersystem crossing to the triplet state [41].

Furthermore, due to the ultraflexible Porphyrin skeleton, the vertical excitation (i.e. the pure electronic transition from the ground state to excited state) tends to be significantly different from the 0–0 transition (i.e. the transition between the vibrational and electronic ground states of the system to the vibrational ground state of the electronic excited state). A basic approximation, such as fixing the orbital occupancies as per Gouterman criteria and performing a constrained relaxation is sufficient to observe the basic forces driving this active Jahn–Teller distortion effect (Fig. 1.14) [46–48].

The vibrational spectra of porphyrin is quite complex, and spectral peaks of interest usually involve a large number of normal modes acting together (see, e.g., [49]). Furthermore, vibrational features in the excited state can be quite distinct from the ground state. This makes analyzing the nature of the vibrational satellites far from straightforward.

The non-radiative processes are not necessarily internal to the porphyrin. When porphyria are in a suitable matrix, or in the close vicinity of other porphyria, there are two prominent mechanisms that transfer energy from/to the molecule. The first one is called the exciton delocalization, where the molecular orbitals perturb to form a new set of excited states that are delocalized over the whole assembly (or in the case of crystal, bands). Exciton delocalization rate is proportional to the inverse cube (r^{-3}) of the distance between the donor and the acceptor, and the effect is more pronounced typically around 1–2 nm of separation. The second is called the Förster or resonance transfer. The Förster transfer occurs when the back transition in the donor induces a transition dipole in the acceptor, transferring the energy non-radiatively. The energy transfer due to this dynamic polarization effect is proportional to inverse sixth of the separation (r^{-6}) and is effective typically at 5–10 nm separation distances [50].

1.4. Some Effects Observed in Assemblies

1.4.1. *Phototautomerization reaction in assemblies, holography, molecular switches*

In free-base porphyrin and its derivatives, light-induced tautomerization reaction leads to persistent spectral hole burning effect (PSHB) when the molecules are anchored in an assembly. One use of this effect is holography (Fig. 1.15). Holography as a means of high-density data storage is regaining

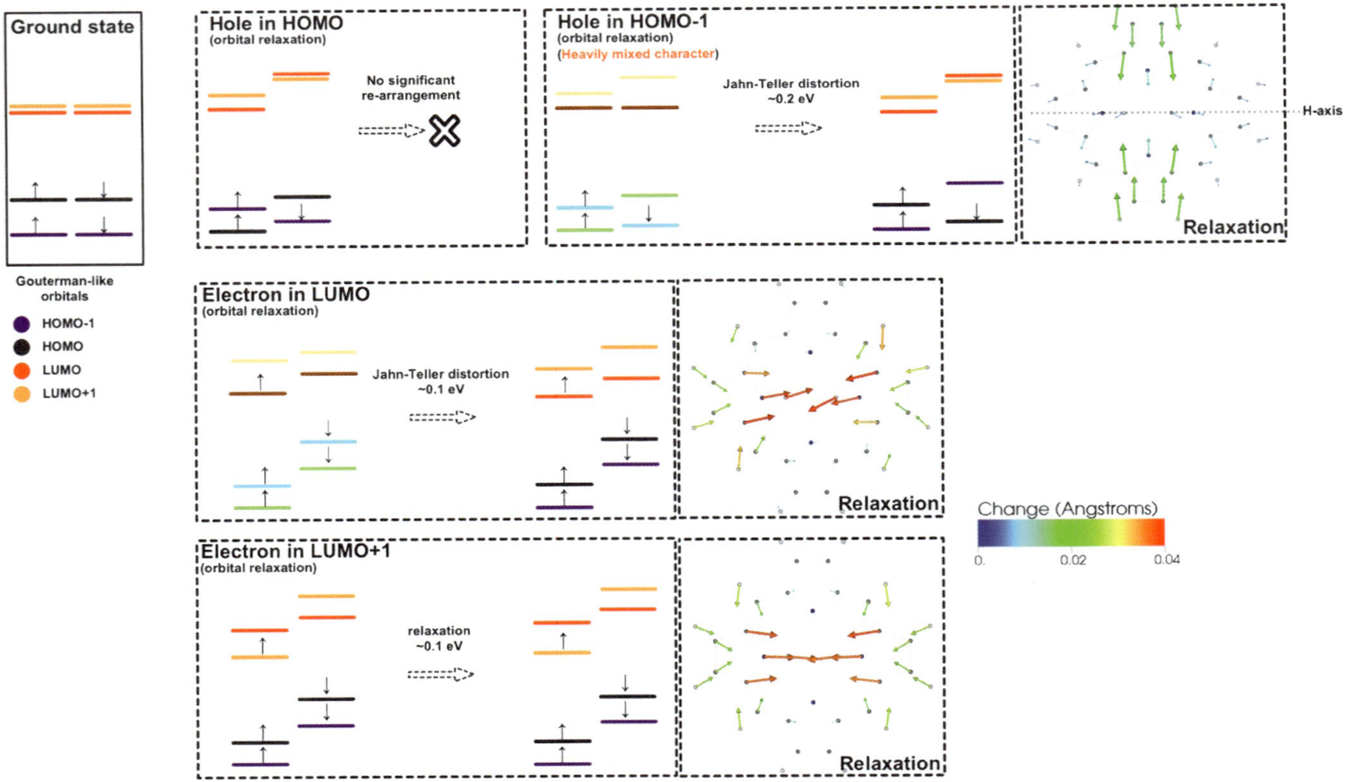

Fig. 1.14: Porphyrin radicals are Jahn–Teller-active, and structural relaxation occurs.

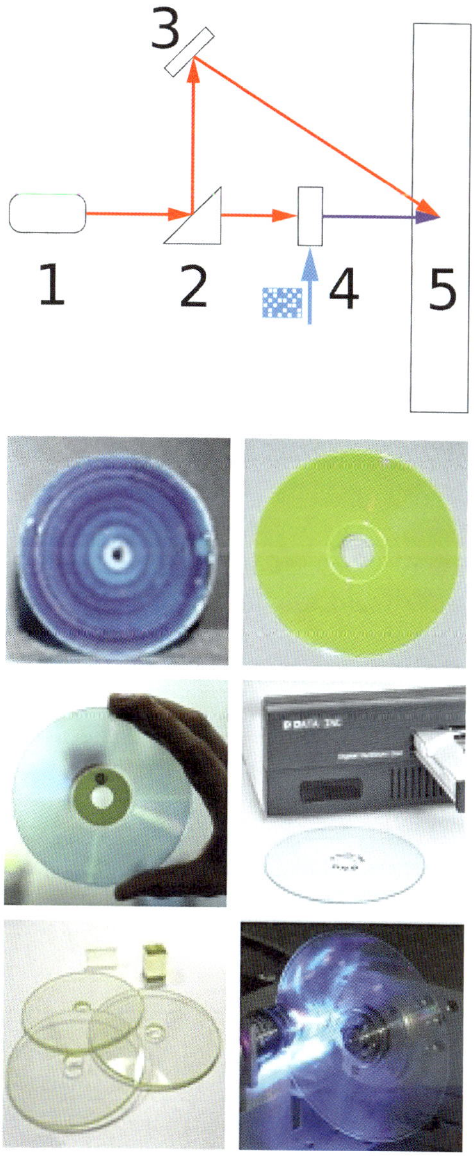

Fig. 1.15: Basic holography, and some high-density data storage devices using the 3D, multi-color, variant of the technology: (1) laser; (2) the beam is split into two sources called the reference beam and the signal beam; (3) mirror; (4) object; (5) the reference beam and the signal beam create an interference pattern in the media. For reading the data, the reference beam alone is adequate. As it passes through the media, due to the index changes in the material that were created during writing, the signal beam where the information is stored is recreated.

attention due to enormous data storage needs of "big data", however, the data write and retrieval rate of such devices is considered as a drawback in modern computing devices. Ultrafast dynamics of the phototautomerization reaction is interesting not only in this aspect but also due to the new way it enables analyzing an optical wave from a source.

Polymer matrices or crystals doped with PSHB materials can capture simultaneously both the spatial as well as the spectral content of a light signal, enabling complete time-and-space-domain storage of optical signals. Ultrafast timescales in free-base porphyrin derivates allow simultaneous spatial and temporal acquisition of an optical wave [51–53].

The displacement between the two tautomer-resonant spectra depends on both dopant molecules and matrix. One observes that a narrow hole always corresponds to a relatively broadened distribution of the photoproduct. The inverse tautomeric process from the product to the educt can occur, provided one irradiates at the product resonant frequency. The detection of the photoproduct in amorphous hosts is more difficult since the spectral shifts of the products from the reactants are usually less than the inhomogeneous width [54].

Holography is not the only use of the ultrafast and stable phototautomerization reaction in the macrocycle of porphyrins. It has been recently shown that when a free-base tetraphenyl-porphyrin molecule is anchored to a silver surface, forming a metal–molecule device assembly, it can function as a molecular conductance switch. Tautomerization leads to change in local conductance levels using electrons from the tip of a scanning tunneling microscope. Furthermore, selective deprotonation leads to a four-level conductance switch [55].

1.4.2. *Host–guest devices and local field effects*

Another source that heavily influences the spectral features of porphyrin is the local variations of the dielectric environment. This effect is visible in light-emitting devices as an unexpected color shift. Neutral free-base and metalated porphyrins have applications in organic light-emitting diodes (OLEDs), as well as sensitizers in dye-sensitized solar cells [56–58]. They have characteristic sharp red emission, moderate fluorescence efficiencies, good film-forming features by means of both solution-based and chemical vapor deposition techniques, and their ease of chemical modification. The state-of-the-art OLEDs based on free-base, Zn-, and

Pt-porphyrin derivatives show moderate luminous efficiencies of around 0.05–0.10 Lum/W [59–63].

A successful and widely used strategy in producing OLEDs is the host–guest strategy. In such a device, electron and/or hole transporter matrices with a wide band gap are used as the host, and the porphyrins are blended into this matrix as the guest. The electroluminescence (EL) mechanism involves the quenching of the emission of the host via an energy transfer process. The guest emits the captured energy.

An OLED device configuration typically involves a multi-layered architecture and thus macroscopic polarization effects in the emitting layer are reduced, as the externally applied electric field is screened out. The host should possess higher triplet energies than those of the guest emitters to prevent reverse energy transfer. Also, the HOMOs and LUMOs of host material should match with those of the neighboring active layers to reduce the hole and electron injection barrier, thus lowering the device driving voltages. The device produces light by radiative decay of triplet excitons (phosphorescence) in the guest. There are a couple of pathways for this to happen (Fig. 1.16). Firstly, the singlet excitons formed in the host under electrical excitation can be transferred to the guest via Förster and Dexter energy transfer, and then singlet excitons in the guest are converted to triplet excitons by fast intersystem crossing. Secondly, the triplet excitons formed in the host can be transferred to phosphor through Dexter energy transfer. Finally, the holes and electrons injected from the anode and the cathode can directly recombine on the phosphorescent guest molecules accompanied by phosphorescence emission.

Although H2TPP shows superior electrochemical and photophysical features compared to ZnTPP as guest, ZnTPP perform better in an OLED device. The latter feature deep red EL, suggesting no change in the nature of the excited emitting state. On the other hand, H2TPP devices exhibit an unexpected yellowish white EL [64]. No degradation upon both charge injection and transport and/or difference in the device mechanism can be observed. Both the shape of the EL spectra and the stark difference in the device performance cannot be rationalized by standard arguments, as both compounds show a similar reversible electrochemical behavior, film morphology, and device mechanism, while the photoluminescence features of the H2TPP in solid state are superior to those of the ZnTPP in terms of, for example, photoluminescence quantum yields (PLQYs). A change in the nature of the emitting excited state in H2TPP devices under electrical

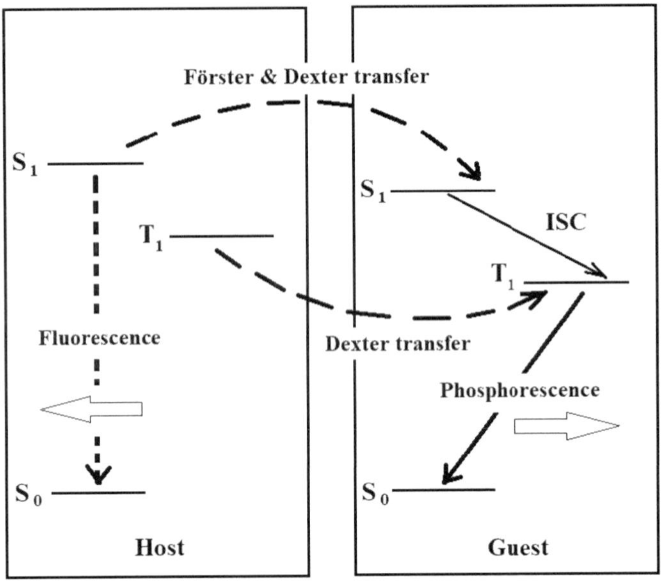

Fig. 1.16: Schematic representation of how a host–guest device converts electrical excitation to light.

stimuli is a more likely explanation, in conjunction with the markedly different dielectric response between the devices. A considerable polarization effect caused by the external applied field modifies the scenario of excited state in H2TPP but not in ZnTPP.

Following the Gouterman argument, H2TPP will have four distinctive Q peaks (two 0–0 and two 0–1 transitions), and ZnTPP will have two (one 0–0 and one 0–1). A major distinction in metal-free porphyrin is the phototautomerization, as discussed previously.

Ab-initio calculations using time-dependent density functional perturbation theory at the static field limit yield [64]:

$$\alpha_R^{\mathrm{H2TPP/C}} = \begin{pmatrix} 463 & 0 & 0 \\ 0 & 460 & 0 \\ 0 & 0 & 201 \end{pmatrix} \quad \alpha_R^{\mathrm{ZnTPP}} = \begin{pmatrix} 470 & 0 & 0 \\ 0 & 466 & 0 \\ 0 & 0 & 205 \end{pmatrix}$$

This is just the electronic contribution. The small variation in microscopic polarizability cannot account for the measured disparity in dielectric constants between H2TPP and ZnTPP devices ($\epsilon_R(0) \approx 11.5$ and $\epsilon_R(0) \approx 6$, respectively) since the molar concentrations used in the devices are the same. It is reasonable to assume linear response for the operation of the

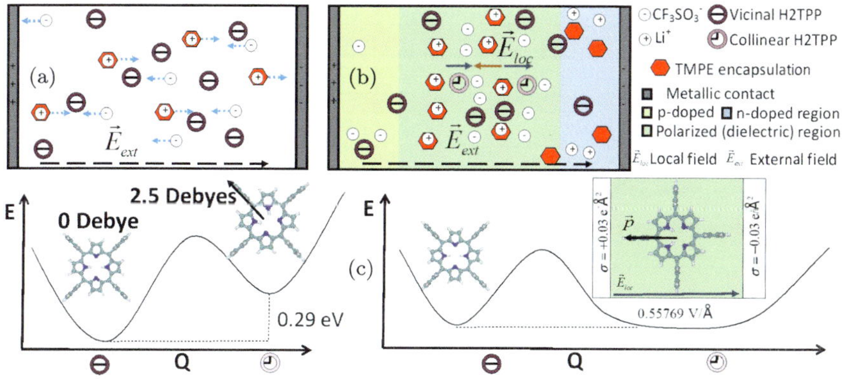

Fig. 1.17: The high dielectric constant H2TPP conformer and the driving mechanism. (a) At the initial field turn on, the encapsulated anions and cations start moving towards the metallic electrodes under the influence of the external electric field. (b) At steady state, p- and n-doped regions form. Some charged species remain trapped to form a polarized dielectric layer. (c) The local fields of the dielectric drive a conformational change in the suitably positioned H2TPP by reducing the energy of the vicinal conformer. The reason why H2TPP device is much more polarizable than the ZnTPP device despite the electronic polarizability of both species being practically the same in the presence of this vicinal conformer.

device. Then, the contributions to the dielectric function can be separated in the following manner:

$$\alpha = \alpha_m + \alpha_e + \alpha_n$$

where α_m is the contribution due to polarizability of the host, α_e is the polarizability of the electronic density of the guests, and α_n is the polarizability due to conformational changes in the guest. Since it is demonstrated that porphyrin species do not react to form an entirely different species (the four Q peaks in high voltage EL spectra is yet another indication of this conclusion). The host in both devices are the same, and the electron density contribution is practically the same as well. This implies a conformational change of the H2TPP guest as a strong candidate that can explain such a large difference (Fig. 1.17).

H2TPP and ZnTPP in their ground state have no molecular dipole moment, but they have relatively large quadrupole moments. The thermal motion of the molecule already leads to a small polarization, but by itself, this effect still cannot account for the huge difference in polarizability. It is energetically not easy to access a ZnTPP conformer with a strong dipole moment, however, the same mobility of macrocycle hydrogen that leads to phototautomerization effect in H2TPP should allow such a re-configuration with relative ease. Indeed, a conformer with a large dipole moment (2.5 Debye per molecule), lies at an energy only ~0.29 eV higher than the ground state (Fig. 1.17(c)). We shall name this as the "vicinal" conformer, where the ground state will be called "collinear" conformer. The collinear conformer has a similar α_e, but the α_n is significantly different due to permanent dipole moment of the conformer.

When an electric field is applied to a dielectric device, the device responds by forming a response dipole. A higher dielectric constant hints at a higher intensity response dipole. In common dielectric materials, a distinction between local fields and the external field is necessary, as the response dipole is usually due to a rolling average of local dipoles, since external field will tend to assemble and organize existing dipoles [65]. The resulting local fields can be orders of magnitude larger than the applied external field, however the macroscopic averaging in a typical experiment will only detect indirect evidence of such large local polarization fields.

Compensating the energy difference between the vicinal and collinear conformer in gas phase solely with an electric field would require a substantial applied potential. However, H2TPP as a guest in this media (as indicated by the ZnTPP device measurement) is most likely exposed to high-intensity local fields.

Electroluminescence spectra of porphyrins has a curious "N-mixing" character [36]. Non-radiative internal processes that move the electron (hole) from higher-order LUMO (lower-order HOMO) orbitals with higher propensity to capture them are suggested as the nature of this character. This character will not be easy to observe in ZnTPP due to symmetry. In Fig. 1.18, we present a schematic representation of the physical process we suggest for the high-voltage device, and the actual shift of the quasi-particle levels, as calculated by sc-GW0 calculations. All the porphyrins, including the collinear H2TPP conformer, has orbitals that would fit well with the Gouterman argument around HOMO and LUMO. However, vicinal H2TPP has orbitals immediately next to Gouterman levels, which extend significantly more along the polarization axis. Actually, these orbitals can be considered to be collinear H2TPP orbitals spread in energy and with higher reactivity. The orbitals of ZnTPP at the same energy range is much more extended, thanks to larger charge accumulation at the metal center.

The vertical spectra of the species are presented in Fig. 1.19. Collinear H2TPP has two vertical Q lines that translate to two 0–0 transitions and two 0–1 transitions. Note that 0–0 transitions can shift significantly with

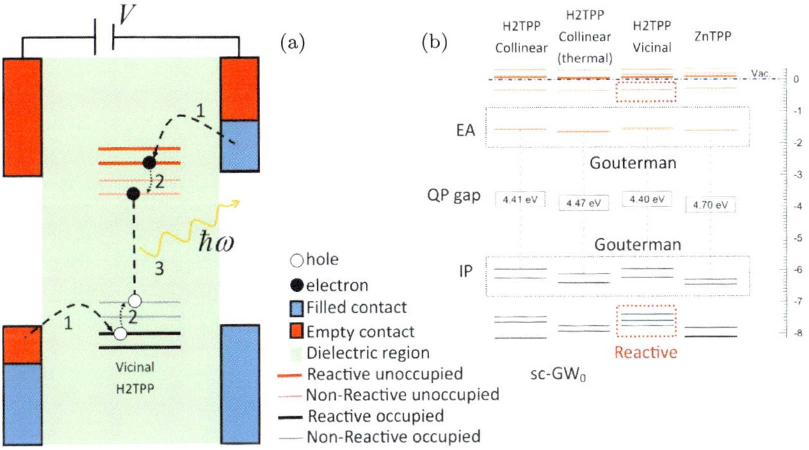

Fig. 1.18: The activation and optical response of a molecule: (a) 1. The propensity to capture an electron (hole) depends on both the level energy alignment and the amount of overlap of the orbital with carrier levels. 2. The electron (hole) captured travels to LUMO (HOMO) via non-radiative processes 3. The electron and hole radiatively recombine to produce a photon. (b) The actual levels of the guest species from scGW$_0$ calculation. The vicinal H2TPP has typical Gouterman levels followed by levels with enhanced overlap. The non-radiative process that mixes N character in luminescence spectra is already known in the literature [36].

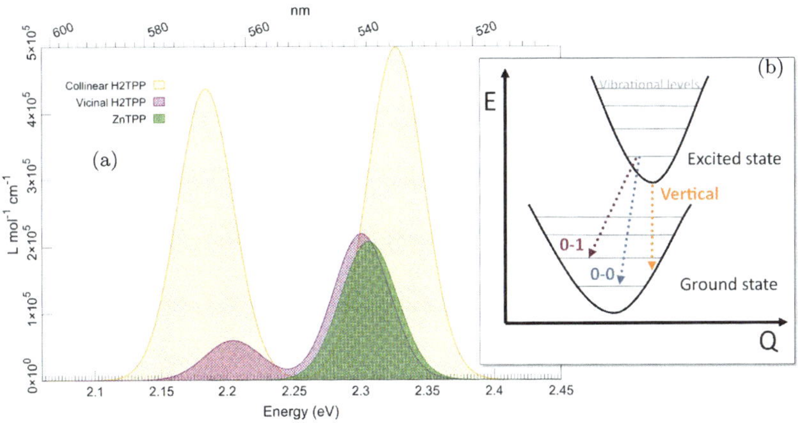

Fig. 1.19: The vertical spectra: (a) Both collinear and vicinal H2TPP have two vertical excitations, which translate to typical 4 Q band fingerprint of metal-free porphyrin in the EL spectra. However, the low-lying Q peak is slightly blue-shifted and lower in intensity, whereas the high-energy Q peak is red-shifted. The sole ZnTPP vertical excitation that will translate to 2 Q bands coincides with the vicinal H2TPP high-energy Q peak. Overall, the quantum efficiency of the vicinal H2TPP is expected to be only marginally better than ZnTPP. (b) Schematic representation for the distinction between vertical and non-vertical transitions.

respect to vertical lines due to excited state relaxation processes. Since the exact shift will depend strongly on the matrix coordination effects and will dilute the focus of the present discussion, we will not discuss them further here. ZnTPP has a sole vertical excitation (two Q peaks), with a lower intensity, that explains the lower quantum efficiency. The vicinal H2TPP has two vertical lines (four Q peaks), but their intensity is much lower than collinear H2TPP. Furthermore, the low-energy Q peak is blue-shifted, whereas the high-energy Q peak is red-shifted. Since this translates to a nonlinear shift in emitted wavelength, the color of the device will also be affected nonlinearly. The quantum efficiency of the collinear H2TPP species appears to be only marginally better than ZnTPP.

1.4.3. *Resonant energy transfer, exciton delocalization, concentration quenching*

Pressure-sensitive paints (PSPs) are used in non-invasive mapping of pressure variations through remote optical detection (Fig. 1.20). They are effective to detect a few orders of magnitude of pressure variations. They have applications ranging from aerodynamic tests to fundamental studies in acoustics [66–69], complex microfluidic and micromechanical devices [70–72], and medicine and biology [73–77].

Many PSPs are composed of probe dyes and coordination compounds immobilized in polymer films. Metalloporphyrins PtOEP and PdOEP (OEP: octaethylporphyrin) display strong room-temperature phosphorescence with high quantum yields and long lifetimes, especially PtOEP is widely used as a PSP probe [78].

In these devices, it is a known effect that a high concentration of the metalloporphyrin dye sensor would lead to quenching of luminescence [79] (this is the exact reason why porphyrins are used in low quantities in host–guest devices discussed previously). This quenching effect becomes significant even at relatively lower mix ratios, around 1.0 g of porphyrin to 60.0 g polymer, and becomes larger at higher ratios of porphyrin to polymer [80]. Some porphyrins are more prone to this effect than others [80]. The effect is related to exciton delocalization and resonant energy transfer mechanisms described in Section 1.3.2.

Although this effect is an adversity for this particular application, it holds a vast potential for other applications. Energy generation is one such case. In bacterial photosynthetic organisms, photosystems 1 and 2 (PS1 and PS2) are arranged into highly effective light-harvesting and

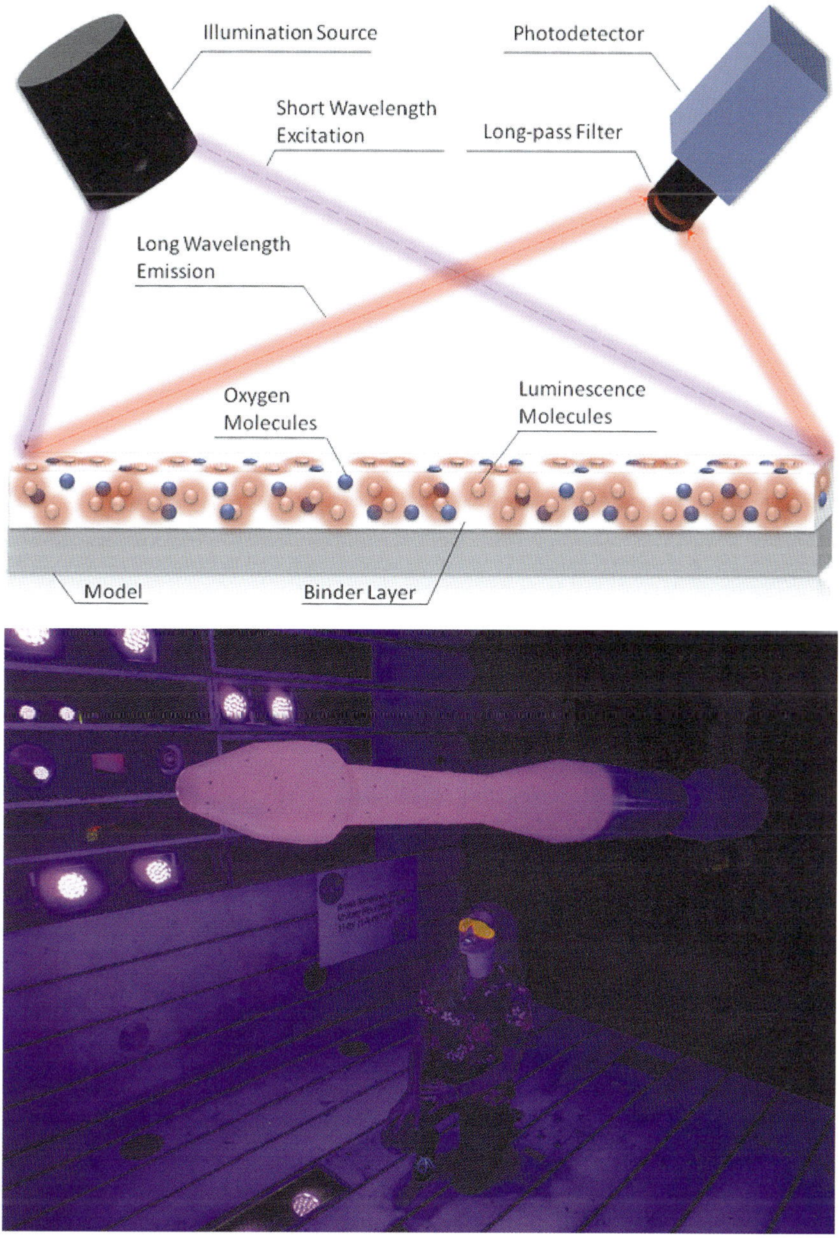

Fig. 1.20: Basic working principle of the pressure-sensitive paint, and the unsteady pressure-sensitive paint applied to generic launch vehicle model in the 11-by-11-foot Transonic Wind Tunnel at Ames Research Center. Credits: NASA Ames.

energy-transferring mechanisms. If one pigment captures a photon, then the excitation energy is transferred to the rest of the array in femtosecond timescale. The trapping center, P700 traps this excitation and uses it to produce a strong reducing agent nicotinamide adenine dinucleotide phosphate (NADPH). This inspired many scientists to model and understand the most significant initial stages of this process [81–83]. Unfortunately, understanding the exact nature of this exciton transfer mechanism is far from trivial, as even model porphyrin systems that are very closely related can exhibit hugely different exciton transfer characteristics. A better understanding of these processes may lead to cheap and efficient solar energy harvesters [84, 85], such as dye sensitized solar cells (DSSCs) [58].

Photovoltaic current generation in a DSSC can be investigated through a four-step process: the absorption of the light by a sensitizer, leading to the electronic excitation of the sensitizer (X) to an excited state (X^*); the dissociation of the generated exciton and injection of excitation energy into the metal oxide, resulting in an oxidized dye (X^+); the restoration of the X^+ back to X by a reducing agent; the regeneration of the reducing agent by photogenerated electrons. The solar-to-electric power conversion efficiency bottlenecks and various technical shortcomings of using liquid electrolytes (such as corrosion of metal contacts) can be overcome using resonant energy transfer mechanisms as a pump acting on the metal oxide directly to produce a "dry DSSC".

Self-metalation of free base porphyrin is a known phenomenon. What is more interesting in this context is the formation of an oxide–porphyrin assembly and the accompanied color change [86, 87]. The complex has many interesting properties for solar energy generation and catalysis (Fig. 1.21).

1.4.4. *Renormalization of energies and band formation in thick films and crystals*

Porphyrin self-assembly and crystallization have been discussed in the previous section. The effect of crystallization on the large, aromatic macrocycle can be approached using archetypal H2TPP and MgTPP crystals (Fig. 1.22). The intimate packing of these aromatic macrocycles results in intriguing photophysical and photochemical effects that are desirable in many technological applications. Mimicking the efficiency of solar energy capture and transfer processes seen in naturally occurring photosystems is one such goal [88–92], along with novel photoactive molecular devices for quantum information processing and/or catalysis.

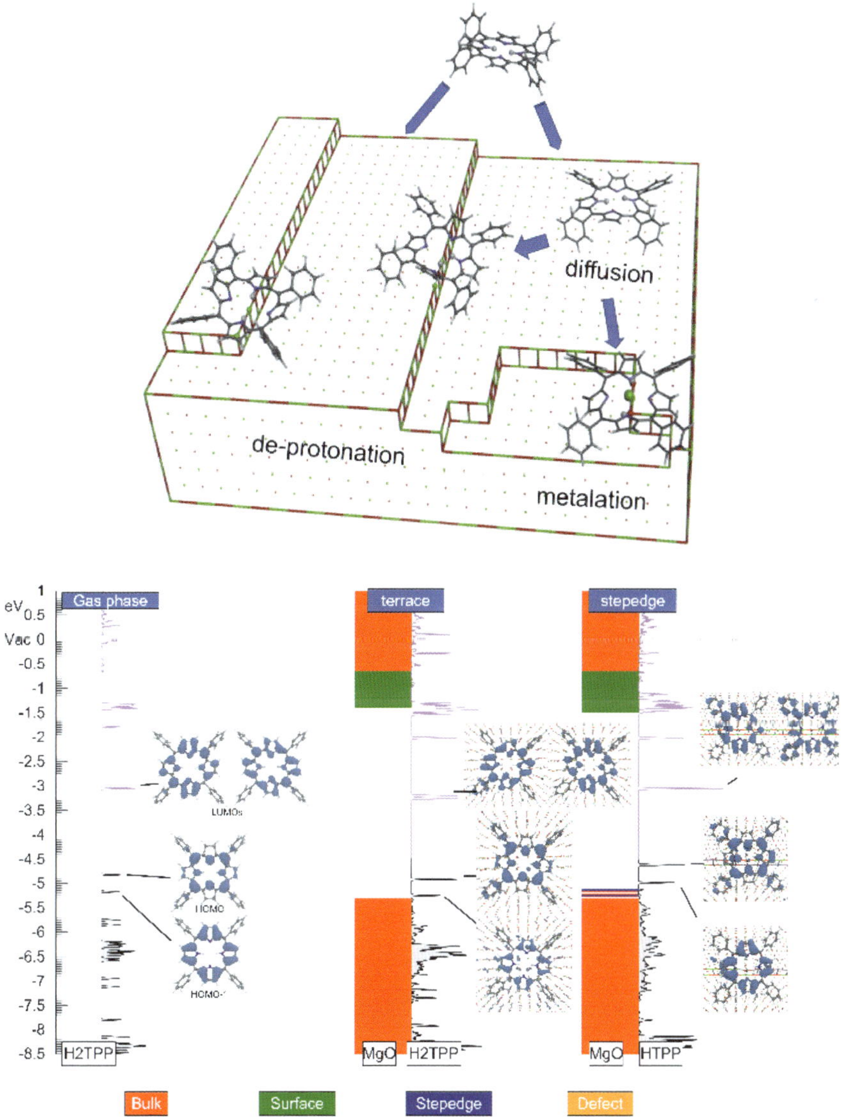

Fig. 1.21: H2TPP metalates and forms anchor themselves into common MgO surface defects. The porphyrin Gouterman levels are situated such that exciton delocalization and resonant energy transfer mechanisms can lead to photovoltaic current generation.

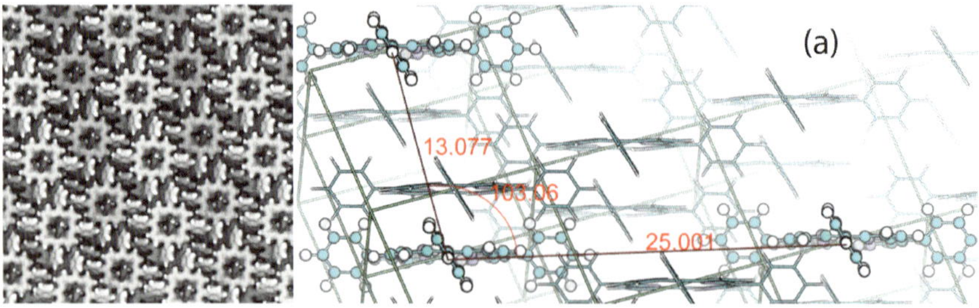

Fig. 1.22: H2TPP crystal [97, 98] (CSD ID:TPHPOR04 [99]).

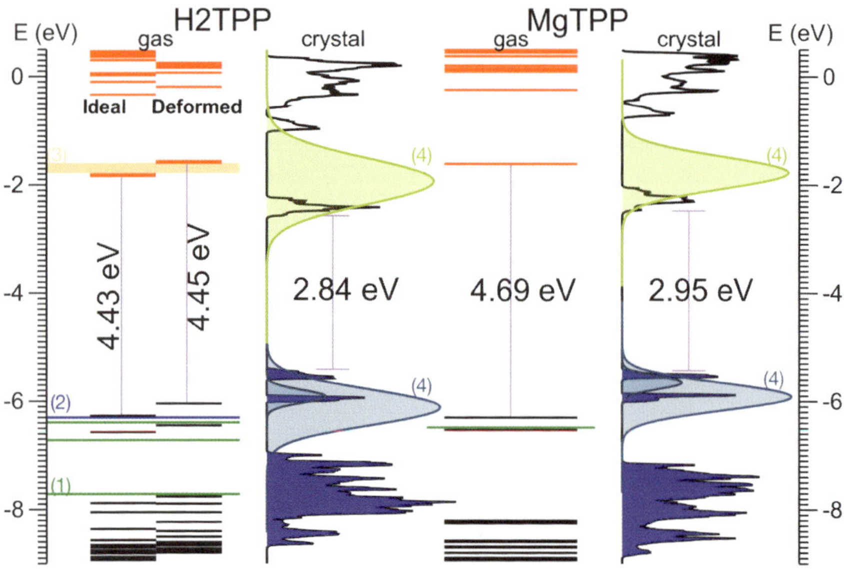

Fig. 1.23: Comparison of DOS of the H2TPP and MgTPP in gas phase and in crystal phase. Theoretical calculations are done at sc-GW0 level and aligned to the vacuum level [98]. Experiments: gas phase: (1) [101], (2) [44], (3) [100]; thick films: (4) [15].

When H2TPP forms thick films, quasi-particle band gap shrinks compared to gas phase [15, 93] (Fig. 1.23). A very similar band shift is also observed in supersonic molecular beam deposition which is known to highly promote crystallinity [94]. This effect is usually overlooked due to robustness of the qualitative features in the DOS of porphyrins. The gas phase DFT density of states [95, 96] qualitatively resembles UPS data, and the large discrepancy is often attributed to the known differences between a Kohn – Sham band and a quasi-particle energy level [94], and various effects such as image states. Hence, an empirical scissor shift to gas phase theory is applied to fit the experiment.

With the recent computational access to more elaborate many-body theories, such as GW approximation of Hedin, it is now possible to attain an NIST-level accuracy for the gas phase H2TPP and MgTPP, and make direct comparison with the experiments [44, 100–103]. Doing so, the gas phase quasi-particle levels do not align well with the experiments involving thick films, with more than 0.5 eV discrepancy for the HOMO itself.

What is the source of this considerable shift? Deformation of the molecule in the crystal or on the surface breaks the symmetry and consequently shifts the Gouterman levels, but the theory indicates this shift is an order of magnitude less and cannot account for the observed shift by itself. Owing to strong dielectric screening of the crystalline porphyrin, substrate-related image charge arguments are weak in the case of a thick film, and the band gap has been reported to be independent of film thickness after a few monolayers anyway.

Instead, theory indicates that the source is the interaction of the orbitals with the neighboring molecules or exciton delocalization, as discussed in the previous sections. It is interesting to note that the effect is significant at GW level, but not that much at long-range coulomb tail-corrected DFT, which suggests a strong contribution from dynamic correlation effects underneath. This observation is in alignment with the experimental observations in the biological systems discussed in the previous section, as exciton delocalization and Förster resonant energy transfer have been identified as the leading energy transfer mechanisms [83].

There are numerous ways the porphyrin can form crystalline structures, however among all the suggested forms in the literature, only the crystalline porphyrin as observed by Kano *et al.* [97] aligns with the experimental data on thick films within reasonable tolerance [14, 93, 94]. This crystalline phase is also significantly more stable energetically than all the other models suggested in the literature [98].

How does the Gouterman model hold in the crystalline phase of these archetypal porphyrins? In Fig. 1.24, a quantitative comparison between the decomposed charge density of the Gouterman molecular orbitals of the isolated molecule and the band-decomposed charge density of the corresponding band states is presented:

$$\Delta\rho_i(\vec{r}) = |\psi_i^g(\vec{r})|^2 - \int_{\mathrm{BZ}} |\psi_{i,\vec{k}}^s(\vec{r})|^2 d^3k$$

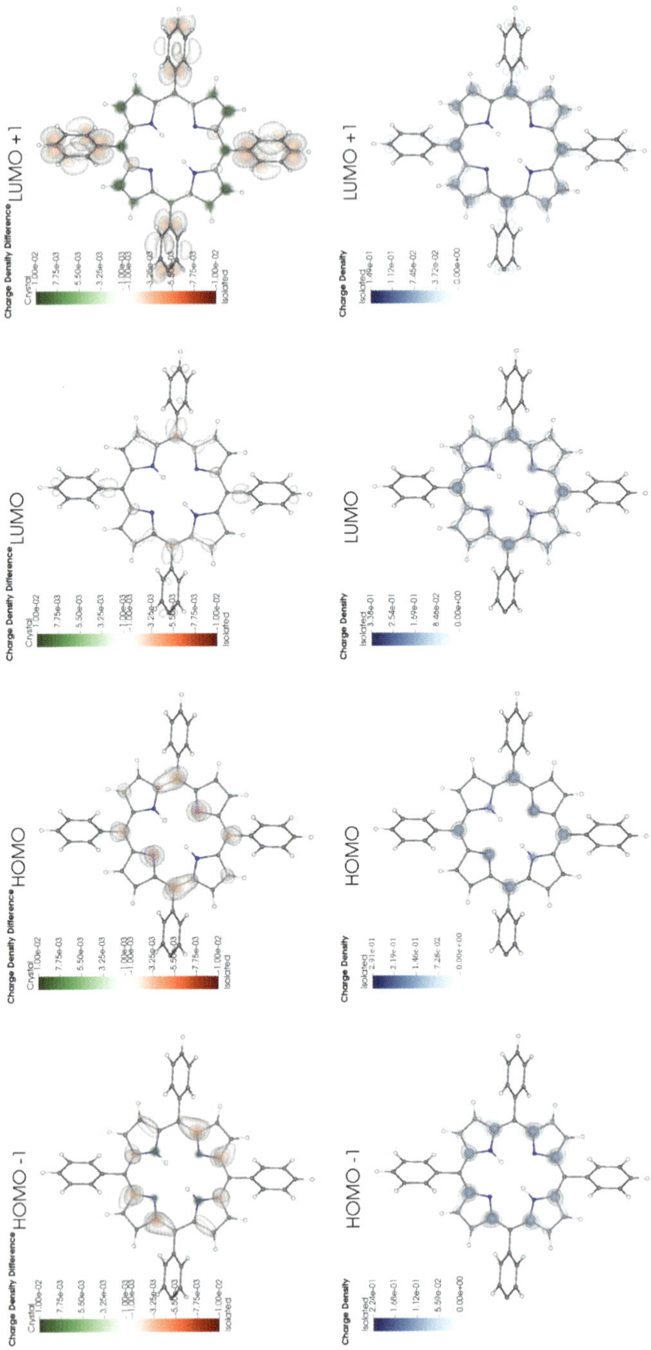

Fig. 1.24: The change in Gouterman orbitals when H2TPP crystallizes [98].

where $\psi_i^g(\vec{r})$ and $\psi_{i,\vec{k}}^s(\vec{r})$ are the wavefunctions of the ith molecular orbital of the isolated molecule and the corresponding band of the crystalline structure, respectively. The band is integrated over the Brillouin zone.

Actually, the small perturbations of the exciton delocalization and resonant mechanisms spread the localized Gouterman levels into bands, but when integrated over the phase, the charge density is almost the same (i.e. the density difference per volume is an order of magnitude less than the electrons per volume in the isolated case). This translates into very similar optical vertical excitation spectra features, thanks to the robustness of the Gouterman model, although the involved wavefunctions now are considerably shifted in energy. This paves the way to band-gap engineering possibilities by substitution [3].

1.4.5. *Polaritons, all organic lasing chambers*

A very interesting recent development is the discovery of a long-range quantum coherence phenomena in photosynthetic systems [104]. The preservation of coherence is attributed to electron–phonon interactions within photosynthetic proteins. In fact, strong exciton–photon coupling at room temperature in an organic cavity based on a porphyrin molecule (4TBP-PZn) blended in a polystyrene has been demonstrated in the pioneering work of David Lidzey and co-workers [105]. Polariton lasing from an organic optical microcavity at room temperature has already been demonstrated [106].

When the light is trapped in a location deep below the optical wavelength, the associated electric field strength increases. This implies enhanced light–matter interaction, strong nonlinearities, large forces, and enhanced emission and absorption probabilities. Polaritons are the eigenstates of the system in the strong coupling regime. They are hybrid light–matter modes involving collective oscillations of electronic charge in the matter. Distinction between the photonic and electronic excitations blurs at this regime. Actually, polaritons are coherent superpositions of the photonic and electronic excitations. Polaritons both exhibit the quantum optical properties of photons and the many-body physics of electronic excitations. Their energy–momentum dispersion relationship is very different from bare uncoupled excitations [2].

There are a number of categories of polaritons [6]. Surface plasmon polaritons, with collective oscillation of surface-bound electrons in metals, are the most well-studied, prominent example. In our context, maybe phonon polaritons formed by atomic vibrations in polar insulators and exciton polaritons in semiconductors are the most relevant.

We have discussed the self-assembly properties of porphyrins previously. With some careful planning, it is possible to tailor microscopic devices that act simultaneously as the active and passive waveguide [4].

Porphyrins represent another class of π-conjugated small molecules that have been shown to form 1D nanostructures. An anionic 5,10,15,20-tetrakis(4-sulfonatophenyl) porphyrin (TPPS) has been employed as a template due to its ability to aggregate into 1D rod-like nanostructure. Diacid form of tetrakis(4-sulfonatophenyl) porphyrin (TPPS) in an acidic solution (pH 0.9–1) self-assembled into nanowires with heights of ~3.8–4.5 nm and bundles with diameters up to 40 nm. A mixture of two porphyrin derivatives were shown to self-assemble into well-defined nanotubes in an acidic solution (pH 2). 1D organic semiconductor nanostructures, such as TPPS, are promising nanowire photodetectors and phototransistors [1].

An all-organic 1D photonic crystal, such as tetraphenylporphyrin (TPP) and octaethylporphyrin (OEP), is demonstrated, which permits the observation of strong exciton–photon coupling at room temperature with Rabi splittings that are an order of magnitude larger than those observed in inorganic analogs. These structures could offer new opportunities for the development of polariton-based devices.

The extinction coefficient of porphyrin at 400–500 nm range ($\approx 2 \times 10^5\,\text{mol}^{-1}\,\text{cm}^{-1}$) [107] is an order of magnitude larger than that of crystalline silicon ($\approx 1 \times 10^4\,\text{mol}^{-1}\,\text{cm}^{-1}$). However, exciton binding energy of a typical III–V semiconductor (≈10 meV) is approximately two orders of magnitude less than the archetypal porphyrins (≈500 meV). No wonder they are preferred in today's solar technology, although they are inefficient in exciton generation, they are quite efficient in breaking the exciton and generating the photovoltaic current. Yet, you don't see III–V semiconductors in plants for solar energy generation. It seems nature has solved the strong exciton binding problem for energy generation by using strong coupling and polaritons [10, 104], and a better understanding of this phenomenon will enable many future technologies, even cheap quantum computers.

References

1. F. S. Kim, G. Ren, and S. A. Jenekhe, One-dimensional nanostructures of π-conjugated molecular systems: Assembly, properties, and applications from photovoltaics, sensors, and nanophotonics to nanoelectronics†. *Chem. Mater.* 23(3), 682–732 (**2011**).

2. G. C. La Rocca, Organic photonics polariton lasing. *Nat. Photon.* 4(6), 343–345 (**2010**).
3. A. Charisiadis, *et al.*, Peripheral substitution of tetraphenyl porphyrins: Fine-tuning self-assembly for enhanced electroluminescence. *ChemPlusChem* 83(4), 254–265 (**2018**).
4. N. Chandrasekhar, *et al.*, Planar active organic waveguide and wavelength filter: Self-assembled meso-tetratolylporphyrin hexagonal nanosheet. *ACS Appl. Mater. Interfaces* 6(3), 1488–1494 (**2014**).
5. D. M. Coles, *et al.*, Polariton-mediated energy transfer between organic dyes in a strongly coupled optical microcavity. *Nat. Mater.* 13(7), 712–9 (**2014**).
6. D. N. Basov, M. M. Fogler, and F. J. Garcia de Abajo, Polaritons in van der Waals materials. *Science* 354(6309), 1992 (**2016**).
7. G. Rojas, *et al.*, Self-assembly and properties of nonmetalated tetraphenyl-porphyrin on metal substrates. *J. Phys. Chem. C* 114(20), 9408–9415 (**2010**).
8. H. Imahori, *et al.*, Substituent effects of porphyrin monolayers on the structure and photoelectrochemical properties of self-assembled monolayers of porphyrin on indium-tin oxide electrode. *J. Phys. Chem B* 108(16), 5018–5025 (**2004**).
9. B. E. Hardin, H. J. Snaith, and M. D. McGehee, The renaissance of dye-sensitized solar cells. *Nat. Photon.* 6(3), 162–169 (**2012**).
10. L. R. Milgrom, *The Colours of Life: An Introduction to the Chemistry of Porphrins and Related Compounds* (Oxford University Press, **1997**).
11. G. C. Ferreira, *et al.*, *The Handbook of Porphyrin Science* Vol. 27 (World Scientific, **2013**).
12. K.-H. Pfoertner and T. Oppenländer, Photochemistr, in *Ullmann's Encyclopedia of Industrial Chemistry* (**2002**).
13. V. C. Zoldan, R. Faccio, and A. A. Pasa, n and p type character of single molecule diodes. *Sci. Rep.* 5(1), 8350 (**2015**).
14. S. J. Silvers and A. Tulinsky, The crystal and molecular structure of triclinic tetraphenylporphyrin. *J. Am. Chem. Soc.* 89(13), 3331–7 (**1967**).
15. A. Classen, *et al.*, Electronic structure of tetraphenylporphyrin layers on Ag(100). *Phys. Rev. B* 95(11), 115414 (**2017**).
16. M. Kielmann and M. O. Senge, Molecular engineering of free-base porphyrins as ligands — The N-H...X binding motif in tetrapyrroles. *Angew. Chem. Int. Ed.* 58(2), 418–441 (**2019**).
17. C. A. Hunter and J. K. M. Sanders, The nature of .pi.–.pi. interactions. *J. Am. Chem. Soc.* 112(14), 5525–5534 (**1990**).
18. J. T. Horeczy, *et al.*, Determination of trace metals in oils. *Anal. Chem.* 27(12), 1899–1903 (**1955**).
19. M. Blumer and G. S. Omenn, Fossil porphyrins — Uncomplexed chlorins in a triassic sediment. *Geochimica Et Cosmochimica Acta* 25(2), 81–90 (**1961**).
20. G. W. Hodgson, *et al.*, Petroleum pigments from recent fresh-water sediments. *Geochimica Et Cosmochimica Acta* 19(4), 272–288 (**1960**).

21. G. W. Hodgson, *et al.*, Geochemistry of porphyrins chlorins and polycyclic aromatics in soils sediments and sedimentary rocks. *Geochimica Et Cosmochimica Acta* 32(7), 737–& (**1968**).
22. N. K. Subbaiyan and F. D'Souza, Studies of a supramolecular photoelectrochemical cell using magnesium tetraphenylporphyrin as photosensitizer. *J. Porphyr. Phthalocyanines* 17(8–9), 733–741 (**2013**).
23. J. P. Mensing, *et al.*, Inkjet-printed sol-gel films containing metal phthalocyanines/porphyrins for opto-electronic nose applications. *Sens. Actuators B-Chem.* 176, 428–436 (**2013**).
24. K. D. Borah and J. Bhuyan, Magnesium porphyrins with relevance to chlorophylls. *Dalton Trans.* 46(20), 6497–6509 (**2017**).
25. N. Amiri, *et al.*, Two novel magnesium(II) meso-tetraphenylporphyrin-based coordination complexes: Syntheses, combined experimental and theoretical structures elucidation, spectroscopy, photophysical properties and antibacterial activity. *J. Solid State Chem.* 258, 477–484 (**2018**).
26. J. A. Simpson, E. S. Weiner, and O. U. Press., *The Oxford English Dictionary*. Vol. 12. (Oxford University Press, **1989**).
27. J. M. Gottfried, Surface chemistry of porphyrins and phthalocyanines. *Surf. Sci. Rep.* 70(3), 259–379 (**2015**).
28. P. Chauhan, Porphyrin derivatives for functional nanomaterials., in *Chemistry* (Padua, **2014**).
29. J.-M. Lehn, *Supramolecular Chemistry: Concepts and Perspectives* (New York: VCH, **1995**).
30. C. J. Medforth, *et al.*, Self-assembled porphyrin nanostructures. *Chem Commun (Camb)* (47), 7261–77 (**2009**).
31. Y. Chen, *et al.*, Porphyrin-based nanostructures for photocatalytic applications. *Nanomaterials* 6(3), 51 (**2016**).
32. J. Cai, *et al.*, Controlled self-assembly and photovoltaic characteristics of porphyrin derivatives on a silicon surface at solid–liquid interfaces. *Soft Matter* 10(15), 2612–2618 (**2014**).
33. S.-P. Wang, *et al.*, Controllable hierarchical self-assembly of porphyrin-derived supra-amphiphiles. *Nat. Commun.* 10(1), 1399 (**2019**).
34. M. Gouterman, Spectra of porphyrins. *J. Mol. Spectrosc.* 6, 138–163 (**1961**).
35. M. Gouterman, G. H. Wagnière, and L. C. Snyder, Spectra of porphyrins. *J. Mol. Spectros.* 11(1–6), 108–127 (**1963**).
36. P. G. Seybold and M. Gouterman, Porphyrins. *J. Mol. Spectrosc.* 31(1–13), 1–13 (**1969**).
37. I. Renge, Symmetry and hole-burning properties of porphyrins. *J. Lumin.* 98(1), 213–220 (**2002**).
38. M.-S. Liao and S. Scheiner, Electronic structure and bonding in metal porphyrins, metal=Fe, Co, Ni, Cu, Zn. *J. Chem. Phys.* 117(1), 205–219 (**2002**).
39. A. L. Maclean, *et al.*, Structural characterization of nickel(II) tetraphenylporphyrin. *Aust. J. Chem.* 49(12), 1273–1278 (**1996**).
40. O. S. M. Volmer, über die Abklingzeit der Fluoreszenz. *Zeitschrift für Physik* 20, (**1919**).

41. J. S. Baskin, H.-Z. Yu, and A. H. Zewail, Ultrafast dynamics of porphyrins in the condensed phase: I. free base tetraphenylporphyrin†. *J. Phys. Chem. A* 106(42), 9837–9844 (**2002**).
42. P. H. Kumar, *et al.*, Ultrafast relaxation dynamics of 5,10,15,20-meso-tetrakis pentafluorophenyl porphyrin studied by fluorescence up-conversion and transient absorption spectroscopy. *J. Phys. Chem. A* 119(8), 1267–78 (**2015**).
43. M. Kruk, *et al.*, Two-photon absorption of tetraphenylporphin free base. *J. Lumin.* 105(1), 45–55 (**2003**).
44. Y. Nakato, K. Abe, and H. Tsubomura, Experimental determination of ionization potentials of tetraphenylporphine and metallotetraphenylporphines. *Chem. Phys. Lett.* 39(2), 358–360 (**1976**).
45. H. Nakatsuji, J.-y. Hasegawa, and M. Hada, Excited and ionized states of free base porphin studied by the symmetry adapted cluster-configuration interaction (SAC-CI) method. *J. Chem.l Phys.* 104(6), 2321 (**1996**).
46. G. V. N. Rajapakse, Photophysical Properties of Metallotetraphenyltetrabenzoporphyrins: Insights from Experimental and Theoretical Studies (Bowling Green State University, **2008**).
47. J. Seth and D. F. Bocian, Electron paramagnetic resonance studies of metalloporphyrin anion radicals. Effects of solvent, counterion, temperature, and isotopic substitution on the Jahn-Teller active 2Eg ground state. *J. Am. Chem. Soc.* 116(1), 143–153 (**1994**).
48. T. G. Spiro, R. S. Czernuszewicz, and X. Y. Li, Metalloporphyrin Structure and Dynamics from Resonance Raman-Spectroscopy. *Coord. Chem. Rev.* 100, 541–571 (**1990**).
49. P. Bour, *et al.*, Vibrational circular dichroism of tetraphenylporphyrin in peptide complexes? A computational study. *Chirality* 12(4), 191–8 (**2000**).
50. T. Förster and O. Sinanoglu, *Modern Quantum Chemistry* Part III Action of Light and Organic Crystals, O. Sinanoglu (ed.) (Academic Press, New York, **1965**).
51. A. Consortini, *Trends in Optics: Research, Developments, and Applications* (Academic Press, **1996**).
52. T. W. Mossberg, Time-domain frequency-selective optical data storage. *Opt. Lett.* 7(2), 77–79 (**1982**).
53. A. Rebane, *et al.*, Photochemical time-domain holography of weak picosecond pulses. *Opt. Commun.* 47(3), 173–176 (**1983**).
54. M. Tian, *et al.*, Phototautomerization and broadband spectral holography. *J. Opt. Soc. Am. B-Opt. Phys.* 15(8), 2216–2225 (**1998**).
55. W. Auwärter, *et al.*, A surface-anchored molecular four-level conductance switch based on single proton transfer. *Nat. Nanotechnol.* 7, 41 (**2011**).
56. T. Higashino and H. Imahori, Porphyrins as excellent dyes for dye-sensitized solar cells: Recent developments and insights. *Dalton Trans.* 44(2), 448–463 (**2015**).
57. B. Tuffy, *Porphyrin Materials for Organic Light Emitting Diodes.* (Brian Tuffy, **2011**).

58. M. Urbani, *et al.*, Meso-substituted porphyrins for dye-sensitized solar cells. *Chem. Rev.* 114(24), 12330–12396 (**2014**).
59. L. J. Zhu, *et al.*, Effect of substituent groups of porphyrins on the electroluminescent properties of porphyrin-doped OLED devices. *J. Phys. Org. Chem.* 23(3), 190–194 (**2010**).
60. A. Ryan, *et al.*, Carbazole-linked porphyrin dimers for organic light emitting diodes: Synthesis and initial photophysical studies. Tetrahedron, 67(43), 8248–8254 (**2011**).
61. J. C. Ostrowski, *et al.*, Near-infrared electroluminescent light-emitting devices based on ethyne-bridged porphyrin fluorophores. *Adv. Mater.* 15(15), 1296–1300 (**2003**).
62. O. Fenwick, *et al.*, Linear and cyclic porphyrin hexamers as near-infrared emitters in organic light-emitting diodes. *Nano Lett.* 11(6), 2451–2456 (**2011**).
63. C. A. Barker, *et al.*, Porphyrin, phthalocyanine and porphyrazine derivatives with multifluorenyl substituents as efficient deep-red emitters. *Chem. — Eur. J.* 13(23), 6710–6717 (**2007**).
64. M. D. Weber, *et al.*, From white to red: Electric-field dependent chromaticity of light-emitting electrochemical cells based on archetypal porphyrins. *Adv. Func.l Mater.* 26(37), 6737–6750 (**2016**).
65. S. Grésillon, *et al.*, Experimental observation of localized optical excitations in random metal-dielectric films. *Phys. Rev. Lett.* 82(22), 4520–4523 (**1999**).
66. O. S. Wolfbeis, Sensor paints. *Adv. Mater.* 20(19), 3759–3763 (**2008**).
67. T. Liu, Pressure- and temperature-sensitive paints, in *Encyclopedia of Aerospace Engineering* W.S. Richard Blockley (ed.) (**2011**).
68. K. J. Disotell, *et al.*, Single-shot temperature- and pressure-sensitive paint measurements on an unsteady helicopter blade. *Exper. Fluids* 55(2), 1671 (**2014**).
69. J. W. Gregory, *et al.*, Fast pressure-sensitive paint for flow and acoustic diagnostics. *Annu. Rev. Fluid Mech.* 46, 303–330 (**2014**).
70. T. C. Corke, *et al.*, *44th AIAA Aerospace Sciences Meeting and Exhibit* (Reno, NV, **2006**).
71. J. W. Gregory, *et al.*, Characterization of the microfluidic oscillator. *AIAA J.* 45(3), 568–576 (**2007**).
72. H. Nagai, *et al.*, Pressure-sensitive paint measurement of pressure distribution in a supersonic micronozzle. *AIAA J.* 46(1), 215–222 (**2008**).
73. F. L. Dickert, H. Besenböck, and M. Tortschanoff, Molecular imprinting through van der Waals interactions: Fluorescence detection of PAHs in water. *Adv. Mater.* 10(2), 149–151 (**1998**).
74. P. Babilas, *et al.*, In vivo phosphorescence imaging of pO2 using planar oxygen sensors. *Microcirculation*, 12(6), 477–487 (**2005**).
75. Wu, M. *et al.*, Fluorescence imaging of the activity of glucose oxidase using a hydrogen-peroxide-sensitive europium probe. *Anal. Biochem.* 340(1), 66–73 (**2005**).
76. M. S. Frederiksen and R. N. Glud, Oxygen dynamics in the rhizosphere of Zostera marina: A two-dimensional planar optode study. *Limnol. Oceanogr.* 51(2), 1072–1083 (**2006**).

77. I. Sánchez-Barragán, *et al.*, Room-temperature phosphorescence (RTP) for optical sensing. *TrAC Trend. Anal. Chem.* 25(10), 958–967 (**2006**).
78. Y. Takeuchi and Y. Amao, *Materials for Luminescent Pressure-Sensitive Paint*, in *Frontiers in Chemical Sensors: Novel Principles and Techniques*, G. Orellana and M.C. Moreno-Bondi, (Eds.) (Springer Berlin Heidelberg: Berlin, Heidelberg), 303–322 (**2005**).
79. M. Gouterman, Oxygen quenching of luminescence of pressure sensitive paint for wind tunnel research. *J. Chem. Educ.* 74(6), 697–702 (**1997**).
80. S. Grenoble, *et al.*, Pressure-sensitive paint (PSP): Concentration quenching of platinum and magnesium porphyrin dyes in polymeric films. *J. Lumin.* 113(1–2), 33–44 (**2005**).
81. K. M. Barkigia, *et al.*, Models of Photosynthetic Chromophores — Molecular-Structure of the Bacteriochlorin (2,3,12,13-Tetrahydro-5,10,15,20-Tetraphenylporphinato)(Pyridine)Zinc(Ii). *Inorg. Chem.* 30(9), 2233–2236 (**1991**).
82. S. L. Miller and L. E. Orgel, *The Origins of Life on the Earth* (Englewood Cliffs, New Jersey. **1974**).
83. R. G. Bennett and R. E. Kellogg, *Progress in Reaction Kinetics.* (Pergamon Press, New York), 215 (**1967**).
84. G. Grassi and D. O. Hall, *Photocatalytic Production of Energy-Rich Compounds* (CRC Press, **1988**).
85. K. K. Rao, *et al.*, Photoelectrochemical responses of photosystem II particles immobilized on dye-derivatized TiO2 films. *J. Photochem. Photobiol. B: Biol.* 5(3), 379–389 (**1990**).
86. J. Schneider, *et al.*, Porphyrin metalation at MgO surfaces: A spectroscopic and quantum mechanical study on complementary model systems. *Chemistry* 22(5), 1744–9 (**2016**).
87. J. Schneider, *et al.*, Porphyrin metalation at the MgO nanocube/toluene interface. *ACS Appl. Mater. Interfaces* 7(41), 22962–22969 (**2015**).
88. J. R. Darwent, *et al.*, Metal phthalocyanines and porphyrins as photosensitizers for reduction of water to hydrogen. *Coord. Chem. Rev.* 44(1), 83–126 (**1982**).
89. M. Grätzel, Artificial photosynthesis: Water cleavage into hydrogen and oxygen by visible light. *Acc. Chem. Res.* 14(12), 376–384 (**1981**).
90. J. K. Kiwi, Kalyanasundaram, and M. Grätzel. Visible light induced cleavage of water into hydrogen and oxygen in colloidal and microheterogeneous systems. in *Solar Energy Materials* (Berlin, Heidelberg: Springer Berlin Heidelberg, **1982**).
91. M. R. Wasielewski, Photoinduced electron transfer in supramolecular systems for artificial photosynthesis. *Chem. Rev.* 92(3), 435–461 (**1992**).
92. M. Grätzel, *Energy Resources Through Photochemistry and Catalysis* (Academic Press inc., **1983**).
93. G. Di Filippo, *et al.*, Interaction of free-base tetraphenylporphyrin with magnesium oxide: Influence of MgO morphology on metalation. *J. Chem. Phys.* 146(6), 064702 (**2017**).

94. M. Nardi, *et al.*, Tetraphenylporphyrin electronic properties: A combined theoretical and experimental study of thin films deposited by SuMBD. *Phys. Chem. Chem. Phys.* 12(4), 871–880 (**2010**).
95. P. Hohenberg and W. Kohn, Inhomogeneous electron gas. *Phys. Rev.* 136(3B), B864–B871 (**1964**).
96. W. Kohn and L. J. Sham, Self-consistent equations including exchange and correlation effects. *Phys. Rev.* 140(4A), A1133–A1138 (**1965**).
97. Kano, K. *et al.*, Factors influencing self-aggregation tendencies of cationic porphyrins in aqueous solution. *J. Am. Chem. Soc.* 122(31), 7494–7502 (**2000**).
98. O. B. Malcioglu, I. Bechis, and M. Bockstedte, Effect of crystallization on the electronic and optical properties of archetypal porphyrins. *Phys. Chem. Chem. Phys.* (**2020**).
99. F. Allen, The Cambridge structural database: A quarter of a million crystal structures and rising. *Acta Crystallographica Section B* 58(3 Part 1), 380–388 (**2002**).
100. H. L. Chen, *et al.*, Gas-phase charge-transfer reactions and electron affinities of macrocyclic, anionic nickel complexes: Ni(SALEN), Ni(tetraphenylporphyrin), and derivatives. *J. Am. Chem. Soci.* 113(7), 2766–2767 (**1991**).
101. S. C. Khandelwal and J. L. Roebber, The photoelectron spectra of tetraphenylporphine and some metallotetraphenylporphyrins. *Chem. Phys. Lett.* 34(2), 355–359 (**1975**).
102. L. Hedin, New method for calculating the one-particle Green's function with application to the electron-gas problem. *Phys. Rev.* 139(3A), A796–A823 (**1965**).
103. X. Blase, C. Attaccalite, and V. Olevano, First-principles GW calculations for fullerenes, porphyrins, phtalocyanine, and other molecules of interest for organic photovoltaic applications. *Phys. Rev. B* 83(11), (**2011**).
104. G. S. Engel, *et al.*, Evidence for wavelike energy transfer through quantum coherence in photosynthetic systems. *Nature* 446(7137), 782–786 (**2007**).
105. D. G. Lidzey, *et al.*, Strong exciton–photon coupling in an organic semiconductor microcavity. *Nature* 395(6697), 53–55 (**1998**).
106. S. Kena-Cohen and S. R. Forrest, Room-temperature polariton lasing in an organic single-crystal microcavity. *Nat. Photon.* 4(6), 371–375 (**2010**).
107. W. M. Horspool and F. Lenci, *CRC Handbook of Organic Photochemistry and Photobiology, Volumes 1 & 2*: CRC press (**2003**).

Chapter 2

Aspects of Time: On the Realistic Treatment of Self-organization Processes

Wolfram Schommers

University of Texas at Arlington, USA
Karlsruhe Institute of Technology (KIT), Germany
wolfram.schommers@partner.kit.edu
gsw.schommers@t-online.de

2.1. Origin of Time: Empirical Facts

The definition of time is essential when we try to describe self-organizing processes. These processes belong to the heart of nanoscience and nanotechnology. However, the definition of "time" in traditional physics is classical in character. A realistic description of a quantum time is needed and is just at the beginning. In this chapter, we summarize new developments in this field.

2.1.1. *Quantum treatment of time*

Within traditional physics, we have one kind of time. It is the "external time", which is shown by our clocks and, in this chapter, we would like to indicate this clock time by the Greek letter τ. The time τ does not reflect the quantum aspect of time but behaves strict classically. This point has often been criticized in literature because a quantum treatment of time is missing.

Within the so-called projection principle (see, for example, Ref. [1]), the basis for a quantum treatment of time are given, which is particularly essential for nanoscience when self-organizing processes are considered. In this connection, an "operator"

$$\hat{C}(\ell, E) \tag{2.1}$$

for time becomes an important description element, and we get a novel expression for the Hamiltonian that can be expressed by

$$\hat{H}\left(\ell, -i\hbar\frac{\partial}{\partial E}\right) \tag{2.2}$$

The parameter ℓ sums up coordinates in the form of operators and the momentums. The variable E is the energy.

An operator for the time could not be introduced in traditional quantum theory. Erwin Schrödinger tried it but without success. In other words, a time operator like $\hat{C}$ is not known in the traditional quantum theory, also not the Hamiltonian $\hat{H}$ in the form of (2.2). The operator$-i\hbar\partial/\partial E$, which appears in $\hat{H}$, is the operator for the time coordinate; also, this kind of operator, i.e., $-i\hbar\partial/\partial E$, is not defined in the traditional quantum theory.

Within the frame of the projection principle, the time, denoted by the letter t, is system-specific. Each physically real object (system) has its own time structure. The clock time τ becomes a reference time. An external time parameter no longer exists.

In this chapter, we would like to discuss some principal facts with respect to space and time. Here, the empirical facts are vital and the entire development with respect to space and time has to be based on the most basic empirical facts.

2.1.2. *Space impressions*

There is an essential difference between space and time: Whereas the space is directly in front of us in the form of a continuum, the time is not directly recognizable. The time in traditional physics is merely defined by our clocks. Nevertheless, in physics, the factor "time" is essential and we have here unique assignments between observed phenomena and time. However, in contrast to our space experience, we have no immediate time experience. The space with the objects is directly before us, the time not. Let us deepen this point.

We express the time feeling through a clock; the clock shows numbers that we want to denote by the Greek letter τ. On the other hand, the space continuum, which we have directly before our eyes, comprises an infinite number of points (space points). In the case of a three-dimensional space, each point consists of three numbers, which we want to denote by x, y, z.

2.1.2.1. *The elements of space and time are not physically real*

There is an essential feature in connection with space and time: The elements of space and time x, y, z, τ cannot be detected. We cannot put a piece of space or a piece of time on the table. Therefore, the following question arises: What kind of reality is behind space and time? Let us discuss this point with respect to typical empirical facts. The following is essential here: The most direct and basic impression about the material world outside is the space structure before us in everyday life.

In Fig. 2.1, a typical empirical procedure is quoted. The world in front of us (before the eyes) reflects "everyday life" experiences and consists at a certain clock time τ of trees, houses, moving and resting cars, animals, etc. This situation is summarized and symbolically expressed in Fig. 2.1 by a geometrical structure denoted by $M(x, y, z)$.

For our life, those structures are of particular interest that vary their structure in space. We connect these structure-variations in space with the

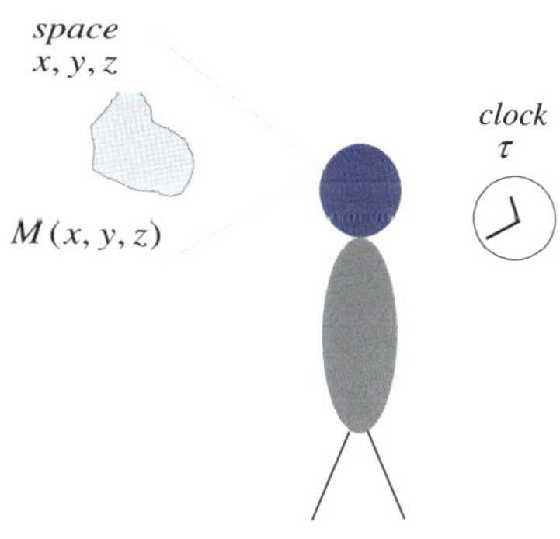

Fig. 2.1: This figure shows a situation, which human beings experience in daily life. The world in front of a human observer (before his eyes) is a space impression that he experiences unconsciously. This space impression reflects "everyday life" experiences and consists at a certain time τ of trees, houses, moving and resting cars, animals, etc. The time τ is the time indicated by a clock; it is the clock time. This situation is symbolically expressed by the geometrical structure denoted by M. The space continuum with the coordinates x, y, z is directly and spontaneously recognizable in connection with the objects, which appear as $M = M(x, y, z)$. The time continuum is not spontaneously recognizable. In traditional physics, the time is exclusively characterized by the clock time τ. The time τ has nothing to do with the systems under investigation summarized by $M(x, y, z)$. In particular, τ is an external quantity, which is indicated by a clock. However, the clock itself does not create the time τ. In particular, the clock is not coupled to the space structure $M(x, y, z)$. We assess the world in daily life by our space impressions that appear spontaneously before us at clock time τ. This conception is basic and is also the basis for physics.

clock time τ, that is, we connect the varying structures in space with an external clock.

Let us indicate the varying space structure by the letters $M_1, M_2, M_3, \ldots$ that are geometrical structures embedded in space, having the coordinates x, y, z. The structure-changes are connected to a clock showing the times $\tau_1, \tau_2, \tau_3, \ldots$. Then, we have the following assignments:

$$\begin{aligned} \tau_1 &\leftrightarrow M_1(x, y, z) \\ \tau_2 &\leftrightarrow M_2(x, y, z) \\ \tau_3 &\leftrightarrow M_3(x, y, z) \\ &\vdots \end{aligned} \tag{2.3}$$

Furthermore, it is assumed in traditional physics that these space structures represent physically real objects, material in character, which are embedded in space.

2.1.2.2. *Space feeling, time feeling and the arrow of time*

A human being experiences the world outside in everyday life as geometrical structure, which we denote by $M(x, y, z)$. The world appears in the form of separate objects, i.e., there are always "extensions". These extensions before the observer's eyes effectuate the "space feeling" of human beings.

The time feeling is based on space structures as well. The structural "changes" effectuate the "time feeling". This situation is expressed by Eq. (2.3). The space structures are fundamental. No time feeling without space (extensions).

The "arrow of time" is also based on space structures. The arrow of time reflects an economic fact. Why? We have to assume that no space structure is repeated in nature. Such a repetition would be superfluous and is therefore avoided in nature. Due to the property "no back to an already known space structure", we get an arrow of time and in particular the "arrow-of-time feeling".

2.1.3. *One-to-one correspondence*

In traditional physics, the following construction is believed to be valid: There is a space–time structure inside a human being which appears however before the eyes of the human being as a picture. It is a picture of the world outside (see in particular Fig. 2.1), and there is a space–time structure outside, which is the material world outside and contains the physically real (material) objects. It is assumed in traditional physics that there is a one-to-one-correspondence between the space–time structures "outside" and the space–time structures "inside".

However, such a view (construction) means the time τ has to be considered as physically real, and this is because we cannot explain how the structure-changes in space come into existence without such a physically real time. In other words, we cannot explain with the tools of traditional physics how the structural changes in space are materialized.

The time τ, which we measure with a clock, actually exists as external entity together with the so-called "stream of time". That is, it has to be assumed within the traditional view that there is a physically real stream of time, which effectuates the structure-changes in space (see Eq. (2.3)).

Note that there is merely the external time τ in traditional physics. No other type of time is defined here.

Thus, the material objects in space and their changes (symbolized by $M_1, M_2, M_3, \ldots$), which appear unconsciously in front of us, are driven somehow through the external time τ. This has to be assumed when we work within traditional physics. However, it is difficult to imagine such a scenario, but the real situation dictates such kind of view. In particular, it presumes that there is somewhere a time-machine that provides the time τ.

The time τ has nothing to do with the system under investigation; it is an external quantity and the clock is not coupled to $M_1, M_2, M_3, \ldots$. But within this view, there is a coupling between the time-machine and the material space structures $M_1, M_2, M_3, \ldots$. This situation is valid for every physically real system in the world. Thus, within this view (construction), the entire world is driven by the external time τ, i.e., it is driven by an external time-machine and the clock marks that.

Everything is equally grasped by the stream of time. This is the case for space structures with and without changes. Trees, houses, animals, moving and resting cars are equally concerned here. There are interacting objects as, for example, the earth with the moon, and their relative structural changes in the course of time τ. These changes are driven by the external time τ. This is the view when we consequently analyze the elements of traditional physics, that is, when we apply the traditional conception consequently.

It has to be emphasized that the appearance of an external time reflects a singularity. This time τ influences everything simultaneously and this behavior follows directly from the formulation of the physical laws. In particular, this time behavior corresponds to the time feeling of human beings. On the other hand, the external time τ cannot be explained by the physical laws itself. In other words, the conception of an external time does not belong to the scientific realm. Its creation remains hidden.

Physics is essentially working on this basis. Newton formulated his mechanics on this conception: Space structures are indexed by an external time defined with our clocks that show the parameter τ. The entire physics after Newton took over this basic conception, also the Theory of Relativity and traditional quantum theory. No doubt, the physical laws have been improved fantastically, but the principal conception remained conserved up to the present day.

2.1.4. *Basic statements*

We may state that the cause for the structural changes $M_1, M_2, M_3, \ldots$ [see Eq. (2.3)] is the time τ. The physically real (material) objects of the world are driven by the time τ. Then, τ becomes a physically real entity. What kind of thing is this entity? It is of course not a clock, but here the often discussed "time-machine" comes into play. A consequent analysis is given in the following.

The time-machine effectuates all kind of structural changes, i.e., the inner structural changes of every material object and the structural changes due to position changes of the objects relative to each other. This is the case in connection with macroscopic objects (trees, houses, cars, etc.), but also with respect to nanosystems.

There must be of course a coupling between the time-machine and the material world:

$$\textit{material world} \quad \xleftrightarrow{\textit{coupling}} \quad \textit{time-machine} \tag{2.4}$$

The structural changes effectuate the "time feeling" that human beings have. This is a qualitative effect and its origin is in fact the structural changes, which human beings have in their observations in everyday life. This time feeling is expressed by clocks, which characterize the time feeling by "numbers" and we get quantitative statements in the form of τ, which marks the clock hands. And behind all that is the time-machine: The time-machine effectuates the structural changes $M_1, M_2, M_3, \ldots$ (see Eq. (2.3)) and through the structural changes the time feeling comes into existence and gives rise to the construction of a clock, which indicates τ, which we called clock time. We have the following scheme:

$$\textit{clock time } \tau \quad \Leftarrow \quad \textit{time-machine} \tag{2.5}$$

What can we say about the time-machine, which is global in character and which influences the entire material world fundamentally and simultaneously?

2.1.5. *The features of the time-machine*

We stated above that the space is directly in front of us in the form of a continuum, but the time is not directly recognizable; the time is merely defined by our clocks. Again, the construction of an external time reflects a singularity. This kind of time influences everything and, in particular, the element τ is used in the formulation of the traditional physical laws.

However, as we already remarked, the external time τ cannot be understood by the physical laws itself. In other words, the conception of an external time and its creation do not belong to the scientific realm. But we work very successfully with this construction. It is, however, only a restricted view and leads to problems when we enter quantum theory.

2.1.5.1. *Couplings*

The time-machine, which is characterized by the stream of time and which is expressed by clocks, is a metaphysical attempt to describe (qualitatively) an "interaction". It is an interaction between the time-machine and the physically real objects in the world. The material world is coupled to the time-machine, which acts globally.

Traditional physics works well with this conception. Newton's mechanics and the Theory of Relativity are in fact decidedly successful. Nevertheless, its epistemological value is rather doubtful. Furthermore, the traditional conception of time seems to be a cul-de-sac road. In fact, no essential step further could be done during the last decades.

2.1.5.2. *The San Francisco cable car*

One of the problems is that we must assume that there is a large time-machine outside. The material objects in space and their changes are symbolized in Fig. 2.1 by M and in connection with Eq. (2.3) by $M_1, M_2, M_3, \ldots$.

All these space structures are directly in front of a human observer in daily life and are experienced unconsciously. The changes themselves are somehow driven through the external time τ. As we have already underlined, it presumes that there is somewhere a time-machine that provides the time τ with its features. However, such an assumption has to be considered as naïve and ridiculous. Nevertheless, this is the ordinary view when we analyze the situation consequently.

There is an instructive example that illustrates the situation unerringly. It is the example with the San Francisco cable car. Simon Fraser remarked [2]: *The ordinary view of time may be represented by an image: that of the underground cable of the San Francisco cable car. Driven by some distant and obscure machinery, the cable remains out of sight; we know that it moves because, at different points, the cable cars get attached to it and are carried along for a ride. Likewise, time in daily life is usually taken to be a universal, cosmic motion of the present – the now – driven by natural or divine powers to which life, matter, man, and society get attached and are carried along for a while.*

The entire world moves from one point in time to another. From Simon Fraser's point of view, it is the "present" that moves. This San Francisco model describes the situation that we believe to have in everyday life, but it is nevertheless a ridiculous, non-physical model. This conception for the clock time τ is not realistic and has to be replaced.

This time conception, also used in usual quantum theory and Newton's mechanics, is not based on a realistic physical background and is therefore metaphysical in character. Such conceptions should not be used in physics and have to be declined. More details are given in the following sections.

Again, the time τ, symbolized by a clock, is an external parameter. The clock time τ can hardly be created through an external time-machine. Such type of machine would not obey the physical laws, which are, on the other hand, valid for all the physically real objects (systems) in the world outside. Therefore, an external time-machine has to be judged as unphysical. Such a conception has to be declined for the final solution of the time problem.

In conclusion, an external time-machine in the sense of the San Francisco cable car, physically detached from the other objects, is an unphysical conception, which is not acceptable from the scientific point of view. The only what we can say about the notion "time" is that it is characterized through clocks All the developments in the scientific realm are based on this time characterization and they do not reflect new ideas concerning the conception of time.

2.1.6. *Switching off the time-machine*

No doubt, to believe that there is somewhere in the world a globally active time-machine is an ill conception. The San Francisco cable car is a nice example for this situation and makes clear that such kind of time mechanism, acting simultaneously on all the physically real systems in the world, has to be excluded. But the conception of an "external time" does hardly admit another possibility.

On the other hand, we observe the systems and their varying structures in space, as indicated by Eq. (2.3). When we work consequently within this conception, these varying structures in space are due to an "interaction" of all the physically real systems in the world with this time-machine. In the case of the San Francisco cable car, it is the interaction of everything with the cable. However, this is merely a descriptive example for this metaphysical and hidden process. As we already remarked, we have to decline such a metaphysical interaction process and we have to find another way to understand the varying structures in space (Eq. (2.3)).

Let us proceed step by step. What happens when we ignore or switch off the time-machine, i.e., when the physically real world and the time-machine, which produces the supposed external time τ, are decoupled? The answer is simple: The structure of the system under investigation remains constant in the course of time. Instead of Eq. (2.3), we obtain

$$\begin{aligned} \tau_1 &\leftrightarrow M(x,y,z) \\ \tau_2 &\leftrightarrow M(x,y,z) \\ \tau_3 &\leftrightarrow M(x,y,z) \\ &\vdots \end{aligned} \tag{2.6}$$

with

$$M(x,y,z) = M_1(x,y,z) = M_2(x,y,z) = \cdots \tag{2.7}$$

In other words, when we switch off or when we decouple the time-machine we obtain for the system under investigation a constant space structure, i.e., it becomes independent on time τ. In this case, the external time-machine is eliminated. However, this cannot be the solution to overcome the external time τ since we actually observe varying structures in space. There can only be one solution: The physically real object itself is not described adequately!

2.1.7. *The system-specific time*

From all that we have to conclude that a modified conception for "time" has to be introduced: The events in space are not driven by a global time-machine or something like that, but the quantity "time" needs a new aspect. The notion "time" has to be defined without an external time-machine, which cannot be specified and merely represents an ad hoc postulated unit.

2.1.7.1. *Physically real interactions create the system-specific time*

In Ref. [1], we argued that the phenomenon "time" should have its basis at the material level itself. The interaction of the globally acting time-machine with the physically real world (with its objects) does not produce the changes in the space structures (Eq. (2.3)), but the material objects themselves create their own time structure. The system-specific time of a material entity comes into existence through interaction processes with other material objects. In other words, a physically real system (object) defines its own time (time structure), i.e., time is no longer a feature that is imposed from outside.

This is valid for all physically real objects in reality. In this case, we get a consistent physical frame without an obscure singularity like the time-machine in analogy to the San Francisco cable car. Again, such a singularity would not be describable by the physical laws, which are, on the other hand, valid for all the other systems in the material world.

This is an important scientific step. Let us therefore repeat: Each system defines its own time structure, that is, the time itself becomes a "system-specific" quantity and is not transferred to the material objects (systems) from outside. A consistent physical frame without singularities like a time-machine is achieved in this way (see in particular Refs. [1] and [3]).

2.1.7.2. *Summary and concluding remarks*

Instead of the external time τ, coupled to an obscure time facility, we have an "internal time" that is system-specific, In the following, we will denote it by the letter t. The time τ, indicated by a clock, remains of course, but the meaning of τ is changed fundamentally. There is longer a time-machine behind the time τ because the time-machine is eliminated. Through the existence of the system-specific time t, the clock time τ becomes a reference time. The situation is summarized in the following scheme:

$$\begin{array}{c} \textit{a global time-machine} \\ \textit{creates the external time } \tau \\ \textit{(indicated by clocks)} \\ \Downarrow \\ \textit{physically real objects create} \\ \textit{the system-specific time } t \\ (\tau \textit{ is a reference time}) \end{array} \tag{2.8}$$

Let us repeat: The interaction of the globally acting time-machine with the physically real world does not produce the changes in the space structures (Eq. (2.3)), but the material objects themselves create their own time structure. The system-specific time of a material object comes into existence through interaction processes with other material objects.

The cosmic time-machine, which is still effective in connection with Eq. (2.3), is eliminated in the case of Eq. (2.6). Thus, there is no global interaction of the physically real systems with a time-machine, which produces the varying structures in space (Eq. (2.3)). When we switch off the time-machine, we obtain for the system under investigation a constant space structure, i.e., it is independent on time τ (see Eq. (2.6)). However, this cannot be the solution for our problem with time because we actually observe varying space structures.

We replace the "external time" τ through an "internal time" t, which is system-specific. This is valid for all physically real objects in reality, i.e., each system in the universe defines a system-specific time t. In this case, we get a consistent physical frame without an obscure singularity like the time-machine, which is not describable by the physical laws that are valid, on the other hand, for all the other systems in the material world.

Not only is the system-specific time t introduced for the description of the physically real system, but the nature of time τ is also changed radically: Since the time-machine is switched off or decoupled, the time τ no longer symbolizes the time-machine. However, the clock as material construction exists further, and τ is restricted to the clock alone; it is the clock time and nothing else.

In other words, in connection with Eq. (2.3) we have the following situation: When the time-machine is switched off (theoretically; the time-machine is an invention of human beings), the clock as material object remains of course conserved; the clock merely expressed the effect of the time-machine, but it does not belong to the time-machine itself. There is no longer a coupling to a time-machine, and this is because we have eliminated the time-machine. Then, the time τ, shown by our clocks, has to be interpreted as a "reference time". Note that the system itself may exist as single object and as an ensemble of objects as well.

Let us underline once again that the quantity "time" gets a new aspect: The material object under investigation defines its own time, i.e., the object itself creates the time and the time is no longer an imposition from outside. This is an important step concerning the nature of time.

The elimination of the artificial time-machine and the idea of a "system-specific" time is an essential step and can be theoretically implemented within the frame of projection theory. But there is a further point within the traditional view, which is doubtful as well: It is the so-called container model for the universe. Within the frame of this model, the space (space–time) is assumed to be a container in which the material objects are embedded. Already Ernst Mach criticized this conception strongly. In fact, the elements of space (space–time) are not observable, i.e., space and time do not belong the category of physically real units.

Therefore, another conception has to be chosen. The time-machine and the container model are both problematic constructions. The necessity for the introduction of the notion "system-specific time" and the elimination of the container model are tightly linked. Let us therefore give some general statements with respect to the container model.

2.2. The Container Model of Traditional Physics

Let us come back to the world of traditional physics, which is often discussed in connection with an external time-machine. This time-machine defines a physically real time that we marked by the letter τ. Let us assume that the time-machine is switched off or decoupled. Then, a physically real time τ is not effective and Eq. (2.6) is valid. The conception of a system-specific time t is not concerned in this subsection.

What does a human being have before his eyes? Is it directly the material world? When the objects before us are assumed to be physically real but without an existing physically real time τ (in analogy to the time-machine), we get a problem because the structural changes (Eq. (2.3)) do not have a physically real cause.

2.2.1. *Confrontation*

Let us work within the traditional frame. Structural changes exist if there are mutual interactions between the material (physically real) objects "and" if there is a physically real stream of time (time-machine). The stream of time as well as the mutual interactions between the physically real objects act simultaneously. Both points are essential factors within the traditional view, starting with Newton' mechanics.

If there is no physically real stream of time effective, i.e., if the material world is not coupled to a stream of time, we may not assume that there are physically real (material) objects embedded in space. In other words, the so-called "container model" has to be excluded in this case.

Another conception has to be introduced. The "projection principle" is obviously the most realistic frame because it considers space and time are treated in accordance with the facts. Within the frame of the projection principle, space and time are compatible with the basic features we actually observe. The projection principle does not need an unrealistic time-machine; here, the material world is projected onto space and time.

Whereas the container model is characterized by the stream of time and mutual interactions between the material objects in space, the projection principle works exclusively with physically real entities; the interactions between the physically real objects are projected onto space and time. In both cases, the observable space structures at clock time τ can be generated. In the case of the container model, we have

$$\{\textit{stream of time (global time-machine)} + \textit{interactions in space}\}$$
$$\rightarrow \textit{space-structure at time } \tau \tag{2.9}$$

On the other hand, the projection principle is based on another conception, which is distinctly different from the container model, and leads to the following scheme:

$$\{\textit{interactions projected onto space and time}\}$$
$$\rightarrow \textit{space-structure at time } \tau \tag{2.10}$$

Again, both views, characterized through scheme (2.9) and scheme (2.10), lead to the observable space structure at clock time (Fig. 2.1).

This new conception, i.e., the projection principle, is discussed in Refs. [1] and [3] and is summarized in Section 2.4.2.

The container model works well in many cases, but it is from the epistemological point of view doubtful. Real scientific progress cannot be expected when the container model in maintained. Another conception has to be introduced. The projection principle is an example. Nevertheless, the container model is used with large success in traditional physics. Let us therefore still discuss and explain some essential facts with respect to the container model.

2.2.2. *Space structures exist twice*

On what facts is the container model based? The common view can be identified that what C.G. Jung wrote [4]: *When one thinks about what consciousness really is, one is deeply impressed of the wonderful fact that an event that takes place in the cosmos outside, produces an inner picture, that the event also takes place inside ....*

What does Jung's statement exactly mean? The traditional point of view can be summarized as follows: We experience the world by our sense organs, that is, the observer interacts with reality outside. Information about the physically real (material) reality outside flows via our senses into the body of the human observer, and the mind forms a picture of it. In this way, we obtain a "picture of reality".

This is definitely a projection. In other words, we have a "reality outside" and an "inner picture" (picture of reality), and we are firmly convinced that the structure of the inner picture is identical with reality outside. In other words, most people are firmly convinced that there is a "one-to-one-correspondence" between reality and the picture. There are geometrical positions in the picture and in reality outside the geometrical positions in the picture are replaced in the world outside by material objects.

This is the opinion of C.G. Jung and a lot of other people believe that too. This standpoint follows directly from the citation above. C.G Jung defines with his statement just that what we have called "container model": We have a picture of reality, which is an inner state, that is, this picture is a state of the mind that is identical with the space impressions before our eyes (see Fig. 2.1). On the other hand, we have a physically real world outside in which material objects are located. Within this view, the reality outside consists of a space that is filled with material objects. This situation defines the container model.

For the container model, the assumption is essential that the structures in the picture are identical with those in the physically real world outside, i.e., it is assumed that there is a one-to-one-correspondence. In particular, space and time are inside the observer as well as in the physically real world outside. But we cannot proof that. In particular, the container model asserts that a human observer is able to grasp the complete world outside that exists without observations.

However, these two points are obviously unrealistic: Space and time cannot be considered as physically real entities (see, for example, Ref. [3]) and only physical real entities define the world outside. Furthermore, the principles of evolution do not allow the detection of the complete information of the world outside [5.6]. From this point of view also, a one-to one-correspondence between the structures in the picture of reality and the actual world outside is an illusion and must be considered as not realistic.

2.2.3. *The container model without time-machine*

Within the frame of the container model, we have a picture of reality, which appears directly before us, and we have a world outside that is physically real. The structures of both parts are identical, that is, we have a one-to-one correspondence.

These features are involved in many formulations. Let us come back to our statement above: "When the objects before our eyes are assumed to be physically real units but without an existing physically real time τ (in analogy to the time-machine), we get a problem because the structural changes (Eq. (2.3)) do not have a physically real cause." This statement is inevitable because the space impressions before us, which are definitely facts, cannot have a physically real background. The so-called "container model" becomes questionable; it is excluded so to speak.

When we say "the objects before our eyes", we mean the situation in the world outside, as if the picture would be the physically real world outside. This is possible in the case of the container model because here the world outside exists in exactly the form of the picture.

When the objects in the world outside are physically real units but without an existing physically real time (see Eq. (2.6), in this case the time-machine is eliminated), we get a problem because the structural changes, expressed by Eq. (2.3), cannot have a physically real cause. Then, the space impressions before us, which are facts, cannot have a physically real background, and the so-called container model has no basis. That is, when we eliminate the obscure time-machine, we eliminate simultaneously the container model. We have to recourse to more realistic views.

2.3. The Realistic Case

To assume that the time τ reflects a physically real entity like matter is not realistic; this conception is interconnected to the existence of a time-machine. Such and similar conceptions are too naive and are not convincing at all. But we do not know how the structural changes (see Eq. (2.3)) come into existence. Note that the effect of structural changes is not an interpretation, but reflects the most direct experiences that human beings have from the physically real (material) material world outside.

The introduction of the system-specific time t opens the door for a deeper understanding of the phenomenon "time". The projection principle is such a frame in which the time and also the space are treated in accordance with what we actually observe. Within this conception, the time-machine is eliminated and simultaneously the time $\boldsymbol{\tau}$ as physically real entity. Nevertheless, the clock time $\boldsymbol{\tau}$ remains in the projection theory an essential element and plays here the role of a reference time.

2.3.1. *Summary of the main facts*

The principles of the projection principle are outlined, for example, in Refs. [1] and [3]. Let us summarize here the main characteristics with respect to the system-specific time t. What are the facts?

Within the projection theory the entire material world "exists" completely at clock time τ. However, only a part is "observable". Human beings observe the x, y, z-structure at clock time $\boldsymbol{\tau}$ (see Fig. 2.1). This is the fundamental fact, which human beings experience unconsciously and which they have before their eyes in everyday life. However, this information is not complete. Let us consider for example a tree. The complete tree is defined through the entire t-spectrum together with the corresponding x, y, z-structures. The complete tree exists at every clock time τ, i.e., it is given at every clock time τ within the x, y, z, t-range for which the tree is expressible. This is in fact the result of the projection theory.

This complete information does not, however, reflect the observed situation: In the x, y, z-structure at clock time τ, the t-spectrum is scanned systematically [1] through time τ, and this scanned information, expressed by $x, y, z, t = \tau$, is observed. It is a selection process. The situation is summarized in the following scheme:

$$\begin{array}{ccc} \tau : x, y, z & & \\ \downarrow & & \\ \tau : x, y, z, t & \xrightarrow{selection} & x, y, z, t = \tau \end{array} \tag{2.11}$$

Clearly, the $x, y, z, t = \tau$-information is identical with the x, y, z-structure at clock time $\boldsymbol{\tau}$:

$$\tau : \quad x, y, z \;\leftrightarrow\; x, y, z, t = \tau \tag{2.12}$$

We have the space structures (x, y, z-structures) directly before the eyes but not the system-specific time t. However, the projection principle teaches that the numerical value of the system-specific time t is identical with the clock time $\boldsymbol{\tau}$.

In this way, the complete set x, y, z, t is obtained; the tree, its past and future with respect to the system-specific time t are existent at every clock time $\boldsymbol{\tau}$.

2.3.2. *Autonomous entities*

Physically real objects (systems) become autonomous when the system-specific time is introduced. The step from the external time τ to the internal time t is seminal: The material objects become "autonomous". It is an autonomy with respect to the description elements, i.e., with respect to the variables. The variables x, y, z, t are considered to be a complete set of description elements. That is, the existence of the physically real world outside by objects is no longer dependent on an external impact through a singular time-machine or a similar formation.

2.4. Basics

Human beings are caught in space and time. Space and time themselves do not belong to the category of physically real entities. Question: How are space and time coupled to the physically real world? The principal facts are based on space and time. The measuring instruments are constructions in space and time. Also, the observations in everyday life are space–time impressions, unconsciously perceived. The realistic theoretical description of the physically real world must therefore be based on space and time, i.e., on the variables x, y, z, t when we work within the frame of the projection principle.

The use of other variables, detached from space and time, is possible but not realistic because the connection to the physically real world is not defined. Physically real are those quantities, which connect the measuring instruments with the theoretical laws. Here, we cannot relinquish on space and time.

2.4.1. *Coupling phenomena*

The coupling phenomena between space and time and the physically real world are essential, especially the elements of space and time are not physically real quantities. This point has to be analyzed in terms of the "projection principle". The details have been discussed in Ref. [1]. Let us first repeat the main facts. After that, we will give statements about the coupling effects.

Space and time are basic elements in the life of human beings and of course in physics. As we have outlined in Section 2.1, the most basic information about the physically real (material) world outside is given by the space structures, which a human being experiences at clock time τ and which appears unconsciously before his eyes.

We underlined in Section 2.1 that space and time do not belong to the category of physically real entities. In particular, we concluded that the space structure at time τ cannot be the material world itself but a picture of it. It is a "space–time picture" that is given in terms of a x, y, z-structure at clock time τ.

In this section, we will outline the details of this view, and we will analyze the situation by means of the projection principle, which is based on quantum-theoretical wave functions.

2.4.2. *The world outside and its picture*

The physically real world is embedded in "basic reality", which we indicated in Ref. [1] by the Greek letter γ. The complex γ does not contain space and time. Only physically real entities occupy basic reality γ. We also stated in Ref. [1] that the principles of evolution are essential when we try to judge the observation possibilities for human beings.

The way from basic reality γ to the space–time picture is straightforward: The tools of evolution pull the human-specific information of the physically real world, and this pulled information has been indicated by the letter α. This information α is transferred to the "mind" of the observing human being and is processed leading to the x, y, z-structure that the human being has at clock time τ before his eyes. The information α is a projection onto space and time. It is outlined in Ref. [1] that the entity a is not dependent on time τ and behaves stationary, i.e., α takes the same value at each time τ.

2.4.2.1. *What can we observe?*

Since the entities α and γ are located outside space and time, they are not observable and, therefore, the entities α and γ have to be considered as "raw information", which cannot be characterized and described further, i.e., no variables are definable for α and γ. Thus, the entities α and γ cannot be furnished with numbers.

In connection with space and time, it is important to know "what do we observe" and "what can we observe". It is the purpose of this section to analyze the situation in this respect. For this analysis, we need the projection principle, expressed in terms of the quantum representation.

2.4.2.2. *The complete information*

Let us consider a physically real (material) object. The object is observed as x, y, z-structure. The entity α contains the "complete" information of this material object. However, the x, y, z-structure varies in the course of time τ. Since these structures are different for different times τ, the information

$$\tau: \quad x_a, y_a, z_a \le x, y, z \le x_b, y_b, z_b \tag{2.13}$$

can only be a part of the complete information α. In (2.13), the sets x_a, y_a, z_a and x_b, y_b, z_b indicate the boundaries of the material object at τ.

Again, the structures for "all" possible clock times τ represent the complete information about the material object under investigation and is expressed through the entity α. Whereas the x, y, z-structures are "system-specific", the clock time τ is not system-specific and has nothing to do with the material object under investigation; it is an external time.

The complex

$$\{(x, y, z\text{-}structure)_i, i = 1, 2, \ldots\} \tag{2.14}$$

defines the material object in "complete" form. Here, the term $(x, y, z\text{-}structures)_i$ is the space structure at clock time τ_i. Not only does the x, y, z-structure at time τ defines the material object, but the object is given for "all" x, y, z-structures in the τ-range of existence. That is, the complex (2.14) is the object and we have

$$\{(x, y, z\text{-}structure)_i, i = 1, 2, \ldots\} = \alpha \tag{2.15}$$

In conclusion, the object is not defined at a specific time but it is at each time τ completely defined through the entity α. From this point of view, the following question is wrong: Is the material object defined at time τ_n or at time τ_k or at another time? The object in complete form is not defined through a specific structure $(x, y, z\text{-}structures)_i$ at a specific time τ_i, but it is at each time τ completely defined through the entity α.

In conclusion, a physically real object is "existent" at every clock time τ, but only a part of it is "observable": The part $(x, y, z\text{-}structures)_i$ at a specific time τ_i. In other words, we have to distinguish between existence and observation.

2.4.2.3. *The time structure of material objects*

The physically real system (object), expressed through (2.14), still contains an "external" element. It is the external clock time τ. However, a material object is individual in character and should therefore be exclusively expressed by system-specific elements. The time τ is a stranger here and has to be eliminated. There are structural changes with respect to the "time". Thus, instead to the external time τ a system-specific time, indicated again by the letter t, has to be introduced. In this way, we come to x, y, z, t-structures.

Like the entity α, the x, y, z, t-*structure* too is not dependent on time τ, i.e., the x, y, z, t-structure behaves stationary with respect to clock time τ. We have at each time τ exactly the same situation for the material object under investigation:

$$\tau: \; x, y, z, t\text{-}structure \tag{2.16}$$

That is, at each clock time τ the complete material object (or even the complete material world outside) is existent. It is at each clock time τ a four-dimensional quantity. Thus, we have

$$x, y, z, t\text{-}structure = \alpha \tag{2.17}$$

With relation (2.15), we immediately get the transition from the traditional view to the projection principle:

$$\{(x, y, z\text{-}structures)_i, i = 1, 2, \ldots\} \Rightarrow x, y, z, t\text{-}structure \tag{2.18}$$

In particular, we get with (2.15) and (2.17)

$$\begin{aligned} &\{(x, y, z\text{-}structures)_i, i = 1, 2, \ldots\} = \\ &\quad x, y, z, t\text{-}structure = \alpha \end{aligned} \tag{2.19}$$

The clock time τ is now eliminated in the characterization of a stationarily existing material object. This is in fact a requirement of the analysis when we judge physical real situations in terms of the entity α, which is directly a result of the projection theory. Instead of structure variations with respect to clock time τ, we have now structure variations in connection with the system-specific time t. The clock time τ remains, but plays now the role of a reference time.

2.4.3. *The general case*

In Section 2.1.7, we discussed the system-specific time t and we compared this new type of time with the clock time τ, which provides the existence of a global, external physically real time-machine. In Section 2.4.2.3, we described the system-specific time by means of the projection principle. This principle makes the existence of the system-specific time t necessary.

2.4.3.1. *Functional connections*

The human-specific basic information α (it can be a single physically real object or even the entire world) has been identified with the complete information we marked by x, y, z, t-*structure*. On the other hand, the information complex x, y, z, t-*structure* is the complete information and summarizes all space structures at various clock times, which is expressed by Eqs. (2.15) and (2.19).

We may bring together the information $\{(x, y, z\text{-}structures)_i, i = 1, 2, \ldots\}$ with the functions $M_1, M_2, M_3, \ldots$, introduced in Sec. 2.1.2. We immediately get

$$\begin{aligned} (x, y, z\text{-}structures)_1 &= M_1(x, y, z) \\ (x, y, z\text{-}structures)_2 &= M_2(x, y, z) \\ &\vdots \end{aligned} \tag{2.20}$$

With the information complex x, y, z, t-*structure* (see Eq. (2.16)) we may define a function $M(x, y, z, t)$, which is required to fulfill

$$M(x, y, z, t) = x, y, z, t\text{-}structure \tag{2.21}$$

The function $M(x, y, z, t)$ describes the entire world or a single material object in human-specific form, i.e., it is identical with the entity α (see also Eq. (2.19)):

$$M(x, y, z, t) = \alpha \tag{2.22}$$

In Section 2.4.4, we will express $M(x, y, z, t)$ through the wave function and treat the characteristics within the frame of the projection principle.

2.4.3.2. *Selections with respect to time*

Human beings do not register the complete space–time structure $M(x, y, z, t)$ at time τ, but only a part of it. The part $M(x, y, z, t = \tau)$ of the complete structure $M(x, y, z, t)$ is selected [1, 3]:

$$M(x, y, z, t) \xrightarrow{selection} M(x, y, z, t = \tau) \tag{2.23}$$

The traditionally defined space structure x, y, z at time τ is then identical with the corresponding structure $M(x, y, z, t = \tau)$:

$$\begin{aligned} M(x, y, z)\, at\, \tau &= (x, y, z\text{-}structure)\, at\, \tau \\ &= M(x, y, z, t = \tau) \end{aligned} \tag{2.24}$$

The material world "exists" in complete form at every time τ [see Eqs. (2.16) and (2.17)], but only a part of can be "observed". This is the message in connection with relation (2.22). A model for the selection process itself can be developed easily and is done in Ref. [3].

2.4.4. *Fictitious realities*

We know nothing about basic reality, which contains all that is physically real. However, basic reality exists. Otherwise, the projection principle would lose its sense. The non-existence of space and time as physically real entities requires the introduction of basic reality.

Even the human-specific part α of the physically real world cannot be observed and described in its basic form without space and time. The description of α is just a space–time picture consisting of geometrical structures. The human observer's mind projects unconsciously the entity α onto space and time [1] having the variables x, y, z, t.

The geometrical points in space and time are correlated. The correlations must be based on physically real interactions, which take place "outside" space and time in basic reality. The reason is obvious: The picture with its elements is not physically real. The space–time elements x, y, z, t are not observable and cannot be measured by measuring instruments (see Section 2.1).

The quantities x, y, z, t are not located in basic reality; only physically real entities are embedded in basic reality. Therefore, the physically real world (basic reality) is not directly accessible to human beings. The projection principle teaches that the impression before our eyes is not the material world itself, but it is a space–time picture [1, 3].

Human beings are caught in space and time. Observations and measurements are exclusively done within space and time, i.e., human beings cannot leave space and time and they can merely recognize "correlations" between the geometrical entities in the picture, but not the physically real interaction processes themselves. The physically real interaction processes exclusively take place in basic reality.

In order to learn something about physically real interaction processes, we need a reality, which corresponds to basic reality or more precisely, which corresponds to the "human-specific" part of basic reality. Again, the principles of evolution only allow to extract the human-specific part of a physically real entity from basic reality. This part can be developed on the basis of space and time. This has been done in the literature (see, for example, Refs. [1] and [3]). In this way, we obtain "fictitious realities". A fictitious reality is a certain kind of virtual reality.

In other words, instead of basic reality, which is not directly accessible and which is the seat of the physically real interactions between the material objects, we construct fictitious realities. For this purpose, we have to introduce variables for the characterization of the fictitious reality. These variables are based on space and time, i.e., on the variables x, y, z, t on which the picture is based.

In Refs. [1] and [3], it is argued that the variables of a fictitious reality are given by the momentum $\mathbf{p} = (p_x, p_y, p_z)$ and the energy E, which are in fact introduced on the basis of space and time and contain the variables x, y, z, t. This situation can be expressed by

$$x, y, z, t \rightarrow p_x, p_y, p_z, E \tag{2.25}$$

i.e., the set p_x, p_y, p_z, E is constructed by the variables x, y, z, t. The picture is located in $(\mathbf{r}, t)$-space with $\mathbf{r} = (x, y, z)$, and fictitious reality is embedded in $(\mathbf{p}, E)$-space.

2.4.4.1. *The quantum-theoretical treatment*

We can describe the general function $M(x, y, z, t)$ in a quantum-theoretical way. Let us denote the quantum information in $(\mathbf{r}, t)$-space by the function $\Psi(x, y, z, t)$, which we called "wave function" in Refs. [1] and [3]. The corresponding wave function in $(\mathbf{p}, E)$-space is marked by $\Psi(p_x, p_y, p_z, E)$.

The function $\Psi(x, y, z, t)$ is the quantum-theoretical information and is an example for the description of the general function $M(x, y, z, t)$:

$$M(x, y, z, t) \leftarrow \Psi(x, y, z, t) \tag{2.26}$$

The function $\Psi(p_x, p_y, p_z, E)$ and the function $\Psi(x, y, z, t)$ belong together and are connected through a Fourier transform, which we express symbolically by

$$\Psi(x, y, z, t) \underset{\longleftarrow}{\xrightarrow{\textit{Fourier transform}}} \Psi(p_x, p_y, p_z, E) \tag{2.27}$$

Both functions contain exactly the same information about the physically real object under investigation.

$\Psi(p_x, p_y, p_z, E)$ reflects reality (fictitious reality) and this entity is projected onto $(\mathbf{r}, t)$-space and we obtain the wave function $\Psi(x, y, z, t)$, which is the picture.

2.4.4.2. *Connections*

Figure 2.1 shows a human being, which has at clock time τ a certain space structure before his eyes. This space structure reflects a geometrical representation (x, y, z-structure) of the material objects of the human being's environment (world outside). The x, y, z-structure is observed at clock time τ and let us denote it by $(x, y, z\text{-}structure)_{observed}$.

The x, y, z-structure is the result of a process in the observer's mind based on the raw information α, which is human-specific and which we have introduced above. The transition from α to the x, y, z-structure is a projection; the x, y, z-structure at time τ reflects a picture of the material objects. We may depict the situation by the following scheme:

$$\begin{gathered} \alpha \\ \downarrow \\ (x, y, z\text{-}structure)_{observed}\, at\, \tau \end{gathered} \tag{2.28}$$

Equation (2.28) reflects the situation with respect to direct observations in everyday life and which is experienced "unconsciously".

The "conscious" treatment of observational facts is based on the wave functions $\Psi(p_x, p_y, p_z, E)$ and $\Psi(x, y, z, t)$. As we have outlined in Ref. [1], the wave function $\Psi(p_x, p_y, p_z, E)$ describes what we called "fictitious reality" and corresponds to the raw information α. Let us repeat: The unit α cannot be described in terms of variables and, therefore, the wave function $\Psi(p_x, p_y, p_z, E)$ is not α, but only corresponds to it:

$$\alpha \hat{=} \Psi(p_x, p_y, p_z, E) \tag{2.29}$$

As we already remarked, the wave function $\Psi(x, y, z, t)$ reflects the picture of $\Psi(p_x, p_y, p_z, E)$; both functions are connected through a Fourier transform (see Eq. (2.27)). The information $\Psi(p_x, p_y, p_z, E)$ is projected onto $(\mathbf{r}, t)$-space. We have

$$\begin{gathered} \Psi(p_x, p_y, p_z, E) \\ \downarrow \\ \Psi(x, y, z, t) \end{gathered} \tag{2.30}$$

As we already outlined, the space structure at clock time τ, expressed by $M(x, y, z, t = \tau)$, is a selection from the complete information $M(x, y, z, t)$ (see Eq. (2.22)). With (2.26), we obtain the quantum-theoretical space structure at τ, i.e., the function $\Psi(x, y, z, t = \tau)$, from the complete information $\Psi(x, y, z, t)$:

$$\Psi(x, y, z, t) \xrightarrow{selection} \Psi(x, y, z, t = \tau) \tag{2.31}$$

Then, Eq. (2.30) combined with (2.31) leads to

$$\begin{array}{ccc} \Psi(p_x, p_y, p_z, E) & & \\ \downarrow & & \\ \Psi(x, y, z, t) & \xrightarrow{selection} & \Psi(x, y, z, t = \tau) \end{array} \tag{2.32}$$

The information $\Psi(x, y, z, t = \tau)$ has to be identified with the unconsciously observed impression $(x, y, z\text{-}structure)_{observed}$ *at* τ and we have with Eq. (2.28) the following connection:

$$(x, y, z\text{-}structure)_{observed}\ at\ \tau \ \leftarrow\ \Psi(x, y, z, t = \tau) \tag{2.33}$$

In other words, if the function $\Psi(x, y, z, t)$ is formulated realistically by a suitable quantum-theoretical model, the complex $(x, y, z\text{-}structure)_{observed}$ at clock time τ, which appears unconsciously before the eyes of a human being, is described.

The unconsciously observed data and the conscious developed theoretical wave functions are connected through scheme (2.28):

$$\begin{array}{ccccc} \alpha & \hat{=} & & & \Psi(p_x, p_y, p_z, E) \\ \downarrow & & & & \downarrow \\ (x, y, z\text{-}structure)_{observed} & \Leftrightarrow & \Psi(x, y, z, t = \tau) & \xleftarrow{selection} & \Psi(x, y, z, t) \\ at\ \tau & & & & \end{array} \tag{2.34}$$

The observed complex $(x, y, z\text{-}structure)_{observed}$ at clock time τ is the impression, which a human being has before his eyes; this impression is "macroscopic" in character. The wave functions $\Psi(x, y, z, t)$ and $\Psi(p_x, p_y, p_z, E)$ in (2.34) are descriptions at the quantum level.

The macroscopic characteristics can be formulated in terms of quantum states, and the quantum states are expressed by the wave functions $\Psi(x, y, z, t)$ and $\Psi(p_x, p_y, p_z, E)$. The transition from the quantum level to the macroscopic level has been discussed in detail in Ref. [1].

The function $M(x, y, z, t)$ [Eq. (2.21)] reflects a general quantity; the system-specific time t, which appears in $M(x, y, z, t)$ is therefore general too, i.e., it is independently from the level at which the physically real object is considered (classical or quantum-theoretical). The appearance of t in $M(x, y, z, t)$ is a requirement of the projection principle.

We introduced the wave functions $\Psi(x, y, z, t)$ and $\Psi(p_x, p_y, p_z, E)$; both functions describe quantum states. $\Psi(x, y, z, t)$ and $\Psi(p_x, p_y, p_z, E)$ have been developed independently from $M(x, y, z, t)$. The details are given, for example, in Refs. [1] and [3]. The system-specific time t in $\Psi(x, y, z, t)$ has be identified with the system-specific time t, which appears in the function $M(x, y, z, t)$. The general function $M(x, y, z, t)$ can be formulated at the quantum level, that is, on the basis of $\Psi(x, y, z, t)$ and $\Psi(p_x, p_y, p_z, E)$, at least in principle. In our discussion, we go from $M(x, y, z, t)$ to $\Psi(x, y, z, t)$ and $\Psi(p_x, p_y, p_z, E)$, i.e., we enter with $\Psi(x, y, z, t)$ and $\Psi(p_x, p_y, p_z, E)$ at the quantum level.:

$$M(x, y, z, t) \to \Psi(x, y, z, t), \Psi(p_x, p_y, p_z, E) \tag{2.35}$$

The quantum information $\Psi(p_x, p_y, p_z, E)$ is located in $(\mathbf{p}, E)$-space and $\Psi(x, y, z, t)$ is embedded in $(\mathbf{r}, t)$-space. The reality (fictitious reality) is described by $\Psi(p_x, p_y, p_z, E)$ and this information is projected onto $(\mathbf{r}, t)$-space and gives $\Psi(x, y, z, t)$.

2.4.5. *Specific properties of the wave functions*

As we remarked several times, in traditional physics, the container-model is used. The physically real world is at every clock time τ embedded in space having the coordinates x, y, z. Within the traditional view, the entire world is driven by the external time τ, i.e., it is driven by an external time-machine and the clock marks that. In other words, behind the clock time, τ hides a global mechanism, which effectuates the structural changes (Eq. (2.3)). Such a global mechanism is rather metaphysical in character.

In contrast to that, within the frame of the projection principle, we work with "two" spaces: $(\mathbf{p}, E)$-space and $(\mathbf{r}, t)$-space. In particular, the projection principle works with the system-specific time t, which has a realistic background and is described through interaction processes with other material objects (Refs. [1] and [3]). The projection principle works with "fictitious realities" and their "pictures of reality". The bases are here the wave functions $\Psi(x, y, z, t)$ and $\Psi(p_x, p_y, p_z, E)$. Again, the quantum information $\Psi(p_x, p_y, p_z, E)$ is located in $(\mathbf{p}, E)$-space (fictitious realty) and $\Psi(x, y, z, t)$ is embedded in $(\mathbf{r}, t)$-space (picture of reality).

As is outlined in Refs. [1] and [3], the wave functions $\Psi(p_x, p_y, p_z, E)$ and $\Psi(x, y, z, t)$ are connected through a Fourier transform and the corresponding inverse Fourier transform

$$\Psi(x,y,z,t) = \frac{1}{(2\pi\hbar)^2}\int_{-\infty}^{\infty}\Psi(p_x,p_y,p_z,E)\exp\left\{i\left[\frac{\mathbf{p}}{\hbar}\cdot\mathbf{r}-\frac{E}{\hbar}t\right]\right\}dp_x dp_y dp_z dE \tag{2.36}$$

and

$$\Psi(p_x,p_y,p_z,E) = \frac{1}{(2\pi\hbar)^2}\int_{-\infty}^{\infty}\Psi(x,y,z,t)\exp\left\{-i\left[\frac{\mathbf{p}}{\hbar}\cdot\mathbf{r}-\frac{E}{\hbar}t\right]\right\}dt dx dy dz \tag{2.37}$$

where $\hbar$ is Planck's constant.

The quantum-theoretical formulation of the theory is the starting point and from this the macroscopic states, as they are observed in everyday life, can be formulated. That is, the macroscopic states and the quantum states are not independent phenomena but the term "macroscopic" is an observation phenomenon, which has its footing at the quantum level.

The conception only makes sense if at least one set of variables is existent at clock time τ. This set is given in connection with $(\mathbf{p}, E)$-space by the four variables p_x, p_y, p_z, E and for $(\mathbf{r}, t)$-space by x, y, z, t:

$$\begin{aligned} \tau &: \ x, y, z, t \\ \tau &: \ p_x, p_y, p_z, E \end{aligned} \tag{2.38}$$

From the Fourier transforms (2.36) and (2.37), we have to conclude that each set x, y, z, t of variables contains the complete p_x, p_y, p_z, E-ensemble (marked by $\{p_x, p_y, p_z, E\}$) of $(\mathbf{p}, E)$-space and, on the other hand, that each set p_x, p_y, p_z, E of the variables contains the complete x, y, z, t-ensemble (marked by $\{x, y, z, t\}$) of $(\mathbf{r}, t)$-space. In other words, at time τ, the entire distributions $\{x, y, z, t\}$ and $\{p_x, p_y, p_z, E\}$ are simultaneously existent. The ranges of the ensemble-values are defined through the wave functions.

In conclusion, due to the features of the Fourier transforms (2.36) and (2.37), we have the following situation:

$$\begin{aligned} \tau &: \ \{p_x, p_y, p_z, E\} \to x, y, z, t \\ \tau &: \ \{x, y, z, t\} \to p_x, p_y, p_z, E \end{aligned} \tag{2.39}$$

In particular, this means that the wave functions $\Psi(x, y, z, t)$ and $\Psi(p_x, p_y, p_z, E)$ are simultaneously existent at time τ:

$$\tau: \ \Psi(x, y, z, t), \Psi(p_x, p_y, p_z, E) \tag{2.40}$$

All these features are reflected by the Fourier transforms (2.36) and (2.37). The projection principle is a conception, which connects basic reality γ with the space–time picture, which is existent at clock time τ as a x, y, z-structure. The picture reflects the human-specific part α, which is pulled from γ.

Also, the wave functions $\Psi(p_x, p_y, p_z, E)$ and $\Psi(x, y, z, t)$ are connected to each other and both functions form a unit and belong together, as in the case α and the observed x, y, z-structure at clock time τ. Both functions, $\Psi(p_x, p_y, p_z, E)$ and $\Psi(x, y, z, t)$, define a projection peculiarity: $\Psi(p_x, p_y, p_z, E)$ describes (fictitious) reality in $(\mathbf{p}, E)$-space and $\Psi(x, y, z, t)$ is the picture in $(\mathbf{r}, t)$-space.

The projection principle, defined by $\Psi(p_x, p_y, p_z, E)$ and $\Psi(x, y, z, t)$, is in principle independent from the projection peculiarities given through the $(\alpha - (x, y, z, \tau\text{-}\textit{structure}))$-unit. However, both conceptions lead to space–time pictures. Thus, the models for $\Psi(x, y, z, t)$ have to be constructed so that they are compatible with the observed x, y, z-structure at clock time τ.

2.4.6. *Physically real objects represented in space and time*

What kind of a thing is a physically real object within the frame of the projection principle? Within this view, a physically real object is existent outside space and time and is located in basic reality. The reality is described by the momentum $\mathbf{p} = (p_x, p_y, p_z)$ and the energy E. In contrast to Newton's theory, the elements $\mathbf{p}$ and E cannot be considered as physically real. The quantities $\mathbf{p}$ and E are introduced on the basis of space and time and contain the variables $x, y, z, t.$, i.e., we have $x, y, z, t \rightarrow p_x, p_y, p_z, E$ (see also Eq. (2.25)). Since the elements x, y, z, t are not physically real, also the quantities p_x, p_y, p_z, E cannot be considered as physically real.

The basic conception of Newton's theory is different. Let us briefly repeat: Within Newton's theory the physically real (material) objects are located in space and move through space in the course of clock time τ. Within this view, it is assumed that the object has a definite mass m_0, a definite momentum $\mathbf{p}$ and of course the energy E. The quantities m_0, $\mathbf{p}$ and E are physically real within this conception.

The projection principle is different and this is of course due to the fact that space and time are not physically real and, therefore, physically real entities cannot be put into such kind of space. In the projection theory, we have the following situation: The processes in connection with observations are physically real and, therefore, they take place in basic reality. But the detector appears in the space picture as well and the human observer registers the x, y, z-pattern at time τ, which are indicated by the detector in the picture. These findings are described by the wave function $\Psi(x, y, z, t)$ for $t = \tau$. The wave function $\Psi(x, y, z, t)$ is not physically real and also not the elements x, y, z, t, but $\Psi(x, y, z, t)$ "describes" physically real facts, i.e., the x, y, z-pattern at time τ which is registered by the detector. The detector itself has to be constructed in accordance with the principle of the theory (projection theory).

Since the functions $\Psi(p_x, p_y, p_z, E)$ and $\Psi(x, y, z, t)$ are coupled [Eqs. (2.36) and (2.37)], also $\Psi(p_x, p_y, p_z, E)$ together with the variables p_x, p_y, p_z, E cannot be considered as physically real quantities.

Nevertheless, $(\mathbf{p}, E)$-space, in which the information $\Psi(p_x, p_y, p_z, E)$ is embedded, has to be considered as a specific form of reality. We called it "fictitious reality" and can be considered as "virtual reality". Reality means that systematic processes are definable. This is in fact the case when we work within the frame of $(\mathbf{p}, E)$-space. The conservation laws for the momentum and for the energy that must be fulfilled dictate the kind of processes, qualitatively and quantitatively. An object interacts with its environment. The interaction is described by $\mathbf{p}, E$-fluctuations for which the conservation laws for the momentum and for the energy are fulfilled. These are $\mathbf{p}, E$-fluctuations between the object under investigation and other objects. A more detailed discussion is given in Section 2.5.3. From our analysis above, we know that in fictitious reality no physically real objects can be embedded, but we may state that fictitious reality can be seen as if it were physically real.

2.4.7. *The traditional view*

Within the traditional treatment of quantum systems, Schrödinger's equation is in the foreground:

$$i\hbar\frac{\partial}{\partial\tau}\varphi(x, y, z, \tau) = -\frac{\hbar^2}{2m_0}\Delta\varphi(x, y, z, \tau) + U(x, y, z)\varphi(x, y, z, \tau) \quad (2.41)$$

We are only interested in principal questions. Therefore, it is sufficient to discuss Schrödinger's equation for the simple case as is given by Eq. (2.40): Here, an object of mass m_0 in an external field $U(x, y, z)$ is formulated.

2.4.7.1. *No quantum aspect of time*

The object, described by the wave function $\varphi(x, y, z, \tau)$, is existent at clock time τ. It is not existent for $\tau < \tau'$ and $\tau > \tau'$. We have the following classification:

$$\begin{aligned} &\textit{physically real object:} \\ &\qquad \textit{existent at } \tau \\ &\qquad \textit{and} \\ &\qquad \textit{not existent and} \tau > \tau' \textit{ and} \tau > \tau' \end{aligned} \tag{2.42}$$

This feature underlines the three-dimensional character of the physically real object within the traditional form of quantum theory.

It is relevant to mention here that Schrödinger's equation (2.41) could not be derived but was assumed on the basis of "reasonable arguments". They can be validated only in connection with experiments. In this respect, the success of Schrödinger's equation is without any doubt impressive. However, we do not really know which physical reality is reflected by this equation and its variations. Thus, it is not or almost not possible to develop these equations systematically further. This is, however, desirable just in connection with the phenomenon "time". The time τ in Schrödinger's equation is still a classical parameter within Schrödinger's theory. There is no quantum aspect of time in usual quantum theory.

2.4.7.2. *Point-like particle in space*

Max Born proposed to interpret the wave function $\varphi = \varphi(x, y, z, \tau)$ not as true material wave but as a "probability wave". According to him, the probability of finding a real quantum object (assumed to be a point-like particle) in the volume element $dV = dxdydz$ at a certain position x, y, z and time τ is given by

$$\varphi^*(x, y, z, \tau)\varphi(x, y, z, \tau)dxdydz \tag{2.43}$$

Thus, $\varphi^*\varphi$ plays the role of a probability density. Born's probability interpretation is one of the fundamentals of usual quantum theory.

In connection with Born's probability interpretation of Schrödinger's equation, a physically real particle has to be assumed. The particle itself reflects a point-like structure and it has the physically real mass m_0. That is, a particle has to be "postulated", which is needed for the understanding of Schrödinger's equation. This particle does not follow from the theory itself. But there is the possibility for its introduction within Schrödinger's theory and its probability interpretation. Nevertheless, it is an artificial construction.

Remark

How do we come to Eq. (2.41)? The way is not based on known physical arguments, but "postulates" are used here. Schrödinger's equations can be formulated in the following way: The classical expression for the energy of an object with mass m_0 in an external field (potential energy) $U(x, y, z)$ is given by

$$E = \frac{1}{2m_0}(p_x^2 + p_y^2 + p_z^2) + U(x, y, z) \tag{2.44}$$

Then, Schrödinger's equation (2.41) can be obtained if specific "postulates" are used: The variables

$$\begin{aligned} &E, \\ &p_x, p_y, p_z, \end{aligned} \tag{2.45}$$

in (2.44) are exchanged by operators and we get

$$\begin{aligned} &\hat{E} = i\hbar\frac{\partial}{\partial \tau} \\ &\hat{p}_x = -i\hbar\frac{\partial}{\partial x}, \hat{p}_y = -i\hbar\frac{\partial}{\partial y}, \hat{p}_z = -i\hbar\frac{\partial}{\partial z}, \\ &x, y, z \end{aligned} \tag{2.46}$$

So, the postulate "keep the coordinates as they are and replace the energy and momenta by *operators*". The source of these two postulates is unknown when we work within usual quantum theory.

2.5. Projection Principle

Within traditional quantum theory, the wave function $\varphi(x, y, z, \tau)$ is of particular interest. This function is determined on the basis of Schrödinger's equation (2.41). We need the interaction $U(x, y, z)$ between a point-like object and its environment. The situation in projection is similar, but the basic equations have to be interpretated quite differently from the corresponding expressions of traditional quantum theory. Let us discuss the situation.

2.5.1. *The basic equations*

As we remarked, within projection theory we work with two spaces, $(\mathbf{p}, E)$-space and $(\mathbf{r}, t)$-space. The notion "time" is here not based on an external time parameter τ, which is identical with that what a clock indicates. Once again, within this view (construction), the entire world is driven by the external time τ, i.e., it is driven by an external time-producing mechanism and the clock marks that. The conception of an external time is eliminated in projection theory and the system-specific t becomes relevant. The time appears in $(\mathbf{r}, t)$-space as variable t and in $(\mathbf{p}, E)$-space as operator

$$\hat{t} = -i\hbar \frac{\partial}{\partial E} \tag{2.47}$$

We obtain equations for the determination of the wave functions $\Psi(p_x, p_y, p_z, E)$ and $\Psi(x, y, z, t)$ in the following way [1, 3]: As in the case of Schrödinger's equation, we start from the classical relation (2.44) and we get on the footing of the two spaces $(\mathbf{p}, E)$-space and $(\mathbf{r}, t)$-space operator rules for the variables.

The momentums p_x, p_y, p_z and the energy E become operators $\hat{p}_x, \hat{p}_y, \hat{p}_z$ in $(\mathbf{r}, t)$-space and we have [1, 3]

$$\hat{p}_x = -i\hbar \frac{\partial}{\partial x}, \hat{p}_y = -i\hbar \frac{\partial}{\partial y}, \hat{p}_z = -i\hbar \frac{\partial}{\partial z}, \hat{E} = i\hbar \frac{\partial}{\partial t} \tag{2.48}$$

The variables x, y, z, t remain as they are in $(\mathbf{r}, t)$-space. In other words, in $(\mathbf{r}, t)$-space, the following elements appear:

$$\begin{aligned} &(\mathbf{r}, t)\text{-}space : x, y, z, t, \\ &-i\hbar \frac{\partial}{\partial x}, -i\hbar \frac{\partial}{\partial y}, -i\hbar \frac{\partial}{\partial z}, i\hbar \frac{\partial}{\partial t} \end{aligned} \tag{2.49}$$

On the other hand, in the case of $(\mathbf{p}, E)$-space, the situation is similar. The coordinates x, y, z and the time t become operators $\hat{x}, \hat{y}, \hat{z}, \hat{t}$ in $(\mathbf{p}, E)$-space and we have [1, 3]

$$\hat{x} = i\hbar \frac{\partial}{\partial p_x}, \hat{y} = i\hbar \frac{\partial}{\partial p_y}, \hat{z} = i\hbar \frac{\partial}{\partial p_z}, \hat{t} = -i\hbar \frac{\partial}{\partial E} \tag{2.50}$$

The variables p_x, p_y, p_z, E remain as they are in $(\mathbf{p}, E)$-space. That is, in $(\mathbf{p}, E)$-space, the following elements appear:

$$\begin{aligned} &(\mathbf{p}, E)\text{-}space : p_x, p_y, p_z, E \\ &i\hbar \frac{\partial}{\partial p_x}, i\hbar \frac{\partial}{\partial p_y}, i\hbar \frac{\partial}{\partial p_z}, -i\hbar \frac{\partial}{\partial E} \end{aligned} \tag{2.51}$$

It is important to underline that in contrast to the usual quantum theory, the operator rules in (2.49) and (2.51) are not postulates, but are an outcome of projection theory, i.e., they are derived here. The operator rules in (2.46) are postulates and cannot be derived in the usual quantum theory.

Using Eq. (2.44) and the rules listed in (2.49) and (2.51), we immediately get relations for the determination of the wave functions $\Psi(p_x, p_y, p_z, E)$ and $\Psi(x, y, z, t)$

($\mathbf{r}$,t)-space:

$$i\hbar\frac{\partial}{\partial t}\Psi(x,y,z,t) = -\frac{\hbar^2}{2m_0}\Delta\Psi(x,y,z,t) + V(x,y,z,t)\Psi(x,y,z,t) \tag{2.52}$$

($\mathbf{p}$,E)-space:

$$\begin{aligned} E\,\Psi(p_x,p_y,p_z,E) &= \frac{p_x^2+p_y^2+p_z^2}{2m_0}\Psi(p_x,p_y,p_z,E) \\ &+ V\left(i\hbar\frac{\partial}{\partial p_x}, i\hbar\frac{\partial}{\partial p_y}, i\hbar\frac{\partial}{\partial p_z}, -i\hbar\frac{\partial}{\partial E}\right)\Psi(p_x,p_y,p_z,E) \end{aligned} \tag{2.53}$$

Instead of the interaction potential $U(x, y, z)$, which is used within the frame of Schrödinger's equation, we have two entities, the function

$$V(x,y,z,t), \quad (\mathbf{r},t)\text{-space} \tag{2.54}$$

and the operator

$$V\left(i\hbar\frac{\partial}{\partial p_x}, i\hbar\frac{\partial}{\partial p_y}, i\hbar\frac{\partial}{\partial p_z}, -i\hbar\frac{\partial}{\partial E}\right), \quad (\mathbf{p},E)\text{-space} \tag{2.55}$$

Both terms, (2.54) and (2.55), describe the same physically real object. Relation (2.54) is the equation for the description in $(\mathbf{r}, t)$-space and relation (2.55) is the equation for the description in $(\mathbf{p}, E)$-space. The situation is summarized in the following.

2.5.2. *Probabilities*

What is the meaning of the wave functions $\Psi(x, y, z, t)$ and $\Psi(p_x, p_y, p_z, E)$, which appear in Eqs. (2.52) and (2.53)?

The wave functions $\Psi(x, y, z, t)$ and $\Psi(p_x, p_y, p_z, E)$ reflect probabilities for the variables. The probability of finding the variables x, y, z, t in the four-dimensional space–time element $dxdydzdt$ around a certain space–time position in $(\mathbf{r}, t)$-space is given by

$$dW_{\mathbf{r},t} = \Psi^*(x,y,z,t)\Psi(x,y,z,t)dxdydzdt \tag{2.56}$$

The quantity $\Psi^*(x, y, z, t)\Psi(x, y, z, t)$ in (2.56) is the probability density. The physically real object takes one set x, y, z, t at clock time τ, i.e., there is one x, y, z, t-event at τ. In the course of time τ, the object creates incessantly x, y, z, t-events within the space–time range for which the probability density is not zero; each of these x, y, z, t-events is given with a certain probability.

The situation in $(\mathbf{p}, E)$-space is similar. The probability of finding the variables p_x, p_y, p_z, E in the element $dp_x dp_y dp_z dE$ around a certain $\mathbf{p}, E$-position is given by

$$dW_{\mathbf{p},E} = \Psi^*(p_x, p_y, p_z, E)\Psi(p_x, p_y, p_z, E)dp_x dp_y dp_z dE \tag{2.57}$$

The quantity $\Psi^*(p_x, p_y, p_z, E)\Psi(p_x, p_y, p_z, E)$ in (2.55) is the probability density in $(\mathbf{p}, E)$-space. The physically real object takes one set p_x, p_y, p_z, E at time τ, i.e., there is one p_x, p_y, p_z, E-event at τ. In the course of time τ the object creates incessantly p_x, p_y, p_z, E-events within the $\mathbf{p}, E$-range for which the probability density is not zero; each of these p_x, p_y, p_z, E-events is given with a certain probability.

The probability behavior within the frame of projection theory reflects a general feature and is valid for each kind of object (system). In Ref. [7], the probability laws are derived, i.e., the statistical behavior of physically real objects is not only an interpretation.

2.5.3. *Interactions*

2.5.3.1. *Constant momentums and energies*

We consider again a physically real object having the momentum $\mathbf{p}$ and the energy E. It is easy to show that in the case of constant momentum $\mathbf{p}$ and constant energy E the object does not exist; if $\mathbf{p} = \mathbf{p}_0$, $E = E_0$, the wave functions $\Psi(p_{x0}, p_{y\beta}, p_{z0}, E_0)$ and $\Psi(x, y, z, t)$ are exactly zero:

$$\begin{gathered}\mathbf{p} = \mathbf{p}_0 = (p_{x0}, p_{y\beta}, p_{z0}) = constant \\ E = E_0 = constant \\ \downarrow \\ \Psi(p_{x0}, p_{y\beta}, p_{z0}, E_0 = 0 \\ \Psi(x, y, z, t) = 0\end{gathered} \tag{2.58}$$

In other words, in the case of (2.58), a physically real object does not exist. There are no changes in $\Delta\mathbf{p}$ with respect to the momentum and there are no changes in ΔE with respect to the energy:

$$\Delta \mathbf{p} = 0, \quad \Delta E = 0 \tag{2.59}$$

Again, in the case of (2.58), the object (system) does not exist. That is, the existence of a physically real object requires the following condition:

$$\Delta \mathbf{p} \neq 0, \quad \Delta E \neq 0 \tag{2.60}$$

This is a basic point. Due to the conservation of momentum and energy, Eq. (2.60) means that there is an exchange of momentum and energy with the environment, at least with one object of the environment. This situation defines the notion "interaction". Within the projection theory, the interaction between two physically real objects is described by an exchange of momentum and energy. The notion "interaction" is defined through $\mathbf{p}, E$-fluctuations.

2.5.3.2. *Momentum-energy fluctuations*

The physically real object is located in $(\mathbf{p}, E)$-space. and the various $(\mathbf{p}, E)$-states are given at clock time τ with a definite probability and the probability density $\Psi^*(p_x, p_y, p_z, E)\Psi(p_x, p_y, p_z, E)$ is relevant. The object under investigation is embedded in a specific environment, which consists of an ensemble of other objects.

The physically real interaction of the object with object k of the environment takes place in $(\mathbf{p}, E)$-space. Here, interaction means that there exist $\mathbf{p}, E$-fluctuations between the object and object k. The interaction process can only take place if the conservation laws for the momentum and the energy are fulfilled. We assume that the object has at time τ the momentum $\mathbf{p}$ and the energy E and we assume that object k has at the same time τ the momentum $\mathbf{p}_k$ and the energy E_k. Then, interaction means that there is an exchange of momentum and energy between both objects. That is, both objects exchange a chunk $\Delta \mathbf{p}, \Delta E$ at time τ. If the object takes the chunk $\Delta \mathbf{p}, \Delta E$, we have

$$object \Rightarrow \left\{ \begin{matrix} \mathbf{p} + \Delta \mathbf{p} = \mathbf{p}' \\ E + \Delta E = E' \end{matrix} \right\} \tag{2.61}$$

Then, due to the conservation of momentum and energy, object k must change as follows:

$$object\, k \left\{ \begin{matrix} \mathbf{p}_k - \Delta \mathbf{p} = \mathbf{p}'_k \\ E_k - \Delta E = E'_k \end{matrix} \right\} \tag{2.62}$$

In conclusion, the physically real interaction process in $(\mathbf{p}, E)$-space, which the object under investigation executes with object k at time τ, is a $\mathbf{p}, E$-fluctuation and we obtain

$$\tau: \quad \left\{ \begin{array}{l} object: \mathbf{p}, E \xrightarrow{\Delta\mathbf{p}, \Delta E\text{-}chunk} \mathbf{p}', E' \\ object\, k: \quad \mathbf{p}_k, E_k \xrightarrow{-\Delta\mathbf{p}, \Delta E\text{-}chunk} \mathbf{p}'_k, E'_k \end{array} \right\} \tag{2.63}$$

The process in connection with the object and object k happens simultaneously and this is because the conservation laws for momentum and energy are only valid if both processes in connection with (2.63) take place simultaneously, i.e., at exactly the same time τ.

The $\Delta\mathbf{p}, \Delta E$-chunk does not move through $(\mathbf{p}, E)$-space.; it does not occupy the $\mathbf{p}_i, E_i$-states between the $\mathbf{p}, E$-state of the object and the $\mathbf{p}_k, E_k$-states of object k. The $\Delta\mathbf{p}, \Delta E$-chunk is created at the $\mathbf{p}, E$-position of the object and the $\Delta\mathbf{p}, \Delta E$-chunk is simultaneously annihilated at the $\mathbf{p}_k, E_k$-position of object k. Both processes take place at exactly the same time τ. Again, the conservation laws for the momentum and the energy require that.

The situation in $(\mathbf{r}, t)$-space is that there are no physically real processes in $(\mathbf{r}, t)$-space and, therefore, there can be no exchange of certain $\mathbf{r}, t$-chunks between the various x, y, z, t-states. In $(\mathbf{r}, t)$-space, there are "correlations" between the x, y, z, t-states. The physically real processes exclusively take place in $(\mathbf{p}, E)$-space.

2.5.3.3. *Object-definition in space and time*

We underlined in Section 2.4.7.2 that within Born's probability interpretation of traditional quantum theory (Schrödinger's equation), a physically real particle has to be assumed. The particle itself has the mass m_0 and is located in space; it reflects a point-like structure. That is, a particle (a physically real object) has to be "postulated", which is needed for the understanding of Schrödinger's equation. This particle does not follow from the theory itself.

Such a postulation of an object is not possible in projection theory. In fact, there can be no physically real objects in $(\mathbf{r}, t)$-space, which could move through $(\mathbf{r}, t)$-space with a certain probability. In other words, the conception cannot be transferred to projection theory.

Objects without $\mathbf{p}, E$-fluctuations, i.e., objects with constant values for $\mathbf{p}$ and E ($\mathbf{p} = \mathbf{p}_0$, $E = E_0$), cannot exist within the projection theory and, therefore, no kind of theory-independent object, point-like or not, can be postulated. Such a free choice of objects would be in contradiction to the theory. On the other hand, the parameter m_0 appears in both basic equations (Eqs. (2.52) and (2.53)). Therefore, individual objects are defined, but they have to be extracted from the theory itself. Again, in the traditional quantum theory, we may postulate a physically real object without violating principal laws. This is in fact done within Born's probability interpretation.

What are the arguments in projection theory? A physically real objects exists if there are $\mathbf{p}, E$-fluctuations in reality, i.e., in $(\mathbf{p}, E)$-space. As we remarked, the $\mathbf{p}, E$-fluctuations reflect an interaction with other objects of the environment. In this case, the condition (2.60) has to be fulfilled. This is the situation with respect to $(\mathbf{p}, E)$-space.

How is this behavior transferred to $(\mathbf{r}, t)$-space where the picture of reality is located? Here, we have to recourse to the basic equation (2.52). The interaction is symbolized by the term $V(x, y, z, t)$. The case

$$V(x, y, z, t) = 0 \tag{2.64}$$

is the interaction-free situation and corresponds to $\Delta\mathbf{p} = 0$, $\Delta E = 0$ (Eq. (2.59)). That is, the object does not exist in $(\mathbf{r}, t)$-space when (2.64) is effective. In other words, the object-existence requires

$$V(x, y, z, t) \neq 0 \tag{2.65}$$

Since a lone object does not exist, the term $V(x, y, z, t)$ correlates the object under investigation with other objects, which we want to symbolize them by the elements

$$x_i, y_i, z_i, t_i, \ i = 1, 2, \ldots \tag{2.66}$$

The object must be at least with one environmental object correlated and let us characterize this other object by x_m, y_m, z_m, t_m. The object itself is characterized by the elements x_0, y_0, z_0, t_0.

The four-dimensional correlations between the object and the other object of the environment are distance-dependent if

$$x = x_0 - x_m, y = y_0 - y_m, z = -z_m, t = -t_m \tag{2.67}$$

This case would mean that the existence of the object under investigation is dependent on the four-dimensional position x_m, y_m, z_m, t_m of the environmental object. However, this case cannot be the basis for an object-definition. The object would have no individuality. Therefore, we explicitly exclude the case (2.67) and we obtain

The correlation with the other object is effective, but we require that this correlation is independent on x_m, y_m, z_m, t_m and we get

$$x = x_0, y = y_0, z = z_0, t = t_0 \tag{2.68}$$

Then, the object in $(\mathbf{r}, t)$-space, i.e., its picture, is defined by the wave function $\Psi(x, y, z, t)$ within x, y, z, t-range where $\Psi(x, y, z, t)$ is not zero:

$$\begin{aligned} &object - definition\, in\, (\mathbf{r}, t)\text{-}space: \\ &\Psi(x, y, z, t), \; -\infty < x, y, z, t < \infty \\ &m_0\, is\, dedicated \end{aligned} \tag{2.69}$$

Since the wave function $\Psi(p_x, p_y, p_z, E)$ is the Fourier transform of wave function $\Psi(x, y, z, t)$, the same object is defined in $(\mathbf{p}, E)$-space by

$$\begin{aligned} &object - definition\, in\, (\mathbf{p}, E)\text{-}space: \\ &\Psi(p_x, p_y, p_z, E), \; -\infty < p_x, p_y, p_z, E < \infty \\ &m_0\, is\, dedicated \end{aligned} \tag{2.70}$$

Both, wave function $\Psi(x, y, z, t)$ as well as the wave function $\Psi(p_x, p_y, p_z, E)$ are not dependent on clock time τ, that is, the object is permanently existent in $(\mathbf{r}, t)$-space and in $(\mathbf{p}, E)$-space.

We may define the object also by means of the probability densities. We have

$$\begin{aligned} &object - definition\ in\ (\mathbf{r}, t)\text{-}space: \\ &\Psi^*(x, y, z, t)\Psi(x, y, z, t), \; -\infty < x, y, z, t < \infty \\ &m_0\, is\, dedicated \end{aligned} \tag{2.71}$$

and

$$\begin{aligned} &object - definition\, in\, (\mathbf{p}, E)\text{-}space: \\ &\Psi^*(p_x, p_y, p_z, E)\Psi(p_x, p_y, p_z, E), \; -\infty < p_x, p_y, p_z, E < \infty \\ &m_0\, is\, dedicated \end{aligned} \tag{2.72}$$

The complete object is defined through (2.69), (2.70) and (2.71), (2.72). It exists permanently in these states. The object, expressed by (2.69), (2.70) and (2.71), (2.72), is existent in these forms at each clock time τ, i.e., the physically real object is τ-independent. The entire future and the entire past with respect to the system-specific time t are simultaneously existent at each clock time τ. The "existence" of the object is concerned here.

However, human beings are caught in the clock time τ, that is, they can only say something about the world outside at clock time τ. As we have stated in Section 2.3.1, the complete object is not observable because the entire t-range is not observable at time τ at once, but only the part for $t = \tau$:

$$\Psi(x, y, z, t = \tau), \quad \Psi^*(x, y, z, t = \tau)\Psi(x, y, z, t = \tau) \tag{2.73}$$

Clearly, the expressions (2.73) reflect only three-dimensional parts of the four-dimensional object, which is described by (2.69) and (2.71).

Explicit limits: The situation in connection with (2.69) and (2.71) can be expressed in terms of explicit limits. If the space limits at τ are x_a, y_a, z_a and x_b, y_b, z_b, we have

$$\begin{aligned} \tau:\ & \Psi(x, y, z, t = \tau) \neq 0 \\ & for \\ & x_a, y_a, z_a \leq x, y, z \leq x_b, y_b, z_b \\ & \Psi(x, y, z, t = \tau) = 0 \\ & for \\ & x, y, z < x_a, y_a, z_a \, and \, x, y, z > x_b, y_b, z_b \end{aligned} \tag{2.74}$$

The limits x_a, y_a, z_a and x_b, y_b, z_b are τ-dependent. Equation (2.74) defines a part of the object. The complete object is a four-dimensional block and is defined for all system-specific times t at every time τ:

$$\tau:\ t_a \leq t \leq t_b \tag{2.75}$$

t_a and t_b are the limits of the system-specific time t. Relation (2.74) together with (2.75) defines the "existence" of the object. The "observations" take place, however, at time τ; the object is in the state t at clock time τ: $t = \tau$. The object can be observed for t-states between t_a and t_b: $\tau_a = t_a$ and $\tau_b = t_b$. Then, we can define the interval $\Delta\tau = \tau_b - \tau_a$ as life-time. However, we have to be careful. The physically real object exists forever if there is no external impact on it. Life-time refers here to the observation.

Again, the physically real object is τ-independent. The entire future and the entire past with respect to the system-specific time t are simultaneously existent at each clock time τ. However, only a part can be observed.

Remark

As we have outlined in Ref. [1], human beings observe in everyday life macroscopic states. The space structure (x, y, z-structure) is not observed at point τ $(\tau \pm 0)$ but the macroscopic states are defined for $\tau \pm \varepsilon$. Here, ε is an arbitrarily small time interval different from zero. The system-specific t-scale is unconsciously scanned by the human observer by $\tau \pm \varepsilon$ and τ moves strictly from the past to the future. But at each point τ, the information within the interval 2ε around τ is registered unconsciously by the human observer, all at once. The physically real events within 2ε are densely packed.

2.5.3.4. *No separated objects*

How is a physically real object defined within the frame of the projection principle? Let us repeat: If there are no $\mathbf{p}$, E-fluctuations, that is, if there is no interaction of the object with its environment, the momentum $\mathbf{p}$ and the energy E are constant: $\mathbf{p} = \mathbf{p}_0$, $E = E_0$. It is relatively easy to show that such kind of isolated object cannot exist; the wave functions $\Psi(p_x, p_y, p_z, E)$ and $\Psi(x, y, z, t)$ are zero in this case (see also Section 2.5.3.1).

Separated objects, detached from everything, do not exist in the projection theory. On the other hand, individual objects exist. What is the solution?

The solutions $\Psi(p_{x0}, p_{y0}, p_{z0}, E_0) = 0$ and $\Psi(x, y, z, t) = 0$ shows that an interaction of the object with other objects is a necessary condition if a separated object reflects a real situation. The other objects have the space–time positions $\{x, y, z, t\}_i, i = 1, 2, \ldots$ at clock time τ.

2.5.4. *Analysis of the interaction terms*

In the transition from the traditional view, expressed by Schrödinger's equation (Eq. (2.41)), to the laws of projection theory, the term $U(x, y, z)$ is replaced by the function $V(x, y, z, t)$ and the operator $V(i\hbar\partial/\partial p_x, i\hbar\partial/\partial p_y, i\hbar\partial/\partial p_z, -i\hbar\partial/\partial E)$. We have

$$U(x,y,z) \;\Rightarrow\; \begin{array}{c} V(x,y,z,t) \\ \updownarrow \\ V\left(i\hbar\dfrac{\partial}{\partial p_x}, i\hbar\dfrac{\partial}{\partial p_y}, i\hbar\dfrac{\partial}{\partial p_z}, -i\hbar\dfrac{\partial}{\partial E}\right) \end{array} \tag{2.76}$$

Let us briefly discuss this point.

2.5.4.1. *The meaning of* $V(i\hbar\partial/\partial p_x, i\hbar\partial/\partial p_y, i\hbar\partial/\partial p_z, -i\hbar\partial/\partial E)$

The physically real object is characterized in $(\mathbf{p}, E)$-space by an ensemble of p_x, p_y, p_z, E-values:

$$\{p_x, p_y, p_z, E\}_{object} \tag{2.77}$$

Object k too is characterized by an ensemble of p_x, p_y, p_z, E-values, also in $(\mathbf{p}, E)$-space:

$$\{p_x, p_y, p_z, E\}_{object\ k} \tag{2.78}$$

The quantity

$$V\left(i\hbar\frac{\partial}{\partial p_x}, i\hbar\frac{\partial}{\partial p_y}, i\hbar\frac{\partial}{\partial p_z}, -i\hbar\frac{\partial}{\partial E}\right) \tag{2.79}$$

in Eq. (2.53) correlates both ensembles through $\mathbf{p}, E$-fluctuations, which are physically real interactions.

$$\{p_x, p_y, p_z, E\}_{object} \xleftrightarrow{\mathbf{p}, E\text{-}fluctuations} \{p_x, p_y, p_z, E\}_{object\ k} \tag{2.80}$$

For the interactions ($\mathbf{p}, E$-fluctuations), the interaction-operator (2.79) is responsible. Since a free object without interaction cannot exist within the projection theory, the object can only be defined in connection with operator (2.79).

In other words, the interaction-operator is responsible for the existence of the object under investigation and this existence is coupled to another object (object k in our case).

But the operator (2.79) is more than that: It determines the properties of the single object, and these properties are expressed by the wave function $\Psi(p_x, p_y, p_z, E)$ and can be calculated by Eq. (2.52).

In conclusion, the operator (2.79) correlates two objects and it also determines the properties of an object. The interaction is necessary for the existence of the object.

It is independent of $\mathbf{p}, E$-distances and it is τ-independent.

2.5.4.2. *The Meaning of* $V(x, y, z, t)$

The physically real object is characterized in $(\mathbf{r}, t)$-space by an ensemble of x, y, z, t-values:

$$\{x, y, z, t\}_{object} \tag{2.81}$$

Object k too is characterized by an ensemble of x, y, z, t-values, also in $(\mathbf{r}, t)$-space:

$$\{x, y, z, t\}_{object\ k} \tag{2.82}$$

The quantity $V(x, y, z, t)$ in Eq. (2.52) correlates both ensembles:

$$\{x, y, z, t\}_{object} \xleftrightarrow{V(x,y,z,t)} \{x, y, z, t\}_{object\ k} \tag{2.83}$$

$V(x, y, z, t)$ is a correlational-interaction. From the $(\mathbf{p}, E)$-space analysis follows that the object cannot exist if $V(x, y, z, t)$ is not effective.

In other words, $V(x, y, z, t)$ is responsible for the existence of the object under investigation and this existence is coupled to another object. It is object k in the case discussed here.

But the quantity $V(x, y, z, t)$ is more than that: It determines the properties of the single object, and these properties are expressed by the wave function $\Psi(x, y, z, t)$ and can be calculated by Eq. (2.52).

In conclusion, $V(x, y, z, t)$ correlates two objects and it determines the properties of an object. The correlational-interaction is necessary for the existence of the object. It is independent of space–time distances and it is τ-independent.

The quantity $V(x, y, z, t)$ is a correlational-interaction and has nothing to do with physically real interactions. $V(x, y, z, t)$ rather reflects a certain kind of order-principle within the space–time picture, i.e., in $(\mathbf{r}, t)$-space, while physically real processes take place in $(\mathbf{p}, E)$-space.

2.5.4.3. *Summary*

Either we model the $\mathbf{p}, E$-fluctuations in $(\mathbf{p}, E)$-space or we develop models for the correlations-interactions in $(\mathbf{r}, t)$-space. This is the situation within the frame of the projection principle. In traditional physics, only the interaction term $U(x, y, z)$ is known and this function describes physically real interaction in space and time. However, as we underlined in Ref, [1], this conception makes problems. No doubt, it works excellently, but the epistemological value of this consideration is questionable.

2.5.5. *Uncertainty relations*

The variables p_x, p_y, p_z, E and x, y, z, t **behave statistically and** are inherently uncertain and we obtain the following uncertainty relations [1, 3]:

$$\delta p_x \delta x \geq \frac{\hbar}{2}, \; \delta p_y \delta y \geq \frac{\hbar}{2}, \; \delta p_z \delta z \geq \frac{\hbar}{2} \tag{2.84}$$

$$\delta E \delta t \geq \frac{\hbar}{2} \tag{2.85}$$

The values

$$\delta x, \delta y, \delta z, \delta t \tag{2.86}$$

and

$$\delta p_x, \delta p_y, \delta p_z, \delta E \tag{2.87}$$

are the uncertainties in the values x, y, z, t and p_x, p_y, p_z, E at the same clock time τ.

The uncertainty relations (2.84) for the momentums and the coordinates are in this form also known in traditional quantum theory, but the uncertainty relation (2.85) for the energy and the system-specific time is not defined in the traditional quantum theory; the system-specific time t is not an element here.

As we pointed out in Refs. [1] and [3], in the traditional quantum theory, there does not exist a proper uncertainty relation for the time and the energy. The time τ appears within Schrödinger's equation (2.41) as an external parameter and in the traditional quantum theory the notion "time" appears still as a classical quantity. But we have uncertainties with respect to the coordinates x, y, z and the momentums p_x, p_y, p_z.

I usual quantum theory, there can be no uncertainty relation for the energy and time which would agree in its physical content with the uncertainty relation for the coordinates and momenta; in the traditional quantum theory, we have the uncertainties $\delta x, \delta y, \delta z$ and $\delta p_x, \delta p_y, \delta p_z$, but the uncertainties δE and δt are not definable. The significance of the well-known relation

$$\Delta E \Delta \tau \geq \frac{\hbar}{2} \tag{2.88}$$

is entirely different from that of Eq. (2.85). This difference is symbolically expressed by the use of Δ instead of δ. The energy E can be measured in usual quantum theory to any degree of accuracy at any instant. The quantity ΔE in Eq. (2.88) is the difference between two exactly measured values of the energy at two different instants and is not the uncertainty in the value of the energy at a given instant (see also the discussion in [8] and [9]). The same is true for the classical time τ: There is definitely no uncertainty with respect to τ. In conclusion, in contrast to the projection theory, there is no quantum aspect of time in the traditional quantum theory.

Mario Bunge's Critique

Since time within conventional quantum theory is still a classical parameter, we should consider the energy–time relation $\Delta E \Delta\tau \geq \hbar/2$ (Eq. (2.88)) as a quasi-classical equation. The value of such an equation is however questionable. In Mario Bunge's opinion the energy–time relation $\Delta E \Delta\tau \geq \hbar/2$ "is a total stranger to quantum theory" [10]. In particular, we find the following in Ref. [10]: *This relation is made plausible by reference to some thought experiments, to radioactive decay, and to line breadths. But unlike the genuine indeterminacy relations, (2.5) has never been proved from first principles. In other words, (C3) does not belong to quantum theory, but is just a piece of doubtful heuristics. The reason for this failure to incorporate (2.5) into quantum mechanics is the following. In this theory, as in every other known theory, time is a "c number" and, more particularly a parameter, not a dynamical variable. Moreover, τ doesn't "belong" (refer) to the system concerned.*

Even in relativistic theories the proper time, though relative to a frame of reference, is not a property of the system on the same footing as its mass or its momentum. In other words, τ does not belong to the family of operators in the Hilbert space associated to every microsystem. Therefore, τ is not a random variable and its scatter vanishes identically ... Consequently, no matter what the scatter in the energy may be, the inequality (C3) does not hold. Also, it does not improve things to regard E, as is sometimes done, as the Hamiltonian of the system the standard deviation of the energy vanishes as well. In conclusion, the so-called fourth indeterminacy (or uncertainty) is a total stranger to the quantum theory although it can be found in works on this theory .....

In short, the fourth scatter relation is not deducible from the principles of the quantum theory, whether relativistic or not. But then why is it sometimes used, for example in the theory of line breadths? The reason is that it is not the same formula: although it has the same typographical form, it has a different content. In particular, "Δ" is not interpreted as a standard deviation (from what?) but as the half-life of the state ψ. But even thus reinterpreted, the formula is not deducible from the postulates of the general *quantum theory. Mind, this has nothing to do with the question whether or not formula (C3) is true under some suitable interpretation. Many other statements are true and yet they do not belong to quantum theory.*

It is in fact adventurous to call relation $\Delta E \Delta \tau \geq \hbar/2$ "uncertainty relation". Relation (2.85) [$\delta E \delta t \geq \hbar/2$] is the adequate expression for the uncertainties with respect to time and energy.

2.5.6. *Energy and time representation*

The function $V(x, y, z, t)$ in (2.52), which has the dimension of an energy, cannot be assigned to specific positions x, y, z, t in $(\mathbf{r}, t)$-space; here, only geometrical structures (space–time positions) are defined and quantities with the dimension of energy cannot be located in $(\mathbf{r}, t)$-space. There are no physically real (material) objects in $(\mathbf{r}, t)$-space, and the potential $V(x, y, z, t)$ would only give sense in connection with material objects. In the projection theory, energies are exclusively positioned in $(\mathbf{p}, E)$-space. Clearly, the situation in connection with the container model is different.

The wave function $\Psi(x, y, z, t)$ is determined by Eq. (2.52). As we have pointed out in Ref. [11], Eq. (2.52) itself cannot be depicted in $(\mathbf{r}, t)$-space, but it is positioned on another "level of reality" which is higher than that where the geometrical structures are representable.

Equation (2.52) is a tool for the determination of $\Psi(x, y, z, t)$ within the energy-representation; other representations for the determination of $\Psi(x, y, z, t)$ can be chosen. Let us briefly discuss the so-called time representation.

When we consider Eqs. (2.52) and (2.53) and formulate the situation within the classical approximation, the uncertainties

$$\delta x, \ldots, \delta t, \delta p_x, \ldots, \delta E \tag{2.89}$$

in the values x, y, z, t, p_x, p_y, p_z, E become zero:

$$\{\delta x, \ldots, \delta t, \delta p_x, \ldots, \delta E\} \rightarrow 0 \tag{2.90}$$

In this case, the variables $x, y, z, t, p_x, p_y, p_z, E$ behave classically without any uncertainty. Since we have in the projection theory the system-specific time t as an additional variable, instead of Eq. (2.44), which can be formulated as $E = E(x, y, z, p_x, p_y, p_z)$, we have

$$E = E(p_x, p_y, p_z, x, y, z, t) \tag{2.91}$$

or the equivalent expression

$$t = t(p_x, p_y, p_z, E, x, y, z) \tag{2.92}$$

Using the quantum-theoretical operator rules, quoted in Sec. 2.5.1 and derived in Refs. [1] and [3], we obtain after minor manipulations the following equation for the determination of the wave function in $(\mathbf{p}, E)$-space:

$$\begin{aligned}&\hat{H}\left(p_x, p_y, p_z, -i\hbar\frac{\partial}{\partial E}, i\hbar\frac{\partial}{\partial p_x}, i\hbar\frac{\partial}{\partial p_y}, i\hbar\frac{\partial}{\partial p_z}\right)\Psi(p_x, p_y, p_z, E)_H \\ &= E\,\Psi(p_x, p_y, p_z, E)_H\end{aligned} \tag{2.93}$$

$\hat{H}$ is the energy-operator and $\Psi(p_x, p_y, p_z, E)_H$ is the wave function in energy-representation. On the other hand, the time representation is an alternative to (2.93). With (2.92) and the corresponding operator rules, we get

$$\begin{aligned}&-i\hbar\frac{\partial}{\partial E}\Psi(p_x, p_y, p_z, E)_C \\ &= \hat{C}\left(p_x, p_y, p_z, E, i\hbar\frac{\partial}{\partial p_x}, i\hbar\frac{\partial}{\partial p_y}, i\hbar\frac{\partial}{\partial p_z}\right)\Psi(p_x, p_y, p_z, E)_C\end{aligned} \tag{2.94}$$

$\hat{C}$ is the time operator and $\Psi(p_x, p_y, p_z, E)_C$ is the wave function in time representation.

The operators

$$\hat{H}\left(p_x, p_y, p_z, -i\hbar\frac{\partial}{\partial E}, i\hbar\frac{\partial}{\partial p_x}, i\hbar\frac{\partial}{\partial p_y}, i\hbar\frac{\partial}{\partial p_z}\right) \tag{2.95}$$

and

$$\hat{C}\left(p_x, p_y, p_z, E, i\hbar\frac{\partial}{\partial p_x}, i\hbar\frac{\partial}{\partial p_y}, i\hbar\frac{\partial}{\partial p_z}\right) \tag{2.96}$$

have to be constructed so that we obtain

$$\Psi(p_x, p_y, p_z, E)_C = \Psi(p_x, p_y, p_z, E)_H \tag{2.97}$$

The time operator $\hat{C}$ has nothing to do with the operator for the time coordinate $\hat{t} = -i\hbar\partial/\partial E$ (Sec. 2.5.1), which is independent on the system under investigation.

Equation (2.94) is a real alternative to Eq. (2.53). While the wave function is determined by Eq. (2.53) within the "energy-representation", it is determined by Eq. (2.94) within the "time representation". The same analysis can be made for $(\mathbf{r}, t)$-space.

2.5.7. *Self-organizing processes*

No doubt, self-organizing processes are important in nanoscience; they belong to the heart of nanoscience (nanotechnology). In general, we have the following situation: We put a system (nanosystem) into a certain environment and both entities, the system and the environment, interact with each other. This situation effectuates that the states of the involved entities will be changed in the course of clock-time τ. This interaction process occurs without influence from outside, i.e., the procedure reflects a self-organizing process.

2.5.7.1. *General definition*

We indicate the initial state of the system under investigation by the letter A and its final state by B, i.e., we have the transition

$$A \underset{\tau}{\longrightarrow} B \tag{2.98}$$

The process starts at time τ_A and it is finished at time τ_B when stationary states are reached. The system and the environment are in stationary states for $\tau < \tau_A$ and $\tau > \tau_B$.

Self-organizing processes have to be considered as "time-sensitive". Just this feature is the problem when we work within traditional quantum theory. Nanosystems behave in most cases quantum-mechanically. As we remarked in the preceding chapters, in traditional quantum theory, there does not exist a quantum aspect of time. The time in traditional quantum theory is given by the external time τ, which behaves classically and is shown by our clocks.

In contrast to this traditional view, within the frame of the projection principle, a quantum aspect of time is defined. We have here a system-specific time t. All these aspects have been discussed in the previous sections.

Let us discuss some characteristics of self-organizing processes within traditional physics, where we have no quantum time. After that, we will treat the problem on the basis of the more realistic projection theory which is characterized through the system-specific quantum time.

2.5.7.2. *Traditional treatment*

In most cases, nanosystems behave quantum-mechanically and we have to apply Schrödinger's equations (Section 2.4.7). A nanosystem consists of a few hundred atoms/molecules. In general, the structure-changes during transition (2.98) are of relevance. Since we are interested in the transition with respect to "time", we may investigate the one-particle Schrödinger equation instead of the more complex Schrödinger equation for the nanosystem consisting of a few hundred objects. Schrödinger's equation for the stationary case is given by Eq. (2.41):

$$i\hbar\frac{\partial\varphi(x,y,z,\tau)}{\partial\tau} = -\frac{\hbar^2}{2m_0}\Delta\varphi(x,y,z,\tau) + U(x,y,z)\varphi(x,y,z,\tau) \qquad (2.99)$$

The non-stationary case is expressed by

$$i\hbar\frac{\partial\varphi(x,y,z,\tau)}{\partial\tau} = -\frac{\hbar^2}{2m_0}\Delta\varphi(x,y,z,\tau) + U(x,y,z,\tau)\varphi(x,y,z,\tau) \quad (2.100)$$

Stationary means that the potential U is not dependent on time τ and we get Eq. (2.100). If the term U is dependent on τ, the non-stationary case (Eq. (2.100)) has to be applied.

In the transition $A \xrightarrow[\tau]{} B$ [scheme (2.98)], the initial state A at $\tau = \tau_A$ and the final state B at $\tau = \tau_B$ are expressed by the stationary equations (see in particular Eq. (2.99)): In particular, we have for $\tau \leq \tau_A$ and $\tau \geq \tau_B$ the following Schrödinger equations:

$\tau \leq \tau_A$:

$$i\hbar\frac{\partial\varphi_A(x,y,z,\tau)}{\partial\tau} = -\frac{\hbar^2}{2m_0}\Delta\varphi_A(x,y,z,\tau) + U_A(x,y,z)\varphi_A(x,y,z,\tau) \qquad (2.101)$$

$\tau \geq \tau_B$:

$$i\hbar\frac{\partial\varphi_B(x,y,z,\tau)}{\partial\tau} = -\frac{\hbar^2}{2m_0}\Delta\varphi_B(x,y,z,\tau) + U_B(x,y,z)\varphi_B(x,y,z,\tau) \qquad (2.102)$$

The transition from A to B is described by Schrödinger's non-stationary equation (see Eq. (2.100)):

$$i\hbar\frac{\partial\varphi_{AB}(x,y,z,\tau)}{\partial\tau} = -\frac{\hbar^2}{2m_0}\Delta\varphi_{AB}(x,y,z,\tau) + U_{AB}(x,y,z,\tau)\varphi_{AB}(x,y,z,\tau) \qquad (2.103)$$

Here, the τ-dependent potential $U_{AB}(x,y,z,\tau)$ has to be used, but not in connection with Eqs. (2.101) and (2.102). An example for a self-organizing process in traditional physics is shown in Fig. 2.2(a).

Remark

For example, in the case of the one-dimensional harmonic oscillator, the potential U is given by $1/2mx^2\omega_0^2 = U(x)$, where m is the mass and ω_0 the frequency of the oscillator. The interaction of this oscillator with a specific environment can lead to the effect that the frequency ω_0 becomes τ-dependent and we have $\omega_0 = \omega_0(\tau)$. Then, in region A, the frequency is given by ω_A and is independent on τ; in region AB the frequency is τ-dependent and we get $\omega_{AB}(\tau)$; in region B, the frequency is given by ω_B and is independent on τ.

The environment and the system (oscillator) influence each other and lead to changes with respect to both the oscillator and the environment. The changes, i.e., the τ-dependent processes, take place during the transition in region AB.

2.5.7.3. *Projection theory*

Scanning Information: In traditional physics, the material objects (systems) exist at time τ and are incessantly produced in the course of clock time τ. However, they do not exist for times $\tau' < \tau$ and $\tau' > \tau$. The view of projection theory is different. At time τ, the physically real system is not produced but selected from the entire information. The entire information of the physically real object exists at each time τ in the form of a four-dimensional indivisible block. Only a part of this block is observable at time τ. This has of course consequences for self-organizing processes. Let us repeat the main facts.

Despite the statistical fluctuations, at τ, the whole of the system-specific time t, the past, present and future, is laid out frozen before us:

$$\begin{aligned} \tau :\ & \Psi^*(x,y,z,t)\Psi(x,y,z,t) \\ & -\infty < x,y,z < \infty,\ -\infty < t < \infty \end{aligned} \tag{2.104}$$

That is, at time τ, the entire information is given all at once. This also means that within the range defined by τ_A and τ_B, where the self-organizing process takes place, the entire information of the wave function is given all of a piece.

The system-specific time t overlaps the τ-range of the self-organizing process and we have

$$-\infty < \tau_A \leq t \leq \tau_B < \infty \tag{2.105}$$

However, we know that the entire information (2.104) is not "observable" at clock-time τ. Instead, we always observe only a certain part of the complete space structure of the system at time τ:

$$\begin{aligned} &\Psi^*(x,y,z,t_C)\Psi(x,y,z,t_C) \\ &-\infty < x,y,z < \infty \end{aligned} \tag{2.106}$$

for $t_C = \tau$. A usual photography represents at a certain time $\tau = t_C$ such a configuration in space.

The curve $\Psi^*(x,y,z,t)\Psi(x,y,z,t)$ is scanned with respect to time t and we observe merely the part at clock-time τ: $\Psi^*(x,y,z,t=\tau)\Psi(x,y,z,t=\tau)$. Scanning processes also take place within the frame of self-organizing processes and this is indicated in Fig. 2.2(b).

We may state quite generally that the information about the wave function $\Psi(x,y,z,t)$ and the probability density $\Psi^*(x,y,z,t)\Psi(x,y,z,t)$, respectively, exists completely at τ not as static unit, but as a fluctuating quantum entity. The fluctuations cannot be dissolved by human beings in daily life and are experienced at time τ as compact block.

Comparison of the Traditional View with the Situation in Projection Theory

In the case of the traditional treatment of self-organizing processes, a piecewise treatment of the three regions, A, AB and B, is demanded and possible. For each region, a specific equation is needed [see Eqs. (2.101)–(2.103)]. The projection principle does not allow such a procedure and this is because we deal here with a four-dimensional indivisible block.

Within this principle the characteristics of self-organizing processes are described by "one" relation, and this is expressed by

$$\begin{aligned} &i\hbar\frac{\partial\Psi_{A,AB,B}(x,y,z,t)}{\partial t} = \\ &\quad -\frac{\hbar^2}{2m_0}\Delta\Psi_{A,AB,B}(x,y,z,t) + V_{A,AB,B}(x,y,z,t)\Psi_{A,AB,B}(x,y,z,t) \end{aligned} \tag{2.107}$$

The entire future and the entire past of the nanosystem with respect to the system-specific time t are simultaneously existent at each clock time τ:

$$\tau : \left\{ \begin{aligned} &\Psi^*_{A,AB,B}(x,y,z,t)\Psi_{A,AB,B}(x,y,z,t) \\ &-\infty < \tau_A \leq t \leq \tau_B < \infty \\ &-\infty < x,y,z < \infty \end{aligned} \right\}_{\substack{existent \\ at\,\tau}} \tag{2.108}$$

However, only a part can be observed; it is just that part for $t = \tau$.

$$\tau : \left\{ \begin{matrix} \Psi^*_{A,AB,B}(x,y,z,t=\tau)\Psi_{A,AB,B}(x,y,z,t=\tau) \\ -\infty < x,y,z < \infty \end{matrix} \right\} \begin{matrix} observable \\ at\,\tau \end{matrix} \tag{2.109}$$

All three regions, A, AB and B, have to be treated all of a piece. The reason is obvious: At clock-time τ with $\tau \leq \tau_A$ quantum states t can be occupied that belongs to region B or to region AB. This property has been discussed in the previous section.

Nevertheless, a human observer experiences the three regions, A, AB and B, as a self-organizing process. He successively scans the curve

$$\Psi^*{}_{A,AB,B}(x,y,z,t)\Psi_{A,AB,B}(x,y,z,t) \tag{2.110}$$

with respect to t and passes through this curve in the course of clock-time τ strictly from the past to the future, i.e., from region A to region B over region AB. This sequence is experienced as a self-organizing process. The scanning process is indicated in Fig. 2.2(b).

The quantum-mechanical description of self-organizing processes in the traditional way has be excluded because the notion "time" is here not adequately treated. Only the clock time τ appears and τ is classical in character. In other words, there is no quantum time involved in the traditional view, as it should be within a proper treatment of time-sensitive processes. Thus, the description of specific self-organizing processes can be utterly wrong when traditional quantum theory is applied.

At time τ_A, we may choose for the probability densities the same values for both approaches (traditional view and projection principle):

$$\varphi^*(x,y,z,\tau_A)\varphi(x,y,z,\tau_A) = \Psi^*_{A,AB,B}(x,y,z,\tau_A)\Psi_{A,AB,B}(x,y,z,\tau_A) \tag{2.111}$$

Since the potential $V_{A,AB,B}(x,y,z,t)$ [Eq. (2.107)] cannot be approximated by the potentials $U_A(x,y,z)$, $U_A(x,y,z)$, $U_{AB}(x,y,z,\tau)$ [Eqs. (2.101)–(2.103)], we have for $\tau > \tau_A$

$$\varphi^*(x,y,z,\tau)\varphi(x,y,z,\tau) \neq \Psi^*_{A,AB,B}(x,y,z,\tau)\Psi_{A,AB,B}(x,y,z,\tau) \tag{2.112}$$

Thus, the description of the self-organizing process with the traditional tools have only a little (or nothing) to do with the more realistic description within the frame of the projection principle. Both views are illustrated in Fig. 2.2.

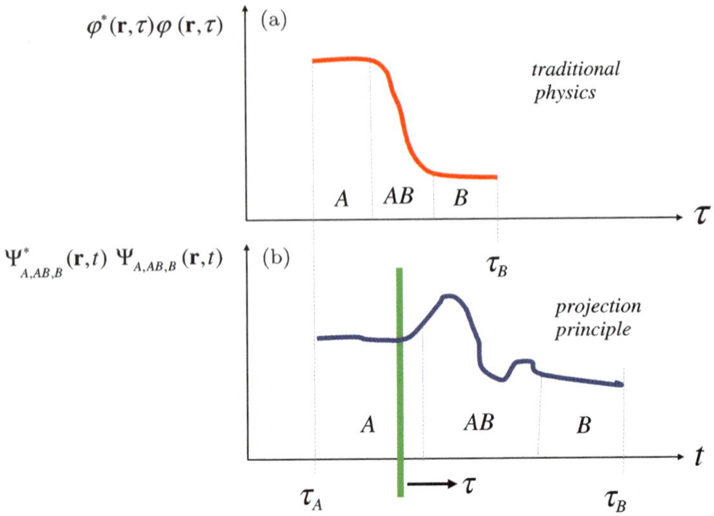

Fig. 2.2: Self-organizing process described within traditional physics. (a) and within the frame of the projection principle in (b). For both views, the probability densities are shown. The boundary conditions at time τ_A are assumed to be identical [see Eq. (2.111)]. The clock time τ has nothing to do with the system-specific time t. Therefore, the curves may differ arbitrarily from each other. In particular, the final states, which is reached at τ_B, can be differ considerably for both approaches. The characteristic three regions, A, AB and B, are located at different positions on the timescales. The scanning process is indicated in (b).

The prediction of self-organizing processes through the traditional way [indicated by Eqs. (2.98)–(2.103)] is not realistic. Just in connection with nanosystems a careful treatment is necessary.

2.6. Summary and Final Remarks

The most direct impression about the material world outside is given by the space structure, which human beings have at clock time τ before their eyes. This is an unconscious experience and is done in everyday life. These are observations in everyday life. This information is the basis for everything in the life of human beings.

Since space and time do not belong to the category of physically real entities, this everyday-life impression can only be a picture of the material world outside. The material world itself is located in basic reality, which is not accessible to human observers.

The clock time τ is often identified with a real global time mechanism, and the entire world is driven by this mechanism. This is an unrealistic point of view and has to be considered as too naive. Here, the entire world is driven by the external time τ, i.e., it is driven by a real external time-machine and the clock marks the numbers of τ. Thus, the clock time τ reflects an external time and has nothing to do with the physically real objects.

Within the frame of the projection principle, the physically real background of the external time τ is eliminated and the conception of the system-specific time t is developed. The clock time τ remains as the reference time. The material objects become four-dimensional entities, which exists permanently.

This has serious consequences for the definition of physically real (material) objects: The entire future and the entire past of a material object with respect to the system-specific time t are simultaneously "existent" at each clock time τ. However, only a part can be "observed"; it is just the part for $t = \tau$.

A human being experiences the world outside at every clock time τ as geometrical structure (x, y, z-structure). The world appears in the form of separate objects, i.e., there are always "extensions". These extensions before the observer's eyes effectuate the "space feeling" of human beings.

On the other hand, the structural "changes" cause the "time feeling". The space structures are fundamental. There is no time feeling without space (extensions).

The "arrow of time" is based on space structures too. We have to assume that nature works economically. That is, in nature, no space structure is "repeated". Such a repetition would be superfluous and is therefore avoided in nature: Due to the property "no back to an already known space structure", we get an arrow of time and in particular the "arrow-of-time feeling".

In projection theory, we have a real uncertainty relation between the energy E and the system-specific time t: $\delta E \delta t \geq \hbar/2$. Such an equation does not exist in the usual form of quantum theory. Since time is still a classical parameter within the usual quantum theory, there is no uncertainty relation of type $\delta E \delta t \geq \hbar/2$ in the traditional quantum theory. However, here, we have the energy–time relation of the form $\Delta E \Delta \tau \geq \hbar/2$. The quantities ΔE and $\Delta \tau$ are differences and they have nothing to do with uncertainties. Relation $\Delta E \Delta \tau \geq \hbar/2$ is a quasi-classical equation. The value of such an equation is, however, questionable. In Mario Bunge's opinion, the energy–time relation $\Delta E \Delta \tau \geq \hbar/2$ "is a total stranger to quantum theory".

In this chapter, we discussed two world views: the "container model" and the "projection principle".

Within the traditional view, the container model is assumed to be valid. Then, the space structures before the eyes of a human being consist of real material objects. A human being touches, for example, a stone and he feels it at once. This direct impact is interpreted in traditional physics as if the material entities of the world outside are directly before the eyes of the observer. This is obviously not realistic.

Let us repeat. The container principle does not fulfill the basic features of space and time: Space and time are here considered as physically real elements, and this is because material objects cannot be embedded in a space–time container, which is not physically real.

The situation is different in the projection theory. Here, the elements of space and time are "not" physically real, and this feature reflects Mach's principle. As we have discussed, this fact is a basic requirement.

Within the projection theory, the objects (systems) "before us" are not material in character but reflect exclusively "geometrical" structures. The physically real material world with all its objects is projected onto space and time and is nothing else than a picture of reality consisting of geometrical structures.

The projection principle is necessary because space and time cannot be considered as physically real elements. The details of the projection principle have been discussed, for example, in Refs. [1] and [11].

The space–time picture reflects a mind-state. Let us consider again a stone, this time within the frame of the projection principle. A human being "sees" the stone and he "feels" it. The observer has the stone and also himself (his fingers) before his eyes. The human being touches the stone in the picture with his fingers and moves it consciously. This is the situation in the picture with its geometrical structures. The corresponding physically real process takes place in basic reality where the stone and the acting human being are embedded as physically real entities. This physically real process in basic reality is also initiated by the human being's mind, but it is an unconscious procedure and this is because the human being has no direct access to basic reality. However, the effect of the physically real process in basic reality is transferred to the mind and the human being feels the stone, simultaneously to the touch in the picture. Details are outlined in Ref. [11].

References

1. W. Schommers, *Expansion of Physics Through Nanoscience* (De Gruyter, Berlin, **2020**).
2. J. T. Fraser, *Die Zeit* (dtv Sachbuch, München, **1987**).
3. W. Schommers, *Quantum Processes* (World Scientific, New Jersey, London, Singapore, **2011**).
4. C. G. Jung, *Synchronizität, Akausalität und Okkultismus* (Deutscher Taschenbuch Verlag GmbH & Co, KG, München, **1990**).
5. H. v. Ditfurth, *Der Geist fiel nicht vom Himmel* (Deutscher Taschenbuch Verlag, **1993**).
6. W. Schommers, *Space and Time. Matter and Mind* (World Scientific, New Jersey, London, Singapore, **1994**).
7. W. Schommers, in preparation.
8. L. de Broglie, *Die Elementarteilchen* (Goverts, Hamburg, **1943**).
9. L. D. Landau and E. M. Lifschitz, *Quantum Mechanics* (Pergamon, Oxford, **1965**).
10. M. Bunge, The so-called fourth indeterminacy relation. *Can J. Phys.* 48, 1410 (**1970**).
11. W. Schommers, *Basic Physics of Nanoscience* (Elsevier, **2019**); W. Schommers, *Mind and Reality* (World Scientific, **2015**).

Chapter 3

Multi-Spinor Nonequilibrium Quantum Transport: Entanglement-Induced Delocalization and Topological Insulators

Felix A. Buot

C&LB Research Institute, Carmen, Cebu 6005, Philippines-Laboratory of Computational Functional Materials, Nanoscience and Nanotechnology, University of San Carlos, Talamban, Cebu City 6000, Philippines

felixa.buot@gmail.com

3.1. Introduction

Originally, the concept of quantum localization and delocalization was restricted to scattering and backscattering in randomly modulated potential at each lattice sites. This is the so-called Anderson localization and delocalization process [1]. A related concept is the so-called Mott insulator and Mott metal–insulator transition [2] due to many-body electron–electron Coulomb interactions. Another closely related concept is the Wigner crystallization [3], whereby potential energy dominates the kinetic energy at low electron densities. Thus, to minimize the potential energy, the electrons form a bcc (body-centered cubic) lattice in 3D, a triangular lattice in 2D and an evenly spaced lattice in 1D. And of course tunneling-restricted localization and delocalization are very well known, which in fact lead to effective chiral dynamics in the case of two tunneling-coupled sites.

Another pioneering idea that was earlier introduced in physics is the concept of quantum entanglement. Originally, the concept of entanglement was restricted to composite systems made up of two spatially separated subsystems or separated pair of spins in a singlet state [4, 5]. However, it has been suggested that nonlocal correlations of similar nature may also exist when considering different degrees of freedom of a single system [6]. It is further claimed that entanglement seems to be a ubiquitous concept that it can apply as well to many different scenarios [7]. It was recently claimed that indeed this should not be unexpected, when considering correlations between different parties such as photons, axial degrees of freedom (as the entanglement of spin and pseudospins [8]) or different sites (as in the orbital entanglement in quantum chemistry bonding [9]). And in the important case of the single-excitation manifold, entanglement is deemed equivalent to coherent delocalization [6].

In this paper, we extend the concept of entanglement, localization and delocalization to entanglement-induced localization and delocalization. In particular, we are referring to entangled degrees of freedom, which has the axial–chiral nature, and present a rounded picture of entanglement localization and delocalization from the nonequilibrium quantum transport phase-space viewpoint. The beauty and power of this approach lie in its simplicity in explaining various experiments including magnetoconductivity measurents and in being able as well to explain different topological phases of matter.

The recent history of condensed matter physics has shown that the study of axial dynamics, namely, vortices, cyclotron orbits, spinors, Berry connections (in older form as Peierls phase factor), Berry curvatures, and Chern numbers has ushered the incarnation of topology in quantum physics [10, 11]. This is earlier manifested in the quantization of quantum Hall effect. Indeed, the importance of several discrete two degrees of freedom, i.e., spinors and their entanglements has emerged as ubiquitous in the physics of new and low-dimensional materials, such as graphene, TMDs, silicene, and topological systems [11, 12]. For example, Pauli–Dirac spin, valley spin (isospin), pseudospin due to low-energy electron–hole symmtery at Dirac points (electron and hole have the same pseudospin), and bilayer pseudospin have gained importance in consideration of the spin quantum transport of two-dimensional materials.

The effect of atomic-layer pseudospin degrees of freedom in bilayer graphene and TMD materials has been vigorously pursued both theoretically and experimentally due to exotic properties, namely, exciton condensation and a new mechanism for the onset of superconductivity [12, 13]. Entangled electron–hole pairs have also been proposed for monolayer graphene [14]. Indeed, in bilayer materials, a new mechansim for the onset of superconductivity has been an intriguing discovery. This is related to the onset of exciton condensate due to the entanglement of layer pseudospin with $\nu = 1$ filling of the lowest Landau levels (LLL), i.e., lowest Landau orbit (LLO) in each of the layers [15, 16], with intralayer quantum Hall effect at $\nu = 1$ of LL0.

In topological insulators, the Anderson type of localization of metallic edge states due to interaction with magnetic impurities causing successive spin flips has also been proposed [17] (in view of our recent findings, this kind of edge-states localization maybe attributed to the entanglement with impurity spin). Spin Kondo effect at the metallic edge states has also been treated [18]. In other words, the edge states have been studied and subjected to the usual treatment of 1D conductors, mostly based on the usual Hamiltonian and exchange interaction context.

In this paper, we present a rounded picture and a new perspective on the physics of topological insulators, QHE-TI, QSHE-TI, and Kondo-TI based on quantum nonlocality as a result of the entanglement of torques induced by the various axial (spinor) degrees of freedom. This is inferred from our results of the generalized nonequilibrium quantum transport equations of spin and pseudospins, and their entanglements. This brings in a novel mechanism for localization and delocalization based on the series of spin and pseudospin torque entanglements. The sort of localization being referred to here is typefied by cyclotron-orbit current localized around orbit center due to either external magnetic field or intrinsic Berry curvature in strongly spin–orbit coupled materials. On the other hand, the sort of delocalization referred to is represented by the metallic edge states in topological insulators [17] as a result of the *quantum nonlocality* brought about by spin and pseudospin entanglements, the type of delocalization responsible for the *resonant quantum tunneling transport* phenomena in resonant tunneling diodes [19, 20].

Here, the point of view in all of these is that of torque or axial entanglement in materials with multi-spinor degrees of freedom. Localization and delocalization of vortices have also found some treatments in the literature that could be interpreted as due to torque interactions [21–23]. In fact, these sorts of localization and *quantum-nonlocality* delocalization effects also shed light on the experimentally observed and ubiquitous phenomena in spin-dependent quantum transport physics. These are the so-called *weak localization* (WL) [24] and *weak antilocalization* (WAL) [25, 26] often referred to in the literature. By virtue of their competing effects in nonequilibrium quantum transport, the crossover from WL to WAL has also been experimentally observed [27].

Our analysis is based on the use of the phase-space formulation of quantum mechanics [28, 29]. The use of position and momentum variables should not be construed as employing a semi-classical analysis. The present analysis is founded on a basic principle, i.e., on the expansion of any quantum mechanical operator in terms of a complete set of orthogonal basis operators or *projectors* [30]. In the phase-space formalism, either discrete or continuum, the orthogonal and complete basis operators are the *phase-space point projectors*, $\hat{\Delta}(p,q)$. Thus, in *operator Hilbert-space* method [30] of quantum theory, any operator $\hat{A}$ can be expanded as the sum or integral of the basis operator, $\hat{\Delta}(p,q)$. As an integral expansion, we have

$$\hat{A} = \left(\frac{1}{2\pi\hbar}\right)^d \int d^d p' d^d q' \; A(p',q')\,\hat{\Delta}(p',q')$$

where d is the dimensionality and the expansion coefficients are the $A(p,q)$'s, where each coefficient is given by the trace of the operator with the *phase-space point projectors*, $\hat{\Delta}(p,q)$,

$$A(p,q) = Tr\left[\hat{\Delta}(p,q)\,\hat{A}\right]$$

The c-function $A(p,q)$ is known as the *lattice Weyl transform* of the operator $\hat{A}$ [29]. Note that the *lattice Weyl transform* operates on matrices. If $\hat{A} = \hat{\rho}$ (the density matrix operator), then $\rho(p,q) \equiv f_w(p,q)$, which is the Wigner distribution function.

It follows that any commutator or anticommutator of $\hat{A}$ and $\hat{B}$ can be treated similarly in the operator Hilbert-space method [30],

$$\left[\hat{A},\hat{B}\right] = \left(\frac{1}{2\pi\hbar}\right)^d \int d^d p' \; d^d q' \; \left[\hat{A},\hat{B}\right](p',q') \;\hat{\Delta}(p',q')$$

where the coefficient of expansion is given by the trace of the operator with $\hat{\Delta}(p,q)$,

$$\left[\hat{A},\hat{B}\right](p,q) = Tr\left(\hat{\Delta}(p,q)\left[\hat{A},\hat{B}\right]\right)$$

and

$$\left\{\hat{A},\hat{B}\right\} = \left(\frac{1}{2\pi\hbar}\right)^d \int d^d p' d^d q' \left\{\hat{A},\hat{B}\right\}(p',q')\,\hat{\Delta}(p',q')$$

where

$$\left\{\hat{A},\hat{B}\right\}(p,q) = Tr\left(\hat{\Delta}(p,q)\left\{\hat{A},\hat{B}\right\}\right)$$

respectively. Note that no semi-classical approximation is involved or needed whatsoever in the above phase-space formalism of operator Hilbert-space method. The exact expressions for $\left[\hat{A},\hat{B}\right](p,q)$ and $\left\{\hat{A},\hat{B}\right\}(p,q)$ are given in terms of powers of Poisson braket (equivalent to expansion to all powers of $\hbar$) or in terms of integral kernels, as shown in Appendix C. It is only when the first power of the Poisson bracket is used that we have a semi-classical approximation. Indeed, the exact quantum nonlocality property of the commutator has been rigorously demonstrated in resonant tunneling diodes, where the theoretical numerical simulation resolves the controversial experimental quantum transport measurements on these nanoodevices [20]. Moreover, the simulation was able to resolve the discrete energy levels of the quantum wells [19] and yields the THz current oscillations responsible for the current plateau in the dc-current measurements [20]. Even in the semi-classical approximation, the quantum nonlocality property of the commutator is already apparent. However, no semi-classical approximation is capable of simulating the current–voltage characteristics of resonant tunneling diodes [20].

Thus, in the context of operator Hilbert-space method, the von Neumann density-matrix evolution equation of quantum statistical dynamics in H-space becomes a super-statevector quantum dynamical equation in L-space or operator Hilbert-space. Thus, the familiar von Neumann density-matrix operator equation in H-space given by

$$i\hbar\frac{\partial}{\partial t}\rho\left(t\right)=\left[\mathcal{H},\rho\right] \tag{3.1}$$

becomes a super-Schrödinger equation for the super-statevector in L-space expressed as

$$i\hbar\frac{\partial}{\partial t}\left|\rho\left(t\right)\right\rangle\!\rangle=\mathcal{L}\left|\rho\left(t\right)\right\rangle\!\rangle \tag{3.2}$$

In Eq. (3.1), $\rho\left(t\right)$ is the density-matrix operator for the whole many-body system in H-space, whereas in Eq. (3.2), $\left|\rho\left(t\right)\right\rangle\!\rangle$ is the corresponding super-statevector in L-space. The superoperator $\mathcal{L}$ corresponds to the commutator $\left[\mathcal{H},\rho\right]$ of Eq. (3.1) and is referred to as the Liouvillian. The corresponding many-body quantum-field operators becomes quantum-superfield operators in L-space. Similar to the zero temperature Green function technique, the super-nonequilibrium Green functions are calculated in the L-space theory. The details in deriving the nonequilibrium quantum transport equations including Cooper pairings are given in the author's book and references therein [30].

3.2. Superfield Quantum Transport Equations

Our starting point is the general quantum transport expressions for fermions, where nonequilibrium quantum superfields and their correlations are the basic variables. These are obtained from the real-time nonequilibrium quantum superfield theoretical transport formulation of Buot [30–32], which can be written by dropping all arguments and discrete indices (undertood as matrix products) as

$$
\begin{aligned}
i\hbar\left(\frac{\partial}{\partial t_1}+\frac{\partial}{\partial t_2}\right)G^{\gtrless} &= \left[\mathcal{H}G^{\gtrless}-G^{\gtrless}\mathcal{H}\right] \\
&\quad + \left[\Sigma^r G^{\gtrless}-G^{\gtrless}\Sigma^a\right]+\left[\Sigma^{\gtrless}G^a-G^r\Sigma^{\gtrless}\right] \\
&\quad + \left[\Delta^r_{hh}g^{\gtrless}_{ee}-g^{\gtrless}_{hh}\Delta^a_{ee}\right] \\
&\quad + \left[\Delta^{\gtrless}_{hh}g^a_{ee}-g^r_{hh}\Delta^{\gtrless}_{ee}\right] \qquad (3.3)
\end{aligned}
$$

The last two brackets account for the Cooper pairings between fermions of the same specie but can differ in some quantum eigenvalues. These do not concern us in the present paper (their corresponding transport equations [30, 31] are important in nonequilibrium superconductivity, where we also need to solve for the nonequilibrium anomalous Green functions). In what follows, we will drop these last two square brackets of the RHS of Eq. (3.3).

In the absence of Cooper pairing of superconductivity, we may also write the general transport equation for '$\gtrless$'-quantities (here, $\{A,B\}$ and $[A,B]$ means anticommutator and commutator, respectively, of operators A and B), as

$$
\begin{aligned}
&i\hbar\left(\frac{\partial}{\partial t_1}+\frac{\partial}{\partial t_2}\right)G^{\lessgtr} \\
&\quad = \left[\tilde{\mathcal{H}},G^{\lessgtr}\right]+\left[\Sigma^{\lessgtr},\mathrm{Re}G^r\right]-\frac{i}{2}\left\{\Gamma,G^{\lessgtr}\right\}+\frac{i}{2}\left\{\Sigma^{\lessgtr},A\right\} \qquad (3.4)
\end{aligned}
$$

where $\Gamma = -\frac{1}{2}\mathrm{Im}\Sigma^r$ is the scattering-out matrix, $A = -\frac{1}{2}\mathrm{Im}\,G^r$ is the spectral function, and the single-particle Hamiltonian is given by

$$
\tilde{\mathcal{H}} = \mathcal{H}_o + \Sigma^\delta + \mathrm{Re}\Sigma^r \qquad (3.5)
$$

where Σ^δ and $\mathrm{Re}\,\Sigma^r$ correspond to the renormalization of the bands or mass terms induced by the self-energy [30, 31]. The appearance of the imaginary i in Eq. (3.4) in the anticommutator is commensurate when viewed in their gradient expansions in terms of Poisson bracket differential operator wherein a commutator of Eq. (3.4) has a factor i whereas the anticommutator does not have. As is well known, in the classical limit, the gradient expansion leads to the Boltzman kinetic transport equation [32].

A more stochastic-oriented quantum transport version is given by the following [30]:

$$
\begin{aligned}
&i\hbar\left(\frac{\partial}{\partial t_1}+\frac{\partial}{\partial t_2}\right)G^{\lessgtr} \\
&\quad=\left[\tilde{\mathcal{H}},G^{\lessgtr}\right]-\frac{i}{2}\left\{\Gamma,G^{\lessgtr}\right\}+\frac{1}{2}A\Gamma+\frac{1}{2}A\mathcal{N} \qquad (3.6)
\end{aligned}
$$

where $\mathcal{N}$ is the noise correlator,

$$
\mathcal{N}=-2\,\mathrm{Im}\,\Sigma^{\mathrm{c}}
$$

with Σ^c being the chronological self-energy correlation. In what follows, we will treat either the two-layer model of chiral Dirac fermions (m,m'), e.g., the bottom and top layers of graphene and TMDs in strong magnetic field or other 2D materials with strong spin–orbit coupling but without magnetic fields [33]. We have generalized the Hamiltonian spinor to account for either of the magnetic field and/or Dresselhaus and/or Rashba spin–orbit coupling [34].

3.3. Transformation to the Eight Tensor Variables

In what follows, we will explicitly consider Pauli–Dirac spin $(k,k'=\{\downarrow,\uparrow\})$, valley indices $(l,l'=\{K,K'\})$, and layer or band indices as the case maybe (e.g., $\{m,m'\}=\{t,b\}$). This means that all quantities in the quantum transport equations are essentially 8×8 matrices, and thus, we have 8×8 matrix equations on the spin and pseudospins indices. As it turns out, the 64 entries in the 'cube' matrix $G^{<}_{k,k'l,l'm,m'}$ can be transformed into eight tensor variables defined in Table 3.1.

Note that in Table 3.1 the eight tensor variables, including the total charge, are mapped to the configurations of three qubits as exhibited by the subcripts of the nonequilibrium tensorial Green function, $G^{<}_{\alpha\beta\gamma}$.

3.3.1. *Products of cube matrices in quantum transport equations*

In a similar manner, the products of 'cube' matrices can be decomposed into scalars, vectors, dyadics, and tensors of third rank, but now

Table 3.1: The eight transport variables.

$G^<_{0,0,0}$	complete scalar, i.e., particle number density or total charge density
$G^<_{k,0,0}$	Pauli–Dirac spin magnetization density
$G^<_{0,l,0}$	isospin magnetization density
$G^<_{0,0,m}$	pseudospin magnetization density
$G^<_{k,l,0}$	entangled or joint Pauli–Dirac spin and isospin magnetization density
$G^<_{k,0,m}$	entangled or joint Pauli–Dirac spin and pseudospin magnetization density
$G^<_{0,l,m}$	entangled or joint isospin–pseudospin magnetization density
$G^<_{k,l,m}$	entangled or joint Pauli–Dirac spin, isospin, pseudospin magnetization density

incorporating torques and torque entanglements. These are the total particle number or total charge density, total Pauli–Dirac spin magnetization density, total isospin magnetization density, total pseudospin magnetization density, total joint Pauli–Dirac spin and isospin magnetization density, total joint Pauli–Dirac spin and pseudospin magnetization density, total joint isospin–pseudospin magnetization density, and total joint Pauli–Dirac spin, isospin, pseudospin magnetization density. The number of each of these terms in a product is dictated by simple combinatoric counting, as shown in the following table.

The quantum transport equations are given in Appendix B. A more detailed derivation is given by Buot *et al.* [35]. In all of the derived coupled spin quantum transport equations, there is no formal difference between the resulting equations of different spins for the same characteristic integer, N_s, which designates the number of spinor degrees of freedom involved in a system. Thus, for the same N_s, the transport equations are mathematically identical and exhibit identical physical behavior, whether Pauli–Dirac spin or any of the pseudospins. The maximal entanglement entropy of the system [36] is given by $S = N_s \ln 2$ by simply associating with the bipartite entanglement of N_s-qubit systems. For carrying out numerical simulations [19, 20] of real highly nonlinear and switching spintronic devices, the prescription for converting all equations to phase space is given in Appendix C.

3.4. Entanglement-Induced Localization and Delocalization

One of the revealing aspects of entanglement in nonequilibrium superfield spinor quantum-transport physics coupled with lattice Weyl transform techniques [19] is the axial–chiral localization and delocalization through the presence of Levi-Civita tensors. The delocalization implies diffusive

Tensors of rank zero	$\left\{\begin{matrix} k=0 \text{ or } \sum_k, \\ l=0 \text{ or } \sum_l, \\ m=0 \text{ or } \sum_m \end{matrix}\right\}$, in three positions, $\Longrightarrow 2^3 = 8$ terms,
Dirac-spin vectors and torques	$\left\{\begin{matrix} l=0 \text{ or } \sum_l, \\ m=0 \text{ or } \sum_m \end{matrix}\right\}, \Longrightarrow \left\{\begin{matrix} 2\times 2^2 = 8 \text{ vectors;} \\ \text{two } 2-\text{DOF for the torques, } 2^2 = 4 \text{ torques} \end{matrix}\right\}$
Isospin vectors and torques	$\left\{\begin{matrix} k=0 \text{ or } \sum_k, \\ m=0 \text{ or } \sum_m \end{matrix}\right\} \Longrightarrow \left\{\begin{matrix} 2\times 2^2 = 8 \text{ vectors;} \\ \text{two 2–DOF for the torques, } 2^2 = 4\text{torques} \end{matrix}\right\}$
Pseudospin vectors and torques	$\left\{\begin{matrix} k=0 \text{ or } \sum_k, \\ l=0 \text{ or } \sum_l \end{matrix}\right\} \Longrightarrow \left\{\begin{matrix} 2\times 2^2 = 8 \text{ vectors;} \\ \text{two 2–DOF for the torques,} 2^2 = 4 \text{ torques} \end{matrix}\right\}$
$\left\{\begin{matrix} \text{Spin–isospin dyadics, torques,} \\ \text{entangled torques} \end{matrix}\right\}$	$\left\{\begin{matrix} m=0 \text{ or } \sum_m \\ k,l \text{ in two ways,} \\ l \text{ and } k \text{ in two ways} \end{matrix}\right\} \Longrightarrow \left\{\begin{matrix} 2^3 = 8 \text{ dyadics, } 2^3 = 8 \text{ torques,} \\ 2 \text{ entangled torques} \end{matrix}\right\}$
$\left\{\begin{matrix} \text{Spin–pseudospin dyadics, torques,} \\ \text{entangled torques} \end{matrix}\right\}$	$\left\{\begin{matrix} l=0, \text{ or } \sum_l, \\ k,m \text{ in two ways,} \\ \text{and } m \text{ and } k \text{ in two ways} \end{matrix}\right\} \Longrightarrow \left\{\begin{matrix} 2^3 = 8 \text{ dyadics, } 2^3 = 8 \text{ torques,} \\ 2 \text{ entangled torques} \end{matrix}\right\}$
$\left\{\begin{matrix} \text{Isopin–pseudospin dyadics, torques,} \\ \text{entangled torques} \end{matrix}\right\}$	$\left\{\begin{matrix} k=0, \text{ or } \sum_k, \\ l,m \text{ in two ways,} \\ \text{and } m \text{ and } l \text{ in two ways} \end{matrix}\right\} \Longrightarrow \left\{\begin{matrix} 2^3 = 8 \text{ dyadics, } 2^3 = 8 \text{ torques,} \\ 2 \text{ entangled torques} \end{matrix}\right\}$
$\left\{\begin{matrix} \text{Spin–isospin-pseudospin tensors,} \\ \text{torques,} \\ \text{entangled 2 and 3 torques} \end{matrix}\right\}$	$\left\{\begin{matrix} k,l, \text{ and } m \text{ considered as three qubits } \Longrightarrow 2^3 = 8 \text{ third-rank tensors,} \\ \text{torques for each spin is } 2^2 = 4 \Longrightarrow \text{ three spins } 3\times 4 = 12 \text{ torques,} \\ \text{entangled torques for each pair of torques is 2–three pairs of torques that can entangle,} \\ \Longrightarrow 3\times 2 = 6 \text{ entangled 2 torques,} \\ \text{Finally we can only have 1 entangled three torques} \end{matrix}\right\}$

processes in phase space or drift-diffusion and quantum tunneling transport in position space. These processes revolve around the lattice Weyl transform of terms, either commutator or anticommutator in the transport equations, multiplied by the Levi-Civita symbols giving the chiral nature of the dynamics.

Indeed, the diffusive property of the commutator is exhibited even in the leading order in gradient expansion. The commutator represents the inherent nonlocality in quantum mechanics and describes diffusive and tunneling motion in phase space of transport kinetics, whereas the anticommutator describes local or nondiffusive events in $(\vec{p}, \vec{q})$ space typefied either by cyclotron-orbit current tied to orbit center or by decay and growth of a phase-space distribution function tied to a point $(\vec{p}, \vec{q})$ in phase space. For example, the nondiffusive character of the scattering anticommutator has been simply approximated by a relaxation-time approximation in Refs. [19, 20], whereas the commutator is entirely responsible for bringing in the nonlocality of quantum transport physics, specifically the quantum tunneling transport in resonant tunneling nanostructures [19, 20] which for the first time resolves the controversial aspects of the current–voltage characteristics of resonant tunneling diodes [20].

For a simple demonstration (see Appendix C for more details), we give here the following lattice Weyl (LW) transform to phase space of a commutator $[A, B]$ and an anticommutator $\{A, B\}$ in terms of Poisson bracket differential operator, Λ, as expansion to all powers of $\hbar$,

$$\begin{aligned} [A,B](p,q) &= \cos\Lambda\,[a(p,q)\,b(p,q) - b(p,q)\,a(p,q)] \\ &\quad - i\sin\Lambda\,\{a(p,q)\,b(p,q) + b(p,q)\,a(p,q)\} \end{aligned} \tag{3.7}$$

$$\begin{aligned} \{A,B\}(p,q) &= \cos\Lambda\ \{a(p,q)\,b(p,q) + b(p,q)\,a(p,q)\} \\ &\quad - i\sin\Lambda\ [a(p,q)\,b(p,q) - b(p,q)\,a(p,q)] \end{aligned} \tag{3.8}$$

where $\Lambda = \frac{\hbar}{2}\left(\frac{\partial^{(a)}}{\partial p}\cdot\frac{\partial^{(b)}}{\partial q} - \frac{\partial^{(a)}}{\partial q}\cdot\frac{\partial^{(b)}}{\partial p}\right)$, the Poisson bracket operator. Thus, if the LW transforms $a(p,q)$ and $b(p,q)$ are not matrices, then to the lowest order, we have

$$[A,B](p,q) = -i\frac{\hbar}{2}\left(\frac{\partial a(p,q)}{\partial p}\cdot\frac{\partial b(p,q)}{\partial q} - \frac{\partial a(p,q)}{\partial q}\cdot\frac{\partial b(p,q)}{\partial p}\right) \tag{3.9}$$

$$\begin{aligned} \{A,B\}(p,q) &= \{a(p,q)\,b(p,q) + b(p,q)\,a(p,q)\} \\ &= 2\,a(p,q)\,b(p,q) \end{aligned} \tag{3.10}$$

clearly showing that even to the lowest order the local and diffusive properties in phase space of anticommutator and commutator, respectively, are apparent. Likewise, in its full integral representation (See Appendix C) which is suitable for carrying numerical simulations, Eq. (3.7) has a nonlocal kernel, whereas Eq. (3.8) has a local kernel of integration. It is precisely the nonlocal character of the kernel given in Eq. (C.11) of the Appendix that is responsible for quantum tunneling in resonant tunneling diodes [19, 20], whose unique current–voltage characteristics cannot be explained by any degree of semi-classical approximations [37].

3.4.1. *Hierarchy of entanglements: Alternating sequence of commutators and anticommutators as function of N_s*

Let us for example start with Pauli–Dirac spinor form of the commutator of A and B which can be written in tensorial form as

$$\begin{aligned}[A,B] &= \frac{1}{4}\left(\left[a_o,\vec{b}\right] + [\vec{a},b_o] + i\left[\vec{a}\times\vec{b} - \vec{b}\times\vec{a}\right]\right)\\ [A,B]_k &= \frac{1}{4}\left([a_o,b_k] + [a_k,b_o] + i\epsilon_{kk_1k_2}\left\{a_{k_1},b_{k_2}\right\}\right)\end{aligned} \tag{3.11}$$

where the *subscript* o indicates total independence with respect to the spin degrees of freedom and $\{a_{k_1}, b_{k_2}\}$ is the anticommutation of a_{k_1} and b_{k_2}. Note the change from original commutator to anticommutator for the term containing one Levi-Civita symbol. In including dependence on other spin degrees of freedom, in all instances the theorem on binary product of 2×2 matrices, Eq. (3.11) is successively applied. Thus, the fully entangled 'torque' (coming from the last term of Eq. (3.11)) now becomes

$$i\epsilon_{ll_1l_2} i\epsilon_{kk_1k_2}\left[a_{k_1,l_1}, b_{k_2,l_2}\right] = -\epsilon_{ll_1l_2}\epsilon_{kk_1k_2}\left[a_{k_1,l_1}, b_{k_2,l_2}\right] \tag{3.12}$$

Upon adding dependence on one more pseudospin degrees of freedom, such as the pseudospin arising either from the symmetry at low energies of the band indices at Dirac points in graphene or from layer pseudospin in bilayer graphene, the entanglement of the torques becomes

$$-i\epsilon_{mm_1m_2}\epsilon_{ll_1l_2}\epsilon_{kk_1k_2}\left\{a_{k_1,l_1,m_1}, b_{k_2,l_2,m_2}\right\} \tag{3.13}$$

which is a third rank tensor. Assuming we have considered only the band pseudospin in Eq. (3.13), if one further considers the pseudospin from layer degrees of freedom in bilayer graphene, for example, then the fully entangled 'torque' becomes

$$\epsilon_{nn_1n_2}\epsilon_{mm_1m_2}\epsilon_{ll_1l_2}\epsilon_{kk_1k_2}\left[a_{k_1,l_1,m_1,n_1}b_{k_2,l_2,m_2,n_2}\right]$$

Thus, starting from the commutation of $[A, B]$, the dependence in spin degrees of freedom results in the alternating commutator and anticommutator of the spin tensor components in the expression for entangled torques, with commutator for even number of $i\epsilon_{ijk}$ (*Levi-Civita* symbols) and anticommutator for odd number of $i\epsilon_{ijk}$. Had we started with $\{A, B\}$, then there will be alternating anticommutator and commutator instead, as the number of spin degrees of freedom grows.

3.4.2. *Local and nonlocal terms*

The commutator rigorously incorporates the quantum nonlocality in phase space or quantum-diffusive motion of transport kinetics in phase space (this is evident even in the semi-classical approximation leading to Boltzmann transport equation, i.e., within the first power of Poisson bracket) and exhibits quantum tunneling across abrupt barriers when all powers of $\hbar$ are taken into account by using the integral kernel [20], whereas the anticommutator describes local or nondiffusive events in $(\vec{p}, \vec{q})$ space as typefied by cyclotron-orbit current tied to orbit center or decay and growth of a phase-space distribution function tied to a point $(\vec{p}, \vec{q})$ in phase space. Moreover, the exact integral form of the commutator and anticommutator exhibits nonlocal kernel and local kernel of integration, respectively. See also the discussion in Appendix C.

One observes that by virtue of Eqs. (3.11)–(3.12), the entangling with a second pseudospin can transform local terms in the transport equation into diffusive or mobile terms, a sort of *delocalization*. This type of delocalization is not surprising as it is exhibited even in the classical case, often used to explain the absence of diamagnetism in classical free electron gas [11]. In the classical case, when localized currents or cyclotron orbits interact with a constriction or opposite boundaries (providing torque interaction), delocalized currents in opposite directions or chirality reside in opposite boundaries, respectively. Thus, the quantum edge states in integer quantum Hall effect (precursor to topological insulators) are often portrayed semi-classically as counterpropagating "skipping orbits" that propagate (delocalize) along the boundary of the system in the sense opposite to that of the Landau orbits, when a particle in a Landau orbit interacts with the boundary and specularly bounces off it (see Figure 3.1) [11].

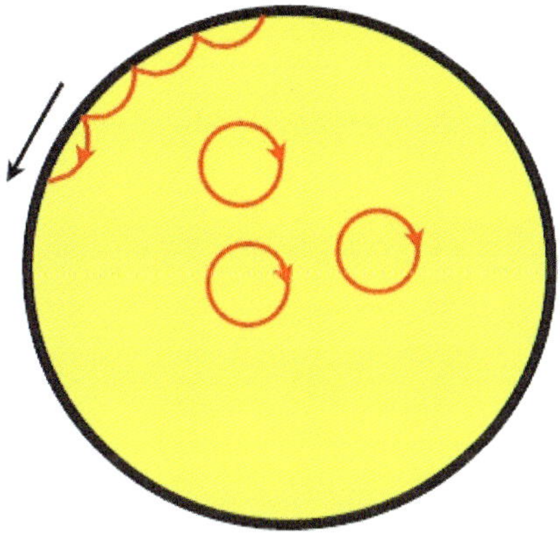

Fig. 3.1: Classical skipping orbits in space, showing the *holographic* nature of topological entanglement, i.e., *bulk-boundary correspondence or* entanglement of $\uparrow$-chirality of the boundary with $\downarrow$-chirality of the bulk. [Reproduced from Ref. [11]].

Thus, consider, for instance, a term in the quantum transport equation for $N_s = 2$ of the form, $i\epsilon_{ll_1l_2}\left\{H_{l_1,o}, G^{\gtrless}_{l_2,o}\right\}$, which is the term coming from the Hamiltonian [our spinor Hamiltonian is generalized to account for the spinor ReΣ^r [40, 41] (which is included in H, Eq. (3.5) as a spinor mass term) and/or magnetic field and/or Dresselhaus and/or Rashba spin orbit coupling] that generates the Landau orbit and is thus a local term by virtue of the anticommutator. However, when the dependence on pseudospin, m, comes into play, this local term is transformed into a nonlocal term, specifically through the term, $i\epsilon_{ll_1l_2}i\epsilon_{mm_1m_2}\left[H_{l_1,m_1}G^{\lessgtr}_{l_2,m_2}\right]$ which is now a nonlocal term, i.e., mobile term in terms of transport point of view, as expressed by the commutator.

3.4.3. *Integer quantum Hall effect topological insulator*

Quantum mechanically speaking, in the specular reflection at the boundary, the incoming and outgoing states present the boundary pseudospin, resulting in the boundary pseudospin-dependent interaction of the Landau orbits. This boundary pseudospin-dependent interaction results in the delocalization of the electron current along the boundary. Since the interaction is elastic and does not cost energy, the delocalized current excitations do not have mass or are zero-excitation modes resulting in the chiral/helical dispersion relations or Dirac point of the excitations as in the case of integer quantum Hall effect. This is the mechanism leading to QHE-TI. In Figure 3.1, one may consider the situation as an entanglement of "+" chirality of the boundary with the "−" chirality of the bulk, a highly localized version of the familiar highly separated entangled pair of spins in a singlet state [4, 5].

Therefore, starting from the commutator $[A, B]$ of Eq. (3.4), the dependence in spinor pseudospin degrees of freedom results in a series of alternating commutator and anticommutator of the spin tensor components as the number of entangled spin degrees of freedom, N_s, grows, with commutator for even number of $i\epsilon_{ijk}$ and anticommutator for odd number of $i\epsilon_{ijk}$. Had we started with $\{A, B\}$, then there will be a series of alternating anticommutator and commutator instead. In other words, entanglement of spins will result in transformations from local to diffusive motion in phase space and vice versa. This novel delocalization mechanism seems to have direct relevance to topological insulators where torques, due either to spin–orbit coupling (Berry curvature) [33, 38] or magnetic fields result in localization or a gapped spectrum in the bulk (*leading* to integer quantum Hall effect in electrical conductivity), are then entangled with pseudospin torque at the boundary consistent with Dirac points, *akin* to classical specular reflection of cyclotron orbits at boundaries. This results in delocalization at the edges (conductive edge quantum states) of 2D systems.

We therefore expect that localization of the edge states can occur if further entanglement with spin-dependent (spinor) impurities (resulting in spin flips) [17] or when layer pseudospin becomes important in bilayer topological insulators, i.e., when three spin degrees of freedom become entangled at the edge. There are actually some evidences on this aspect of edge-state localization in bilayer topological quantum Hall 2D systems where edge states are not even mentioned in the discussion of transport [39]. On the other hand, the intralayer gapped bulk states of the LLO in bilayer systems become mobile due to the interaction with layer pseudospin, in fact in pseudospin ferromagnetic state, this becomes superconducting mediated by the condensation of bilayer bosons of excitons [16].

The inherent nonlocality property of commutators is fundamentally rooted in the uncertainty principle of quantum mechanics where the smallest nonlocality in phase space is measured by the Planck constant $\hbar$. Indeed, the quantum uncertainty principle naturally follows from the formalism discussed in Section 3.1.

3.4.4. *Metallic edge states and bilayer superconductivity*

If the pseudospin m is the boundary pseudospin, then the delocalization will result in the so-called edge metallic states. Similarly, in the bilayer system with each layer in their LLO localized state, the entangling with the layer pseudospin will delocalize the LLO states into nonlocal conducting states. In fact, in the layer-pseudospin ferromagnetic state, experiments have shown that the electron–hole boson condensate at nonzero momentum leads to a new mechanism for the onset of superconductivity.

3.4.5. *Localization and delocalization*

Thus, we expect, in the absence of magnetic field, localization from Dresselhaus and Rashba spin–orbit coupling as incorporated in the Hamiltonian, H, yielding the *orbit term* $i\epsilon_{mm_1m_2}\left\{H_{m_1}, G^{\gtrless}_{m_2}\right\}$, whereas delocalization contribution is expected from the Dyakonov–Perel and Elliott–Yafet mechanism as incorporated in the *scattering-out* Γ-term, and *scattering-in* A-term. These are the commutator terms containing one Levi-Civita symbol, which describe a complex motion similar to the motion and deformation of strong magnetic field-induced Landau orbits, due to the torque exerted by the electric field [42], resulting in the build-up of Hall voltage [43]. There are indeed theoretical and experimental works which support this assertion, the effect is related to what is often referred to in the literature as weak localization due to Dresselhaus and Rashba spin–orbit couplings [24] and weak antilocalization or supression of scattering rates due to the Dyakonov–Perel and Elliott–Yafet scattering mechanisms [25, 26] or entanglement of spin–orbit couplings with another torque (spinor) degrees of freedom. Delocalization or nonlocal scattering terms may also occur where the scattering centers have a pseudospin character, such as the two-level dependence of the scattering in Wigner-function quantum transport formalism recently studied by Rossi *et al.* [44, 45].

3.4.5.1. *Crossover from weak localization to weak antilocalization*

In the literature, the terms "weak localization" (WL) and "weak antilocalization" (WAL) often specifically refers to the presence of both spin–orbit coupling and magnetic field in a 2D system. The effect of the magnetic field is to induce weak localization where carriers execute Wilson loops (i.e., attributed to $i\epsilon_{kk_1k_2}\left\{H_{k_1,o}, G^{\gtrless}_{k_2,o}\right\}$ anticommutator term in transport equation) along its path, but the effect of spin–orbit coupling is to induce weak antilocalization (WAL) by trying to modify the Wilson loops into a quantum diffusive term (i.e., from the $i\epsilon_{kk_1k_2}i\epsilon_{mm_1m_2}\left[H_{k_1,m_1}G^{\lessgtr}_{k_2,m_2}\right]$ commutator term). These contrasting and competing effects as well as their respective dependence on magnetic-field strength for Wislon localization loops, and gated electric-field strength or thickness of a 2D layer system for the spin–orbit delocalization effects, can result in an observed crossover between WL and WAL in the magnetoconductivity measurements [27] as a function of gate voltage or magnetic field. The observation of weak antilocalization implies that a strong spin–orbit coupling is the dominant effect [46].

In exprimental conditions, we can assume that statistically an Ω number of particles are executing Wilson loops in their transport, i.e., $i\epsilon_{kk_1k_2}\left\{H_{k_1,o}, G^{\gtrless}_{k_2,o}\right\}$ due to the effects of applied magnetic fields, whereas $(N-\Omega)$ number of particles already felt as well the applied gate-electric field, resulting in a stronger spin–orbit coupling leading to $i\epsilon_{kk_1k_2}i\epsilon_{mm_1m_2}\left[H_{k_1,m_1}G^{\lessgtr}_{k_2,m_2}\right]$ torque-entanglement modification. At constant gate voltage, the magnetoconductivity is a decreasing function of the magnetic field. However by increasing the gate voltage, the spin–orbit coupling start to become dominant in counteracting the effects of the Wilson weak-localization loops and therefore a crossover point in the magnetoconductivity measurent is expcted to occur at $\Omega = 0.5N$ after which the magnetoconductivity will start to increase with the applied gate voltage. Indeed, the crossover point or minimum point of the magnetoconductivity has been observed in the experiments. The terminology '*weak*' as used in the experimental observations is due to the fact that there are other important terms that contribute to quantum nonlocality and hence magnetoconductivity transport measurements, e.g., nonlocality of spin-dependent scattering terms [43].

3.4.6. *Quantum spin Hall effect topological insulator*

We remark on quantum spin Hall effect topological insulator (QSHE-TI) as it relates to the above discussion on QHE-TI. In time-reversal symmetric systems with strong spin–orbit coupling, Kramer's degenerate states play an important role. Since spin–orbit coupling does not break time-reversal symmetry, Kramer's degenerate pair remains intact and moves in opposite directions with opposite spins. Thus, the effective spin–orbit magnetic field rotates Kramer's degenerate pair in opposite sense, say clockwise and counterclockwise rotating pair. When this pair interacts with the boundary pseudospin, in the same sense as discussed above, this oppositely moving Kramer's Landau orbit pair forms metallic edge states of oppositely moving currents having opposite spins, i.e., helical edge states. This is the mechanism leading to QSHE-TI.

3.4.7. *Topological Kondo insulator*

It is worthwhile to add comments on topological Kondo insulator [47], which represents the intersection of topology and strongly correlated and heavy fermion systems. Within the quantum transport perspective, the hybridization of localized f and mobile d-band, which gives rise to a hybridization gap insulator, is within the domain of quantum transport localization term expressed by the '*orbit term*' $i\epsilon_{kk_1k_2}\left\{H_{k_1}, G_{k_2}^{\gtrless}\right\}$ driven by strong spin–orbit interaction of the f-electrons [47], as discussed above. Likewise, the metallic edge states are described by the delocalization term $i\epsilon_{kk_1k_2}i\epsilon_{mm_1m_2}\left[H_{k_1,m_1}G_{k_2,m_2}^{\lessgtr}\right]$ similar to the mechanism leading to QSHE-TI with helical edge or surface states. It should be pointed out that since the orbit centers are the atoms themselves, the delocalization at the surface could results in the restructuring of the atomic surface atoms resulting in twisting or warping, as exhibited by some Kondo insulator materials, such as $CeNiSn$ possessing nonsymmorphic symmetries, i.e., containing three glide reflections, three screw rotations and an inversion symmetry [48]. Here, nonsymmorphic symmetries give rise to a momentum-dependent twist that enables the surface states to be detached from the bulk on the glide plane, leading to the so-called Möbius-twisted surface states [48].

3.5. Concluding Remarks

In summary, it is worth noting that although topological phases are often discussed in the Hamiltonian context, it has also been shown that associated topological protection and phenomena can also emerge in open quantum systems with engineered dissipation [49]. In this paper, we have shown that a rounded picture and intuitive aspects of topological phases and bilayer superconductivity arise from the entanglement of different pseudospins in highly nonlinear and nonequilibrium quantum transport equations of spin and pseudospins. These mechanisms also shed light in the so-called weak localization (WL) and delocalization (WAL) as observed experimentally [27]. Two entangled torques from two spin degrees of freedom, one in the bulk extending to the boundary and one at the boundary, are needed to characterize topological insulators. Moreover, in most experiments on WL and WAL, two entangled torques, one from the applied magnetic field and one from spin–orbit coupling, are also needed to characterize the results of experiments on WL versus WAL and their crossover point.

The nonequilibrium quantum transport results of this paper also serve as springboard to applications in the emerging field of spincaloritronics and pseudospincaloritronics [50, 51]. The formalism employed here can be extended to the inclusion of the spinor form of the electron–phonon/plasmon and spinon/magnon self-energies in accounting for the coupling of the equations obtained here to the spin-dependent nonequilibrium quantum transport of phonons and plasmons [31]. This will have relevance concerning heat flow in spincaloritronics and pseudospincaloritronics.

Acknowledgements

The author is grateful to Prof. Wolfram Schommers for his invitation to contribute to this book series.

Appendix

A. Binary Product of 'Cube Matrices'

The result of any binary product of matrices characterized by Pauli–Dirac spin indices, valley or isospin indices, and energy bands or pseudospin indices can be classified into *eight* tensorial groups, obtained by *successive iteration* of the spinor transformation of 2×2 matrices given in Eq. (3.11). We obtained the following expressions, where as before repeated indices are summed up.

A.1. *Total scalars or tensors of rank zero*

Here, we have $\left\{k = 0 \text{ or } \sum\limits_k,\ l = 0 \text{ or } \sum\limits_l,\ m = 0 \text{ or } \sum\limits_m\right\}$, i.e., three 2-degrees of freedom, yielding $2^3 = 8$ terms,

$$S_{o,o,o} = \left(\frac{1}{4}\right)^3 \left\{ \begin{array}{l} \left[\bar{a}_{o,o,o}\bar{b}_{o,o,o} + a_{k,o,o}b_{k,o,o} + \bar{a}_{o,o,m}\bar{b}_{o,o,m}\right] \\ + \bar{a}_{o,l,o}\bar{b}_{o,l,o} + a_{k,l,o}b_{k,l,o} + \bar{a}_{o,l,m}\bar{b}_{o,l,m} \\ + a_{k,o,m}b_{k,o,m} + a_{k,l,m}b_{k,l,m} \end{array} \right\}$$

A.2. *Dirac spin vectors and torques*

Here, we have $[l = 0 \text{ or } \sum\limits_l,\ m = 0 \text{ or } \sum\limits_m]$ on each factor, yielding $2 \times 2^2 = 8$ vectors, and two 2-degrees of freedom for the torques, $2^2 = 4$ torques:

$$V_{k,o,o} = \left(\frac{1}{4}\right)^3 \times \begin{bmatrix} \bar{a}_{o,o,o}b_{k,o,o} + a_{k,o,o}\bar{b}_{o,o,o} + i\epsilon_{kk_1k_2}\left(a_{k_1,o,o}b_{k_2,o,o}\right) \\ + \bar{a}_{o,l,o}b_{k,l,o} + a_{k,l,o}\bar{b}_{o,l,o} + i\epsilon_{kk_1k_2}\left(a_{k_1,l,o}b_{k_2,l,o}\right) \\ + \bar{a}_{o,o,m}b_{k,o,m} + a_{k,o,m}\bar{b}_{o,o,m} + i\epsilon_{kk_1k_2}\left(a_{k_1,o,m}b_{k_2,o,m}\right) \\ + \bar{a}_{o,l,m}b_{k,l,m} + a_{k.l,m}\bar{b}_{o,l,m} + i\epsilon_{kk_1k_2}\left(a_{k_1,l,m}b_{k_2,l,m}\right) \end{bmatrix} \tag{A.1}$$

A.3. *Isospin vectors and torques*

Here, we have [$k = 0$ or $\sum_k$, $m = 0$ or $\sum_m$] on each factor, yielding $2 \times 2^2 = 8$ vectors, and two 2-degrees of freedom for the torques, $2^2 = 4$ torques:

$$V_{o,l,o} = \left(\frac{1}{4}\right)^3 \begin{Bmatrix} \left[\bar{a}_{o,o,o}\bar{b}_{o,l,o} + \bar{a}_{o,l,o}\bar{b}_{o,o,o} + i\epsilon_{ll_1l_2}\left(\bar{a}_{o,l_1,o}\bar{b}_{o,l_2,o}\right)\right] \\ + \bar{a}_{o,o,m}\bar{b}_{o,l,m} + \bar{a}_{o,l,m}\bar{b}_{o,o,m} + i\epsilon_{ll_1l_2}\left(\bar{a}_{o,l_1,m}\bar{b}_{o,l_2,m}\right) \\ + \left[a_{k,o,o}b_{k,l,o} + a_{k,l,o}b_{k,o,o} + i\epsilon_{ll_1l_2}\left(a_{k,l_1,o}b_{k,l_2,o}\right)\right] \\ + a_{k,o,m}b_{k,l,m} + a_{k.l,m}b_{k,o,m} + i\epsilon_{ll_1l_2}\left(a_{k,l_1,m}b_{k,l_2,m}\right) \end{Bmatrix}. \tag{A.2}$$

A.4. *Pseudospin vectors and torques*

Similarly, here, we have [$k = 0$ or $\sum_k$, $l = 0$ or $\sum_l$] on each factor, yielding $2 \times 2^2 = 8$ vectors, and two 2-degrees of freedom for the torques, $2^2 = 4$ torques:

$$V_{o,o,m} = \left(\frac{1}{4}\right)^3 \begin{Bmatrix} \left[\bar{a}_{o,o,o}\bar{b}_{o,o,m} + \bar{a}_{o,o,m}\bar{b}_{o,o,o} + i\epsilon_{mm_1m_2}\left(\bar{a}_{o,o,m_1}\bar{b}_{o,o,m_2}\right)\right] \\ + \left[a_{k,o,o}b_{k,o,m} + a_{k,o,m}b_{k,o,o} + i\epsilon_{mm_1m_2}\left(a_{k,o,m_1}b_{k,o,m_2}\right)\right] \\ + \left[\bar{a}_{o,l,o}\bar{b}_{o,l,m} + \bar{a}_{o,l,m}\bar{b}_{o,l,o} + i\epsilon_{mm_1m_2}\left(\bar{a}_{o,l,m_1}\bar{b}_{o,l,m_2}\right)\right] \\ + \left[a_{k,l,o}b_{k,l,m} + a_{k,l,m}b_{k,l,o} + i\epsilon_{mm_1m_2}\left(a_{k,l,m_1}b_{k,l,m_2}\right)\right] \end{Bmatrix}$$

A.5. *Dirac spin–isospin dyadics, torques and entangled torques*

Here, we have [$m = 0$ or $\sum_m$, k, l can be distributed in two ways, and l and k can be distributed in two ways], yielding $2^3 = 8$ dyadics, $2^3 = 8$ torques, and $2^1 = 2$ entangled torques from $m = 0$ or $\sum_m$ only. We have

$$T_{k,l} = \left(\frac{1}{4}\right)^3 \times \left\{ \begin{bmatrix} \bar{a}_{o,o,o}b_{k,l,o} + a_{k,l,o}\bar{b}_{o,o,o} + \bar{a}_{o,l,o}b_{k,o,o} + a_{k,o,o}\bar{b}_{o,l,o} \\ + i\epsilon_{kk_1k_2}\left(a_{k_1,o,o}b_{k_2,l,o}\right) + i\epsilon_{kk_1k_2}\left(a_{k_1,l,o}b_{k_2,o,o}\right) \\ + \left[i\epsilon_{ll_1l_2}\left(\bar{a}_{o,l_1,o}b_{k,l_2,o}\right)\right] + i\epsilon_{ll_1l_2}\left(a_{k,l_1,o}\bar{b}_{o,l_2,o}\right) \\ + \frac{1}{2}\left[i\epsilon_{kk_1k_2}i\epsilon_{ll_1l_2}\left(a_{k_1,l_1,o}b_{k_2,l_2,o}\right)\right] \\ + \frac{1}{2}i\epsilon_{kk_1k_2}i\epsilon_{ll_1l_2}\left(a_{k_1,l_1,o}b_{k_2,l_2,o}\right) \end{bmatrix} + \begin{bmatrix} + \bar{a}_{o,o,m}b_{k,l,m} + a_{k.l,m}\bar{b}_{o,o,m} + \bar{a}_{o,l,m}b_{k,o,m} + a_{k,o,m}\bar{b}_{o,l,m} \\ + i\epsilon_{kk_1k_2}\left(a_{k_1,l,m}b_{k_2,o,m}\right) + i\epsilon_{kk_1k_2}\left(a_{k_1,o,m}b_{k_2,l,m}\right) \\ + i\epsilon_{ll_1l_2}\left(a_{k,l_1,m}\bar{b}_{o,l_2,m}\right) + \left[i\epsilon_{ll_1l_2}\left(\bar{a}_{o,l_1,m}b_{k,l_2,m}\right)\right] \\ + \frac{1}{2}\left[i\epsilon_{kk_1k_2}i\epsilon_{ll_1l_2}\left(a_{k_1,l_1,m}b_{k_2,l_2,m}\right)\right] \\ + \frac{1}{2}\left[i\epsilon_{kk_1k_2}i\epsilon_{ll_1l_2}\left(a_{k_1,l_1,m}b_{k_2,l_2,m}\right)\right] \end{bmatrix} \right\}$$

A.6. *Dirac spin–pseudospin dyadics, torques and entangled torques*

Here, we have [$l = 0$, or $\sum_l$, k, m can be distributed in two ways, and m and k can be distributed in two ways], yielding $2^3 = 8$ dyadics, $2^3 = 8$ torques, and $2^1 = 2$ entangled torques from $l = 0$, or $\sum_l$ only. We have

$$T_{k,m} = \left(\frac{1}{4}\right)^3 \begin{bmatrix} \bar{a}_{o,o,o}b_{k,o,m} + a_{k,o,m}\bar{b}_{o,o,o} + \bar{a}_{o,o,m}b_{k,o,o} + a_{k,o,o}\bar{b}_{o,o,m} \\ + i\epsilon_{kk_1k_2}\left(a_{k_1,o,o}b_{k_2,o,m}\right) + i\epsilon_{kk_1k_2}\left(a_{k_1,o,m}b_{k_2,o,o}\right) \\ + i\epsilon_{mm_1m_2}\left(\bar{a}_{o,o,m_1}b_{k,o,m_2}\right) + i\epsilon_{mm_1m_2}\left(a_{k,o,m_1}\bar{b}_{o,o,m_2}\right) \\ + \frac{1}{2}i\epsilon_{kk_1k_2}i\epsilon_{mm_1m_2}\left(a_{k_1,o,m_1}b_{k_2,o,m_2}\right) \\ + \frac{1}{2}i\epsilon_{kk_1k_2}i\epsilon_{mm_1m_2}\left(a_{k_1,o,m_1}b_{k_2,o,m_2}\right) \end{bmatrix}$$

$$+ \left(\frac{1}{4}\right)^3 \begin{bmatrix} \bar{a}_{o,l,o}b_{k,l,m} + a_{k.l,m}\bar{b}_{o,l,o} + a_{k,l,o}\bar{b}_{o,l,m} + \bar{a}_{o,l,m}b_{k,l,o} \\ + i\epsilon_{kk_1k_2}\left(a_{k_1,l,o}b_{k_2,l,m}\right) + i\epsilon_{kk_1k_2}\left(a_{k_1,l,m}b_{k_2,l,o}\right) \\ + i\epsilon_{mm_1m_2}\left(\bar{a}_{o,l,m_1}b_{k,l,m2}\right) + i\epsilon_{mm_1m_2}\left(a_{k.l,m_1}\bar{b}_{o,l,m_2}\right) \\ + \frac{1}{2}i\epsilon_{kk_1k_2}i\epsilon_{mm_1m_2}\left(a_{k_1,l,m_1}b_{k_2,l,m_2}\right) \\ + \frac{1}{2}i\epsilon_{kk_1k_2}i\epsilon_{mm_1m_2}\left(a_{k_1,o,m_1}b_{k_2,o,m_2}\right) \end{bmatrix}$$

A.7. *Isospin–pseudospin dyadics, torques and entangled torques*

Here, we have [$k = 0$ or $\sum_k$, l, m can be distributed in two ways, and m and l can be distributed in two ways], yielding $2^3 = 8$ dyadics, $2^3 = 8$ torques, and $2^1 = 2$ entangled torques from $k = 0$ or $\sum_k$ only. We have

$$T_{l,m} = \left(\frac{1}{4}\right)^3 \begin{bmatrix} \bar{a}_{o,o,o}\bar{b}_{o,l,m} + \bar{a}_{o,l,m}\bar{b}_{o,o,o} + \bar{a}_{o,l,o}\bar{b}_{o,o,m} + \bar{a}_{o,o,m}\bar{b}_{o,l,o} \\ + i\epsilon_{mm_1m_2}\left(\bar{a}_{o,o,m_1}\bar{b}_{o,l,m_2}\right) + i\epsilon_{mm_1m_2}\left(\bar{a}_{o,l,m_1}\bar{b}_{o,o,m_2}\right) \\ + i\epsilon_{ll_1l_2}\left(\bar{a}_{o,l_1,o}\bar{b}_{o,l_2,m}\right) + i\epsilon_{ll_1l_2}\left(\bar{a}_{o,l_1,m}\bar{b}_{o,l_2,o}\right) \\ + \frac{1}{2}i\epsilon_{ll_1l_2}i\epsilon_{mm_1m_2}\left(\bar{a}_{o,l_1,m_1}\bar{b}_{o,l_2,m_2}\right) \\ + \frac{1}{2}i\epsilon_{ll_1l_2}i\epsilon_{mm_1m_2}\left(\bar{a}_{o,l_1,m_1}\bar{b}_{o,l_2,m_2}\right) \end{bmatrix}$$

$$+ \left(\frac{1}{4}\right)^3 \begin{bmatrix} a_{k,l,o}b_{k,o,m} + a_{k,o,m}b_{k,l,o} + a_{k.l,m}b_{k,o,o} + a_{k,o,o}b_{k,l,m} \\ + i\epsilon_{mm_1m_2}\left(a_{k.l,m_1}b_{k,o,m_2}\right) + i\epsilon_{mm_1m_2}\left(a_{k,o,m_1}b_{k,l,m_2}\right) \\ + i\epsilon_{ll_1l_2}\left(a_{k,l_1,m}b_{k,l_2,o}\right) + i\epsilon_{ll_1l_2}\left(a_{k,l_1,o}b_{k,l_2,m}\right) \\ + \frac{1}{2}i\epsilon_{ll_1l_2}i\epsilon_{mm_1m_2}\left(a_{k,l_1,m_1}b_{k,l_2,m_2}\right) \\ + \frac{1}{2}i\epsilon_{ll_1l_2}i\epsilon_{mm_1m_2}\left(a_{k,l_1,m_1}b_{k,l_2,m_2}\right) \end{bmatrix}.$$

A.8. *Dirac spin–isospin–pseudospin tensors, torques, and entangled 2 and 3 torques*

Here, we have [k, l, and m can be considered as three qubits], yielding $2^3 = 8$ tensors of third rank, the number of torques for each spin is $2^2 = 4$, yielding for three spins $3 \times 4 = 12$ torques, the number of entngled torques for each pair of torques is 2 and since there are three pairs of torques that can entangle, the results is $3 \times 2 = 6$ entangled torques. Finally, we can only have one entangled three torques. We have the result,

$$T_{k,l,m}$$

$$= \left(\frac{1}{4}\right)^3 \begin{bmatrix} a_{k,l,m}b_{0,0,0} + a_{k,l,0}b_{0,0,m} + a_{k,0,m}b_{0,l,0} + a_{0,l,m}b_{k,0,0} \\ + a_{k,0,0}b_{0,l,m} + a_{0,0,m}b_{k,l,0} + a_{0,l,0}b_{k,0,m} + a_{o,o,o}b_{k,l,m} \end{bmatrix}$$

$$+\left(\frac{1}{4}\right)^3\begin{bmatrix}\epsilon_{kk_1k_2}a_{k_1,0,0}b_{k_2,l,m}+\epsilon_{kk_1k_2}a_{k_1,l,0}b_{k_2,0,m}\\+\epsilon_{kk_1k_2}a_{k_1,0,m}b_{k_2,l,0}+\epsilon_{kk_1k_2}a_{k_1,l,m}b_{k_2,0,0}\\+\epsilon_{ll_1l_2}a_{0,l_1,0}b_{k,l_2,m}+\epsilon_{ll_1l_2}a_{k,l_1,0}b_{0,l_2,m}\\+\epsilon_{ll_1l_2}a_{0,l_1,m}b_{k,l_2,0}+\epsilon_{ll_1l_2}a_{k,l_1,m}b_{0,l_2,0}\\+\epsilon_{m;m_1,m_2}a_{0,0.m_1}b_{k,l,m_2}+\epsilon_{m;m_1,m_2}a_{0,l.m_1}b_{k,0,m_2}\\+\epsilon_{m;m_1,m_2}a_{k,0.m_1}b_{0,l,m_2}+\epsilon_{m;m_1,m_2}a_{k,l.m_1}b_{0,0,m_2}\end{bmatrix}$$

$$+\left(\frac{1}{4}\right)^3\begin{bmatrix}i\epsilon_{kk_1k_2}i\epsilon_{ll_1l_2}\left(a_{k_1,l_1,o}b_{k_2,l_2,m}+a_{k_1,l_1,m}b_{k_2,l_2,o}\right)\\+i\epsilon_{kk_1k_2}i\epsilon_{mm_1m_2}\left(a_{k_1,o,m_1}b_{k_2,l,m_2}+a_{k_1,l,m_1}b_{k_2,o,m_2}\right)\\+i\epsilon_{ll_1l_2}i\epsilon_{mm_1m_2}\left(a_{o,l_1,m_1}b_{k,l_2,m_2}+a_{k,l_1,m_1}b_{o,l_2,m_2}\right)\\+i\epsilon_{kk_1k_2}i\epsilon_{ll_1l_2}i\epsilon_{mm_1m_2}\left(a_{k_1,l_1,m_1}b_{k_2,l_2,m_2}\right)\end{bmatrix}.$$

B. Quantum Transport Equations for $N_s = 3$

We have for the total scalar,

$$2^{N_s}i\hbar\left(\frac{\partial}{\partial t_1}+\frac{\partial}{\partial t_2}\right)G^{\gtrless}_{o,o,o}$$
$$=\left[\tilde{\mathcal{H}},G^{\lessgtr}\right]_{o,o,o}+\left[\Sigma^{\lessgtr},\mathrm{Re}G^{r}\right]_{o,o,o}-\frac{i}{2}\left\{\Gamma,G^{\lessgtr}\right\}_{o,o,o}+\frac{i}{2}\left\{\Sigma^{\lessgtr},A\right\}_{o,o,o} \tag{B.1}$$

where

$$\left[\tilde{\mathcal{H}},G^{\lessgtr}\right]_{o,o,o}$$
$$=\left[H_{o,o,o},G^{\gtrless}_{o,o,o}\right]+\left[H_{o,o,m},G^{\gtrless}_{o,o,m}\right]+\left[H_{o,l,o},G^{\gtrless}_{o,l,o}\right]$$
$$+\left[H_{o,l,m},G^{\gtrless}_{o,l,m}\right]+\left[H_{k,o,o},G^{\gtrless}_{k,o,o}\right]+\left[H_{k,o,m},G^{\lessgtr}_{k,o,m}\right]$$
$$+\left[H_{k,l,o},G^{\gtrless}_{k,l,o}\right]+\left[H_{k,l,m},G^{\gtrless}_{k,l,m}\right] \tag{B.2}$$

$$\left[\Sigma^{\lessgtr},\mathrm{Re}G^{r}\right]_{o,o,o}$$
$$=\left[\Sigma^{\gtrless}_{o,o,o},\mathrm{Re}G^{r}_{o,o,o}\right]+\left[\Sigma^{\lessgtr}_{o,o,m},\mathrm{Re}G^{r}_{o,o,m}\right]+\left[\Sigma^{\gtrless}_{o,l,o},\mathrm{Re}G^{r}_{o,l,o}\right]$$
$$+\left[\Sigma^{\gtrless}_{o,l,m},\mathrm{Re}G^{r}_{o,l,m}\right]+\left[\Sigma^{\lessgtr}_{k,o,o},\mathrm{Re}G^{r}_{k,o,o}\right]+\left[\Sigma^{\gtrless}_{k,o,m},\mathrm{Re}G^{r}_{k,o,m}\right]$$
$$+\left[\Sigma^{\lessgtr}_{k,l,o},\mathrm{Re}G^{r}_{k,l,o}\right]+\left[\Sigma^{\lessgtr}_{k,l,m},\mathrm{Re}G^{r}_{k,l,m}\right] \tag{B.3}$$

$$\begin{aligned}
&\left\{\Gamma, G^{\lessgtr}\right\}_{o,o,o} \\
&= \left\{\Gamma_{o,o,o}, G^{\gtrless}_{o,o,o}\right\} + \left\{\Gamma_{o,o,m}, G^{\gtrless}_{o,o,m}\right\} + \left\{\Gamma_{o,l,o}, G^{\gtrless}_{o,l,o}\right\} \\
&\quad + \left\{\Gamma_{o,l,m}, G^{\gtrless}_{o,l,m}\right\} + \left\{\Gamma_{k,o,o}, G^{\gtrless}_{k,o,o}\right\} + \left\{\Gamma_{k,o,m}, G^{\lessgtr}_{k,o,m}\right\} \\
&\quad + \left\{\Gamma_{k,l,o}, G^{\gtrless}_{k,l,o}\right\} + \left\{\Gamma_{k,l,m}, G^{\gtrless}_{k,l,m}\right\}
\end{aligned} \tag{B.4}$$

$$\begin{aligned}
&\left\{\Sigma^{\lessgtr}, A\right\}_{o,o,o} \\
&= \left\{\Sigma^{\gtrless}_{o,o,o}, A_{o,o,o}\right\} + \left\{\Sigma^{\lessgtr}_{o,o,m}, A_{o,o,m}\right\} + \left\{\Sigma^{\gtrless}_{o,l,o}, A_{o,l,o}\right\} \\
&\quad + \left\{\Sigma^{\gtrless}_{o,l,m}, A_{o,l,m}\right\} + \left\{\Sigma^{\lessgtr}_{k,o,o}, A_{k,o,o}\right\} + \left\{\Sigma^{\gtrless}_{k,o,m}, A_{k,o,m}\right\} \\
&\quad + \left\{\Sigma^{\lessgtr}_{k,l,o}, A_{k,l,o}\right\} + \left\{\Sigma^{\lessgtr}_{k,l,m}, A_{k,l,m}\right\}
\end{aligned} \tag{B.5}$$

We have for the Pauli–Dirac spin,

$$\begin{aligned}
&2^{N_s} i\hbar \left(\frac{\partial}{\partial t_1} + \frac{\partial}{\partial t_2}\right) G^{\gtrless}_{k,o,o} \\
&= \left[\tilde{\mathcal{H}}, G^{\lessgtr}\right]_{k,o,o} + \left[\Sigma^{\lessgtr}, \mathrm{Re}G^r\right]_{k,o,o} - \frac{i}{2}\left\{\Gamma, G^{\lessgtr}\right\}_{k,o,o} + \frac{i}{2}\left\{\Sigma^{\lessgtr}, A\right\}_{k,o,o}
\end{aligned} \tag{B.6}$$

where

$$\begin{aligned}
&\left[\tilde{\mathcal{H}}, G^{\lessgtr}\right]_{k,o,o} \\
&= \left[H_{o,o,o}, G^{\gtrless}_{k,o,o}\right] + \left[H_{o,o,m}, G^{\gtrless}_{k,o,m}\right] + i\epsilon_{kk_1k_2}\left\{H_{k_1,o,o}, G^{\gtrless}_{k_2,o,o}\right\} \\
&\quad + \left[H_{o,l,o}, G^{\gtrless}_{k,l,o}\right] + \left[H_{o,l,m}, G^{\gtrless}_{k,l,m}\right] + i\epsilon_{kk_1k_2}\left\{H_{k_1,l,o}, G^{\gtrless}_{k_2,l,o}\right\} \\
&\quad + \left[H_{k,o,o}, G^{\gtrless}_{o,o,o}\right] + \left[H_{k,o,m}, G^{\gtrless}_{o,o,m}\right] + i\epsilon_{kk_1k_2}\left\{H_{k_1,o,m}, G^{\gtrless}_{k_2,o,m}\right\} \\
&\quad + \left[H_{k,l,o}, G^{\gtrless}_{o,l,o}\right] + \left[H_{k.l,m}, G^{\gtrless}_{o,l,m}\right] + i\epsilon_{kk_1k_2}\left\{H_{k_1,l,m}, G^{\gtrless}_{k_2,l,m}\right\}
\end{aligned} \tag{B.7}$$

$$\begin{aligned}
&\left[\Sigma^{\lessgtr}, \mathrm{Re}G^{r}\right]_{k,o,o} \\
&\quad = \left[\Sigma^{\lessgtr}_{o,o,o}, \mathrm{Re}G^{r}_{k,o,o}\right] + \left[\Sigma^{\lessgtr}_{o,o,m}, \mathrm{Re}G^{r}_{k,o,m}\right] + i\epsilon_{kk_1k_2}\left\{\Sigma^{\lessgtr}_{k_1,o,o}, \mathrm{Re}G^{r}_{k_2,o,o}\right\} \\
&\qquad + \left[\Sigma^{\lessgtr}_{o,l,o}, \mathrm{Re}G^{r}_{k,l,o}\right] + \left[\Sigma^{\lessgtr}_{o,l,m}, \mathrm{Re}G^{r}_{k,l,m}\right] + i\epsilon_{kk_1k_2}\left\{\Sigma^{\lessgtr}_{k_1,l,o}, \mathrm{Re}G^{r}_{k_2,l,o}\right\} \\
&\qquad + \left[\Sigma^{\lessgtr}_{k,o,o}, \mathrm{Re}G^{r}_{o,o,o}\right] + \left[\Sigma^{\lessgtr}_{k,o,m}, \mathrm{Re}G^{r}_{o,o,m}\right] \\
&\qquad + i\epsilon_{kk_1k_2}\left\{\Sigma^{\lessgtr}_{k_1,o,m}, \mathrm{Re}G^{r}_{k_2,o,m}\right\} + \left[\Sigma^{\lessgtr}_{k,l,o}, \mathrm{Re}G^{r}_{o,l,o}\right] \\
&\qquad + \left[\Sigma^{\lessgtr}_{k.l,m}, \mathrm{Re}G^{r}_{o,l,m}\right] + i\epsilon_{kk_1k_2}\left\{\Sigma^{\lessgtr}_{k_1,l,m}, \mathrm{Re}G^{r}_{k_2,l,m}\right\}
\end{aligned} \tag{B.8}$$

$$\begin{aligned}
&\left\{\Gamma, G^{\lessgtr}\right\}_{k,o,o} \\
&\quad = \left\{\Gamma_{o,o,o}, G^{\gtrless}_{k,o,o}\right\} + \left\{\Gamma_{o,o,m}, G^{\gtrless}_{k,o,m}\right\} + i\epsilon_{kk_1k_2}\left[\Gamma_{k_1,o,o}, G^{\gtrless}_{k_2,o,o}\right] \\
&\qquad + \left\{\Gamma_{o,l,o}, G^{\gtrless}_{k,l,o}\right\} + \left\{\Gamma_{o,l,m}, G^{\gtrless}_{k,l,m}\right\} + i\epsilon_{kk_1k_2}\left[\Gamma_{k_1,l,o}, G^{\gtrless}_{k_2,l,o}\right] \\
&\qquad + \left\{\Gamma_{k,o,o}, G^{\gtrless}_{o,o,o}\right\} + \left\{\Gamma_{k,o,m}, G^{\gtrless}_{o,o,m}\right\} + i\epsilon_{kk_1k_2}\left[\Gamma_{k_1,o,m}, G^{\gtrless}_{k_2,o,m}\right] \\
&\qquad + \left\{\Gamma_{k,l,o}, G^{\gtrless}_{o,l,o}\right\} + \left\{\Gamma_{k.l,m}, G^{\gtrless}_{o,l,m}\right\} + i\epsilon_{kk_1k_2}\left[\Gamma_{k_1,l,m}, G^{\gtrless}_{k_2,l,m}\right]
\end{aligned} \tag{B.9}$$

$$\begin{aligned}
&\left\{\Sigma^{\lessgtr}, A\right\}_{k,o,o} \\
&\quad = \left\{\Sigma^{\lessgtr}_{o,o,o}, A_{k,o,o}\right\} + \left\{\Sigma^{\lessgtr}_{o,o,m}, A_{k,o,m}\right\} + i\epsilon_{kk_1k_2}\left[\Sigma^{\lessgtr}_{k_1,o,o}, A_{k_2,o,o}\right] \\
&\qquad + \left\{\Sigma^{\lessgtr}_{o,l,o}, A_{k,l,o}\right\} + \left\{\Sigma^{\lessgtr}_{o,l,m}, A_{k,l,m}\right\} + i\epsilon_{kk_1k_2}\left[\Sigma^{\lessgtr}_{k_1,l,o}, A_{k_2,l,o}\right] \\
&\qquad + \left\{\Sigma^{\lessgtr}_{k,o,o}, A_{o,o,o}\right\} + \left\{\Sigma^{\lessgtr}_{k,o,m}, A_{o,o,m}\right\} + i\epsilon_{kk_1k_2}\left[\Sigma^{\lessgtr}_{k_1,o,m}, A_{k_2,o,m}\right] \\
&\qquad + \left\{\Sigma^{\lessgtr}_{k,l,o}, A_{o,l,o}\right\} + \left\{\Sigma^{\lessgtr}_{k.l,m}, A_{o,l,m}\right\} + i\epsilon_{kk_1k_2}\left[\Sigma^{\lessgtr}_{k_1,l,m}, A_{k_2,l,m}\right]
\end{aligned} \tag{B.10}$$

We have for the valley spin,

$$\begin{aligned}
&2^{N_s} i\hbar\left(\frac{\partial}{\partial t_1} + \frac{\partial}{\partial t_2}\right) G^{\gtrless}_{o,l,o} \\
&\quad = \left[\tilde{\mathcal{H}}, G^{\lessgtr}\right]_{o,l,o} + \left[\Sigma^{\lessgtr}, \mathrm{Re}G^{r}\right]_{o,l,o} - \frac{i}{2}\left\{\Gamma, G^{\lessgtr}\right\}_{o,l,o} + \frac{i}{2}\left\{\Sigma^{\lessgtr}, A\right\}_{o,l,o}
\end{aligned} \tag{B.11}$$

where

$$\begin{aligned}
&\left[\tilde{\mathcal{H}}, G^{\lessgtr}\right]_{o,l,o} \\
&\quad = \left[H_{o,o,o}, G^{\gtrless}_{o,l,o}\right] + \left[H_{o,o,m}, G^{\gtrless}_{o,l,m}\right] + i\epsilon_{ll_1l_2}\left\{H_{o,l_1,o}, G^{\gtrless}_{o,l_2,o}\right\} \\
&\qquad + \left[H_{o,l,o}, G^{\lessgtr}_{o,o,o}\right] + \left[H_{o,l,m}, G^{\gtrless}_{o,o,m}\right] + i\epsilon_{ll_1l_2}\left\{H_{o,l_1,m}, G^{\gtrless}_{o,l_2,m}\right\} \\
&\qquad + \left[H_{k,o,o}, G^{\gtrless}_{k,l,o}\right] + \left[H_{k,o,m}, G^{\gtrless}_{k,l,m}\right] + i\epsilon_{ll_1l_2}\left\{H_{k,l_1,o}, G^{\gtrless}_{k,l_2,o}\right\} \\
&\qquad + \left[H_{k,l,o}, G^{\gtrless}_{k,o,o}\right] + \left[H_{k.l,m}, G^{\gtrless}_{k,o,m}\right] + i\epsilon_{ll_1l_2}\left\{H_{k,l_1,m}, G^{\gtrless}_{k,l_2,m}\right\}
\end{aligned} \tag{B.12}$$

$$\begin{aligned}
&\left[\Sigma^{\lessgtr}, \mathrm{Re}G^{r}\right]_{o,l,o} \\
&\quad = \left[\Sigma^{\gtrless}_{o,o,o}, \mathrm{Re}G^{r}_{o,l,o}\right] + \left[\Sigma^{\gtrless}_{o,o,m}, \mathrm{Re}G^{r}_{o,l,m}\right] + i\epsilon_{ll_1l_2}\left\{\Sigma^{\gtrless}_{o,l_1,o}, \mathrm{Re}G^{r}_{o,l_2,o}\right\} \\
&\qquad + \left[\Sigma^{\gtrless}_{o,l,o}, \mathrm{Re}G^{r}_{o,o,o}\right] + \left[\Sigma^{\gtrless}_{o,l,m}, \mathrm{Re}G^{r}_{o,o,m}\right] + i\epsilon_{ll_1l_2}\left\{\Sigma^{\gtrless}_{o,l_1,m}, \mathrm{Re}G^{r}_{o,l_2,m}\right\} \\
&\qquad + \left[\Sigma^{\gtrless}_{k,o,o}, \mathrm{Re}G^{r}_{k,l,o}\right] + \left[\Sigma^{\gtrless}_{k,o,m}, \mathrm{Re}G^{r}_{k,l,m}\right] + i\epsilon_{ll_1l_2}\left\{\Sigma^{\gtrless}_{k,l_1,o}, \mathrm{Re}G^{r}_{k,l_2,o}\right\} \\
&\qquad + \left[\Sigma^{\gtrless}_{k,l,o}, \mathrm{Re}G^{r}_{k,o,o}\right] + \left[\Sigma^{\gtrless}_{k.l,m}, \mathrm{Re}G^{r}_{k,o,m}\right] + i\epsilon_{ll_1l_2}\left\{\Sigma^{\gtrless}_{k,l_1,m}, \mathrm{Re}G^{r}_{k,l_2,m}\right\}
\end{aligned} \tag{B.13}$$

$$\begin{aligned}
&\left\{\Gamma, G^{\lessgtr}\right\}_{o,l,o} \\
&\quad = \left\{\Gamma_{o,o,o}, G^{\gtrless}_{o,l,o}\right\} + \left\{\Gamma_{o,o,m}, G^{\gtrless}_{o,l,m}\right\} + i\epsilon_{ll_1l_2}\left[\Gamma_{o,l_1,o}, G^{\gtrless}_{o,l_2,o}\right] \\
&\qquad + \left\{\Gamma_{o,l,o}, G^{\lessgtr}_{o,o,o}\right\} + \left\{\Gamma_{o,l,m}, G^{\gtrless}_{o,o,m}\right\} + i\epsilon_{ll_1l_2}\left[\Gamma_{o,l_1,m}, G^{\gtrless}_{o,l_2,m}\right] \\
&\qquad + \left\{\Gamma_{k,o,o}, G^{\gtrless}_{k,l,o}\right\} + \left\{\Gamma_{k,o,m}, G^{\gtrless}_{k,l,m}\right\} + i\epsilon_{ll_1l_2}\left[\Gamma_{k,l_1,o}, G^{\gtrless}_{k,l_2,o}\right] \\
&\qquad + \left\{\Gamma_{k,l,o}, G^{\gtrless}_{k,o,o}\right\} + \left\{\Gamma_{k.l,m}, G^{\gtrless}_{k,o,m}\right\} + i\epsilon_{ll_1l_2}\left[\Gamma_{k,l_1,m}, G^{\gtrless}_{k,l_2,m}\right]
\end{aligned} \tag{B.14}$$

$$
\begin{aligned}
&\left\{\Sigma^{\lessgtr}, A\right\}_{o,l,o} \\
&\quad = \left\{\Sigma^{\gtrless}_{o,o,o}, A_{o,l,o}\right\} + \left\{\Sigma^{\gtrless}_{o,o,m}, A_{o,l,m}\right\} + i\epsilon_{ll_1l_2}\left[\Sigma^{\gtrless}_{o,l_1,o}, A_{o,l_2,o}\right] \\
&\qquad + \left\{\Sigma^{\gtrless}_{o,l,o}, A_{o,o,o}\right\} + \left\{\Sigma^{\gtrless}_{o,l,m}, A_{o,o,m}\right\} + i\epsilon_{ll_1l_2}\left[\Sigma^{\gtrless}_{o,l_1,m}, A_{o,l_2,m}\right] \\
&\qquad + \left\{\Sigma^{\gtrless}_{k,o,o}, A_{k,l,o}\right\} + \left\{\Sigma^{\gtrless}_{k,o,m}, A_{k,l,m}\right\} + i\epsilon_{ll_1l_2}\left[\Sigma^{\gtrless}_{k,l_1,o}, A_{k,l_2,o}\right] \\
&\qquad + \left\{\Sigma^{\gtrless}_{k,l,o}, A_{k,o,o}\right\} + \left\{\Sigma^{\gtrless}_{k.l,m}, A_{k,o,m}\right\} + i\epsilon_{ll_1l_2}\left[\Sigma^{\gtrless}_{k,l_1,m}, A_{k,l_2,m}\right]
\end{aligned} \tag{B.15}
$$

We have for the pseudospin,

$$
\begin{aligned}
&2^{N_s} i\hbar\left(\frac{\partial}{\partial t_1} + \frac{\partial}{\partial t_2}\right) G^{\gtrless}_{o,o,m} \\
&\quad = \left[\tilde{\mathcal{H}}, G^{\lessgtr}\right]_{o,o,m} + \left[\Sigma^{\lessgtr}, \mathrm{Re}G^{r}\right]_{o,o,m} - \frac{i}{2}\left\{\Gamma, G^{\lessgtr}\right\}_{o,o,m} + \frac{i}{2}\left\{\Sigma^{\lessgtr}, A\right\}_{o,o,m}
\end{aligned} \tag{B.16}
$$

where

$$
\begin{aligned}
&\left[\tilde{\mathcal{H}}, G^{\lessgtr}\right]_{o,o,m} \\
&\quad = \left[H_{o,o,o}, G^{\gtrless}_{o,o,m}\right] + \left[H_{o,o,m}, G^{\gtrless}_{o,o,o}\right] + i\epsilon_{mm_1m_2}\left\{H_{o,o,m_1}, G^{\gtrless}_{o,o,m_2}\right\} \\
&\qquad + \left[H_{o,l,o}, G^{\gtrless}_{o,l,m}\right] + \left[H_{o,l,m}, G^{\gtrless}_{o,l,o}\right] + i\epsilon_{mm_1m_2}\left\{H_{o,l,m_1}, G^{\gtrless}_{o,l,m_2}\right\} \\
&\qquad + \left[H_{k,o,o}, G^{\gtrless}_{k,o,m}\right] + \left[H_{k,o,m}, G^{\gtrless}_{k,o,o}\right] + i\epsilon_{mm_1m_2}\left\{H_{k,o,m_1}, G^{\gtrless}_{k,o,m_2}\right\} \\
&\qquad + \left[H_{k,l,o}, G^{\gtrless}_{k,l,m}\right] + \left[H_{k,l,m}, G^{\gtrless}_{k,l,o}\right] + i\epsilon_{mm_1m_2}\left\{H_{k,l,m_1}, G^{\gtrless}_{k,l,m_2}\right\}
\end{aligned} \tag{B.17}
$$

$$
\begin{aligned}
&\left[\Sigma^{\lessgtr}, \mathrm{Re}G^{r}\right]_{o,o,m} \\
&\quad = \left[\Sigma^{\lessgtr}_{o,o,o}, \mathrm{Re}G^{r}_{o,o,m}\right] + \left[\Sigma^{\lessgtr}_{o,o,m}, \mathrm{Re}G^{r}_{o,o,o}\right] \\
&\qquad + i\epsilon_{mm_1m_2}\left\{\Sigma^{\lessgtr}_{o,o,m_1}, \mathrm{Re}G^{r}_{o,o,m_2}\right\} \\
&\qquad + \left[\Sigma^{\lessgtr}_{o,l,o}, \mathrm{Re}G^{r}_{o,l,m}\right] + \left[\Sigma^{\lessgtr}_{o,l,m}, \mathrm{Re}G^{r}_{o,l,o}\right] \\
&\qquad + i\epsilon_{mm_1m_2}\left\{\Sigma^{\lessgtr}_{o,l,m_1}, \mathrm{Re}G^{r}_{o,l,m_2}\right\}
\end{aligned}
$$

$$+\left[\Sigma^{\lessgtr}_{k,o,o}, \mathrm{Re}G^{r}_{k,o,m}\right] + \left[\Sigma^{\lessgtr}_{k,o,m}, \mathrm{Re}G^{r}_{k,o,o}\right]$$
$$+ i\epsilon_{mm_1m_2}\left\{\Sigma^{\lessgtr}_{k,o,m_1}, \mathrm{Re}G^{r}_{k,o,m_2}\right\}$$
$$+\left[\Sigma^{\lessgtr}_{k,l,o}, \mathrm{Re}G^{r}_{k,l,m}\right] + \left[\Sigma^{\lessgtr}_{k,l,m}, \mathrm{Re}G^{r}_{k,l,o}\right]$$
$$+ i\epsilon_{mm_1m_2}\left\{\Sigma^{\lessgtr}_{k,l,m_1}, \mathrm{Re}G^{r}_{k,l,m_2}\right\} \tag{B.18}$$

$$\left\{\Gamma, G^{\lessgtr}\right\}_{o,o,m}$$
$$= \left\{\Gamma_{o,o,o}, G^{\gtrless}_{o,o,m}\right\} + \left\{\Gamma_{o,o,m}, G^{\gtrless}_{o,o,o}\right\} + i\epsilon_{mm_1m_2}\left[\Gamma_{o,o,m_1}, G^{\gtrless}_{o,o,m_2}\right]$$
$$+ \left\{\Gamma_{o,l,o}, G^{\gtrless}_{o,l,m}\right\} + \left\{\Gamma_{o,l,m}, G^{\gtrless}_{o,l,o}\right\} + i\epsilon_{mm_1m_2}\left[\Gamma_{o,l,m_1}, G^{\gtrless}_{o,l,m_2}\right]$$
$$+ \left\{\Gamma_{k,o,o}, G^{\gtrless}_{k,o,m}\right\} + \left\{\Gamma_{k,o,m}, G^{\gtrless}_{k,o,o}\right\} + i\epsilon_{mm_1m_2}\left[\Gamma_{k,o,m_1}, G^{\gtrless}_{k,o,m_2}\right]$$
$$+ \left\{\Gamma_{k,l,o}, G^{\gtrless}_{k,l,m}\right\} + \left\{\Gamma_{k,l,m}, G^{\gtrless}_{k,l,o}\right\} + i\epsilon_{mm_1m_2}\left[\Gamma_{k,l,m_1}, G^{\gtrless}_{k,l,m_2}\right] \tag{B.19}$$

$$\left\{\Sigma^{\lessgtr}, A\right\}_{o,o,m}$$
$$= \left\{\Sigma^{\lessgtr}_{o,o,o}, A_{o,o,m}\right\} + \left\{\Sigma^{\lessgtr}_{o,o,m}, A_{o,o,o}\right\} + i\epsilon_{mm_1m_2}\left[\Sigma^{\lessgtr}_{o,o,m_1}, A_{o,o,m_2}\right]$$
$$+ \left\{\Sigma^{\lessgtr}_{o,l,o}, A_{o,l,m}\right\} + \left\{\Sigma^{\lessgtr}_{o,l,m}, A_{o,l,o}\right\} + i\epsilon_{mm_1m_2}\left[\Sigma^{\lessgtr}_{o,l,m_1}, A_{o,l,m_2}\right]$$
$$+ \left\{\Sigma^{\lessgtr}_{k,o,o}, A_{k,o,m}\right\} + \left\{\Sigma^{\lessgtr}_{k,o,m}, A_{k,o,o}\right\} + i\epsilon_{mm_1m_2}\left[\Sigma^{\lessgtr}_{k,o,m_1}, A_{k,o,m_2}\right]$$
$$+ \left\{\Sigma^{\lessgtr}_{k,l,o}, A_{k,l,m}\right\} + \left\{\Sigma^{\lessgtr}_{k,l,m}, A_{k,l,o}\right\} + i\epsilon_{mm_1m_2}\left[\Sigma^{\lessgtr}_{k,l,m_1}, A_{k,l,m_2}\right] \tag{B.20}$$

We have for the entangled Dirac spin and valley spin,

$$2^{N_s} i\hbar\left(\frac{\partial}{\partial t_1} + \frac{\partial}{\partial t_2}\right) G^{\gtrless}_{k,l,o}$$
$$= \left[\tilde{\mathcal{H}}, G^{\lessgtr}\right]_{k,l,o} + \left[\Sigma^{\lessgtr}, \mathrm{Re}G^{r}\right]_{k,l,o} - \frac{i}{2}\left\{\Gamma, G^{\lessgtr}\right\}_{k,l,o} + \frac{i}{2}\left\{\Sigma^{\lessgtr}, A\right\}_{k,l,o} \tag{B.21}$$

where

$$\begin{aligned}
&\left[\tilde{\mathcal{H}}, G^{\lessgtr}\right]_{k,l,o} \\
&= \left[H_{o,o,o}, G^{\lessgtr}_{k,l,o}\right] + \left[H_{o,o,m}, G^{\lessgtr}_{k,l,m}\right] + i\epsilon_{kk_1k_2}\left\{H_{k_1,o,o}, G^{\lessgtr}_{k_2,l,o}\right\} \\
&\quad + i\epsilon_{ll_1l_2}\left\{H_{o,l_1,o}, G^{\lessgtr}_{k,l_2,o}\right\} + \frac{1}{2}i\epsilon_{kk_1k_2}i\epsilon_{ll_1l_2}\left[H_{k_1,l_1,o}, G^{\lessgtr}_{k_2,l_2,o}\right] \\
&\quad + \left[H_{o,l,o}, G^{\lessgtr}_{k,o,o}\right] + \left[H_{o,l,m}, G^{\lessgtr}_{k,o,m}\right] + i\epsilon_{kk_1k_2}\left\{H_{k_1,l,o}, G^{\lessgtr}_{k_2,o,o}\right\} \\
&\quad + i\epsilon_{ll_1l_2}\left\{H_{k,l_1,o}G^{\lessgtr}_{o,l_2,o}\right\} + \frac{1}{2}i\epsilon_{kk_1k_2}i\epsilon_{ll_1l_2}\left[H_{k_1,l_1,o}, G^{\lessgtr}_{k_2,l_2,o}\right] \\
&\quad + \left[H_{k,l,o}, G^{\lessgtr}_{o,o,o}\right] + \left[H_{k.l,m}, G^{\lessgtr}_{o,o,m}\right] + i\epsilon_{kk_1k_2}\left\{H_{k_1,l,m}, G^{\lessgtr}_{k_2,o,m}\right\} \\
&\quad + i\epsilon_{ll_1l_2}\left\{H_{k,l_1,m}, G^{\lessgtr}_{o,l_2,m}\right\} + \frac{1}{2}i\epsilon_{kk_1k_2}i\epsilon_{ll_1l_2}\left[H_{k_1,l_1,m}, G^{\lessgtr}_{k_2,l_2,m}\right] \\
&\quad + \left[H_{k,o,o}, G^{\lessgtr}_{o,l,o}\right] + \left[H_{k,o,m}, G^{\lessgtr}_{o,l,m}\right] + i\epsilon_{kk_1k_2}\left\{H_{k_1,o,m}, G^{\lessgtr}_{k_2,l,m}\right\} \\
&\quad + i\epsilon_{ll_1l_2}\left\{H_{o,l_1,m}, G^{\lessgtr}_{k,l_2,m}\right\} + \frac{1}{2}i\epsilon_{kk_1k_2}i\epsilon_{ll_1l_2}\left[H_{k_1,l_1,m}, G^{\lessgtr}_{k_2,l_2,m}\right]
\end{aligned} \tag{B.22}$$

$$\begin{aligned}
&\left[\Sigma^{\lessgtr}, \mathrm{Re}G^r\right]_{k,l,o} \\
&= \left[\Sigma^{\lessgtr}_{o,o,o}, \mathrm{Re}G^r_{k,l,o}\right] + \left[\Sigma^{\lessgtr}_{o,o,m}, \mathrm{Re}G^r_{k,l,m}\right] + i\epsilon_{kk_1k_2}\left\{\Sigma^{\lessgtr}_{k_1,o,o}, \mathrm{Re}G^r_{k_2,l,o}\right\} \\
&\quad + i\epsilon_{ll_1l_2}\left\{\Sigma^{\lessgtr}_{o,l_1,o}, \mathrm{Re}G^r_{k,l_2,o}\right\} + \frac{1}{2}i\epsilon_{kk_1k_2}i\epsilon_{ll_1l_2}\left[\Sigma^{\lessgtr}_{k_1,l_1,o}, \mathrm{Re}G^r_{k_2,l_2,o}\right] \\
&\quad + \left[\Sigma^{\lessgtr}_{o,l,o}, \mathrm{Re}G^r_{k,o,o}\right] + \left[\Sigma^{\lessgtr}_{o,l,m}, \mathrm{Re}G^r_{k,o,m}\right] + i\epsilon_{kk_1k_2}\left\{\Sigma^{\lessgtr}_{k_1,l,o}, \mathrm{Re}G^r_{k_2,o,o}\right\} \\
&\quad + i\epsilon_{ll_1l_2}\left\{\Sigma^{\lessgtr}_{k,l_1,o}\mathrm{Re}G^r_{o,l_2,o}\right\} + \frac{1}{2}i\epsilon_{kk_1k_2}i\epsilon_{ll_1l_2}\left[\Sigma^{\lessgtr}_{k_1,l_1,o}, \mathrm{Re}G^r_{k_2,l_2,o}\right] \\
&\quad + \left[\Sigma^{\lessgtr}_{k,l,o}, \mathrm{Re}G^r_{o,o,o}\right] + \left[\Sigma^{\lessgtr}_{k.l,m}, \mathrm{Re}G^r_{o,o,m}\right] \\
&\quad + i\epsilon_{kk_1k_2}\left\{\Sigma^{\lessgtr}_{k_1,l,m}, \mathrm{Re}G^r_{k_2,o,m}\right\} + i\epsilon_{ll_1l_2}\left\{\Sigma^{\lessgtr}_{k,l_1,m}, \mathrm{Re}G^r_{o,l_2,m}\right\} \\
&\quad + \frac{1}{2}i\epsilon_{kk_1k_2}i\epsilon_{ll_1l_2}\left[\Sigma^{\lessgtr}_{k_1,l_1,m}, \mathrm{Re}G^r_{k_2,l_2,m}\right] + \left[\Sigma^{\lessgtr}_{k,o,o}, \mathrm{Re}G^r_{o,l,o}\right] \\
&\quad + \left[\Sigma^{\lessgtr}_{k,o,m}, \mathrm{Re}G^r_{o,l,m}\right] + i\epsilon_{kk_1k_2}\left\{\Sigma^{\lessgtr}_{k_1,o,m}, \mathrm{Re}G^r_{k_2,l,m}\right\} \\
&\quad + i\epsilon_{ll_1l_2}\left\{\Sigma^{\lessgtr}_{o,l_1,m}, \mathrm{Re}G^r_{k,l_2,m}\right\} + \frac{1}{2}i\epsilon_{kk_1k_2}i\epsilon_{ll_1l_2}\left[\Sigma^{\lessgtr}_{k_1,l_1,m}, \mathrm{Re}G^r_{k_2,l_2,m}\right]
\end{aligned} \tag{B.23}$$

$$
\begin{aligned}
&\left\{\Gamma, G^{\lessgtr}\right\}_{k,l,o} \\
&= \left\{\Gamma_{o,o,o}, G^{\lessgtr}_{k,l,o}\right\} + \left\{\Gamma_{o,o,m}, G^{\lessgtr}_{k,l,m}\right\} + i\epsilon_{kk_1k_2}\left[\Gamma_{k_1,o,o}, G^{\lessgtr}_{k_2,l,o}\right] \\
&\quad + i\epsilon_{ll_1l_2}\left[\Gamma_{o,l_1,o}, G^{\lessgtr}_{k,l_2,o}\right] + \frac{1}{2} i\epsilon_{kk_1k_2} i\epsilon_{ll_1l_2}\left\{\Gamma_{k_1,l_1,o}, G^{\lessgtr}_{k_2,l_2,o}\right\} \\
&\quad + \left\{\Gamma_{o,l,o}, G^{\lessgtr}_{k,o,o}\right\} + \left\{\Gamma_{o,l,m}, G^{\lessgtr}_{k,o,m}\right\} + i\epsilon_{kk_1k_2}\left[\Gamma_{k_1,l,o}, G^{\lessgtr}_{k_2,o,o}\right] \\
&\quad + i\epsilon_{ll_1l_2}\left[\Gamma_{k,l_1,o} G^{\lessgtr}_{o,l_2,o}\right] + \frac{1}{2} i\epsilon_{kk_1k_2} i\epsilon_{ll_1l_2}\left\{\Gamma_{k_1,l_1,o}, G^{\lessgtr}_{k_2,l_2,o}\right\} \\
&\quad + \left\{\Gamma_{k,l,o}, G^{\lessgtr}_{o,o,o}\right\} + \left\{\Gamma_{k.l,m}, G^{\lessgtr}_{o,o,m}\right\} + i\epsilon_{kk_1k_2}\left[\Gamma_{k_1,l,m}, G^{\lessgtr}_{k_2,o,m}\right] \\
&\quad + i\epsilon_{ll_1l_2}\left[\Gamma_{k,l_1,m}, G^{\lessgtr}_{o,l_2,m}\right] + \frac{1}{2} i\epsilon_{kk_1k_2} i\epsilon_{ll_1l_2}\left\{\Gamma_{k_1,l_1,m}, G^{\lessgtr}_{k_2,l_2,m}\right\} \\
&\quad + \left\{\Gamma_{k,o,o}, G^{\lessgtr}_{o,l,o}\right\} + \left\{\Gamma_{k,o,m}, G^{\lessgtr}_{o,l,m}\right\} + i\epsilon_{kk_1k_2}\left[\Gamma_{k_1,o,m}, G^{\lessgtr}_{k_2,l,m}\right] \\
&\quad + i\epsilon_{ll_1l_2}\left[\Gamma_{o,l_1,m}, G^{\lessgtr}_{k,l_2,m}\right] + \frac{1}{2} i\epsilon_{kk_1k_2} i\epsilon_{ll_1l_2}\left\{\Gamma_{k_1,l_1,m}, G^{\lessgtr}_{k_2,l_2,m}\right\}
\end{aligned} \tag{B.24}
$$

$$
\begin{aligned}
&\left\{\Sigma^{\lessgtr}, A\right\}_{k,l,o} \\
&= \left\{\Sigma^{\lessgtr}_{o,o,o}, A_{k,l,o}\right\} + \left\{\Sigma^{\lessgtr}_{o,o,m}, A_{k,l,m}\right\} + i\epsilon_{kk_1k_2}\left[\Sigma^{\lessgtr}_{k_1,o,o}, A_{k_2,l,o}\right] \\
&\quad + i\epsilon_{ll_1l_2}\left[\Sigma^{\lessgtr}_{o,l_1,o}, A_{k,l_2,o}\right] + \frac{1}{2} i\epsilon_{kk_1k_2} i\epsilon_{ll_1l_2}\left\{\Sigma^{\lessgtr}_{k_1,l_1,o}, A_{k_2,l_2,o}\right\} \\
&\quad + \left\{\Sigma^{\lessgtr}_{o,l,o}, A_{k,o,o}\right\} + \left\{\Sigma^{\lessgtr}_{o,l,m}, A_{k,o,m}\right\} + i\epsilon_{kk_1k_2}\left[\Sigma^{\lessgtr}_{k_1,l,o}, A_{k_2,o,o}\right] \\
&\quad + i\epsilon_{ll_1l_2}\left[\Sigma^{\lessgtr}_{k,l_1,o} A_{o,l_2,o}\right] + \frac{1}{2} i\epsilon_{kk_1k_2} i\epsilon_{ll_1l_2}\left\{\Sigma^{\lessgtr}_{k_1,l_1,o}, A_{k_2,l_2,o}\right\} \\
&\quad + \left\{\Sigma^{\lessgtr}_{k,l,o}, A_{o,o,o}\right\} + \left\{\Sigma^{\lessgtr}_{k.l,m}, A_{o,o,m}\right\} + i\epsilon_{kk_1k_2}\left[\Sigma^{\lessgtr}_{k_1,l,m}, A_{k_2,o,m}\right] \\
&\quad + i\epsilon_{ll_1l_2}\left[\Sigma^{\lessgtr}_{k,l_1,m}, A_{o,l_2,m}\right] + \frac{1}{2} i\epsilon_{kk_1k_2} i\epsilon_{ll_1l_2}\left\{\Sigma^{\lessgtr}_{k_1,l_1,m}, A_{k_2,l_2,m}\right\} \\
&\quad + \left\{\Sigma^{\lessgtr}_{k,o,o}, A_{o,l,o}\right\} + \left\{\Sigma^{\lessgtr}_{k,o,m}, A_{o,l,m}\right\} + i\epsilon_{kk_1k_2}\left[\Sigma^{\lessgtr}_{k_1,o,m}, A_{k_2,l,m}\right] \\
&\quad + i\epsilon_{ll_1l_2}\left[\Sigma^{\lessgtr}_{o,l_1,m}, A_{k,l_2,m}\right] + \frac{1}{2} i\epsilon_{kk_1k_2} i\epsilon_{ll_1l_2}\left\{\Sigma^{\lessgtr}_{k_1,l_1,m}, A_{k_2,l_2,m}\right\}
\end{aligned} \tag{B.25}
$$

We have for the entangled Dirac spin and pseudospin,

$$2^{N_s} i\hbar \left(\frac{\partial}{\partial t_1} + \frac{\partial}{\partial t_2}\right) G^{\gtrless}_{k,o,m}$$

$$= \left[\tilde{\mathcal{H}}, G^{\lessgtr}\right]_{k,o,m} + \left[\Sigma^{\lessgtr}, \mathrm{Re}G^{r}\right]_{k,o,m} - \frac{i}{2}\left\{\Gamma, G^{\lessgtr}\right\}_{k,o,m}$$

$$+ \frac{i}{2}\left\{\Sigma^{\lessgtr}, A\right\}_{k,o,m} \tag{B.26}$$

where

$$\left[\tilde{\mathcal{H}}, G^{\lessgtr}\right]_{k,o,m}$$

$$= \left[H_{o,o,o}, G^{\lessgtr}_{k,o,m}\right] + \left[H_{o,o,m}, G^{\lessgtr}_{k,o,o}\right] + i\epsilon_{kk_1k_2}\left\{H_{k_1,o,o}, G^{\lessgtr}_{k_2,o,m}\right\}$$

$$+ i\epsilon_{mm_1m_2}\left\{H_{o,o,m_1}, G^{\lessgtr}_{k,o,m_2}\right\} + \frac{1}{2}i\epsilon_{kk_1k_2}i\epsilon_{mm_1m_2}\left[H_{k_1,o,m_1}G^{\lessgtr}_{k_2,o,m_2}\right]$$

$$+ \left[H_{o,l,o}, G^{\lessgtr}_{k,l,m}\right] + \left[H_{o,l,m}, G^{\lessgtr}_{k,l,o}\right] + i\epsilon_{kk_1k_2}\left\{H_{k_1,l,o}, G^{\lessgtr}_{k_2,l,m}\right\}$$

$$+ i\epsilon_{mm_1m_2}\left\{H_{o,l,m_1}, G^{\lessgtr}_{k,l,m2}\right\} + \frac{1}{2}i\epsilon_{kk_1k_2}i\epsilon_{mm_1m_2}\left[H_{k_1,l,m_1}, G^{\lessgtr}_{k_2,l,m_2}\right]$$

$$+ \left[H_{k,o,o}, G^{\lessgtr}_{o,o,m}\right] + \left[H_{k,o,m}, G^{\lessgtr}_{o,o,o}\right] + i\epsilon_{kk_1k_2}\left\{H_{k_1,o,m}, G^{\lessgtr}_{k_2,o,o}\right\}$$

$$+ i\epsilon_{mm_1m_2}\left\{H_{k,o,m_1}, G^{\lessgtr}_{o,o,m_2}\right\} + \frac{1}{2}i\epsilon_{kk_1k_2}i\epsilon_{mm_1m_2}\left[H_{k_1,o,m_1}, G^{\lessgtr}_{k_2,o,m_2}\right]$$

$$\left[H_{k,l,o}, G^{\lessgtr}_{o,l,m}\right] + \left[H_{k.l,m}, G^{\lessgtr}_{o,l,o}\right] + i\epsilon_{kk_1k_2}\left\{H_{k_1,l,m}, G^{\lessgtr}_{k_2,l,o}\right\}$$

$$+ i\epsilon_{mm_1m_2}\left\{H_{k.l,m_1}, G^{\lessgtr}_{o,l,m_2}\right\} + \frac{1}{2}i\epsilon_{kk_1k_2}i\epsilon_{mm_1m_2}\left[H_{k_1,o,m_1}, G^{\lessgtr}_{k_2,o,m_2}\right] \tag{B.27}$$

$$\left[\Sigma^{\lessgtr}, \mathrm{Re}G^{r}\right]_{k,o,m}$$

$$= \left[\Sigma^{\lessgtr}_{o,o,o}, \mathrm{Re}G^{r}_{k,o,m}\right] + \left[\Sigma^{\lessgtr}_{o,o,m}, \mathrm{Re}G^{r}_{k,o,o}\right] + i\epsilon_{kk_1k_2}\left\{\Sigma^{\lessgtr}_{k_1,o,o}, \mathrm{Re}G^{r}_{k_2,o,m}\right\}$$

$$+ i\epsilon_{mm_1m_2}\left\{\Sigma^{\lessgtr}_{o,o,m_1}, \mathrm{Re}G^{r}_{k,o,m_2}\right\} + \frac{1}{2}i\epsilon_{kk_1k_2}i\epsilon_{mm_1m_2}$$

$$\times \left[\Sigma^{\lessgtr}_{k_1,o,m_1}\mathrm{Re}G^{r}_{k_2,o,m_2}\right]\left[\Sigma^{\lessgtr}_{o,l,o}, \mathrm{Re}G^{r}_{k,l,m}\right] + \left[\Sigma^{\lessgtr}_{o,l,m}, \mathrm{Re}G^{r}_{k,l,o}\right]$$

$$
\begin{aligned}
&+ i\epsilon_{kk_1k_2}\left\{\Sigma^{\lessgtr}_{k_1,l,o}, \mathrm{Re}G^r_{k_2,l,m}\right\} + i\epsilon_{mm_1m_2}\left\{\Sigma^{\lessgtr}_{o,l,m_1}, \mathrm{Re}G^r_{k,l,m2}\right\} \\
&+ \frac{1}{2} i\epsilon_{kk_1k_2} i\epsilon_{mm_1m_2}\left[\Sigma^{\lessgtr}_{k_1,l,m_1}, \mathrm{Re}G^r_{k_2,l,m_2}\right] \\
&+ \left[\Sigma^{\lessgtr}_{k,o,o}, \mathrm{Re}G^r_{o,o,m}\right] + \left[\Sigma^{\lessgtr}_{k,o,m}, \mathrm{Re}G^r_{o,o,o}\right] \\
&+ i\epsilon_{kk_1k_2}\left\{\Sigma^{\lessgtr}_{k_1,o,m}, \mathrm{Re}G^r_{k_2,o,o}\right\} + i\epsilon_{mm_1m_2}\left\{\Sigma^{\lessgtr}_{k,o,m_1}, \mathrm{Re}G^r_{o,o,m_2}\right\} \\
&+ \frac{1}{2} i\epsilon_{kk_1k_2} i\epsilon_{mm_1m_2}\left[\Sigma^{\lessgtr}_{k_1,o,m_1}, \mathrm{Re}G^r_{k_2,o,m_2}\right] \\
&\times \left[\Sigma^{\lessgtr}_{k,l,o}, \mathrm{Re}G^r_{o,l,m}\right] + \left[\Sigma^{\lessgtr}_{k.l,m}, \mathrm{Re}G^r_{o,l,o}\right] + i\epsilon_{kk_1k_2}\left\{\Sigma^{\lessgtr}_{k_1,l,m}, \mathrm{Re}G^r_{k_2,l,o}\right\} \\
&+ i\epsilon_{mm_1m_2}\left\{\Sigma^{\lessgtr}_{k.l,m_1}, \mathrm{Re}G^r_{o,l,m_2}\right\} + \frac{1}{2} i\epsilon_{kk_1k_2} i\epsilon_{mm_1m_2} \\
&\times \left[\Sigma^{\lessgtr}_{k_1,o,m_1}, \mathrm{Re}G^r_{k_2,o,m_2}\right]
\end{aligned}
\tag{B.28}
$$

$$
\begin{aligned}
&\left\{\Gamma, G^{\lessgtr}\right\}_{k,o,m} \\
&= \left\{\Gamma_{o,o,o}, G^{\lessgtr}_{k,o,m}\right\} + \left\{\Gamma_{o,o,m}, G^{\lessgtr}_{k,o,o}\right\} + i\epsilon_{kk_1k_2}\left[\Gamma_{k_1,o,o}, G^{\lessgtr}_{k_2,o,m}\right] \\
&+ i\epsilon_{mm_1m_2}\left[\Gamma_{o,o,m_1}, G^{\lessgtr}_{k,o,m_2}\right] + \frac{1}{2} i\epsilon_{kk_1k_2} i\epsilon_{mm_1m_2}\left\{\Gamma_{k_1,o,m_1} G^{\lessgtr}_{k_2,o,m_2}\right\} \\
&+ \left\{\Gamma_{o,l,o}, G^{\lessgtr}_{k,l,m}\right\} + \left\{\Gamma_{o,l,m}, G^{\lessgtr}_{k,l,o}\right\} + i\epsilon_{kk_1k_2}\left[\Gamma_{k_1,l,o}, G^{\lessgtr}_{k_2,l,m}\right] \\
&+ i\epsilon_{mm_1m_2}\left[\Gamma_{o,l,m_1}, G^{\lessgtr}_{k,l,m2}\right] + \frac{1}{2} i\epsilon_{kk_1k_2} i\epsilon_{mm_1m_2}\left\{\Gamma_{k_1,l,m_1}, G^{\lessgtr}_{k_2,l,m_2}\right\} \\
&+ \left\{\Gamma_{k,o,o}, G^{\lessgtr}_{o,o,m}\right\} + \left\{\Gamma_{k,o,m}, G^{\lessgtr}_{o,o,o}\right\} + i\epsilon_{kk_1k_2}\left[\Gamma_{k_1,o,m}, G^{\lessgtr}_{k_2,o,o}\right] \\
&+ i\epsilon_{mm_1m_2}\left[\Gamma_{k,o,m_1}, G^{\lessgtr}_{o,o,m_2}\right] + \frac{1}{2} i\epsilon_{kk_1k_2} i\epsilon_{mm_1m_2}\left\{\Gamma_{k_1,o,m_1}, G^{\lessgtr}_{k_2,o,m_2}\right\} \\
&+ \left\{\Gamma_{k,l,o}, G^{\lessgtr}_{o,l,m}\right\} + \left\{\Gamma_{k.l,m}, G^{\lessgtr}_{o,l,o}\right\} + i\epsilon_{kk_1k_2}\left[\Gamma_{k_1,l,m}, G^{\lessgtr}_{k_2,l,o}\right] \\
&+ i\epsilon_{mm_1m_2}\left[\Gamma_{k.l,m_1}, G^{\lessgtr}_{o,l,m_2}\right] + \frac{1}{2} i\epsilon_{kk_1k_2} i\epsilon_{mm_1m_2}\left\{\Gamma_{k_1,o,m_1}, G^{\lessgtr}_{k_2,o,m_2}\right\}
\end{aligned}
\tag{B.29}
$$

$$\begin{aligned}
&\left\{\Sigma^{\lessgtr}, A\right\}_{k,o,m} \\
&= \left\{\Sigma^{\lessgtr}_{o,o,o}, A_{k,o,m}\right\} + \left\{\Sigma^{\lessgtr}_{o,o,m}, A_{k,o,o}\right\} + i\epsilon_{kk_1k_2}\left[\Sigma^{\lessgtr}_{k_1,o,o}, A_{k_2,o,m}\right] \\
&\quad + i\epsilon_{mm_1m_2}\left[\Sigma^{\lessgtr}_{o,o,m_1}, A_{k,o,m_2}\right] + \frac{1}{2} i\epsilon_{kk_1k_2} i\epsilon_{mm_1m_2}\left\{\Sigma^{\lessgtr}_{k_1,o,m_1} A_{k_2,o,m_2}\right\} \\
&\quad + \left\{\Sigma^{\lessgtr}_{o,l,o}, A_{k,l,m}\right\} + \left\{\Sigma^{\lessgtr}_{o,l,m}, A_{k,l,o}\right\} + i\epsilon_{kk_1k_2}\left[\Sigma^{\lessgtr}_{k_1,l,o}, A_{k_2,l,m}\right] \\
&\quad + i\epsilon_{mm_1m_2}\left[\Sigma^{\lessgtr}_{o,l,m_1}, A_{k,l,m2}\right] + \frac{1}{2} i\epsilon_{kk_1k_2} i\epsilon_{mm_1m_2}\left\{\Sigma^{\lessgtr}_{k_1,l,m_1}, A_{k_2,l,m_2}\right\} \\
&\quad + \left\{\Sigma^{\lessgtr}_{k,o,o}, A_{o,o,m}\right\} + \left\{\Sigma^{\lessgtr}_{k,o,m}, A_{o,o,o}\right\} + i\epsilon_{kk_1k_2}\left[\Sigma^{\lessgtr}_{k_1,o,m}, A_{k_2,o,o}\right] \\
&\quad + i\epsilon_{mm_1m_2}\left[\Sigma^{\lessgtr}_{k,o,m_1}, A_{o,o,m_2}\right] + \frac{1}{2} i\epsilon_{kk_1k_2} i\epsilon_{mm_1m_2}\left\{\Sigma^{\lessgtr}_{k_1,o,m_1}, A_{k_2,o,m_2}\right\} \\
&\quad + \left\{\Sigma^{\lessgtr}_{k,l,o}, A_{o,l,m}\right\} + \left\{\Sigma^{\lessgtr}_{k.l,m}, A_{o,l,o}\right\} + i\epsilon_{kk_1k_2}\left[\Sigma^{\lessgtr}_{k_1,l,m}, A_{k_2,l,o}\right] \\
&\quad + i\epsilon_{mm_1m_2}\left[\Sigma^{\lessgtr}_{k.l,m_1}, A_{o,l,m_2}\right] + \frac{1}{2} i\epsilon_{kk_1k_2} i\epsilon_{mm_1m_2}\left\{\Sigma^{\lessgtr}_{k_1,o,m_1}, A_{k_2,o,m_2}\right\}
\end{aligned} \tag{B.30}$$

We have for the entangled valley spin and pseudospin,

$$\begin{aligned}
&2^{N_s} i\hbar\left(\frac{\partial}{\partial t_1} + \frac{\partial}{\partial t_2}\right) G^{\gtrless}_{o,l,m} \\
&= \left[\tilde{\mathcal{H}}, G^{\lessgtr}\right]_{o,l,m} + \left[\Sigma^{\lessgtr}, \mathrm{Re}G^{r}\right]_{o,l,m} - \frac{i}{2}\left\{\Gamma, G^{\lessgtr}\right\}_{o,l,m} + \frac{i}{2}\left\{\Sigma^{\lessgtr}, A\right\}_{o,l,m}
\end{aligned} \tag{B.31}$$

where,

$$\begin{aligned}
&\left[\tilde{\mathcal{H}}, G^{\lessgtr}\right]_{o,l,m} \\
&= \left[H_{o,o,o}, G^{\lessgtr}_{o,l,m}\right] + \left[H_{o,o,m}, G^{\lessgtr}_{o,l,o}\right] + i\epsilon_{mm_1m_2}\left\{H_{o,o,m_1}, G^{\lessgtr}_{o,l,m_2}\right\} \\
&\quad + i\epsilon_{ll_1l_2}\left\{H_{o,l_1,o}, G^{\lessgtr}_{o,l_2,m}\right\} + \frac{1}{2} i\epsilon_{ll_1l_2} i\epsilon_{mm_1m_2}\left[H_{o,l_1,m_1}, G^{\lessgtr}_{o,l_2,m_2}\right] \\
&\quad + \left[H_{o,l,o}, G^{\lessgtr}_{o,o,m}\right] + \left[H_{o,l,m}, G^{\lessgtr}_{o,o,o}\right] + i\epsilon_{mm_1m_2}\left\{H_{o,l,m_1}, G^{\lessgtr}_{o,o,m_2}\right\}
\end{aligned}$$

$$
\begin{aligned}
&+ i\epsilon_{ll_1l_2}\left\{H_{o,l_1,m}, G^{\lessgtr}_{o,l_2,o}\right\} + \frac{1}{2}i\epsilon_{ll_1l_2}i\epsilon_{mm_1m_2}\left[H_{o,l_1,m_1}, G^{\lessgtr}_{o,l_2,m_2}\right] \\
&+ \left[H_{k,l,o}, G^{\lessgtr}_{k,o,m}\right] + \left[H_{k.l,m}, G^{\lessgtr}_{k,o,o}\right] + i\epsilon_{mm_1m_2}\left\{H_{k.l,m_1}, G^{\lessgtr}_{k,o,m_2}\right\} \\
&+ i\epsilon_{ll_1l_2}\left\{H_{k,l_1,m}, G^{\lessgtr}_{k,l_2,o}\right\} + \frac{1}{2}i\epsilon_{ll_1l_2}i\epsilon_{mm_1m_2}\left[H_{k,l_1,m_1}, G^{\lessgtr}_{k,l_2,m_2}\right] \\
&+ \left[H_{k,o,o}, G^{\lessgtr}_{k,l,m}\right] + \left[H_{k,o,m}, G^{\lessgtr}_{k,l,o}\right] + i\epsilon_{mm_1m_2}\left\{H_{k,o,m_1}, G^{\lessgtr}_{k,l,m_2}\right\} \\
&+ i\epsilon_{ll_1l_2}\left\{H_{k,l_1,o}, G^{\lessgtr}_{k,l_2,m}\right\} + \frac{1}{2}i\epsilon_{ll_1l_2}i\epsilon_{mm_1m_2}\left[H_{k,l_1,m_1}, G^{\lessgtr}_{k,l_2,m_2}\right]
\end{aligned}
\tag{B.32}
$$

$$
\begin{aligned}
&\left[\Sigma^{\lessgtr}, \mathrm{Re}G^r\right]_{o,l,m} \\
&\quad= \left[\Sigma^{\lessgtr}_{o,o,o}, \mathrm{Re}G^r_{o,l,m}\right] + \left[\Sigma^{\lessgtr}_{o,o,m}, \mathrm{Re}G^r_{o,l,o}\right] \\
&\qquad+ i\epsilon_{mm_1m_2}\left\{\Sigma^{\lessgtr}_{o,o,m_1}, \mathrm{Re}G^r_{o,l,m_2}\right\} \\
&\qquad+ i\epsilon_{ll_1l_2}\left\{\Sigma^{\lessgtr}_{o,l_1,o}, \mathrm{Re}G^r_{o,l_2,m}\right\} \\
&\qquad+ \frac{1}{2}i\epsilon_{ll_1l_2}i\epsilon_{mm_1m_2}\left[\Sigma^{\lessgtr}_{o,l_1,m_1}, \mathrm{Re}G^r_{o,l_2,m_2}\right] \\
&\qquad+ \left[\Sigma^{\lessgtr}_{o,l,o}, \mathrm{Re}G^r_{o,o,m}\right] + \left[\Sigma^{\lessgtr}_{o,l,m}, \mathrm{Re}G^r_{o,o,o}\right] \\
&\qquad+ i\epsilon_{mm_1m_2}\left\{\Sigma^{\lessgtr}_{o,l,m_1}, \mathrm{Re}G^r_{o,o,m_2}\right\} \\
&\qquad+ i\epsilon_{ll_1l_2}\left\{\Sigma^{\lessgtr}_{o,l_1,m}, \mathrm{Re}G^r_{o,l_2,o}\right\} \\
&\qquad+ \frac{1}{2}i\epsilon_{ll_1l_2}i\epsilon_{mm_1m_2}\left[\Sigma^{\lessgtr}_{o,l_1,m_1}, \mathrm{Re}G^r_{o,l_2,m_2}\right] \\
&\qquad+ \left[\Sigma^{\lessgtr}_{k,l,o}, \mathrm{Re}G^r_{k,o,m}\right] + \left[\Sigma^{\lessgtr}_{k.l,m}, \mathrm{Re}G^r_{k,o,o}\right] \\
&\qquad+ i\epsilon_{mm_1m_2}\left\{\Sigma^{\lessgtr}_{k.l,m_1}, \mathrm{Re}G^r_{k,o,m_2}\right\} \\
&\qquad+ i\epsilon_{ll_1l_2}\left\{\Sigma^{\lessgtr}_{k,l_1,m}, \mathrm{Re}G^r_{k,l_2,o}\right\} \\
&\qquad+ \frac{1}{2}i\epsilon_{ll_1l_2}i\epsilon_{mm_1m_2}\left[\Sigma^{\lessgtr}_{k,l_1,m_1}, \mathrm{Re}G^r_{k,l_2,m_2}\right] \\
&\qquad+ \left[\Sigma^{\lessgtr}_{k,o,o}, \mathrm{Re}G^r_{k,l,m}\right] + \left[\Sigma^{\lessgtr}_{k,o,m}, \mathrm{Re}G^r_{k,l,o}\right]
\end{aligned}
$$

$$+ i\epsilon_{mm_1m_2}\left\{\Sigma^{\lessgtr}_{k,o,m_1}, \mathrm{Re}G^r_{k,l,m_2}\right\}$$
$$+ i\epsilon_{ll_1l_2}\left\{\Sigma^{\lessgtr}_{k,l_1,o}, \mathrm{Re}G^r_{k,l_2,m}\right\}$$
$$+ \frac{1}{2}i\epsilon_{ll_1l_2}i\epsilon_{mm_1m_2}\left[\Sigma^{\lessgtr}_{k,l_1,m_1}, \mathrm{Re}G^r_{k,l_2,m_2}\right] \tag{B.33}$$

$$\begin{aligned}
&\left\{\Gamma, G^{\lessgtr}\right\}_{o,l,m}\\
&= \left\{\Gamma_{o,o,o}, G^{\lessgtr}_{o,l,m}\right\} + \left\{\Gamma_{o,o,m}, G^{\lessgtr}_{o,l,o}\right\} + i\epsilon_{mm_1m_2}\left[\Gamma_{o,o,m_1}, G^{\lessgtr}_{o,l,m_2}\right]\\
&\quad + i\epsilon_{ll_1l_2}\left[\Gamma_{o,l_1,o}, G^{\lessgtr}_{o,l_2,m}\right] + \frac{1}{2}i\epsilon_{ll_1l_2}i\epsilon_{mm_1m_2}\left\{\Gamma_{o,l_1,m_1}, G^{\lessgtr}_{o,l_2,m_2}\right\}\\
&\quad + \left\{\Gamma_{o,l,o}, G^{\lessgtr}_{o,o,m}\right\} + \left\{\Gamma_{o,l,m}, G^{\lessgtr}_{o,o,o}\right\} + i\epsilon_{mm_1m_2}\left[\Gamma_{o,l,m_1}, G^{\lessgtr}_{o,o,m_2}\right]\\
&\quad + i\epsilon_{ll_1l_2}\left[\Gamma_{o,l_1,m}, G^{\lessgtr}_{o,l_2,o}\right] + \frac{1}{2}i\epsilon_{ll_1l_2}i\epsilon_{mm_1m_2}\left\{\Gamma_{o,l_1,m_1}, G^{\lessgtr}_{o,l_2,m_2}\right\}\\
&\quad + \left\{\Gamma_{k,l,o}, G^{\lessgtr}_{k,o,m}\right\} + \left\{\Gamma_{k.l,m}, G^{\lessgtr}_{k,o,o}\right\} + i\epsilon_{mm_1m_2}\left[\Gamma_{k.l,m_1}, G^{\lessgtr}_{k,o,m_2}\right]\\
&\quad + i\epsilon_{ll_1l_2}\left[\Gamma_{k,l_1,m}, G^{\lessgtr}_{k,l_2,o}\right] + \frac{1}{2}i\epsilon_{ll_1l_2}i\epsilon_{mm_1m_2}\left\{\Gamma_{k,l_1,m_1}, G^{\lessgtr}_{k,l_2,m_2}\right\}\\
&\quad + \left\{\Gamma_{k,o,o}, G^{\lessgtr}_{k,l,m}\right\} + \left\{\Gamma_{k,o,m}, G^{\lessgtr}_{k,l,o}\right\} + i\epsilon_{mm_1m_2}\left[\Gamma_{k,o,m_1}, G^{\lessgtr}_{k,l,m_2}\right]\\
&\quad + i\epsilon_{ll_1l_2}\left[\Gamma_{k,l_1,o}, G^{\lessgtr}_{k,l_2,m}\right] + \frac{1}{2}i\epsilon_{ll_1l_2}i\epsilon_{mm_1m_2}\left\{\Gamma_{k,l_1,m_1}, G^{\lessgtr}_{k,l_2,m_2}\right\}
\end{aligned} \tag{B.34}$$

$$\begin{aligned}
&\left\{\Sigma^{\lessgtr}, A\right\}_{o,l,m}\\
&= \left\{\Sigma^{\lessgtr}_{o,o,o}, A_{o,l,m}\right\} + \left\{\Sigma^{\lessgtr}_{o,o,m}, A_{o,l,o}\right\} + i\epsilon_{mm_1m_2}\left[\Sigma^{\lessgtr}_{o,o,m_1}, A_{o,l,m_2}\right]\\
&\quad + i\epsilon_{ll_1l_2}\left[\Sigma^{\lessgtr}_{o,l_1,o}, A_{o,l_2,m}\right] + \frac{1}{2}i\epsilon_{ll_1l_2}i\epsilon_{mm_1m_2}\left\{\Sigma^{\lessgtr}_{o,l_1,m_1}, A_{o,l_2,m_2}\right\}\\
&\quad + \left[\Sigma^{\lessgtr}_{o,l,o}, A_{o,o,m}\right] + \left[\Sigma^{\lessgtr}_{o,l,m}, A_{o,o,o}\right] + i\epsilon_{mm_1m_2}\left\{\Sigma^{\lessgtr}_{o,l,m_1}, A_{o,o,m_2}\right\}\\
&\quad + i\epsilon_{ll_1l_2}\left\{\Sigma^{\lessgtr}_{o,l_1,m}, A_{o,l_2,o}\right\} + \frac{1}{2}i\epsilon_{ll_1l_2}i\epsilon_{mm_1m_2}\left[\Sigma^{\lessgtr}_{o,l_1,m_1}, A_{o,l_2,m_2}\right]\\
&\quad + \left\{\Sigma^{\lessgtr}_{k,l,o}, A_{k,o,m}\right\} + \left\{\Sigma^{\lessgtr}_{k.l,m}, A_{k,o,o}\right\} + i\epsilon_{mm_1m_2}\left[\Sigma^{\lessgtr}_{k.l,m_1}, A_{k,o,m_2}\right]
\end{aligned}$$

$$+ i\epsilon_{ll_1l_2}\left[\Sigma^{\lessgtr}_{k,l_1,m}, A_{k,l_2,o}\right] + \frac{1}{2}i\epsilon_{ll_1l_2}i\epsilon_{mm_1m_2}\left\{\Sigma^{\lessgtr}_{k,l_1,m_1}, A_{k,l_2,m_2}\right\}$$
$$+ \left[\Sigma^{\lessgtr}_{k,o,o}, A_{k,l,m}\right] + \left[\Sigma^{\lessgtr}_{k,o,m}, A_{k,l,o}\right] + i\epsilon_{mm_1m_2}\left\{\Sigma^{\lessgtr}_{k,o,m_1}, A_{k,l,m_2}\right\}$$
$$+ i\epsilon_{ll_1l_2}\left\{\Sigma^{\lessgtr}_{k,l_1,o}, A_{k,l_2,m}\right\} + \frac{1}{2}i\epsilon_{ll_1l_2}i\epsilon_{mm_1m_2}\left[\Sigma^{\lessgtr}_{k,l_1,m_1}, A_{k,l_2,m_2}\right] \tag{B.35}$$

We have for the entangled Dirac spin, valley spin, and pseudospin,

$$2^{N_s}i\hbar\left(\frac{\partial}{\partial t_1} + \frac{\partial}{\partial t_2}\right)G^{\gtrless}_{k,l,m}$$
$$= \left[\tilde{\mathcal{H}}, G^{\lessgtr}\right]_{k,l,m} + \left[\Sigma^{\lessgtr}, \mathrm{Re}G^r\right]_{k,l,m} - \frac{i}{2}\left\{\Gamma, G^{\lessgtr}\right\}_{k,l,m} + \frac{i}{2}\left\{\Sigma^{\lessgtr}, A\right\}_{k,l,m} \tag{B.36}$$

where,

$$\left[\tilde{\mathcal{H}}, G^{\lessgtr}\right]_{k,l,m}$$
$$= \begin{bmatrix} \left[H_{k,l,m},\ G^{\lessgtr}_{0,0,0}\right] + \left[H_{k,l,0},\ G^{\lessgtr}_{0,0,m}\right] \\ + \left[H_{k,0,m},\ G^{\lessgtr}_{0,l,0}\right] + \left[H_{0,l,m},\ G^{\lessgtr}_{k,0,0}\right] \\ + \left[H_{k,0,0},\ G^{\lessgtr}_{0,l,m}\right] + \left[H_{0,0,m},\ G^{\lessgtr}_{k,l,0}\right] \\ + \left[H_{0,l,0},\ G^{\lessgtr}_{k,0,m}\right] + \left[H_{o,o,o},\ G^{\lessgtr}_{k.l.m}\right] \end{bmatrix}$$
$$+ \begin{bmatrix} \epsilon_{kk_1k_2}\left\{H_{k_1,0,0},\ G^{\lessgtr}_{k_2,l,m}\right\} + \epsilon_{kk_1k_2}\left\{H_{k_1,l,0},\ G^{\lessgtr}_{k_2,0,m}\right\} \\ + \epsilon_{kk_1k_2}\left\{H_{k_1,0,m},\ G^{\lessgtr}_{k_2,l,0}\right\} + \epsilon_{kk_1k_2}\left\{H_{k_1,l,m},\ G^{\lessgtr}_{k_2,0,0}\right\} \\ + \epsilon_{ll_1l_2}\left\{H_{0,l_1,0},\ G^{\lessgtr}_{k,l_2,m}\right\} + \epsilon_{ll_1l_2}\left\{H_{k,l_1,0},\ G^{\lessgtr}_{0,l_2,m}\right\} \\ + \epsilon_{ll_1l_2}\left\{H_{0,l_1,m},\ G^{\lessgtr}_{k,l_2,0}\right\} + \epsilon_{ll_1l_2}\left\{H_{k,l_1,m},\ G^{\lessgtr}_{0,l_2,0}\right\} \\ + \epsilon_{m;m_1,m_2}\left\{H_{0,0.m_1},\ G^{\lessgtr}_{k,l,m_2}\right\} + \epsilon_{m;m_1,m_2}\left\{H_{0,l.m_1},\ G^{\lessgtr}_{k,0,m_2}\right\} \\ + \epsilon_{m;m_1,m_2}\left\{H_{k,0.m_1},\ G^{\lessgtr}_{0,l,m_2}\right\} + \epsilon_{m;m_1,m_2}\left\{H_{k,l.m_1},\ G^{\lessgtr}_{0,0,m_2}\right\} \end{bmatrix}$$

$$
+\left[\begin{array}{l} i\epsilon_{kk_1k_2}i\epsilon_{ll_1l_2}\left(\left[H_{k_1,l_1,o},G^{\lessgtr}_{k_2,l_2,m}\right]+\left[H_{k_1,l_1,m},G^{\lessgtr}_{k_2,l_2,o}\right]\right)\\ +i\epsilon_{kk_1k_2}i\epsilon_{mm_1m_2}\left(\left[H_{k_1,o,m_1},G^{\lessgtr}_{k_2,l,m_2}\right]+\left[H_{k_1,l,m_1},G^{\lessgtr}_{k_2,o,m_2}\right]\right)\\ +i\epsilon_{ll_1l_2}i\epsilon_{mm_1m_2}\left(\left[H_{o,l_1,m_1},G^{\lessgtr}_{k,l_2,m_2}\right]+\left[H_{k,l_1,m_1},G^{\lessgtr}_{o,l_2,m_2}\right]\right)\\ +i\epsilon_{kk_1k_2}i\epsilon_{ll_1l_2}i\epsilon_{mm_1m_2}\left\{H_{k_1,l_1,m_1},G^{\lessgtr}_{k_2,l_2,m_2}\right\} \end{array}\right] \tag{B.37}
$$

$$
\begin{aligned}
&\left[\Sigma^{\lessgtr},\mathrm{Re}G^r\right]_{k,l,m}\\
&=\left[\begin{array}{l} \left[\Sigma^{\lessgtr}_{k,l,m},\ \mathrm{Re}G^r_{0,0,0}\right]+\left[\Sigma^{\lessgtr}_{k,l,0},\ \mathrm{Re}G^r_{0,0,m}\right]\\ +\left[\Sigma^{\lessgtr}_{k,0,m},\ \mathrm{Re}G^r_{0,l,0}\right]+\left[\Sigma^{\lessgtr}_{0,l,m},\ \mathrm{Re}G^r_{k,0,0}\right]\\ +\left[\Sigma^{\lessgtr}_{k,0,0},\ \mathrm{Re}G^r_{0,l,m}\right]+\left[\Sigma^{\lessgtr}_{0,0,m},\ \mathrm{Re}G^r_{k,l,0}\right]\\ +\left[\Sigma^{\lessgtr}_{0,l,0},\ \mathrm{Re}G^r_{k,0,m}\right]+\left[\Sigma^{\lessgtr}_{o,o,o},\ \mathrm{Re}G^r_{k.l.m}\right] \end{array}\right]\\
&+\left[\begin{array}{l} \epsilon_{kk_1k_2}\left\{\Sigma^{\lessgtr}_{k_1,0,0},\ \mathrm{Re}G^r_{k_2,l,m}\right\}+\epsilon_{kk_1k_2}\left\{\Sigma^{\lessgtr}_{k_1,l,0},\ \mathrm{Re}G^r_{k_2,0,m}\right\}\\ +\epsilon_{kk_1k_2}\left\{\Sigma^{\lessgtr}_{k_1,0,m},\ \mathrm{Re}G^r_{k_2,l,0}\right\}+\epsilon_{kk_1k_2}\left\{\Sigma^{\lessgtr}_{k_1,l,m},\ \mathrm{Re}G^r_{k_2,0,0}\right\}\\ +\epsilon_{ll_1l_2}\left\{\Sigma^{\lessgtr}_{0,l_1,0},\ \mathrm{Re}G^r_{k,l_2,m}\right\}+\epsilon_{ll_1l_2}\left\{\Sigma^{\lessgtr}_{k,l_1,0},\ \mathrm{Re}G^r_{0,l_2,m}\right\}\\ +\epsilon_{ll_1l_2}\left\{\Sigma^{\lessgtr}_{0,l_1,m},\ \mathrm{Re}G^r_{k,l_2,0}\right\}+\epsilon_{ll_1l_2}\left\{\Sigma^{\lessgtr}_{k,l_1,m},\ \mathrm{Re}G^r_{0,l_2,0}\right\}\\ +\epsilon_{m;m_1,m_2}\left\{\Sigma^{\lessgtr}_{0,0.m_1},\ \mathrm{Re}G^r_{k,l,m_2}\right\}\\ +\epsilon_{m;m_1,m_2}\left\{\Sigma^{\lessgtr}_{0,l.m_1},\ \mathrm{Re}G^r_{k,0,m_2}\right\}\\ +\epsilon_{m;m_1,m_2}\left\{\Sigma^{\lessgtr}_{k,0.m_1},\ \mathrm{Re}G^r_{0,l,m_2}\right\}\\ +\epsilon_{m;m_1,m_2}\left\{\Sigma^{\lessgtr}_{k,l.m_1},\ \mathrm{Re}G^r_{0,0,m_2}\right\} \end{array}\right]\\
&+\left[\begin{array}{l} i\epsilon_{kk_1k_2}i\epsilon_{ll_1l_2}\left(\left[\Sigma^{\lessgtr}_{k_1,l_1,o},\ \mathrm{Re}G^r_{k_2,l_2,m}\right]\right.\\ \left.+\left[\Sigma^{\lessgtr}_{k_1,l_1,m},\ \mathrm{Re}G^r_{k_2,l_2,o}\right]\right)\\ +i\epsilon_{kk_1k_2}i\epsilon_{mm_1m_2}\left(\left[\Sigma^{\lessgtr}_{k_1,o,m_1},\ \mathrm{Re}G^r_{k_2,l,m_2}\right]\right.\\ \left.+\left[\Sigma^{\lessgtr}_{k_1,l,m_1},\ \mathrm{Re}G^r_{k_2,o,m_2}\right]\right)\\ +i\epsilon_{ll_1l_2}i\epsilon_{mm_1m_2}\left(\left[\Sigma^{\lessgtr}_{o,l_1,m_1},\ \mathrm{Re}G^r_{k,l_2,m_2}\right]\right.\\ \left.+\left[\Sigma^{\lessgtr}_{k,l_1,m_1},\ \mathrm{Re}G^r_{o,l_2,m_2}\right]\right)\\ +i\epsilon_{kk_1k_2}i\epsilon_{ll_1l_2}i\epsilon_{mm_1m_2}\left\{\Sigma^{\lessgtr}_{k_1,l_1,m_1},\ \mathrm{Re}G^r_{k_2,l_2,m_2}\right\} \end{array}\right]
\end{aligned} \tag{B.38}
$$

$$
\begin{aligned}
&\left\{\Gamma, G^{\lessgtr}\right\}_{k,l,m} \\
&= \begin{bmatrix} \left\{\Gamma_{k,l,m},\ G^{\lessgtr}_{0,0,0}\right\} + \left\{\Gamma_{k,l,0},\ G^{\lessgtr}_{0,0,m}\right\} \\ + \left\{\Gamma_{k,0,m},\ G^{\lessgtr}_{0,l,0}\right\} + \left\{\Gamma_{0,l,m},\ G^{\lessgtr}_{k,0,0}\right\} \\ + \left\{\Gamma_{k,0,0},\ G^{\lessgtr}_{0,l,m}\right\} + \left\{\Gamma_{0,0,m},\ G^{\lessgtr}_{k,l,0}\right\} \\ + \left\{\Gamma_{0,l,0},\ G^{\lessgtr}_{k,0,m}\right\} + \left\{\Gamma_{o,o,o},\ G^{\lessgtr}_{k.l.m}\right\} \end{bmatrix} \\
&\quad + \begin{bmatrix} \epsilon_{kk_1k_2}\left[\Gamma_{k_1,0,0},\ G^{\lessgtr}_{k_2,l,m}\right] + \epsilon_{kk_1k_2}\left[\Gamma_{k_1,l,0},\ G^{\lessgtr}_{k_2,0,m}\right] \\ + \epsilon_{kk_1k_2}\left[\Gamma_{k_1,0,m},\ G^{\lessgtr}_{k_2,l,0}\right] + \epsilon_{kk_1k_2}\left[\Gamma_{k_1,l,m},\ G^{\lessgtr}_{k_2,0,0}\right] \\ + \epsilon_{ll_1l_2}\left[\Gamma_{0,l_1,0},\ G^{\lessgtr}_{k,l_2,m}\right] + \epsilon_{ll_1l_2}\left[\Gamma_{k,l_1,0},\ G^{\lessgtr}_{0,l_2,m}\right] \\ + \epsilon_{ll_1l_2}\left[\Gamma_{0,l_1,m},\ G^{\lessgtr}_{k,l_2,0}\right] + \epsilon_{ll_1l_2}\left[\Gamma_{k,l_1,m},\ G^{\lessgtr}_{0,l_2,0}\right] \\ + \epsilon_{m;m_1,m_2}\left[\Gamma_{0,0.m_1},\ G^{\lessgtr}_{k,l,m_2}\right] + \epsilon_{m;m_1,m_2}\left[\Gamma_{0,l.m_1},\ G^{\lessgtr}_{k,0,m_2}\right] \\ + \epsilon_{m;m_1,m_2}\left[\Gamma_{k,0.m_1},\ G^{\lessgtr}_{0,l,m_2}\right] + \epsilon_{m;m_1,m_2}\left[\Gamma_{k,l.m_1},\ G^{\lessgtr}_{0,0,m_2}\right] \end{bmatrix} \\
&\quad + \begin{bmatrix} i\epsilon_{kk_1k_2} i\epsilon_{ll_1l_2}\left(\left\{\Gamma_{k_1,l_1,o},\ G^{\lessgtr}_{k_2,l_2,m}\right\} + \left\{\Gamma_{k_1,l_1,m},\ G^{\lessgtr}_{k_2,l_2,o}\right\}\right) \\ + i\epsilon_{kk_1k_2} i\epsilon_{mm_1m_2}\left(\left\{\Gamma_{k_1,o,m_1},\ G^{\lessgtr}_{k_2,l,m_2}\right\}\right. \\ \left. + \left\{\Gamma_{k_1,l,m_1},\ G^{\lessgtr}_{k_2,o,m_2}\right\}\right) \\ + i\epsilon_{ll_1l_2} i\epsilon_{mm_1m_2}\left(\left\{\Gamma_{o,l_1,m_1},\ G^{\lessgtr}_{k,l_2,m_2}\right\} + \left\{\Gamma_{k,l_1,m_1},\ G^{\lessgtr}_{o,l_2,m_2}\right\}\right) \\ + i\epsilon_{kk_1k_2} i\epsilon_{ll_1l_2} i\epsilon_{mm_1m_2}\left[\Gamma_{k_1,l_1,m_1},\ G^{\lessgtr}_{k_2,l_2,m_2}\right] \end{bmatrix}
\end{aligned} \tag{B.39}
$$

$$
\begin{aligned}
&\left\{\Sigma^{\lessgtr}, A\right\}_{k,l,m} \\
&= \begin{bmatrix} \left\{\Sigma^{\lessgtr}_{k,l,m},\ A_{0,0,0}\right\} + \left\{\Sigma^{\lessgtr}_{k,l,0},\ A_{0,0,m}\right\} \\ + \left\{\Sigma^{\lessgtr}_{k,0,m},\ A_{0,l,0}\right\} + \left\{\Sigma^{\lessgtr}_{0,l,m},\ A_{k,0,0}\right\} \\ + \left\{\Sigma^{\lessgtr}_{k,0,0},\ A_{0,l,m}\right\} + \left\{\Sigma^{\lessgtr}_{0,0,m},\ A_{k,l,0}\right\} \\ + \left\{\Sigma^{\lessgtr}_{0,l,0},\ A_{k,0,m}\right\} + \left\{\Sigma^{\lessgtr}_{o,o,o},\ A_{k.l.m}\right\} \end{bmatrix}
\end{aligned}
$$

$$
+\left[\begin{array}{l}
\epsilon_{kk_1k_2}\left[\Sigma^{\lessgtr}_{k_1,0,0},\ A_{k_2,l,m}\right]+\epsilon_{kk_1k_2}\left[\Sigma^{\lessgtr}_{k_1,l,0},\ A_{k_2,0,m}\right]\\
+\epsilon_{kk_1k_2}\left[\Sigma^{\lessgtr}_{k_1,0,m},\ A_{k_2,l,0}\right]+\epsilon_{kk_1k_2}\left[\Sigma^{\lessgtr}_{k_1,l,m},\ A_{k_2,0,0}\right]\\
+\epsilon_{ll_1l_2}\left[\Sigma^{\lessgtr}_{0,l_1,0},\ A_{k,l_2,m}\right]+\epsilon_{ll_1l_2}\left[\Sigma^{\lessgtr}_{k,l_1,0},\ A_{0,l_2,m}\right]\\
+\epsilon_{ll_1l_2}\left[\Sigma^{\lessgtr}_{0,l_1,m},\ A_{k,l_2,0}\right]+\epsilon_{ll_1l_2}\left[\Sigma^{\lessgtr}_{k,l_1,m},\ A_{0,l_2,0}\right]\\
+\epsilon_{m;m_1,m_2}\left[\Sigma^{\lessgtr}_{0,0.m_1},\ A_{k,l,m_2}\right]+\epsilon_{m;m_1,m_2}\left[\Sigma^{\lessgtr}_{0,l.m_1},\ A_{k,0,m_2}\right]\\
+\epsilon_{m;m_1,m_2}\left[\Sigma^{\lessgtr}_{k,0.m_1},\ A_{0,l,m_2}\right]+\epsilon_{m;m_1,m_2}\left[\Sigma^{\lessgtr}_{k,l.m_1},\ A_{0,0,m_2}\right]
\end{array}\right]
$$

$$
+\left[\begin{array}{l}
i\epsilon_{kk_1k_2}i\epsilon_{ll_1l_2}\left(\left\{\Sigma^{\lessgtr}_{k_1,l_1,o},\ A_{k_2,l_2,m}\right\}+\left\{\Sigma^{\lessgtr}_{k_1,l_1,m},\ A_{k_2,l_2,o}\right\}\right)\\
+i\epsilon_{kk_1k_2}i\epsilon_{mm_1m_2}\left(\left\{\Sigma^{\lessgtr}_{k_1,l_1,o},\ A_{k_2,l_2,m}\right\}+\left\{\Sigma^{\lessgtr}_{k_1,l_1,m},\ A_{k_2,l_2,o}\right\}\right)\\
+i\epsilon_{ll_1l_2}i\epsilon_{mm_1m_2}\left(\left\{\Sigma^{\lessgtr}_{k_1,l_1,o},\ A_{k_2,l_2,m}\right\}+\left\{\Sigma^{\lessgtr}_{k_1,l_1,m},\ A_{k_2,l_2,o}\right\}\right)\\
+i\epsilon_{kk_1k_2}i\epsilon_{ll_1l_2}i\epsilon_{mm_1m_2}\left[\Sigma^{\lessgtr}_{k_1,l_1,m_1},\ A_{k_2,l_2,m_2}\right]
\end{array}\right]
\tag{B.40}
$$

C. Transport Equations in Phase Space

The transport equations in phase space are obtained by applying the "lattice" Weyl transformation (although continuum approximation is often employed in this paper, this is not essential and we use the word "lattice" when referring to solid-state problems [19, 20, 29, 30]) of the propagator equations for the respective excitations by using the identities detailed in the following sections.

C.1. *"Lattice" Weyl transform of differential operators*

$$
i\hbar\left(\frac{\partial}{\partial t_1}+\frac{\partial}{\partial t_2}\right)Y\left(t_1,t_2\right)\Leftrightarrow i\hbar\frac{\partial}{\partial t}Y\left(E,t\right) \tag{C.1}
$$

$$
\left(\frac{\partial^2}{\partial t_1^2}-\frac{\partial^2}{\partial t_2^2}\right)Y\left(t_1,t_2\right)\Leftrightarrow \frac{2i}{\hbar}E\frac{\partial}{\partial t}Y\left(E,t\right) \tag{C.2}
$$

$$
\left(\nabla_1^2-\nabla_2^2\right)Y\left(r_1,r_2\right)\Leftrightarrow \frac{2i}{\hbar}p\cdot\nabla_q Y\left(p,q\right) \tag{C.3}
$$

C.2. *Lattice Weyl transform of product of arbitrary operators*

C.2.1. *Poisson bracket expansion*

In terms of differential "Poisson bracket" operator,

$$AB\,(p,q) = \exp\frac{\hbar}{i}\left(\frac{\partial^{(a)}}{\partial p}\cdot\frac{\partial^{(b)}}{\partial q} - \frac{\partial^{(a)}}{\partial q}\cdot\frac{\partial^{(b)}}{\partial p}\right)\ a\,(p,q)\ b\,(p,q) \tag{C.4}$$

Thus, we are led to the lattice Weyl transform of a commutator, $[A,B]\,(p,q)$, and an anticommutator, $\{A,B\}\,(p,q)$, as

$$\begin{aligned}[A,B]\,(p,q) &= \cos cos\Lambda\,[a\,(p,q)\ b\,(p,q) - b\,(p,q)\,a\,(p,q)]\\ &\quad - i\sin\Lambda\,[a\,(p,q)\ b\,(p,q) + b\,(p,q)\,a\,(p,q)] \qquad \text{(C.5)}\\ \{A,B\}\,(p,q) &= \cos cos\Lambda\,[a\,(p,q)\ b\,(p,q) + b\,(p,q)\,a\,(p,q)]\\ &\quad - i\sin\Lambda\,[a\,(p,q)\ b\,(p,q) - b\,(p,q)\,a\,(p,q)] \qquad \text{(C.6)}\end{aligned}$$

where $\Lambda = \frac{\hbar}{2}\left(\frac{\partial^{(a)}}{\partial p}\cdot\frac{\partial^{(b)}}{\partial q} - \frac{\partial^{(a)}}{\partial q}\cdot\frac{\partial^{(b)}}{\partial p}\right)$. Note that so far no semi-classical approximation is involved in the derivation.

C.2.2. *Integral representation*

In terms of integral operators, we have

$$\begin{aligned}AB\,(p,q) &= \frac{1}{(\hbar^4)^2}\int dp'\ dq'\ K_A^{+}\,(p,q;p'q')\ b\,(p',q')\\ &= \frac{1}{(\hbar^4)^2}\int dp'\ dq'\ a\,(p',q')\ K_B^{-}\,(p,q;p'q')\end{aligned} \tag{C.7}$$

where integral kernels are defined by

$$\begin{aligned}&K_Y^{\pm}\,(p,q;p'q')\\ &= \int dudv\exp\left\{\frac{i}{\hbar}\left[(p-p')\cdot v + (q-q')\cdot u\right]\right\} y\left(p\pm\frac{u}{2}, q\mp\frac{v}{2}\right)\end{aligned} \tag{C.8}$$

Therefore, in terms of integral operators,

$$[A,B](p,q) = \frac{1}{(\hbar^4)^2}\int dp'\,dq'\,\left[K_A^+(p,q;p'q')\,b(p',q')\right.$$
$$\left.-b(p',q')\,K_A^-(p,q;p'q')\right] \tag{C.9}$$

$$\{A,B\}(p,q) = \frac{1}{(\hbar^4)^2}\int dp'dq'\left[K_A^+(p,q;p'q')\,b(p',q')\right.$$
$$\left.+b(p',q')\,K_A^-(p,q;p'q')\right] \tag{C.10}$$

C.2.3. *Local and nonlocal integral kernels*

The above expressions simplify considerably when the lattice Weyl transform $a(p,q)$ and $b(p,q)$ are scalar functions. We have for the integral representations,

$$[A,B](p,q) = \frac{1}{(\hbar^4)^2}\int dp'\,dq'dudv\exp\left\{\frac{i}{\hbar}\left[(p-p')\cdot v+(q-q')\cdot u\right]\right\}$$
$$\times\begin{bmatrix} a\left(p+\frac{u}{2},q-\frac{v}{2}\right) \\ -a\left(p-\frac{u}{2},q+\frac{v}{2}\right)\end{bmatrix} b(p',q') \tag{C.11}$$

$$\{A,B\}(p,q) = \frac{1}{(\hbar^4)^2}\int dp'\,dq'dudv\exp\left\{\frac{i}{\hbar}\left[(p-p')\cdot v+(q-q')\cdot u\right]\right\}$$
$$\times\begin{bmatrix} a\left(p+\frac{u}{2},q-\frac{v}{2}\right) \\ +a\left(p-\frac{u}{2},q+\frac{v}{2}\right)\end{bmatrix} b(p',q') \tag{C.12}$$

Thus, in its integral forms, which are better suited for carrying out numerical simulations, Eq. (C.11) exhibits a *nonlocal* kernel for the commutator, $[A,B](p,q)$, whereas Eq. (C.12) shows an '*averaging*' type of kernel tied at point (p,q) or *local* kernel. This is already shown for the anticommutator $\{A,B\}(p,q)$ to a leading order. Indeed, it is precisely the nonlocal character of the kernel in Eq. (C.11) that is entirely responsible for quantum tunneling transport in resonant tunneling diodes [19, 20] which exhibit entirely quantum nonlocality characteristics.

References

1. P. W. Anderson, Absence of diffusion in certain random lattices, *Phys. Rev.* 109(5), 1492–1505 (**1958**).

2. N. F. Mott, The basis of the electron theory of metals, with special reference to the transition metals, *Proc. Phys. Society. Series A.* 62(7), 416–422 (**1949**).
3. E. Wigner, On the interaction of electrons in metals, *Phys. Rev.* 46(11), 1002–1011 (**1934**).
4. J. Bell, On the Einstein–Poldolsky–Rosen paradox. *Physics* 1(3), 195–200, (**1964**).
5. D. Deutsch and P. Hayden, How quantum teleportation actually works, *Proc. R. Soc. Lond. A* 456, 1759–1774 (**2000**).
6. R. J. C. Spreeuw, *Found. Phys.* 28, 361 (**1998**).
7. R. de J. León-Montiel, A. Vallés, H. M. Moya-Cessa, and J. P. Torres, Coherent delocalization: Views of entanglement in different scenarios, *Laser Phys. Lett.* 12, 085204(6pp) (**2015**).
8. B. Doucot, M. O. Goerbig, P. Lederer, and R. Moessner, *Entanglement Skyrmions in multicomponent quantum hall systems*, arXiv:0806.0229v2 [cond.mat.mess-Hall] (November 6, 2008).
9. K. Boguslawski and P. l Tecmer, *Orbital Entanglement in Quantum Chemistry*, arXiv:1409.8017v2 [physics.chem-ph] (October 4, 2016).
10. J. M. Kosterlitz, Nobel Lecture: Topological defects and phase transitions, *Rev. Mod. Phys.* 89, 040501-1 (**2017**).
11. F. D. M. Haldane, Nobel lecture: Topological quantum matter, *Rev. Mod. Phys.* 89, 040502-2 (**2017**).
12. X. Xu, W. Yao, D. Xiao, and T. F. Heinz, Spin and pseudospins in layered transition metal dichalcogenides, *Nat. Phys.* 10, 343 (**2014**).
13. M. S. Fuhrer and A. R. Hamilton, Chasing the exciton condensate, *Physics Viewpoint, Physics* 9, 80, July18, (**2016**).
14. M. Kindermann, *Pseudospin entanglement and Bell test in graphene*, arXiv:0807.0217v2 [cond-mat.mess-hall] (July 3, 2008).
15. X. Liu, K. Watanabe, T. Taniguchi, B. L. Halperin, and P. Kim, Quantum hall drag of exciton condensate in graphene, *Nat. Phys.* 13, 746 (**2017**).
16. I. Sodemann, H. Chen, and A. H. MacDonald, *Theory of transport phenomena in coherent quantum hall bilayers*, arXiv: 1411.0008v1 [cond-mat.mess-hall] (October 31, 2014).
17. B. Altshuler, Anderson Localization at 1D edge of 2D Topological Insulator, *Phys. Rev. Lett.* 111, 08641 (**2013**).
18. J. Maciejko, C. Liu, Y. Oreg, X.-L. Qi, C. Wu, and S.-C. Zhang, Kondo effect in helical edge liquid of the quantum spin Hall state, *Phys. Rev. Lett.* **102**, 256803 (2009). See also, R. Bonazola, J. A. Andrade, J. I. Facio, D. J. Garc, and P. S. Cornaglia, *SU(4) Kondo entanglement in double quantum dot devices*, arXiv:1705.00750v3 [cond-mat.mess-hall] (August 15, 2017).
19. F. A. Buot and K. L. Jensen, Lattice Weyl-Wigner formulation of exact many-body quantum transport theory and applications to novel solid-state quantum-based devices, *Phys. Rev. B* 42, 9429 (**1990**).
20. K. L. Jensen and F.A. Buot, Numerical simulation of intrinsic bistability and high-frequency current oscillations in resonant tunneling structures, *Phys. Rev. Lett.* 66, 1078 (**1991**).

21. A. van Oudenaarden, S. J. K. Várdy, and J. E. Mooij, One-dimensional localization of quantum vortices in disordered Josephson junction arrays, *Phys. Rev. Lett.* 77, 4257 (**1996**).
22. E. Babaev, J. Jäykkä, and M. Speight, *Magnetic Field Delocalization and Flux Inversion in Fractional Vortices in Two-component Superconductors*, arXiv:0903.3339v2 [cond-mat.supr-con] (December 8, 2009).
23. C. J. Olson Reichhardt and M. B. Hastings, Do vortices entangle?, *Phys. Rev. Lett.* 92(15) 157002 (**2004**).
24. J.-L. Ge, T.-R. Wu, M. Gao, Z.-B. Bai, L. Cao, X.-F. Wang, Y.-Y. Qin, and F.-Q. Song, Weak localization of bismuth cluster-decorated graphene and its spin–orbit interaction, *Front. Phys.* 12, 127210 (**2017**).
25. W. Knapp, G. E. Pikus, F. G. Pikus, S. V. Iordanskii, V. Mosser, K. Zekentes, and Yu. B. Lyanda-Geller, *Weak Antilocalization and Spin Precession in Quantum Wells*, arXiv:cond-mat/9602068v1 (1996).
26. M. Kammermeier, P. Wenk, J. Schiliemann, S. Heedt, and T. Schäpers, Weak (anti)localization in tubular semiconductor nanowires with spin–orbit coupling, *Phys. Rev B* 93, 205306 (**2016**).
27. S. I. Dorozhkin, A. A. Kapustin, and S. S. Murzin, Observation of crossover from weak localization to antilocalization in the temperature dependence of the resistance of a two-dimensional system with spin–orbit interaction, *JETP Lett.* 97, 149–154 (**2013**).
28. F. A. Buot, Discrete phase-space model for quantum mechanics, in M. Kafatos, Ed., *Bell's Theorem,Quantum Theory, and Conceptions of the Universe* (Kluwer, NY, 1989, Fundamental Physics Series), pp. 159–162.
29. F. A. Buot, Method for calculating Tr $\mathcal{H}^n$ in solid state theory, *Phys. Rev. B* 10, 3700 (**1974**).
30. Felix A. Buot, *Nonequilibrium Quantum Transport Physics in Nanosystems: Foundation of Computational Nonequilibrium Physics in Nanoscience and Nanotechnology* (World Scientific, NJ, 2009).
31. F. A. Buot, General theory of quantum distribution function transport equations: Superfluid systems and ultrafast dynamics of optically excited semiconductors, *La Rivista del Nuovo Cimento* 20(9), 1–75 (**1997**).
32. F. A. Buot, R. A. Loberternos, D. L. Villarin, and R. E. S. Otadoy, Spin magnetization quantum transport equations: Coupled nonequilibrium spin and charge correlations, *Quantum Matter* 2, 1–7 (**2013**).
33. M. Büttiker, Edge state without magnetic fields, *Science* 325, 278 (**2009**).
34. G. Baym and T. Ozawa, Condensation of bosons with Rashba-Dresselhaus spin–orbit coupling, *J. Phys.: Conf. Ser.* 529, 012006 (**2014**). See also: *Slides on spin–orbit coupling in solids*, physik.uni-regensburg.de.
35. Felix A. Buot, Karla B. Rivero, and Roland E. S. Otadoy, Generalized nonequilibrium quantum transport of spin and pseudospins: Entanglements and topological phases, *Physica B: Cond. Mat.* 559, 42–61 (**2019**).
36. F. Wilczek, *Basic Notions of Entropy and Entanglement, Lecture notes*, Center for Theoretical Physics, MIT, Cambridge MA 02139 USA (March 3, 2014).
37. F. A. Buot, Y. Jiang, and A.I. Fedoseyev, Novel quantum hydrodynamic equations and quantum-hierarchy decoupling scheme, *Phys. Rev. E* 66, 066119 (**2002**).

38. S. Murakami, Two-dimensional topological insulators and their edge states, *J. Phys.: Conf. Series* 302, 012019 (**2011**).
39. E. D. K. Finck, J. P. Eisenstein, L. N. Pfeifer, and K. W. West, Exciton transport and andreev reflection in a bilayer quantum hall system, *Phys. Rev. Lett.* 106, 236807 (**2011**).
40. A. K. Rajagopal, Spin-orbit interactions in the many-body theory of magnetic electron systems, *Phys. Rev. B* 57, 11582 (**1998**).
41. A. K. Rajagopal and M. Mochena, *Broken-Symmetry-Adapted Green's Function Theory of Condensed Matter: Towards a Vector Spin Density Functional Theory*, cond-mat/9907089, (July 6, 1999).
42. O. D. Jefimenko, Torque exerted by a moving electric charge on a stationary electric charge distribution, *J. Phys. A: Math. Gen.* 35, 5305–5314 (**2002**).
43. M. I. D'yakonov and V. I. Perel', Possibility of orienting electron spins with current, *ZhETF Pis. Red.* 13(11), 657–660 (**1971**).
44. R. C. Iotti, F. Dolcini, and F. Rossi, Wigner-function formalism applied to semiconductor quantum devices: Need for nonlocal scattering models, *Phys. Rev. B* 96, 115420 (**2017**).
45. R. C. Iotti and F. Rossi, Microscopic theory of energy dissipation and decoherence in solid-state quantum devices: Need for nonlocal scattering models, *Entropy* 20, 726 (**2018**).
46. P. J. Newton, R. Mansell, S. N. Holmes, M. Myronov, and C. H. W. Barnes, Weak localization and weak antilocalization in doped germanium epilayers, *App. Phys. Lett.* 110, 062101 (**2017**).
47. M. Dzero, J. Xia, V. Galitski, and P. Coleman, Topological Kondo Insulators, *Annual Rev. Cond. Matter Phys.* 7, 249–280 (**2016**).
48. P.-Y. Chang, O. Erten, and P. Coleman, Möbius Kondo insulators, *Nat. Phys.* 13, 794–798 (**2017**).
49. S. Diehl, E. Rico, M. A. Baranov, and P. Zoller, Topology by dissipation in atomic quantum wires, *Nat. Phys.* 7, 971 (**2011**).
50. X. Chen, L. Zhang, and Hong Guo, Valley caloritronics and its realization by graphene nanoribbons, *Phys. Rev. B* 92, 155427 (**2015**).
51. H. Adachi, K. Uchida, E. Saitoh, and S. Maekawa, Theory of the spin Seebeck effect, *Rep. Prog. Phys.* 76, 036501 (**2013**).

Chapter 4

Layered Nanodiamond Hybrid Materials

Markus Mohr* **and Hans J. Fecht†**

Institute of Functional Nanosystems, Ulm University,
Albert-Einstein-Allee 47 89081 Ulm, Germany
**markus.mohr@uni-ulm.de*
†hans.fecht@uni-ulm.de

4.1. Introduction

The physical and chemical property changes of materials occurring on the nanoscale are of great interest for science and technology [1, 2]. The modifications allow the perfect tuning of a macroscopic material's properties if it consists of nanoscale building blocks with controlled size and composition.

Basically, every property has a critical length scale. If a nanoscale building block is made smaller than this critical length scale, the fundamental physics of that property, such as the mechanical, electronic, magnetic, and optical properties as well as corrosion behavior and diffusion, changes. By altering the sizes of these building blocks, controlling their internal and surface chemistry, their atomic structure, and their assembly, nanostructured materials can nowadays be produced in large quantities, as predicted and shown in Fig. 4.1 [3]. Thus, it becomes possible to engineer the properties and functionalities of materials and components in completely new ways.

In all these cases, interfaces and surfaces that separate the different particles, layers, and crystalline or non-crystalline domains from each other play a crucial role in controlling the nanostructures' properties and stability. Of particular importance for engineering applications are thin layers and coatings such as micromechanical systems, energy systems, and health and medical applications.

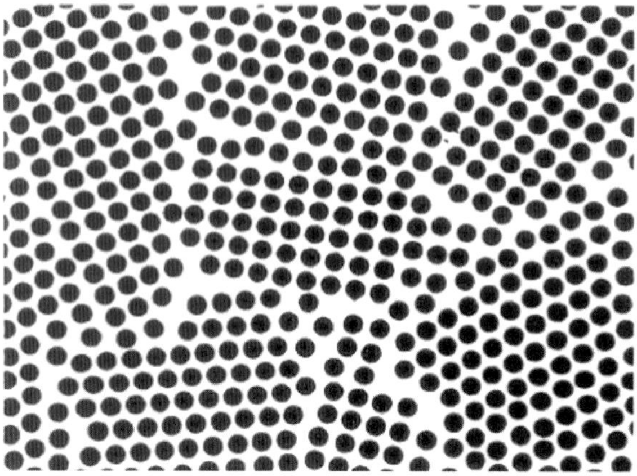

Fig. 4.1: Schematic depiction of a rigid body nanocrystalline material consisting of grains and grain boundaries [3]. Reprinted from [3] with permission from Elsevier.

There are major unsolved questions and new phenomena observed in the field of interface-controlled nanostructures when the grain size range approaches a few nanometers. This applies, for example, to the mechanical properties of engineering materials.

Some of the basic principles in improving the mechanical strength of a material have already been developed in early steelmaking back in 1500 BC, for example, Damascene steel. For metals, the mechanical strength strongly depends on the grain size and is generally described by the Hall–Petch relation, where the strength increases as the inverse square root of the grain size.

This behavior is connected with the formation, movement, and pile-up of dislocations on obstacles, such as grain boundaries when applied to mechanical stresses [3].

For small grain diameters in the range of 10–20 nm, when these dislocations-based plastic deformation mechanisms are exhausted, other mechanisms become activated. Examples are grain boundary sliding, creep, and similar. These mechanisms are known for high-temperature deformation behavior for brittle materials such as ceramics and intermetallics and have led to the development of high-strength metallic materials and composites. For the smallest grain sizes achieved, the mechanical strength can be improved by a factor of four to five and therefore considerably enhance performance and lifetime [2].

Table 4.1: Physical properties of single- or polycrystalline diamond — Superior to other materials like Si, steel or SiO_2.

Property	**Diamond**	**Si**	**Steel**	**SiO_2**
Young's Modulus (GPa)	700–1143	110–190	190–215	60–95
Fracture Strength (GPa)	4–10	1.4	0.4–1.1	0.4–1.1
Density (g/cm^3)	3.51	2.32	6–8	2.5
CTE ($10^6/K$)	1.1	2.6	10–20	0.5–0.9
Thermal Conductivity (W/mK)	$1–2\ 10^3$	105–145	10–45	1–1.6

For non-metals and ceramics, the situation is quite different. Here, the principles of plasticity by dislocation activity are generally not relevant due to the different nature of the covalent bond. In the case of ceramics, typically produced through sintering processes, the plastic deformation — if any, due to the inherent brittleness — is rather controlled by processes like grain boundary sliding, creep, and alike. Nowadays, the mechanical behavior of ceramics can be improved by interface and grain boundary engineering but often at the cost of mechanical strength.

Considered in general as a ceramic material, diamond is a (metastable) allotrope of carbon. The carbon atoms are arranged in a variation of the face-centered cubic crystal structure — the diamond lattice. Natural diamonds are generally formed at high temperature and pressure deep in the Earth's mantle. Carbon-containing minerals provide the carbon source, but growth occurs over long periods of time.

In Table 4.1, the main material properties of diamond are given in comparison with other materials, such as semiconductors (silicon), metals (steel), and ceramics (silicon-dioxide). Thus, diamond is the hardest of all known materials, making it attractive for wear-stressed parts and microparts.

The high fracture strength and Young's modulus make diamond parts very robust and dimensionally stable. The high thermal conductivity combined with the low coefficient of thermal expansion and the low heat capacity enable thermally stressed applications like heat spreaders, laser windows, or thermal microactuators. Undoped diamond is electrically insulating. However, diamond might be *p*- and *n*-doped at different levels to become semiconducting, and even metal-like, at high doping concentrations [1].

Therefore, diamond is a very attractive material for many applications due to its versatile properties. However, since natural diamond is rare, expensive, and unavailable in sizes and shapes required for most engineering applications, extensive research has been undertaken to produce it synthetically.

The first successful approach in the 1950s was based on duplicating nature's method of producing pieces of diamond [4]. Using conditions similar to great depth within the earth, i.e., high pressure of typically several GPa and temperatures in excess of 1200°C, synthetic diamond can be produced on a large scale. This technique is referred to as "high-pressure high-temperature" (HPHT) method and is well established in today's synthetic diamond industry, producing hundreds of tons each year. Due to its restriction in shape and size and infeasibility of integration into standard manufacturing processes, the use of HPHT is mainly found in mechanical abrasion applications, such as cutting, grinding and polishing.

A more versatile alternative method to synthesize diamond is based on a low-pressure chemical vapor deposition (CVD) technique developed in 1962 [5]. In contrast to the HPHT method, yielding single-crystalline diamond stones, CVD diamonds typically form a continuous film consisting of various crystallites. To give an overview of the wide pressure and temperature range to synthesize diamond, Fig. 4.2 exhibits the formation conditions of different C-based phases, including CVD-grown diamond [6].

4.2. CVD Techniques for Diamond Growth

4.2.1. *Basic principles of CVD*

First of all, a feed gas, typically composed of hydrogen (H_2) and methane (CH_4) as a source for carbon, needs to be activated to start the diamond formation process. Three techniques with different activation techniques are well established in CVD diamond growth. In the microwave plasma CVD (MWPCVD), the energy is supplied by electric discharge, generating a plasma. The combustion CVD (CCVD) uses an oxyacetylene torch, and in the hot filament CVD (HFCVD), the gas is thermally activated by hot filaments. While the first two methods are widely used for nanocrystalline diamond deposition, the combustion CVD technique, known for its high growth rates, is only used to produce polycrystalline diamond [7].

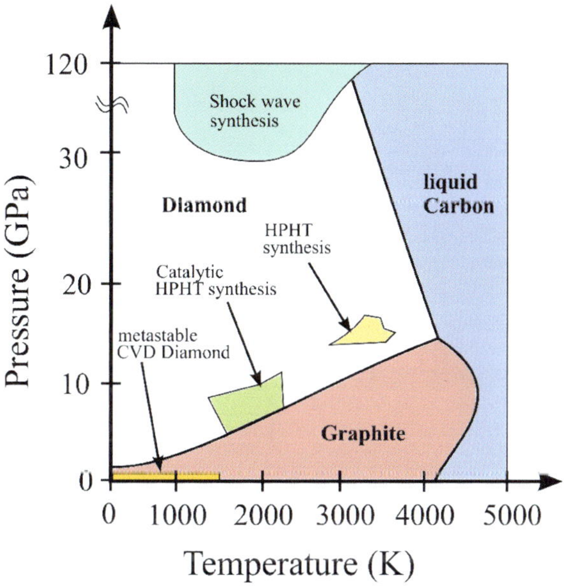

Fig. 4.2: The carbon phase diagram with variable pressure and temperature indicating the stability range of different C-phases produced by different methods (after [6]).

The respective activation zone provides enough energy for a complex set of reactions. The electric discharge or heat dissociates the feed gas into reactive radicals, such as atomic hydrogen and hydrocarbons (CH_3, CH_2, etc.). Additionally, collisions of atomic hydrogen with hydrocarbons lead to a series of CH_x radicals.

Beyond the activation region, these reactive fragments continue to mix and undergo a complex set of chemical reactions with different growth species, such as C, CH, C_2, C_2H, CH_3, and C_2H_2. Former experimental work and modeling studies have identified the methyl radical (CH_3) and acetylene (C_2H_2) as the only two species being sufficiently reactive and present at high concentrations to serve as a precursor for diamond growth. According to widely accepted theories, CH_3 is the most likely growth species for diamond.

At this point, the species can either absorb and react with the surface, being desorbed back into the gas phase, or diffuse around close to the surface until an appropriate reaction site is found. If a surface reaction occurs and if all the conditions are suitable, one possible outcome is the formation of diamond.

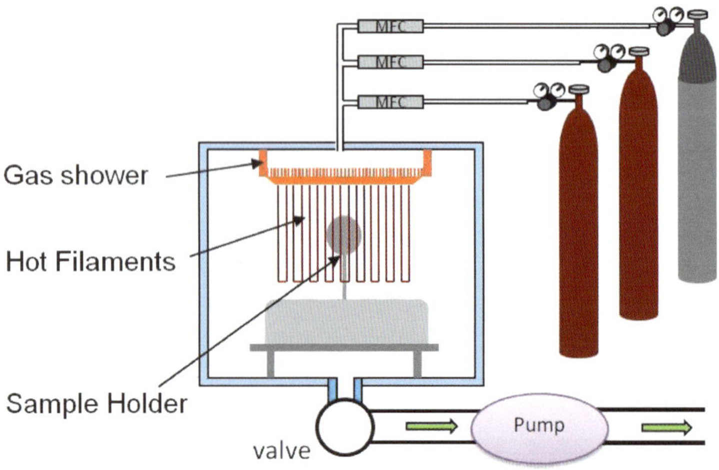

Fig. 4.3: Schematic illustration of the HFCVD diamond process.

4.2.2. *Hot filament CVD*

The hot filament CVD is the most used technique for large-scale deposition of polycrystalline diamond films, especially on three-dimensional substrates. A sketch of the typical HFCVD reactor principles is presented in Fig. 4.3.

The process uses a vacuum chamber, and initial to deposition, the vacuum is typically in the range of 10^{-2}–10^{-4} mbar. The energy for the activation of the gas phase is provided by the temperature of filament. These filaments are typically made of a refractory material, such as tungsten (W), tantalum (Ta) or rhenium (Re).

During the deposition process, the filaments are electrically heated to 1800–2100°C, supplying the energy for the hydrogen dissociation. Filaments made of tungsten or other carbide-forming metals require a carburization step before they can be effectively used for diamond deposition.

4.2.2.1. *Seeding and substrate pretreatment*

The growth of diamond on non-diamond materials, irrespective of the method and crystallographic structure, implies an initial substrate pretreatment to promote nucleation. Due to the high deposition temperatures around 700–750°C, only substrates can be used which are thermally stable in that range and chemically compatible with carbon. In this respect, silicon substrates in the form of single-crystal wafer material are perfect and

Fig. 4.4: View inside the hot processing chamber with multiple tungsten filaments heated to about 2000°C. Six 6-inch silicon wafers can be processed in one step at the same time.

WC–Co hard metal can be used as well, when properly pretreated on the surface before deposition.

A combination of the high surface energy, low precursor sticking probability, and a chemical competition between the carbon phases hampers primary diamond nucleation. The growth of diamond nuclei starts when individual carbon atoms on the substrate's surface begin to form sp^3 tetrahedral lattices. Since non-diamond substrates initially do not promote such a template for the tetrahedral structure, carbon atoms reaching the surface are immediately etched by the highly reactive atomic hydrogen. Consequently, diamond formation is not supposed to take place on ideal surfaces. In reality, a low nucleation density of 10^3–$10^5\,cm^{-2}$ is detected.

Even though spontaneous surface nucleation occurs on non-seeded substrates, the film evolution in terms of film uniformity, film adhesion, and roughness is considerably controlled by the initial nucleation density. Additionally, in order to ensure a minimum thickness at film coalescence, the nucleation density should be as high as possible.

Using a proper seeding procedure using nanocrystalline diamond precursor particles, the formation of diamond crystals in the early stages of diamond deposition is illustrated in Fig. 4.5. After a short growth step of about 5 min, the primary diamond seeds are overgrown by nanosized secondary crystallites. The formation of nanometric grains at the active sites are clearly seen in the high-resolution SEM images. Since the deposition conditions favor a high rate of secondary nucleation, the newly formed diamond crystals have a typical size of 10 nm under optimum growth conditions.

Fig. 4.5: SEM images of NCD cluster formation around primary nanodiamond seeds after 5 min growth (M. Wiora, private communication).

A relatively homogenous nucleation density in the order of 10^{11} cm^{-2} over the three-inch wafer and over a batch of several wafers could be achieved, as shown in Fig. 4.5.

Diamond growth involves numerous critical deposition parameters, such as gas pressure, gas composition, filament material, filament arrangement, substrate temperature, and reactor geometry. These parameters significantly determine growth rate, microstructure, and diamond quality. To minimize the grain size, a high rate of secondary nucleation is crucial. Previous research using HFCVD showed the significance of several growth parameters on the microstructure and diamond quality [1].

4.2.2.2. *Microcrystalline vs. nanocrystalline diamond*

In contrast to the HPHT technique, CVD diamond growth results in polycrystalline films. Depending on the distinct growth parameters, the resulting coatings can have different microstructures [8–11].

The (average) crystal size can be controlled by, first, the nucleation step and, second, the growth conditions from several micrometers down to the nanometer range. Regardless of the size, the single grains are separated by grain boundaries, usually consisting of non-diamond components, such as sp^2- and sp^3-hybridized carbon. Accordingly, the properties can differ significantly depending on the microstructure. Polycrystalline diamond films can be classified into three categories according to their crystal sizes:

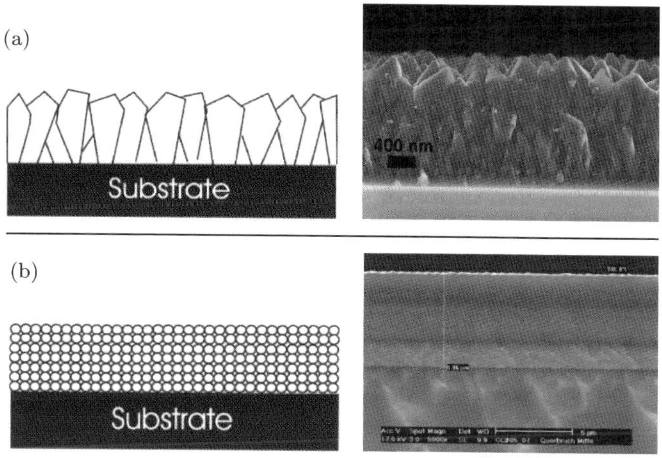

Fig. 4.6: (a) Schematic illustration and SEM image in cross-section of a microcrystalline (MCD) diamond film. (b) Schematic illustration and SEM image in cross-section of a nanocrystalline (NCD) diamond film.

- microcrystalline diamond (MCD) (>100 nm);
- nanocrystalline diamond (NCD) (10–20 nm);
- ultra-nanocrystalline diamond (UNCD) (3–5 nm).

Each category offers distinctly different properties, which are consequently controlled and dependent on the grain size. The microstructural differences of the films are schematically illustrated in Fig. 4.6. Microcrystalline diamond films can be seen together with SEM images in cross-section, clearly distinguishing the different structures.

As the name implies, the film consists of large and facetted crystals, usually several micrometers in size. A distinct characteristic of MCD films is the non-uniform and anisotropic evolution of the crystallites. As displayed in Fig. 4.6, MCD films typically develop a fiber-like grain texture with increasing crystallite size while the film thickness increases. This texture evolution, also sometimes called "columnar growth", can be described using the "van der Drift growth selection mechanism" [12].

The van der Drift selection mechanism considers the competitive growth between crystallites having different texture orientations and growth velocities. In the early beginning of the growth phase, the crystallites are randomly orientated. However, with increasing growth time, the crystallites having the most rapid growth direction perpendicular to the substrate (i.e., for diamond (110) direction) will gradually overgrow their neighboring

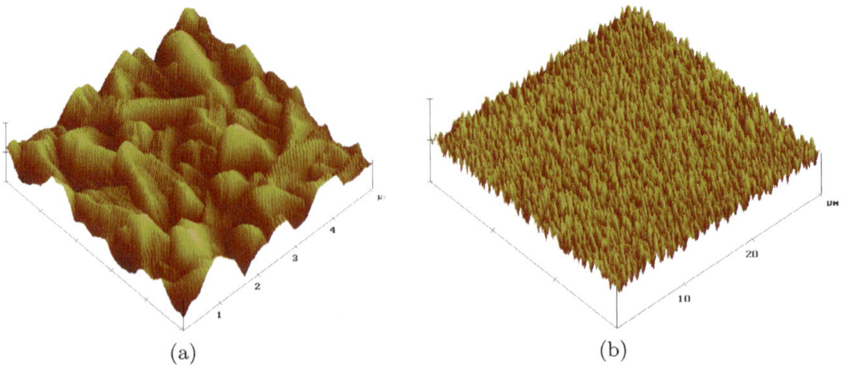

Fig. 4.7: (a) AFM surface image of a microcrystalline diamond (MCD) layer. (b) AFM surface image of an ultra-smooth nanocrystalline diamond (NCD) layer.

slow-growing grains and hence develop a columnar structure with a preferential orientation (110) normal to the substrate. The crystallites of MCD films are typically equal to the film thickness.

This evolution phenomenon of MCD films describing an anisotropic growth has an essential influence on the material properties. The most specific characteristic of MCD is the high surface roughness, which originates from the large and facetted crystallites, which are typically proportional to the film thickness. Roughness values ranging from hundreds of nanometers up to several microns are reported. As such, in Fig. 4.7, AFM images of rough polycrystalline diamond and ultra-smooth nanocrystalline diamond surfaces are shown.

4.2.2.3. *Optimization of process conditions*

Polycrystalline diamond films having grain sizes considerably less than 100 nm are often classified as nanocrystalline diamond films. A sub-category is ultra-nanocrystalline diamond films with a crystal size of 3–8 nm and atomically abrupt ($\sim$0.5 nm) high-energy grain boundaries.

The most attractive attribute of both categories is the independence of the grain size from the film thickness, i.e., the diamond layer does not grow in the "van der Drift mode" but by renucleation steps, resulting in homogeneous equiaxed nanometer-sized grains. Thus, the key for NCD and UNCD growth is to offer a very high and constant rate of secondary nucleation entirely through the deposition procedure, thus preventing the formation of large crystals. In order for this to happen, the processing conditions need to be optimized by adopting the relevant parameters, such as gas mixture, pressure, and substrate temperature.

The gas pressure is an important parameter for CVD diamond growth. The gas pressure sensitively changes the gas phase fluid dynamics and gas phase chemical reaction kinetics among various gas species in a CVD reactor. According to the kinetic theory of gases, the lower the pressure, the lower the mean free path, and more active species or radicals (atomic H, CH_3, C_2H_2 radical, etc.) can survive collisions and reach the substrate surface. In comparison, the higher the pressure, the more active species can be generated by the filament at a constant filament temperature, however, the mean free path is shorter, and consequently, the collision of hydrogen with each other as well as with other molecules are more frequent. Hence, the probability of hydrogen atom recombination and consumption is greater. Furthermore, an increase in gas pressure leads to an increase of convective heat dissipation of the filaments and the substrate.

A high collision rate of gas species with the filaments additionally cools down the filaments. This thermal issue, in turn, lowers the substrate temperature and finally leads to a decrease in growth rate. These contradicting effects of pressure on determining the concentration of active species at the substrates surface obviously need balancing.

Furthermore, selection of the correct gas composition is one of the key factors for the successful growth of high-quality diamond films. A typical gas mixture of 1% CH_4 in H_2 results in polycrystalline films with grain sizes in the micron or tens of micron range. Changing the hydrogen/methane ratio or even introducing additional gases has promising potential for both enhancing the diamond quality and suppressing the growth of large crystallites, however, at the cost of the overall growth rate.

The technologically most feasible method to control the grown diamond microstructure relies on the variation of CH_4/H_2 in the gas mixture. A low CH_4 concentration (<0.5%) leads to large and strongly facetted crystallites. The microstructure of these films is highly thickness-dependent, resulting in columnar grain growth and large surface roughness relating to the van der Drift growth mode [12]. Increasing the CH_4 concentration in the CH_4/H_2 mixture to a few percent reveals a change in surface morphology and considerable reduction in average grain size due to renucleation events (see Figs. 4.8(a) and 4.8(d)). However, the lower hydrogen content and the less effective sp^2-etching mechanism cause a high degree of carbon supersaturation on the surface and a rather poor film quality with a high degree of non-diamond carbon when CH_4 percentage is exceeding typical

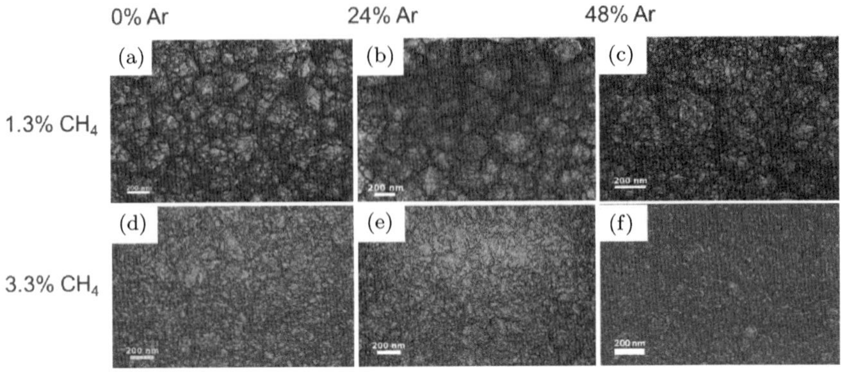

Fig. 4.8: High-resolution SEM images with varying average grain sizes dependent on the CH_4/H_2 ratio and the Ar/H_2 ratio.

values around 4%. Another option to vary the growth behavior is the indirect manipulation of the heat transport in the HFCVD by adding an inert gas with relatively low thermal conductivity, such as Argon. This leads to a reduced grain size, as can be seen in Figs. 4.8(a)–(f).

The addition of oxygen into the gas mixture has the potential to suppress the growth of non-diamond components and thus to increase the growth rate. Oxygen dissociates into O ions and can react with surrounding hydrogen and hydrocarbon into OH, H_2O, CO and CO_2. It is well known that the highly reactive OH radicals can etch non-diamond components more rapidly than atomic hydrogen by directly oxidizing carbon and removing them as CO or CO_2. Hence, the addition of oxygen into the CH_4/H_2 gas phase of an HFCVD system is a common approach for producing high-quality microcrystalline diamond films at high growth rates.

The established process with a gas mixture of methane (3%) in hydrogen (96%) and additional oxygen (1%) has been used as a base at a gas pressure set to 5 mbar. The process with the highest O_2 concentration clearly evolves large and facetted crystals with a pronounced growth texture. Despite the rather high methane content of ∼3% in the gas phase, no degradation effect of the film is observed. It can be assumed that the enhanced etching rate of the hydroxyl radicals OH assists the diamond growth evidently by compensating the non-diamond formation and promotion of secondary nucleation. With a reduction of the oxygen concentration, the microstructure evolves from strongly facetted to non-facetted small grains, as depicted in the TEM analysis in Figs. 4.9(a)–(d).

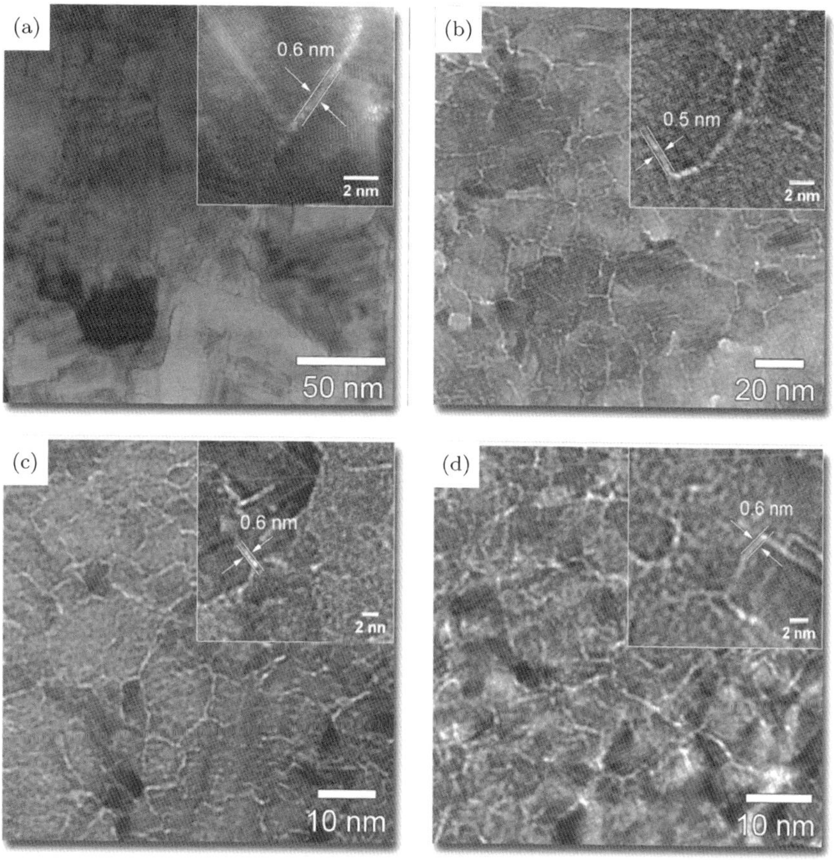

Fig. 4.9: TEM images of samples with variation of oxygen content in the gas phase mixture between 0.81% (a), 0.54% (b), 0.11% (c) and 0% (d) [7].

The average grain size of the films produced is generally determined using X-ray diffraction in a standard Bragg–Brentano configuration and applying Scherrer's formula or similar ones. The peak broadening of the diamond (111) diffraction peak is used to roughly estimate the average grain size, neglecting peak broadening from microstrains.

The XRD measurements are presented in Figs. 4.10 and 4.11. The UNCD sample shows no apparent texture (see Fig. 4.11(a)), and the average grain size was found to be $<D> = 6\,\text{nm}$ (see Fig. 4.10). The NCD sample exhibits a weak (110) fiber texture, confirmed by measurement of pole figures for the (111) diffraction peak (see Fig. 4.11(b)).

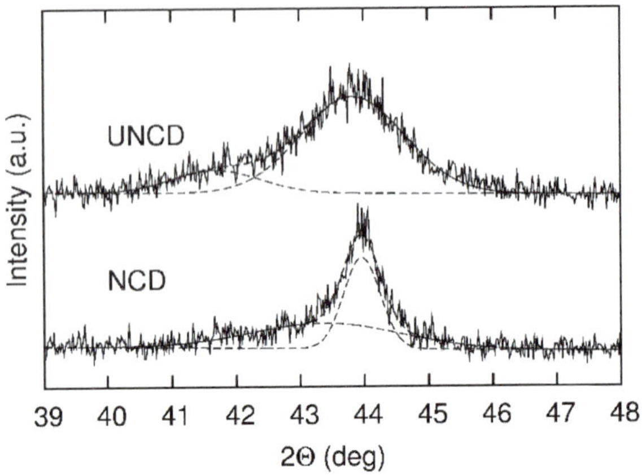

Fig. 4.10: X-ray diffraction peaks (111) for sample UNCD and NCD. An additional small halo in the background (dashed line), which is attributed to stacking faults, can be seen. Reprinted from [17], with the permission of AIP Publishing.

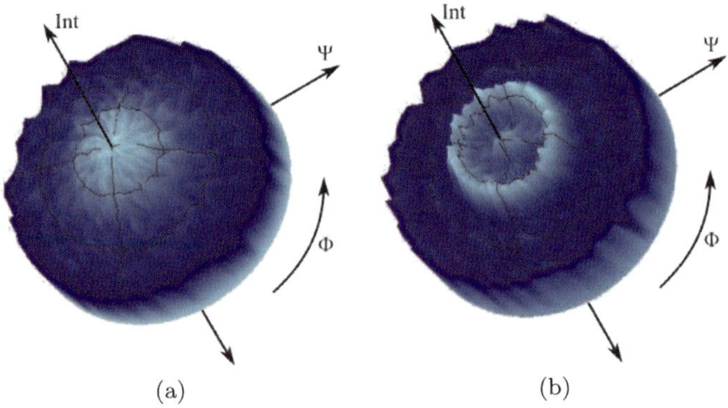

Fig. 4.11: (a) Pole figure of the (111) peak of sample UNCD, showing no apparent texture. (b) Pole figure of (111) peak of sample NCD, showing maximum intensities at tilt angle $\psi = 35°$, indicating a weak monoaxial (110) fiber texture. Reprinted from [17], with the permission of AIP Publishing.

In this case, the average grain size $<D>$ was found to be about 15 nm (see Fig. 4.10).

4.3. Physical and Mechanical Properties

Due to the large crystallites in MCD, the mechanical, electric, thermal and optical properties are rather similar to a single-crystalline diamond.

However, the implications of the anisotropic microstructure, i.e., the increasing grain size with film thickness and the texture evolution, is noticeable. Such a texture can lead to significantly different elastic coefficients depending on the nature of the preferred orientation. Due to the polycrystalline nature, the hardness in MCD can slightly vary between 80 and 100 GPa, which can be attributed to the orientation of the individual crystals and grain boundary-related weak bonds [13]. Since the overall bonding environment is dominated by sp^3-bonded carbon, the hardness as well as Young's modulus are rather similar to single crystals but strongly depend on the growth conditions and related levels of defects. With its extreme hardness and stiffness, diamond is a rather brittle material; however, it still surpasses the most common materials (Si, SiC, SiO_2) used for micromechanical applications. Especially unfavorable for MCD films is the columnar growth of the grains leading to a predefined crack path through the film along the grain boundaries. The fracture strength of MCD diamond is reported to be between 0.5 and 1.8 GPa [14].

In the nanocrystalline state, Young's modulus of metallic and nonmetallic materials is found to decrease in general with decreasing grain size due to the different atomic coordination and excess volume within the grain boundaries [3]. However, the reported range of values for the elastic moduli of diamond with grain sizes in the range of 4–100 nanometers is quite substantial [1]. This large spread is often attributed to the high variability of morphology and structure of the investigated nanocrystalline and ultra-nanocrystalline diamond films. Computer simulations have shown that three- and four-coordinated atoms coexist in the grain boundaries of diamond [15]. In contrast, most of the atoms in silicon grain boundaries are four-coordinated.

The mechanical properties of UNCD and NCD are generally different from the conventional MCD and single-crystalline materials due to the large number of grain boundaries present in the materials volume. Typical values for the fracture strength, Young's modulus, shear modulus, Poisson's ratio are summarized in Table 4.2, while the properties of natural diamond are considerably different and the values are 1143 GPa for Young's modulus (theoretical value), 534 GPa for the shear modulus and 0.07 for Poisson's ratio. From the large variation of these properties as a function of grain size and the excellent thermal stability of nanocrystalline diamond up to more than 1200°C, it becomes evident that this is the ideal material that can be tailored in an extremely wide range of physical properties and up to very high temperatures.

Table 4.2: Mechanical properties of the investigated UNCD and NCD films, the values for Young's, shear modulus, and Poisson's ratio are weighted averages with their standard deviation.

Property	**UNCD**	**NCD**
Fracture strength (GPa)	1.8 +/− 0.6	5.0 +/− 0.2
Young's modulus (GPa)	463 +/− 11	752 +/− 22
Shear modulus (GPa)	195 +/− 3	367 +/− 8
Poisson's ratio	0.201 +/− 0.041	0.034 +/− 0.017
Density (g/cm^3)	2.92 +/− 0.02	3.27/−0.02

Furthermore, the electrical (and thermal) conductivity in highly doped nanocrystalline material is highly sensitive to grain boundaries and thus to the average grain size due to pronounced electron scattering. The grain boundaries generally have a considerable excess volume with atomic coordination different from the crystalline parts [16], resulting in a decrease of the transport properties by several orders of magnitude.

4.3.1. *Fracture strength*

A nanoindenter was used as an instrument to bend microbeams with a length of 500 μm, a width of 200 μm, and a thickness of 17 μm. The fabrication of the microbeams is described in more detail in the following chapter. A special stainless steel tip was employed to apply a line load at the end of the microbeams (see Fig. 4.12). The force F and deflection d were recorded simultaneously until fracture occurred.

The stress at fracture σ_f was then calculated according to

$$\sigma_f = \frac{t}{2} \frac{L \cdot d_{\max}}{I} \frac{\partial F}{\partial d}$$

where t is the thickness of the beam, $d_{\max}$ the deflection of the microbeam at fracture, and L the beam length. The probability distribution of stress at fracture was then fitted with a Weibull distribution function, assuming the weakest link model for failure [17].

Figure 4.13 exhibits the Weibull plots of the UNCD and NCD diamond films, respectively. For the UNCD samples, a value for the fracture stress σ_0 of 1.8 ± 0.6 GPa and a Weibull modulus of 5.2 ± 0.2 while for the NCD samples σ_0 of 5.0 ± 0.2 GPa and a Weibull modulus of 6.1 ± 0.4 have been determined, as given in Table 4.2.

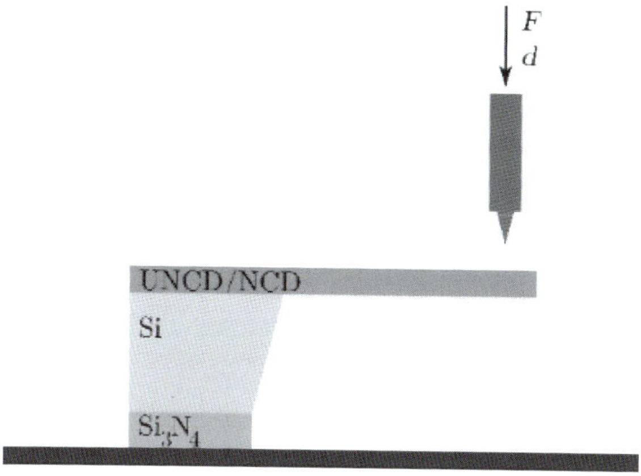

Fig. 4.12: Schematic diagram of the experimental setup for the fracture experiment. Reprinted from [17], with the permission of AIP Publishing.

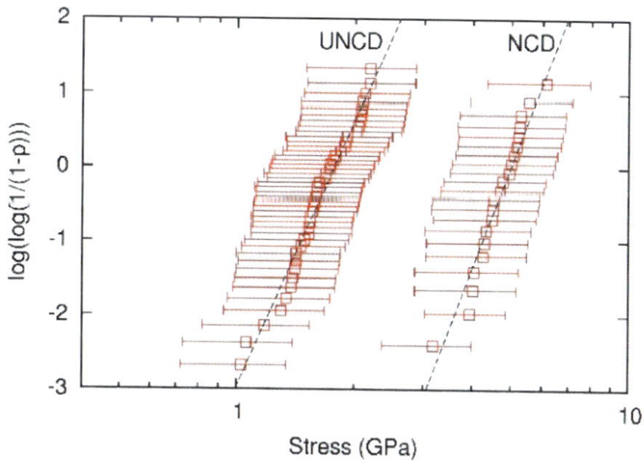

Fig. 4.13: Weibull plots for sample UNCD and sample NCD. Reprinted from [17], with the permission of AIP Publishing.

4.3.2. *Young's modulus*

Bending several microbeams (1500 μm length) at 10 different distances l (1320–1490 μm) from the beam fixture and recording the force F and displacement d, Young's modulus E of the microbeams can be determined. From the measured bending stiffness k of the microbeams given as

$$k = \frac{\partial F}{\partial d} = \frac{3EI}{l^3}$$

the Young's modulus of the microbeams is calculated and given in Table 4.2. Furthermore, for an entire range of microbeams with grain sizes varying between 6 and 15 nm, such measurements have been performed [19], as shown in Fig. 4.14.

The results of samples numbered S1 to S8, as shown in Fig. 4.14, indicate the decrease of Young's modulus E from 1143 GPa of single-crystalline diamond to about 450 GPa due to grain boundary-induced softening. While this reduction of about 60% is considerable, the resulting material properties are still exceptional and correspond to double or triple the value of a typical steel.

4.3.3. *Poisson's ratio*

The Poisson's ratio ν is a fundamental measure to compare material properties, and it generally reflects the resistance of a material against volume change per shape change [20]. The ratio of shear modulus G to bulk modulus B describes the shear stability of a material:

$$\frac{G}{B} = \frac{3}{2}\frac{(1 - 2v)}{1 + v} \tag{4.1}$$

The G/B ratio for single-crystalline diamond is 1.201 [18] and therefore larger than that for other ceramics, glasses, and semiconductors. The Poisson's ratio for single-crystalline diamond is dependent on the crystal direction and varies accordingly between 0.00786 and 0.115 [18] having an average value (assuming randomly oriented grains and neglecting grain boundary effects) of 0.0691. For polycrystalline diamond with (110) fiber

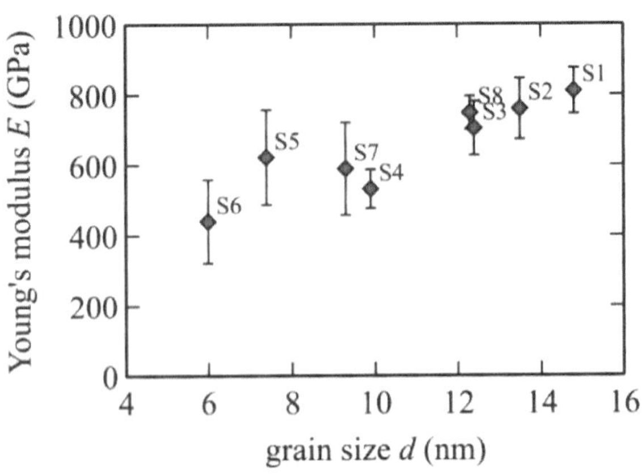

Fig. 4.14: Young's modulus E as a function of average grain size [19]. Reprinted from [19] with permission from Elsevier.

texture for Poisson's ratio, again neglecting effects from grain boundaries, averaged values of 0.0730 for longitudinal orthogonal elongation and 0.0592 for transversal orthogonal elongation have been reported.

To determine Poisson's ratio for microbeams, bending and torsion oscillations of nanocrystalline and ultra-nanocrystalline diamond microbeams ($w = 200\,\mu$m, $L = 1500\,\mu$m) were excited using a transversal acoustical transducer and were recorded by monitoring the dynamic lateral and vertical beam deflection by means of the optical deflection system of a dedicated AFM. The resonance frequencies of the nth bending and torsion modes were extracted from amplitude spectra by fitting with a Lorentzian function. The elastic moduli E and G were calculated from the bending and torsion resonance frequencies of the nth modes, and thus, Poisson's ratio was obtained, as given in Table 4.2.

4.3.4. *Wear behavior*

Diamond is well known for having excellent tribological properties owing to its high hardness and stiffness. However, many factors can adversely affect friction and wear behavior. The high surface roughness in particular downgrades the friction coefficient and the wear performance. Figure 4.15 shows a collection of friction values as a function of roughness ranging from 0.03 up to 0.5 (for very rough microcrystalline films) [21].

Tribology tests on the NCD samples were performed using a ball-on-disk tribometer from CSM Instruments (see Fig. 4.16). In a dry environment, Si_3N_4 balls coated with NCD and having a radius of 1 mm were loaded against the NCD films, with the normal load fixed at 1 N. Considering a

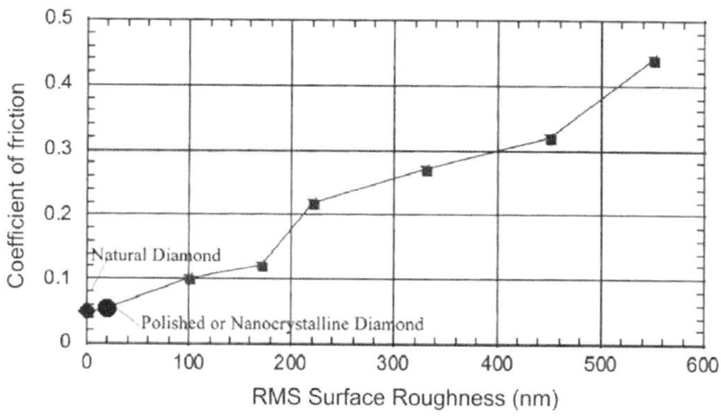

Fig. 4.15: Relationship between coefficient of friction of diamond films and their surface roughness [21].

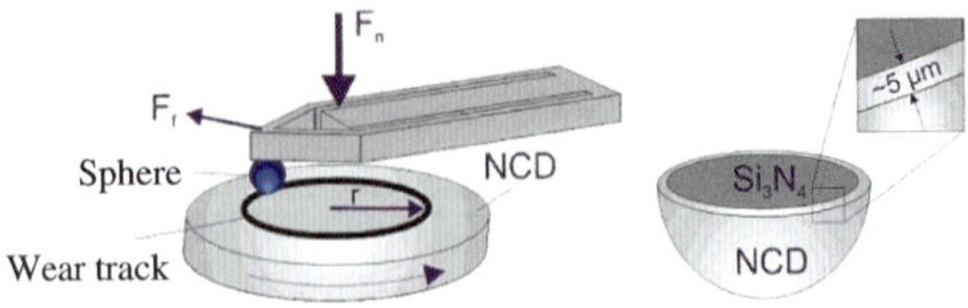

Fig. 4.16: Schematic diagram of the ball-on-disk tribometer setup together with the NCD coated Si_3N_4 sphere as wear counterpart.

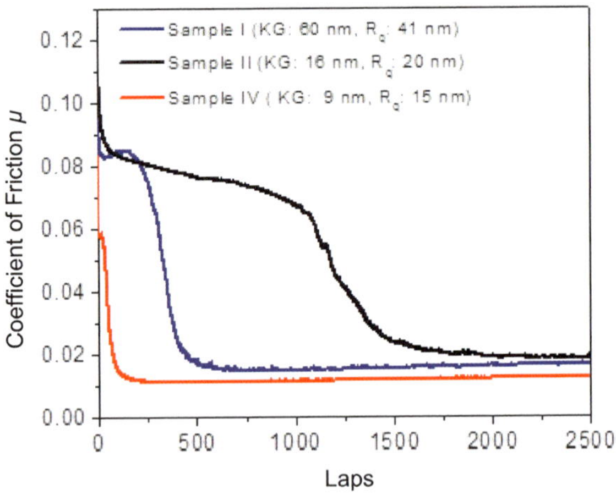

Fig. 4.17: Measured coefficients of friction using the device shown in Fig. 4.16 for three different sample systems I, II, and IV, with varying grain size (KG) between 60 and 9 nm, and thus varying surface roughness R_O.

Hertzian elastic contact for the Si_3N_4 balls coated with NCD, the sphere ($E = 300\,\text{GPa}$, $\nu = 0.25$) pressed against a flat silicon wafer ($E = 130\,\text{GPa}$, $\nu = 0.28$), we calculated that the applied loads used in this study produced initial peak pressures of ∼1 GPa (mean contact pressure 0.64 GPa). The tests were performed in ambient atmosphere (room humidity of 30–60%) at room temperature, a constant speed of 0.5 cm/s, a sliding radius of 1.5 mm and with different sliding distances.

Representative results of the sliding experiments of film systems I, II, and IV are displayed in Fig. 4.17. All three samples show a typical friction evolution of polycrystalline diamond films. After initial high values of the coefficient of friction μ, also defined as "running-in", the curves are decreasing quickly from the starting value between 0.06 and 0.09, reaching a steady-state friction coefficient between 0.02 and 0.01. Such extremely low friction coefficients between nanocrystalline diamond contacts in a dry and ambient atmosphere are among the lowest known and close to perfectly cleaved single crystals.

Considering the early friction stage, one can observe a significant difference between the three film systems. Figure 4.17 clearly displays that the coarse-grained sample I, having the highest roughness, needs a considerably longer phase (~2000 laps) to reach the low steady-state friction value, whereas the smoother samples II and IV have a distinctly shorter run-in period of 650 and 220 laps, respectively. From the SEM surface analysis, it becomes obvious that the initial high friction is largely a result of plowing of the surface by asperities. Plastic asperity deformation takes place on both the disk and the ball until the surface is super-polished. Consequently, the coefficient of friction in the initial stage of the diamond films is largely dependent on the surface morphology. The film system I, with an average grain size of 60 nm having large asperities with high surface roughness, needs a significantly longer time in order to wear down the protruding grains, consequently preventing asperity locking. The smoother the initial surface, the faster the asperity plowing event is completed, and low steady-state friction is possible, such as for film system IV.

The described mechanism also has strong implications on the wear behavior of the nanodiamond coating. While during the pin-on-disk testing, a super-polished diamond surface is achieved, wear events almost completely disappeared, relating mostly to environmental effects. This observation led to the development of new cutting tools. As such, NCD layers have been deposited on hard metal surfaces, i.e., WC with Co binder, a standard cutting tool material. In studying the cutting performance of gels and foils containing super-abrasive TiO_2 nanoparticles for whitening design purposes, a superior performance has been found.

While the tool life of conventional cutting tools made of hard metal and sintered ceramics is considered as standard in the cutting industry, the tool life of the NCD-coated and plasma-sharpened WC tool (Diamaze PSD) excelled the standard materials by a factor of 25 and more without failure.

4.4. Microfabrication Process Technology

Microfabrication technology is the process technology for the fabrication of structures on the micrometer scales and smaller [2]. Historically, the earliest microfabrication processes were used for integrated electronic circuit fabrication, also known as "semiconductor manufacturing" or

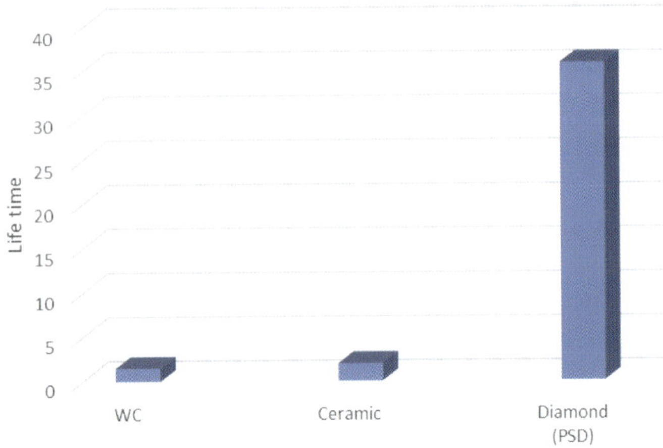

Fig. 4.18: Tooling lifetime of difference cutting materials including hard metal WC-based, ceramics (Si_3N_4), and plasma sharpened diamond (Diamaze PSD) in hours cutting TiO_2-containing foils (courtesy GFD).

"semiconductor device fabrication". More recently, microelectromechanical systems (MEMS) and microsystems technology and their extension into nanoscale have reused, adapted or extended microfabrication methods. The major concepts and principles that are used in microfabrication technologies are lithography, doping, thin films, etching, bonding, and polishing. The microfabrication technologies have their origin in the microelectronics industry, where the devices are usually made on silicon wafers as well as on substrates of other semiconductor materials.

4.4.1. *General aspects*

To fabricate a microdevice, many process steps must be performed, including the deposition of a film, patterning of the film with the desired microstructures, and the removal (etching) of film portions. Process control along the individual process steps involves thin-film metrology methods to ensure that the film structure has the desired characteristics in terms of thickness, surface topography, optical properties, and other coefficients for suitable device behavior.

Often, the film has to be patterned in order to form openings in some of the layers. These features are usually on the micrometer or nanometer scale. To this end, a 'mask' is used to define portions of the film which will be removed. Examples of patterning techniques include photolithography and shadow masking followed by etching.

Etching is the removal of some part of the thin film or substrate. To this purpose, the thin film or substrate is exposed to an etching medium (acids or plasmas), which chemically and/or physically removes the parts of film until the desired geometry is achieved. Physical etching techniques are based on dry etching (Plasma etching), such as reactive-ion etching (RIE) or deep reactive-ion etching (DRIE) using appropriate media.

4.4.2. *Diamond microtechnology and RIE micromachining*

As described briefly before, and as is well established for silicon microtechnology, basically three different steps are being used. In the first step, a thin diamond film is grown on a single crystalline and well-defined silicon substrate with diameters up to six inches. This process is performed using a hot filament chemical vapor deposition (HFCVD) system (CEMECON Lisa 800). With appropriate process parameters established as described above, nanocrystalline diamond layers are grown in a very reproducible and reliable manner with appropriate thickness.

In the second step, the shape and geometry of the objects to be produced are designed and defined in a dedicated mask, for example, for a 1D microbeam, a 2D cogwheel, or a 3D microscalpel, or any other complex shape and intricate geometry.

This mask is then transferred onto the diamond film using standard photolithography. This technique allows the fabrication of complex shapes at a high lateral resolution of better than $0.5\,\mu$m. A photoresist is used to transfer the shape of the object in the previously deposited mask by reactive ion etching (RIE). The shape is then transferred into the diamond film using an anisotropic etching process. The etching process stops at the diamond silicon interface. The result is a relief-structured wafer, where the residual diamond film defines the shape of the object.

In the third and last step, the diamond wafer is plasma polished. The process is very similar to plasma etching. Here, the diamond film is removed atomic layer by atomic layer, resulting in the desired geometry of the part to be manufactured. The process is self-adjusting and therefore insensitive against overetch. Thus, the polishing (or etching) procedure represents the physical limit, utilizing diamond as the hardest material and an exact achievement of complex geometries with an accuracy of a few nanometers. Finally, any remaining parts of the silicon substrate can be etched away

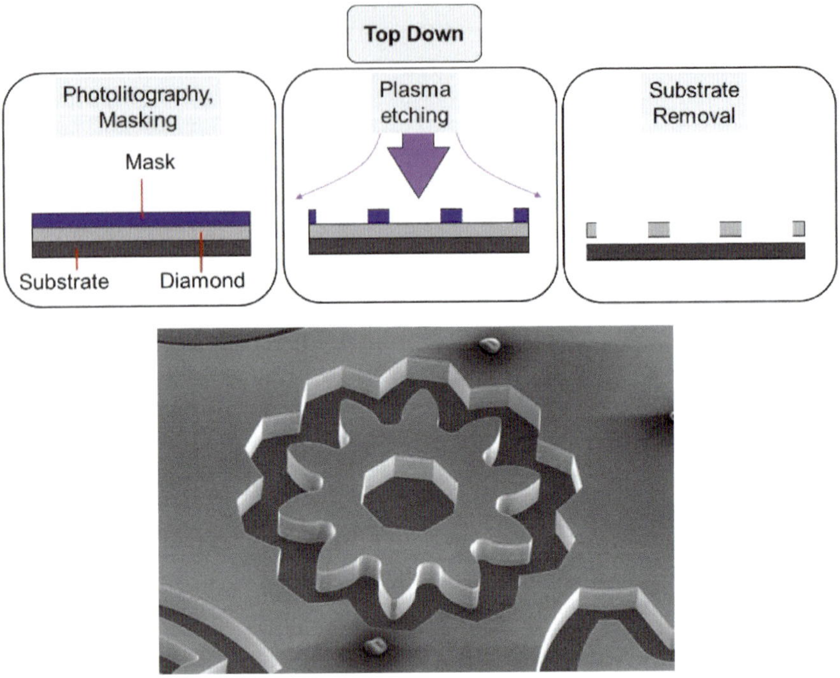

Fig. 4.19: Schematic diagram of the nanodiamond-based top-down process and a resulting free-standing full-diamond microwheel together with etched-off areas in black (courtesy GFD).

chemically, for example, in a KOH solution. Thus, this so-called *top-down process* allows producing diamond microparts and components with special shapes and geometries, as shown in Fig. 4.19.

Alternatively, the full range of established silicon-based microtechnology can be used to produce appropriate and intentionally designed precursor silicon microparts, which can further be coated by a thin protecting diamond layer. When all the conditions of seeding, nucleation, and growth under the appropriate conditions in the CVD reactor are fulfilled, nanocrystalline diamond layers can be grown on 3D structured silicon objects guaranteeing all the advantageous properties of diamond, such as high hardness, ultra-smooth and thus lubrication-free surfaces.

This so-called *bottom-up process* is schematically shown in Fig. 4.19 as well as a resulting microcogwheel made of silicon while a homogeneous surface layer typically 5–10 μm in thickness is fully sufficient to provide all the advantages of diamond coatings. The excellent mechanical properties

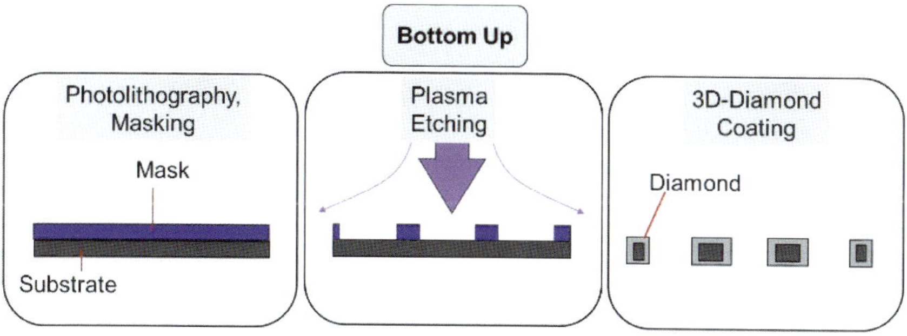

Fig. 4.20: Bottom-up process in coating complex silicon microparts with a relatively thin protecting and mechanically stable nanocrystalline diamond layer with an ultra-smooth surface (courtesy GFD).

in combination with the low coefficient of friction and the low thermal expansion coefficient of diamond may allow running microdrives without lubricants at high efficiency.

The SEM analysis of a seeded 3D microgear showed similar results as on planar Si-Wafers. Figure 4.20 displays as an insert a cross-section of a single cog of the gear after seeding and diamond growth. Detailed SEM analysis shows a high and homogenous nucleation density ($\sim 10^{11}\,\text{cm}^{-2}$) at all relevant positions. For commercial purposes in comparison with laboratory achievements, upscaling procedures are an absolute necessity for high and fast throughput. In the case of the diamond microfabrication, both top-down and bottom-up processes can be scaled up because of the advantageous CVD technology. Based on the current design, six silicon wafers with six-inch diameters can be used as substrates for the diamond film deposition inside the processing chamber. This fully optimized situation is

Fig. 4.21: Upscaling the fabrication of microcogwheels for micromechanical applications.

unique and allows the fabrication of a multitude of microcogwheels on one wafer at once, as demonstrated in Fig. 4.21.

This scale-up procedure might pinpoint the direction for future developments. Thus, a lubricant-free operation of diamond microgears at high revolution speeds, wear and lubrication-free, low moment of inertia, high efficiency, and high reliability can be realized for a manifold of applications.

4.5. Microengineering Applications

4.5.1. *Nanosharp cutting tools and scalpels*

Following a similar direction, ultra-precise diamond tools can be designed and developed for modern surgical methods. In order to decrease the recovery times of patients, there is an obvious trend towards ambulatory surgical techniques. However, these treatments require less traumatic wounds. Especially in the field of microsurgery, like eye- and neurosurgery, the demand for ultra-sharp cutting tools is therefore continuously increasing [22].

The first major step towards this trend was the introduction of diamond keratomes for eye surgery made from natural gemstones approximately 35 years ago. Since diamond is the hardest of all known materials and has outstanding mechanical properties, it was possible to fabricate microcutting tools that have an outstanding and long-lasting sharpness. Although these types of knifes have the highest functionality, only a small percentage of all cuts are performed with diamond scalpels mainly due to the high price.

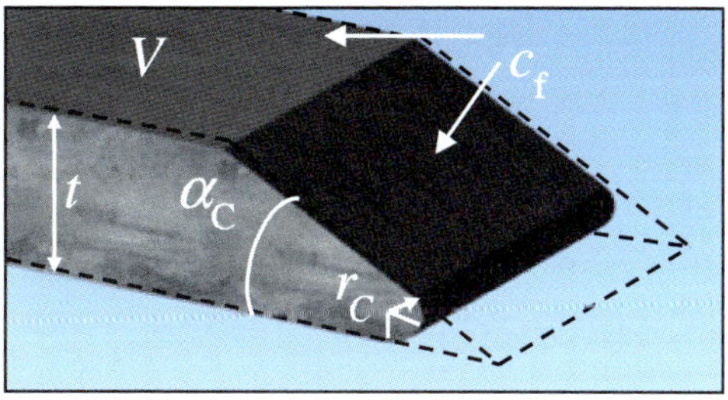

Fig. 4.22: Schematic drawing of a cutting tool with the cutting edge being the most critical part indicated by its radius of curvature.

In order to provide ultra-sharp scalpels for an affordable price, a novel diamond scalpel has been developed, utilizing a thin CVD diamond film on a silicon backing. Furthermore, the shape and sharpness of the scalpel is no longer realized by manual polishing. A novel plasma process as described above has been developed, which results in radii of curvatures of the diamond blade tip of less than 5 nm.

The main cutting parameters for any blade are V the volume, t the thickness, α_c the apex angle, c_f the coefficient of friction and r_c the radius of curvature of the blade, as shown in Fig. 4.22. The sharpness of the blade is mainly given by the radius of curvature r_c at the tip of the blade. Since stainless steel is a relatively soft material in contrast to diamond, even a very small radius of curvature will increase very fast during the cutting of even soft materials, like, for instance, a human cornea. As a consequence, most metal blades are dulling very fast and are therefore used only once (the so-called single-use or disposable products). Furthermore, the residual surface of a corneal cut and therefore the wound sealing and the wound healing are strongly influenced by the radius of curvature. The radius of curvature of a mechanically polished blade of medical grade stainless steel is typically in the range of 3 μm and is expected to increase significantly during the first cut into a human cornea.

The larger the radius of curvature r_c of a blade, the more cutting defects in the cornea are likely to occur and the higher the residual corneal surface roughness is expected. Therefore, a key to further cutting improvement lies in the decrease of the radius of curvature to a minimum value, which is at the very least in the range of the radius of one atom. In order to avoid dulling of the cutting edge, diamond as the hardest known material should be chosen.

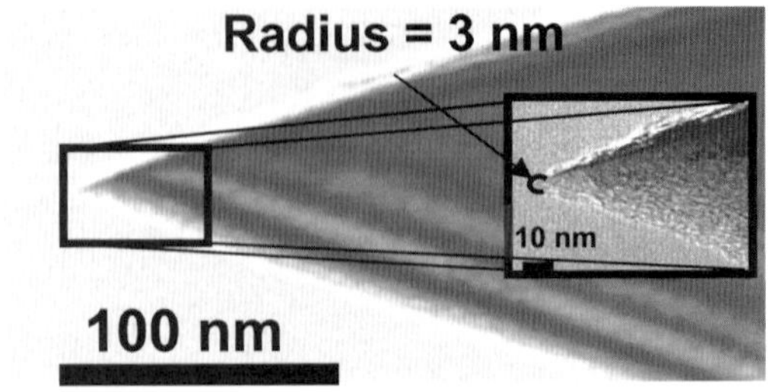

Fig. 4.23: Diamond cutting edge for eye surgery with a radius of curvature of only 3 nm (courtesy GFD).

As such, a new design for scalpel blades for particular purposes in eye- or neurosurgery or others can be designed and fabricated.

In order to overcome state of the art in terms of mechanical grinding procedures, the plasma-based diamond polishing process has been applied for the fabrication of diamond cutting edges, resulting in a radius of curvature of less than a few nanometers. Due to the high mechanical strength and very low surface roughness of NCD, full advantage of the nanometer cutting edge can be taken advantage of, such as very low cutting forces, little damage to tissue, and long-lasting sharpness. The TEM image in Fig. 4.23 exhibits a nanodiamond cutting blade with a radius of curvature of 3 nm — the sharpest cutting edge and best performance ever achieved in any material.

4.5.2. *Micromechanical sensors*

Industrial applications require more and more temperature stable and durable pressure sensors. Silicon is so far the material of choice in present piezoresistive sensors. Nonetheless, in harsh environments, silicon needs to be protected from the media, and other materials need to be used.

As such, diamond offers special properties that make it interesting for micromechanical devices like pressure sensors in harsh environments, especially the chemical inertness, bandgap engineering, and high thermal conductivity are beneficial for such applications. An important advantage is the possibility to integrate (ultra-)nanocrystralline diamond films into the common silicon technology. The ability to grow electrically conductive ultra-nanocrystalline diamond films that exhibit a piezoresistive effect allow one to fabricate hybrid pressure sensors composed of diamond and silicon. Here, the successful fabrication and characterization of a membrane pressure sensor made of nanocrystalline diamond grown on a silicon substrate is described.

The sensor consists basically of a nanocrystalline diamond membrane with ultra-nanocrystalline piezoresistive diamond sensor elements. The piezoresistive ultra-nanocrystalline diamond (UNCD) films exhibit electrical conductivity at room temperature [23, 24]. The occurrence of this conductivity is often explained by the ordering of sp^2 bonds in the grain boundaries, influenced by nitrogen addition to the gas phase and enhanced substrate temperature during growth. A major advantage of UNCD films is the absence of permanent degradation of its properties. Even after annealing at temperatures up to 900°C, the electrical conductivity is unaltered.

The growth of (ultra-)nanocrystalline diamond films by hot filament chemical vapor deposition (HFCVD) is advantageous, considering the wide variety of possible substrate materials as well as the high scalability of the method. Making use of both, the highly developed silicon technology as well as the emerging nanocrystalline diamond technology, a hybrid diamond-silicon technology, can be established.

For initial testing purposes of the piezoresistive effect of nanodiamond, firstly microcantilever beams have been designed and fabricated, as exhibited in Fig. 4.24.

The free-standing cantilevers of nanocrystalline diamond were then deflected using a micromanipulator. The change in electrical resistivity was measured to give the piezoresistive coefficient (gauge factor)

$$k = \frac{\Delta R/R}{\varepsilon}$$

while the mechanical (tensile) strain is determined as

$$\varepsilon = \frac{3}{2}s\left(l - \frac{x}{2}\right)\frac{h}{l^3}$$

For the measurement of the gauge factor, the free end of the cantilever is deflected downwards in 10 μm steps using a needle with a sharp tip. Precise positioning of the needle is possible with the help of a tiny hole at the cantilever end. Before fracture, the needle is pulled backwards in the same intervals. Four piezoresistors are arranged in a Wheatstone bridge configuration; one of them is exposed to the applied strain. The change in resistance is measured simultaneously. The results are shown in Fig. 4.25 where the slope gives a gauge factor k equals 7.

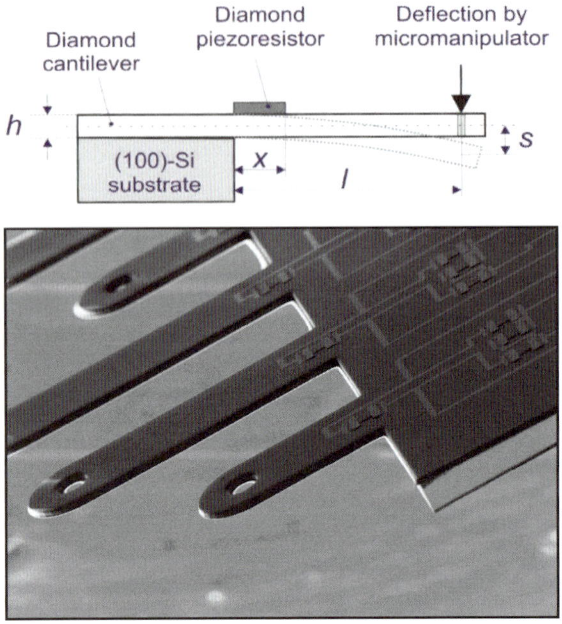

Fig. 4.24: Schematic drawing of a microcantilever and a micrograph of microcantilevers of nanocrystalline diamond with a thickness of typically 10 μm and a length between 500 and 1000 μm. Reprinted from [23] with permission from Elsevier.

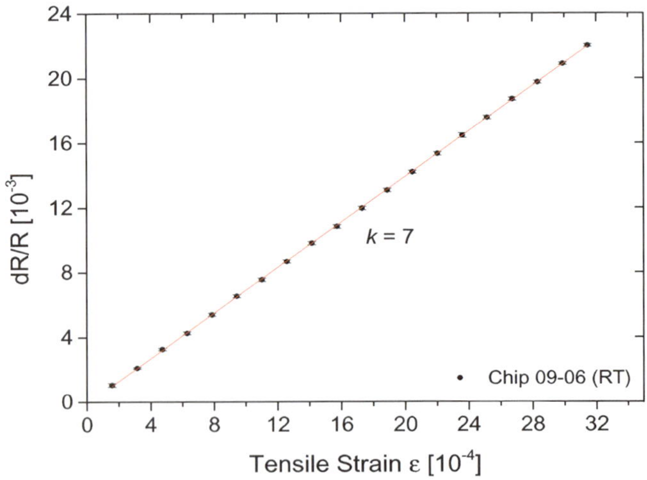

Fig. 4.25: Measurement of the change of resistance of a UNCD microcantilever as a function of tensile strain indicating a gauge factor k of 7.

For the fabrication of a pressure sensor, silicon wafers with a diameter of six inches, a (100) surface and a thin Si_3N_4 layer on the backside were used in the first step to grow a nominally intrinsic nanocrystalline diamond layer of around 15 μm thickness using a gas mixture of $CH_4/H_2/N_2$ at filament temperatures of about 1900°C. In the second step, a thin layer of around 1 μm of conductive ultra-nanocrystalline diamond was deposited on top using a gas mixture of $CH_4/H_2/NH_3$. The filament temperature was around 2000°C. The schematic layer stack is shown in Fig. 4.26(a). In the next step, the backside was masked with photoresist and using standard contact lithography and etching in a CF_4/O_2 plasma, squares were etched into the Si_3N_4 layer (Fig. 4.26(b)). Subsequently, the front side of the wafers were covered with an aluminum mask, which was structured by photolithography to mask the structures for the sensitive elements. Next, the UNCD layer on the front side was etched until the NCD layer in a reactive ion etching process. The aluminum etching mask was removed afterwards by wet chemical etching (Fig. 4.26(c)). In the next step, the wafers were immersed in a KOH (Potassium Hydroxide) solution in order to be etched into the silicon. The KOH etching leads to the formation of the diamond membrane, Fig. 4.26(d).

In the last step, a structured metallization was sputtered (Ta/Pt/Au) to connect the resistive elements with contact pads, Fig. 4.26(e).

The sensor is designed as a Wheatstone bridge circuit, having one element on the edge of the deformable membrane and the others on the undeformed substrate. The applied source voltage at the Wheatstone bridge V_s was 30 V and the voltage signal V_p was measured using a millivolt meter. See Fig. 4.27 for illustration.

A photograph of a completed sensor is shown in Fig. 4.27. As can be seen, in all designs, the strain acting on the resistive element is parallel to the electric field, therefore only the longitudinal piezoresistive effect is investigated.

In order to characterize the sensor response, a pressure difference was applied to the membrane pressure sensor in a controlled manner. That way, the sensitivity, $S = d(V_p/V_s)/d\Delta p$, was determined for three pressure sensors with slightly different electrical resistivity and k-factor of the sensitive elements. Figure 4.28 shows the output signal (V_p/V_s) as a function of the applied pressure difference Δp. This first analysis confirms a good sensitivity of the sensor in the range between $4 \cdot 10^{-6}\,\text{mbar}^{-1}$ and $6 \cdot 10^{-6}\,\text{mbar}^{-1}$.

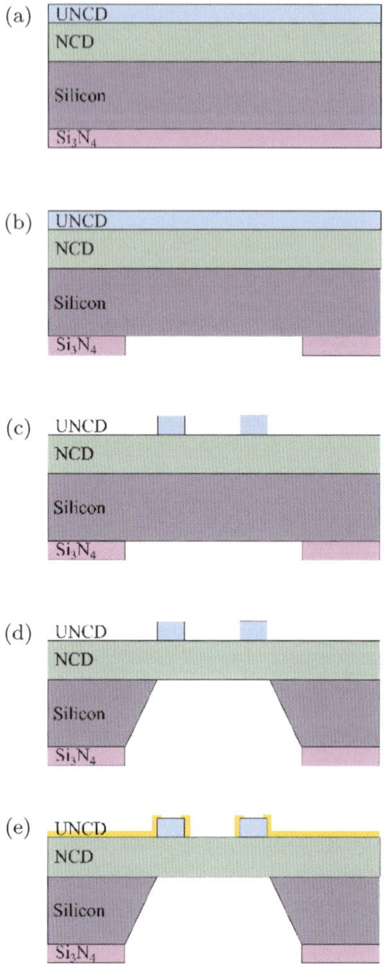

Fig. 4.26: Schematic drawing of the process flow, which was used to produce a piezoresistive, diamond-based pressure sensor.

4.5.3. *Bioapplications and future developments*

Materials which are similar in scale to biologic molecules and systems are allowing for the realization of architectures on biorelevant scales, including the nanoscale, microscale and macroscale [25]. This opens the door to exciting new applications in biomedical engineering and nanomedicine. Specifically, chemically inert substrates with surface feature sizes matching

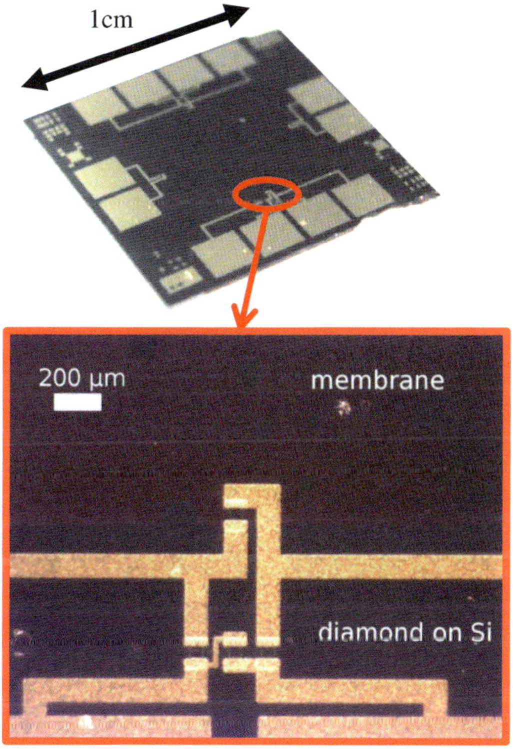

Fig. 4.27: Photograph of the produced membrane pressure sensor (top). One resistor is located on the edge of the membrane, while three resistors of the Wheatstone-bridge circuit reside on the substrate (bottom).

cell receptors, such as microcavities adjusted to the dimension of otocysts or embryoid bodies are in the center of attention.

In this regard, methods like microthermoforming and microreplication have recently been developed as a basic platform for investigations in life science based on three-dimensional cell cultivation [26]. Microreplication techniques, such as hot embossing techniques, microinjection molding or nanoimprinting can be used in order to produce appropriate micropatterns using thermoplastic materials. While these methods have produced important results [27] during long time experiments, polymer materials tend to react chemically with body fluids or their substitutes that are chemically very aggressive [28].

For these reasons and with regard to the stability of the interface between polymer (polystyrene) and chemically aggressive biological fluids,

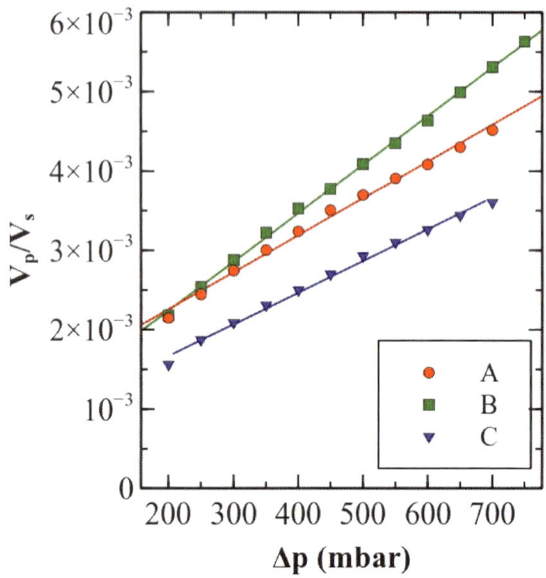

Fig. 4.28: Output signal as a function of pressure difference for three different pressure sensors.

an alternative method has been developed to etch cavities or other geometries directly in thick diamond layers or, alternatively, into silicon which can be coated with nanocrystalline diamond, as described above.

This way, a system of array structures was designed and fabricated to grow spherical cell aggregates called embryoid bodies, thus acting as a 3D patterned chemically inert Petri dish. This now allows one to implement the complete set of determinants of biocompatibility into one biomimetic model inspired by the characteristic periodic niche topography on the surface of strawberries. The structure that resembles the surface of a strawberry (the strawberry seeds are protected in cavities on the surface) provides a setting to reproducibly culture spherical cell aggregates in suspension. The spheroids are standardly used to study embryogenesis and tumorigenesis *in vitro* [29]. Routinely, embryoid bodies are produced by the hanging drop method and cultured in polystyrene Petri dishes. A prototype of the 3D Petri dish is displayed in Fig. 4.29 (left). Nanoscopic interfacial water layers, masking the moderately hydrogenated diamond surfaces, prevent the attachment of the predominantly adherent cells to the substrate during events of first contact, thereby accentuating their binding to each other — a condition for the formation of embryoid bodies. A precondition for their

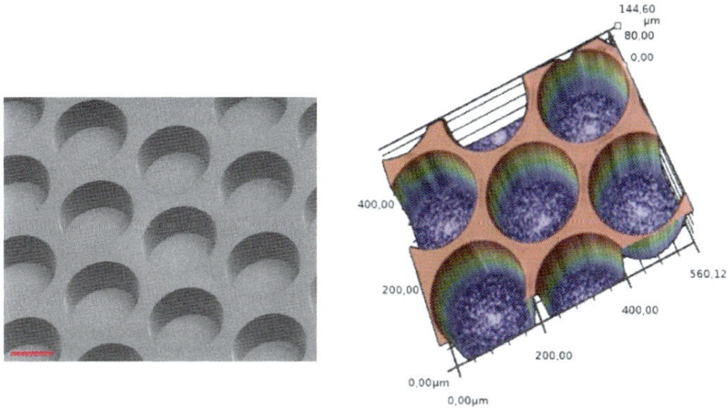

Fig. 4.29: SEM (left) and digital microscope images (right) with nanocrystalline diamond deposited at the bottom of the cavities with 200 μm diameter.

formation is a certain critical cell concentration below which cells tend to attach to the surface of the microcavities where they form monolayers (Fig. 4.29, right). Embryoid bodies start to form at higher cell concentrations, i.e. where competitive bindings favor cell–attachment relative to cell–substrate attachment [30].

The confocal laser scanning microscopy (CLSM) image shows a monolayer of P19 cells at the bottom of the diamond-coated cavities. The diameter of cavities is ~200 μm (see Fig. 4.30).

More recently, the effect of the diamond-coated Petri dishes on the formation of spherical aggregates of pluripotent stem cells (embryoid bodies) has been investigated. Notably, the diamond surfaces promoted embryoid body formation, whereas embryoid bodies do not form on polystyrene.

One major advantage of the microcavity method (3D Petri dish), when compared with the conventional "hanging drop method", is that the spheroids are immobilized in the cavities. In this way, it is possible to change the cell culture medium (adjusting the correct pH and safeguarding constant nutrient supply which is essential for growth) without disturbing the embryoid bodies or washing them away before maturity, i.e. desired harvest size. As such, embryoid body formation on nanocrystalline diamond-coated microcavities is now a simple and straightforward process; it protects from potentially oxidative stress caused by centrifugation or by the substrate itself and as such, microstructured NCD is becoming the "gold standard" for investigations of biomedical cell cultures with far-reaching results [31] and consequences.

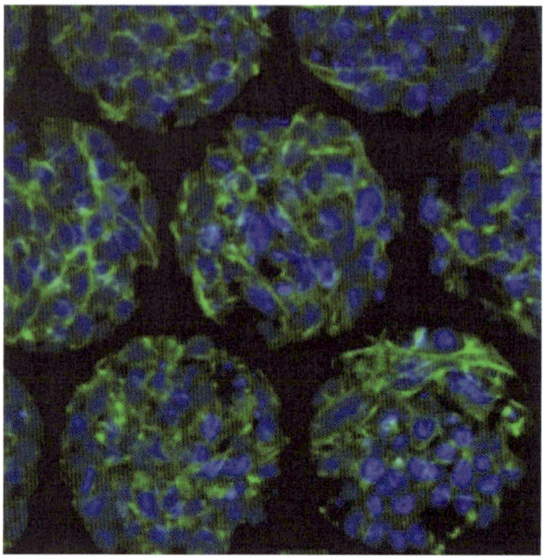

Fig. 4.30: Confocal laser scanning microscopy (CLSM) image shows a monolayer of P19 cells at the bottom of the diamond-coated cavities.

Acknowledgements

The authors gratefully acknowledge the financial support by the German Ministry of Education and Research (BMBF) within the project "DiTect" (FKZ 03XP0192B) and the expert support by P. Gluche, M. Mertens (GFD Ulm), A. P. Sommer (UUlm), and K. Brühne (UUlm).

References

1. H.-J. Fecht, K. Brühne, und P. Gluche, *Carbon-based Nanomaterials and Hybrids — Synthesis, Properties and Commercial Applications* (Singapore: Pan Stanford Publishing, **2014**).
2. M. Van de Voorde, M. Werner, and H.-J. Fecht, *The Nano-Micro Interface — Bridging the Micro and Nanoworlds*, 2nd edition (Wiley-VCH, **2015**).
3. H. Gleiter, *Nanocryst. Mater.* 33, 223–315 (**1989**).
4. F. P. Bundy, H. T. Hall, H. M. Strong, and R. H. Wentorf, *Nature* 176, 51–55 (**1955**).
5. W. G. Eversole, Synthesis of diamond, US Patent 3,3030,1880 (**1962**).
6. F. P. Bundy, *J. Geophys. Res.* 85(B12), 6930 (**1980**).
7. M. Wiora, PhD thesis (2013), Ulm University and M. Wiora in Ref. [1].
8. D. M. Gruen, *Annu. Rev. Mater. Sci.* 29 (**1999**); O. Auciello and A. V. Sumant, Diamond *Relat. Mater.* 19(7–9), 699–718 (**2010**).
9. M. Werner, P. Gluche, M. Adamschik, E. Kohn, and H.-J. Fecht, *IEEE International Symposium on Industrial Electronics*, 1998. Proceedings. ISIE '98, (**1998**).

10. S. O. Kucheyev, J. Biener, J. W. Tringe, Y. M. Wang, P. B. Mirkarimi, T. van Buuren, S. L. Baker, A. V. Hamza, K. Brühne, and H.-J. Fecht, *Appl. Phys. Lett.* 86, (**2005**).
11. M. Wiora, K. Brühne, A. Flöter, P. Gluche, T. M. Willey, S. O. Kucheyev, A. W. van Buuren, A. V. Hamza, J. Biener, and H.-J. Fecht, *Diam. Relat. Mater.* 18(5–8), (**2009**).
12. A. van der Drift, *Philips Res. Rep.* 22, 21 (**1967**).
13. F. Szuecs, M. Werner, R. S. Sussmann, C. S. J. Pickles, and H.-J. Fecht, *J. Appl. Phys.* 86(11), 6010–6017 (**1999**).
14. P. Hollman, A. Alahelisten, M. Olsson, and S. Hogmark, *Thin Solid Films* 270(1–2), 137–142 (**1995**).
15. P. Keblinski, D. Wolf, S. R. Phillpot, and H. Gleiter, *J. Mater. Res.* 13(8), 2077–2100 (**1998**).
16. V. Mitic, M. Ljubiša, L. Kocić, S. Tidrow, and H.-J. Fecht, *Nanotechnology, Energy and Fractals Nature, Nanotechnology for Energy Sustainability,* B. Raj, M. Van de Vorde, Y. Mahajan (eds.) (Wiley-VCH, **2016** in print).
17. M. Mohr, A. Caron, P. Herbeck-Engel, R. Bennewitz, P. Gluche, K. Brühne, and H.-J. Fecht, *J. Appl. Phys.* 116, 124308 (**2014**).
18. P. Hess, *J. Appl. Phys.* 111, 051101 (**2012**).
19. M. Mohr, L. Daccache, S. Horvat, K. Brühne, T. Jacob, and H.-J. Fecht, *Acta Materialia* 122, 92–98 (**2017**).
20. G. N. Greeves, A. L. Greer, R. S. Lakes, and T. Rouxel, *Nat. Mater.* 10, 823–835 (**2011**).
21. A. Erdemir and C. Donnet, Tribology of diamond, diamond-like carbon and related films. In *Modern Tribology Handbook*, B. Bhushan (ed.) (**2000**)
22. P. Gluche in: H.-J. Fecht and M. Werner (eds.), *The Nano-Micro Interface — Bridging the Micro and Nano Worlds* (Wiley-VCH, **2004**).
23. N. Wiora, M. Mertens, M. Mohr, K. Brühne, and H.-J. Fecht, *Diamond and Related Materials* 70, 145–150 (**2016**).
24. M. Mohr, P. Gluche, K. Brühne, and H.-J. Fecht, *Quantum Matter* 3 (**2014**).
25. A. P. Sommer, M. Wiora, and H.-J. Fecht, Biomimetric Hierarchies in Diamond-Based Architectures, in Ref. 2 (**2015**).
26. S. Giselbrecht, T. Gietzelt, E. Gottwald, C. Trautmann, R. Truckenmüller, K.-F. Weibezahn, and A. Welle, 3D tissue culture substrates produced by microthermoforming of pre-processed polymer films, *Biomed. Microdev.* 8, 191–199 (**2006**).
27. R. Truckenmüller, K.-F. Weibezahn, and A. Welle, Highly adaptable microstructured 3D cell culture platform in the 96 well format for stem cell differentiation and characterization Proc. 9th Int. Conf. on Miniaturized Systems for Chemistry and Life Sciences, *Micro Total Analysis Systems, μTAS* (Boston, MA, **2005**) pp. 376–378
28. A. P. Sommer, D. Zhu, D. F. Gagsteiger, and H.-J. Fecht, Online Proceedings Library of the Materials Research Society, 1511 (**2013**). DOI: 10.1557/opl.2013.168.

29. H. Kurosawa, Methods for inducing embryoid body formation: In vitro differentiation system of embryonic stem cells. *J. Biosci. Bioeng.* **103**(5), 389–398 (**2007**).
30. A. P. Sommer, D. Zhu, T. Scharnweber, and H.-J. Fecht, *J. Bionic Eng.* 6(1), 14–17 (**2009**).
31. A. P. Sommer, M. Haddad, and H.-J. Fecht, *Sci. Rep.* 8(5), 12029 (**2015**).

Chapter 5

Heat Transport and Thermal Characterization of Nanoscale Materials and Devices

Xiaomeng Wang* and Jin He†

Shenzhen SoC Key Lab., Peking University Shenzhen Institute and PKU_HKUST Shenzhen_Hongkong Institution, P.R. China

**wangxm@ier.org.cn*

†*frankhe@pku.edu.cn*

5.1. Introduction

The size of typical electronic devices is being reduced to as small as deep nanometer magnitude, on a fast approach to the feature wavelength of many particles or quasi-particles, such as electron, phonon, with the continuous advancement of the micro- and nanofabrication technology and process achievement. In this realm, we can expect devices to be much smaller and faster while exhibiting new unexpected phenomena of scientific interest and technological importance. Due to the changing of size in electronics and optoelectronics devices from micro to nanoscale, there is much more interest to look at the properties of these materials and structures for their outstanding applications. At small-sized scales, the number of atoms or the number of electrons in the system becomes small, so that continuum mechanics and elastic continuum models have to be replaced by models that take into account the discrete nature of the electronic and vibrational states and their distribution in energy band. Under such a technical background, the thermal transport at the nanometer scale is expected to be significantly different from classical or macroscopic transport in these materials and structures. In fact, thermal transport of nanoengineered materials gets enhanced or depressed dramatically in comparison to their bulk counterparts. So, the thermal transport at nanoscale has drawn much attention among research community due to the emergence of various nanostructured materials and their promises to address energy issues. The ability to control thermal conductive property of objects is of great interest in diverse areas such as solar and thermo-photo-voltaic energy conversion, thermoelectric applications. This chapter will mainly focus on nanoscale energy transport due to surface phonon scattering through three parts: background, typical application and future development.

5.2. Background

Thermal transport is one of the fundamental characteristics that determine the applications of materials. In the thermal transport process, the wave packets of lattice vibrations can be represented by the local concentration of phonons, which can be treated as pseudo-classical particles. Based on the kinetic theory [1], thermal conductivity (k) can be determined by

$$k = \frac{1}{3} C v \Lambda \tag{5.1}$$

As seen from Eq. (5.1), for a phonon-dominated thermal transport process, thermal conductivity is related to specific heat C, phonon group velocity v, and phonon mean free path Λ. Thus, thermal conductivity can be tuned by changing these three parameters. However, changing specific heat of a chosen material is difficult, unless the size of the crystal is on the atomic scale so that lattice vibrations are confined and the dispersion relation is modified. One limit for phonon group velocity is the speed of sound. As shown in Eq. (5.2), the speed of sound v in a crystal is related to elastic stiffness S and mass density ρ_0 and can be expressed by

$$v = \sqrt{\frac{S}{\rho_0}} \tag{5.2}$$

Therefore, materials with light atoms and strong bonding usually possess high thermal conductivity, such as diamond. On the contrary, materials with heavy atoms and weak bonding, such as polymers with van der Waals bonding, usually have low thermal conductivity. For a chosen material, the most practical way to manipulate thermal conductivity is by changing the phonon mean free path Λ.

The phonon mean free path is determined by collision events during the transport process, which is called phonon scattering. In general, there are three types of phonon scattering: phonon–phonon scattering, impurity scattering, and boundary scattering.

1. Phonon–phonon scattering: Phonon–phonon scattering is caused by the anharmonicity of chemical bonds between atoms. In a harmonic spring-mass system, the spring constant is independent of the spring deformation. While for chemical bonds, the spring constant will change when it is deformed. When one phonon propagates through a plane of atoms, these atoms will shift away from their equilibrium positions, so spring constants between these atoms will change. If another phonon propagates on the same plane, it will meet these different spring constants and get scattered.
2. Impurity scattering: In a crystal, alloy atoms, dopants, isotope variations, and point defects would cause point inhomogeneity which scatters phonons and reduces thermal conductivity. Usually, these impurities are on the atomic scale, so phonon-impurity scattering always lies in the Rayleigh scattering regime. For Rayleigh scattering, short wavelength phonons are scattered more effectively than middle-to-long wavelength phonons [2].
3. Boundary scattering: Phonons also get scattered at boundaries of a crystal or grain boundaries of polycrystalline materials, which reduces the phonon mean free path and hence reduces the thermal conductivity. Boundary scattering is the dominant scattering process for single crystalline materials at low temperature, since the phonon mean free path increases with decreasing temperature. For polycrystalline materials, the phonon mean free path could be much smaller than the intrinsic phonon mean free path in the single crystalline materials, which will result in a much lower thermal conductivity.

With the size reduction of nanoscale electronic devices, the heat generated by the unit area in integrated circuits will be increasing exponentially, and consequently the thermal management in these devices is a very important issue. So, the thermal transport properties of nanoscale systems and nanomaterials have attracted increasing attention in both experiments and calculations. Generally, representative nanomaterials comprise semiconductor or metallic nanostructures (one-dimensional nanostructures [3, 4], two-dimensional nanostructures [5, 6], nanocomposites [7, 8]), graphene [9, 10], carbon nanotubes (CNT) [11, 12], and nanofluids [13]. There is a great interest in the preparation and application of nanometer-sized materials since they show dramatic changes at nanoscale range, which exhibits new properties of industrial interest. For instance, an extremely low thermal conductivity is highly expected to attain a high efficiency for nanoscale thermoelectric application [14]. A good example comes from quasi-one-dimensional (1D) core-shell nanowires, which renders phonon thermal conductance $G \approx 0.2 \times 10^{-9}\,\text{W/K}$ by means of strong phonon-boundary scattering and enhanced thermoelectric properties owing to a quantum confinement effect [15]. Meanwhile, an intriguing issue faced by nanostructured materials is how to rapidly dissipate the generated heat for device-scale usage based on these materials. Many attempts have been made to develop novel nanostructures with ultra-high thermal conductivity (κ), such as a suspended single-layer graphene with superb κ up to $5300\,\text{W/m} \cdot \text{K}$ [10], and a stretched carbon nanotube fiber with κ of around $770\,\text{W/mK}$ [14]. Undoubtedly, understanding the thermal transport properties, especially their dependence on the nanostructures for these nanomaterials are essential to the potential applications.

5.2.1. *Thermal transport property at nanoscale*

The nature of thermal transport is the propagation of energy carriers in a medium in the form of wave packets [16]. Figure 5.1 shows the schematic diagram of a wave packet with wavelength λ propagating through a medium with different characteristic lengths L. During the propagation process, the wave packet collides with defects, grain boundaries, and other wave packets in the medium. These collision events will cause the scattering of the wave packet, changing of its energy and propagation direction, and thus impede the propagation process. The average distance between two successive collision events is called "mean free path", denoted by Λ. In general, the mean free path is much larger than the wavelength. At macroscale, since $\lambda \ll \Lambda \ll L$, the wave packet can be treated as a particle and its wave nature can be neglected, as shown in Fig. 5.1(b). At micro/nanoscale, when $\lambda \ll L \leq \Lambda$, wave packets will collide more frequently with the system boundary, which will further impede the propagation process. It is known as the classical size effect, as shown in Fig. 5.1(c). If $L$is further reduced to be comparable to λ, the wave packet can no longer be treated as a particle, and the wave interference effect must be considered [17, 18], which is known as quantum size effect as shown in Fig. 5.1(d).

As described above, thermal transport process at micro/nanoscale is different from that at macroscale because the phonon transport in nanostructured materials is different from the counterpart in bulk materials. Due

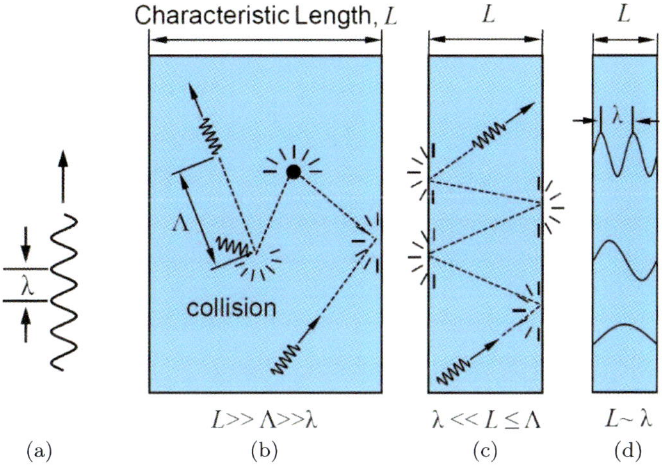

Fig. 5.1: Schematic diagram of the wave packet propagation in a medium with different characteristic lengths. (a) A wave packet with a wavelength λ; (b) Bulk behavior, $\lambda \ll \Lambda \ll L$; (c) Classical size effect, $\lambda \ll L \leq \Lambda$; (d) Quantum size effect, $L \sim \lambda$ [16].

to the classical size effect, thermal conductivity of nanostructured materials could be significantly lower than the bulk value. Traditional phenomenological laws, such as the Fourier's law for heat conduction, are also challenged. When the characteristic length of a system is comparable to the phonon mean free path, the increased phonon boundary scattering can reduce thermal conductivity significantly [19–21]. While when the characteristic length is comparable to phonon wavelength, the coherent wave interference effects need to be considered [17, 18]. Understanding thermal transport process at nanoscale is of great importance. Next, thermal transport in four kinds of nanostructured materials including thin films, nanowires, nanocomposites and carbon allotropes will be reviewed.

5.2.1.1. *One-dimensional nanostructure*

1D nanomaterials such as various kinds of nanowires, nanobelts, nanofibers, and nanotubes, made of metals, semiconductors, insulators, and organic compounds, have attracted tremendous interest due to their potential applications in electronic and energy conversion devices, magnetic devices, optical and chemical sensors, and biological labels [22–24].

Thermal energy transport in 1D materials is generally controlled by two types of energy carriers, i.e., phonons and electrons. Electrons dominate heat transport in good electrical conductors like metals, while phonons are in charge of heat transport in semiconductors and insulators. The thermal conductivity κ of electrical conductors and semiconductors consists of two parts, i.e., $\kappa = \kappa_l + \kappa_e$, where κ_l is the phonon or lattice thermal conductivity and κ_e is the electron thermal conductivity. Both phonon transport and electron transport are affected by various scattering mechanisms. For example, for phonon transport, scattering mechanisms include phonon–phonon scattering, phonon–electron scattering, phonon-impurity scattering, and phonon-boundary scattering, among which the phonon–phonon scattering mechanism is usually the dominant one to determine thermal conductivity of the bulk samples at room temperature. Since phonons collide with the boundary more frequently in 1D nanostructures than in bulk materials, the phonon-boundary scattering causes higher resistance to the phonon transport, which results in lower thermal conductivity than their bulk counterparts. On the other hand, the phonon mean free path could range from 10 nm to 1 μm, while the electron mean free path is ~10 nm at room temperature. The large difference between the phonon and electron mean free paths provides an opportunity to tune phonon transport and electron transport separately in some materials. If the dimension of 1D nanostructures is smaller than the phonon mean free path but larger than that of charge carriers, then the lattice (phonon) thermal conductivity will be significantly reduced but electron thermal transport will not be affected much.

To date, lots of work have been focused on the theoretical calculation of thermal conductivity of 1D materials using the Boltzmann transport equation, molecular dynamics (MD) simulation, Monte Carlo simulation, and the first-principle calculation. For example, by solving the Boltzmann transport equation, Walkauskas *et al.* calculated the lattice thermal conductivity of individual GaAs nanowires, which is much lower than the bulk value and decreases with decreasing the wire diameter due to the enhanced phonon-boundary scattering [25]. The Monte Carlo method was adopted to simulate the phonon transport in Si nanowires by Chen *et al.* For nanowires with a diameter larger than 37 nm, simulated results illustrate that thermal conductivity depends on the diameter of nanowires strongly, which can be ascribed to the enhanced phonon-boundary scattering and possible phonon spectrum modification [26].

In contrast with extensive theoretical studies on phonon transport in 1D nanostructures, experimental studies were only initiated when Tighe *et al.* designed and fabricated a GaAs-based heterostructure [27]. With this design, thermal conductivity presents strong diameter dependence and

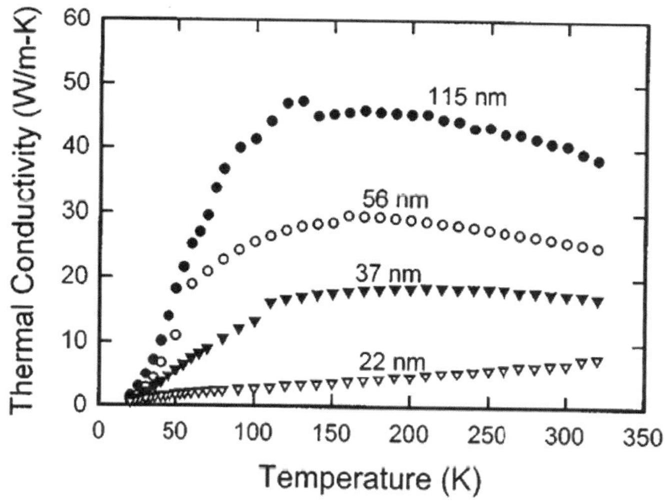

Fig. 5.2: Measured thermal conductivity of Si nanowires with different diameters [28].

the thermal conductivity value of the wire with a diameter of 22 nm is more than two orders of magnitude lower than the bulk value as shown in Fig. 5.2 [28]. Boukai *et al.* [29] also found that thermal conductivity decreases with decreasing the nanowire diameter due to the phonon-boundary scattering and reported that the rough nanowire surface could scatter phonons effectively.

5.2.1.2. *Two-dimensional nanostructure*

Two-dimensional (2D) materials, such as graphene, phosphorene, and transition metal dichalcogenides (e.g., MoS2 and WS2), have attracted a great deal of attention recently due to their extraordinary structural, mechanical, and physical properties. Thermal conductivity of 2D nanomaterials has shown great potential in the design and operation of microelectronic devices, thermal management and thermoelectric energy generation. In recent decades, thermal transport in thin solid films with thickness ranging from tens of nanometers to micrometers has become more and more important [21]. In this thickness range, thermal conductivity will be reduced by boundary scattering, which might be a problem for heat dissipation. In order to understand the fundamental physics of heat conduction in thin films, previous researchers have conducted a great amount of works, both experimentally [30–33] and theoretically [34–38].

For thermal transport perpendicular to a thin film, according to the gray model, the thin film thermal conductivity can be related to the bulk value using the following relation [39]:

$$k_{film} = k_{bulk} B_t \left(\frac{\Lambda_{bulk}}{t} \right) \tag{5.3}$$

where B_t is the boundary scattering function that depends on the ratio of Λ bulk and t.

The effective phonon mean free path along the cross-plane of thin film, Λ film, can be calculated using the Matthiessen's rule as [8]

$$\frac{1}{\Lambda_{film}} = \frac{1}{\Lambda_{bulk}} + \frac{1}{t} \tag{5.4}$$

The boundary scattering function B_t can be determined from Eq. (5.4) as

$$B_t \left(\frac{\Lambda_{bulk}}{t} \right) \equiv \frac{\Lambda_{film}}{\Lambda_{bulk}} = \frac{1}{1 + \frac{\Lambda_{bulk}}{t}} \tag{5.5}$$

Substituting Eq. (5.5) into Eq. (5.3), we obtain

$$k_{film} = k_{bulk} \frac{t}{t + \Lambda_{bulk}} \tag{5.6}$$

Layered materials with very different inter-atomic potentials in the in-plane and cross-plane directions exhibit highly anisotropic properties. For example, the in-plane thermal conductivity of highly oriented pyrolytic graphite (HOPG) is about 2000 $W/m \cdot$K at room temperature; while the value in the c-axis direction is only around 6.8 W/m · K, which is more than two orders of magnitude lower than the in-plane value [40]. The low c-axis thermal conductivity of graphite has been attributed to the weak van der Waals interactions between atomic layers.

For thermal transport along the in-plane direction of a thin film, Fuchs and Sondheimer derived an analytical solution based on the Boltzmann transport equation [41, 42]. The equation was first derived for electron transport and was then extended to phonon thermal transport. The Fuchs–Sondheimer solution assumes an average phonon mean free path Λ bulk, which enables the calculation of the thin film thermal conductivity as a function of the film thickness t,

$$k_{film} = k_{bulk} \left(1 - \frac{3}{2} \frac{\Lambda_{bulk}}{t} \int_1^{\infty} \left(\frac{1}{x^3} - \frac{1}{x^5} \right) \left(1 - \exp\left(-xt/\Lambda_{bulk} \right) \right) dx \right) \tag{5.7}$$

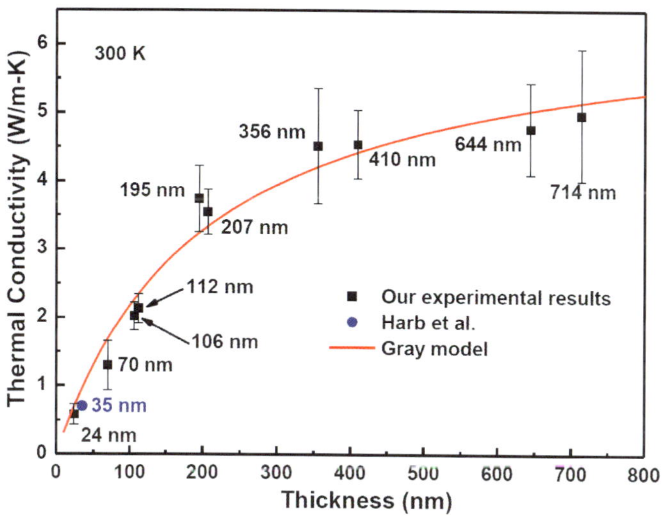

Fig. 5.3: The c-axis thermal conductivity of graphite films at 300 K shows a strong dependence on the film thickness. The red line is the fitting curve using the Gray model [46].

Asheghi *et al.* [33] measured the in-plane thermal conductivity of silicon films using Joule heating and electrical-resistance thermometry method, and found an obvious thermal conductivity reduction at low temperature for the film thickness ranging from 0.4 to 1.6 μm. Later, Ju and Goodson [32] measured the in-plane thermal conductivity of silicon films with a thickness down to 74 nm and found a thermal conductivity reduction up to 50% at room temperature.

Experimental studies on thickness dependence of the cross-plane thermal conductivity of thin films are more challenging since the effects of interfacial thermal resistance and in-plane heat spreading on the measurements are usually difficult to eliminate at nanoscale [43–45]. Recently, Fu *et al.* [46] measured the cross-plane thermal conductivity of thin graphite flakes with various thicknesses by using a differential three omega method. Results show that as the film thickness increases from 24 nm to 714 nm, the measured *c*-axis thermal conductivity also follows a clear increasing trend (Fig. 5.3) and eventually approaches the bulk value as the film becomes much thicker than phonon mean free path (204 nm at room temperature) and the boundary effect vanishes [46].

So, when the film thickness is smaller than the phonon mean free path, thermal conductivity of thin films in both in-plane and cross plane direction should be much lower than the bulk value due to the enhanced phonon-boundary scattering, the phonon mean free path is limited by the film thickness.

5.2.1.3. *Nanocomposite*

Nanocomposites refer to the format of nanoinclusions such as nanoparticles or nanowires embedded in a host matrix or a heterostructure geometry with nanostructures adjacent to each other. Nanocomposites that do not require any specific morphology are practical for mass fabrication and easy to incorporate into commercial devices. In fact, nanograined materials from the same type usually show desirable thermal conductivity reduction, although these materials often possess additional internal structures such as nanoprecipitates [47].

Particle-host nanocomposite: One kind of nanocomposites is the particle-host nanocomposite, which consists of nanostructures embedded in a host material. The reduced thermal conductivity in particle-host-type nanocomposite was first reported by Worlockin in 1966 [48], who synthesized NaCl containing AgCl nanodots by chemical reaction method. An obvious thermal conductivity reduction was found at low temperature even with a very low concentration of AgCl, 2.5×10^{-5} (mole fraction). The author claimed that the embedded AgCl nanodots can scatter phonons in geometrical regime and hence reduce thermal conductivity at low temperature where boundary scattering dominates and phonons have a large mean free path. Recently, with the development in materials synthesis technologies, high quality materials (host) with embedded nanostructures (particle) can be synthesized. Kim *et al.* studied thermal conductivity of $In_{0.53}Ga_{0.47}As$ with embedded ErAs nanoparticles grown by molecular beam epitaxy (MBE) and they found that thermal conductivity can be effectively reduced for the whole temperature range due to the existence of ErAs nanoparticles, as shown in Fig. 5.4. Based on theoretical analyses, the atomic scale alloy atoms only affect short wavelength phonons according to the Rayleigh scattering theorem [49]. But embedded nanoparticles can scatter mid-long wavelength phonons more efficiently [50], and the shape, size, and concentration of nanostructures all play important roles in phonon scattering processes. This additional scattering mechanism can reduce thermal conductivity of particle-host nanocomposites below the alloy limit, which provides a new route to tune thermal conductivity of crystalline solids.

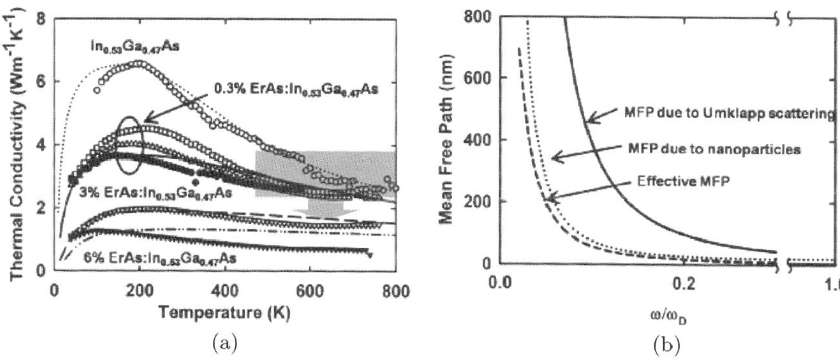

Fig. 5.4: (a) Temperature dependence of thermal conductivity of 6% and 3% ErAs in $In_{0.53}Ga_{0.47}As$ (solid and open downward triangles, respectively) is lower than that 0.3% ErAs and without ErAs in $In_{0.53}Ga_{0.47}As$ [51]. Dotted, solid, and dashed lines are based on theoretical analysis. (b) The phonon mean free path (MFP) versus normalized frequency of 3% ErAs in $In_{0.53}Ga_{0.47}As$ at 600 K [49].

Particle–particle nanocomposites: Another type of nanocomposites are particle–particle nanocomposites, which are polycrystalline materials, usually with grain sizes smaller than 100 nm. They can be synthesized by a combined ball milling and hot pressing process [52–54]. As long as the nanograin size is smaller than the phonon mean free path, it can effectively reduce thermal conductivity. The high-density grain boundaries can scatter phonons with all wavelengths [55, 56]. Due to the increased phonon scattering, all these nanostructured materials show reduced thermal conductivity. Particle–particle nanocomposites have been proven to be an effective and practical approach to improve the efficiency of thermoelectric materials by engineering materials into nanocomposites.

5.2.1.4. *Carbon allotropes*

Carbon allotropes and their derivatives occupy a unique place in terms of their ability to conduct heat. The room-temperature thermal conductivity of carbon materials spans an extraordinary large range — of over five orders of magnitude — from the lowest in amorphous carbons to the highest in graphene and carbon nanotubes, as shown in Fig. 5.5 [57]. Undoubtedly, carbon nanotube (CNT) and graphene are the most successful alternatives for nanoelectronics [58–60].

Carbon nanotube (CNT): CNTs were first observed and studied by Iijima [61] two decades ago. CNTs are hollow cylinders composed of one

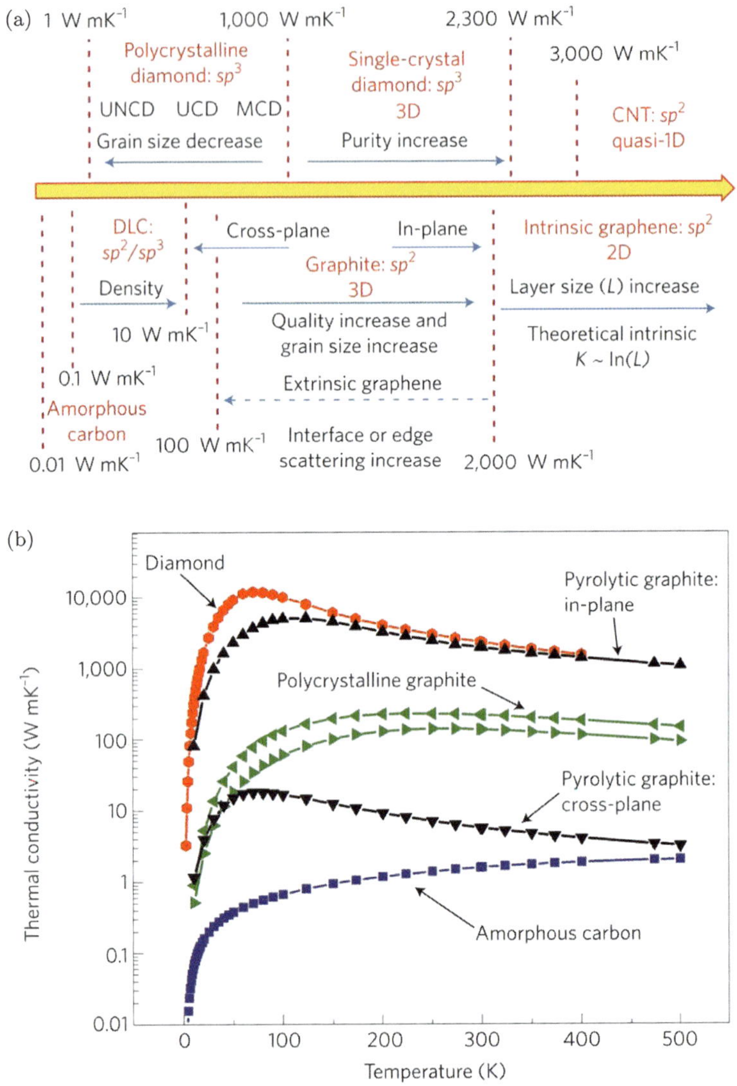

Fig. 5.5: (a) Diagram based on average values reported in the literature. The axis is not to scale. (b) Thermal conductivity of bulk carbon allotropes as a function of T [57].

or more concentric layers of carbon atoms in a honeycomb lattice arrangement [62, 63]. CNTs include both single-walled and multiwalled structures, as shown in Fig. 5.6. Due to their unique crystalline structure (the high sound velocity and long phonon mean free path), boundary scattering

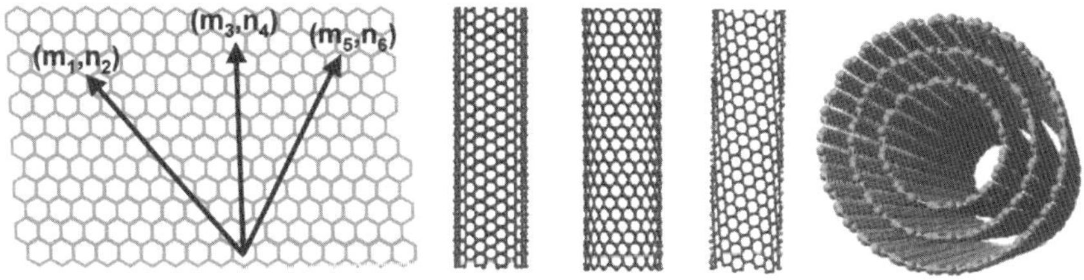

(a) (b) (c) (d) (e)

Fig. 5.6: Schematic of the honeycomb structure of a graphene sheet (a). SWCNTs can be formed by folding the sheet along the shown lattice vectors leading to armchair (b), zigzag (c), and chiral (d) tubes, respectively. The graphene sheets rolled up into concentric cylinders form MWCNT (e) [73].

is nearly absent in CNTs [63, 64], so that CNTs provide the high thermal conductivity [65, 66] with the experimentally determined RT value $k \approx 3000\,\mathrm{W/m\cdot K}$ for an individual multiwall carbon nanotube (MW-CNT) [66] and $k \approx 3500\,\mathrm{W/m\cdot K}$ for an individual single-wall carbon nanotube (SW-CNT) [67]. The room temperature (RT) thermal conductivity in the range 1750–5800 W/m · K was reported for the crystalline "ropes" of SW CNTs [68]. These values exceed those of the best bulk crystalline thermal conductor, diamond, which has the thermal conductivity in the range $k = 1000$–$2200\,\mathrm{W/m\cdot K}$. [69]. Thus, CNTs show high thermal and electrical conductivity and this makes CNTs an ideal candidate for promising applications in nanoelectronics, which is expected to act like heat pipes due to their high thermal conductivity [70], field-effect transistors (FETs) due to their excellent electrical properties [71] and nanobioelectronics [72].

Graphene: As a single atomic plane of carbon, graphene can be wrapped up into other graphitic materials such as fullerene, carbon nanotubes and thin graphene films [74]. Due to its strong in-plane covalent bonds (in-plane phonon group velocities and phonon density) formed by sp^2-hybridized carbon arrangement, graphene exhibits many advantages compared with traditional materials, like super high electrical conductivity (high mobility of charge carriers) [75] and superior thermal conductivity [10, 57]. For thermal property, Balandin *et al.* [10] discovered that individual single-layer graphene manifests an extraordinary high RT thermal conductivity of up to 5300 W/m · K, suggesting that graphene can outperform carbon nanotubes (CNTs) in heat conduction. Based on the dispersion calculations in graphitic materials [76], the phonon group velocity in graphene is higher than that in CNTs [77], which leads to larger k. While single-layer graphene

(SLG) is hard to produce, graphene multilayers are much cheaper and are expected to retain the heat conducting property.

5.2.2. *Characterization methods of nanoscale materials*

Thermal conductivity can be defined as the rate at which heat is transferred by conduction through a unit cross-section area of a material, when a temperature gradient exits perpendicular to the area [78, 79]. For bulk materials, thermal conductivity measurement techniques have been developed since 1960s. And most thermal conductivity measurement techniques for bulk involve creating a temperature gradient along a bulk sample and monitoring the temperatures by thermocouples at different places [80]. The sample may also be placed between standard materials with known thermal conductivities. So, by comparing the temperature gradient, the sample thermal conductivity can be derived. However conventional thermal measurement techniques do not have sufficient spatial resolution to resolve temperature and thermophysical properties of submicron structures. There are several challenges to characterize thermal transport at micro/nanoscales. So, the experimental evidence of thermal transport phenomena in microelectronics and nanostructures is unique and critical.

Thermal conductivity (k) of solid materials can be determined by

$$k = -\frac{Q/A}{\Delta T/L} \tag{5.8}$$

where Q is the amount of heat conducted through a cross section (A), and ΔT is the induced temperature difference over a distance L. Thermal conductivity measurement involves determination of the heat flux (Q/A) and the temperature gradient ($\Delta T/L$) across the material. Over the past decade, the establishment of several nanoscale thermal transport measurement techniques has made it possible to use nanostructures as model systems to test the phonon transport theories, each possessing their own limitations. Broadly speaking, thermal property measurement techniques can be divided into contact and non-contact methods, as well as steady-state and transient methods. Contact methods typically involve incorporating the sample of interest into an electronic circuit, the temperature-dependent behavior of which is assessed and subsequently related to an unknown thermal property of the sample through appropriate modeling. Techniques in this category include the 3ω method [81], bolometry [82], suspended thermal bridge and electrical pulse heating [83]. Non-contact techniques, in contrast, involve measurement of some temperature-dependent property that does not require sample incorporation into an electric circuit [84]. Raman thermometry [85, 86], the photo-acoustic method [87], time-domain thermoreflectance (TDTR) [88], dilatometry [89], and the laser flash method [90] are the most commonly used non-contact techniques.

Steady-state measurement: In steady-state measurements, the thermal conductivity and the interfacial thermal conductance are determined by measuring the temperature difference (ΔT) under a steady-state heat flow (Q) through the sample (thickness) [91]. The thermal conductivity is given then by the slope of the power versus the temperature gradient between both sides of the sample [78]. All steady-state methods are directly based on Fourier's law (Eq. (5.8)) [92]. These methods measure directly thermal conductivity and are suitable for low thermal conductivity and composite materials. However, steady-state techniques do require relatively large samples, and longer time tests than transient methods. They also present other drawbacks such as parasitic heat losses and contact resistance because of the use of temperature sensors [78]. Steady-state methods have the advantage of one-directional heat flow over a larger area which accounts for measurement of samples with anisotropic thermal conductivity and composites. It normally requires well-designed experimental setup and along waiting time to reach thermal equilibrium. So, for the steady-state method, the temperature difference does not change with time and thermal conductivity can be determined by measuring ΔT, Q, A, and L.

Transient measurement: Transient measurement of thermal properties in measurement involves using transient heat source with a temperature probe that can be localized spatially. These methods determine the thermal diffusivity (α) through a transitional thermal flow during the heating process. For the transient method, thermal conductivity measurements are performed during the heating process and it is much faster compared to the steady-state method. Thermal diffusivity is measured in parallel to thermal conductivity, and both of them can be highly sensitive to the frequency of ac heat source. It uses small sample amounts and can work at high temperature and pressure. However, the accuracies are lower than the ones for steady-state methods. To obtain thermal conductivity values, the specific heat capacity and density of the studied samples [93] need to be known/measured, which increases the error of the measurement.

In the following, several widely used methods for thermal conductivity measurement of nanoscale materials will be introduced.

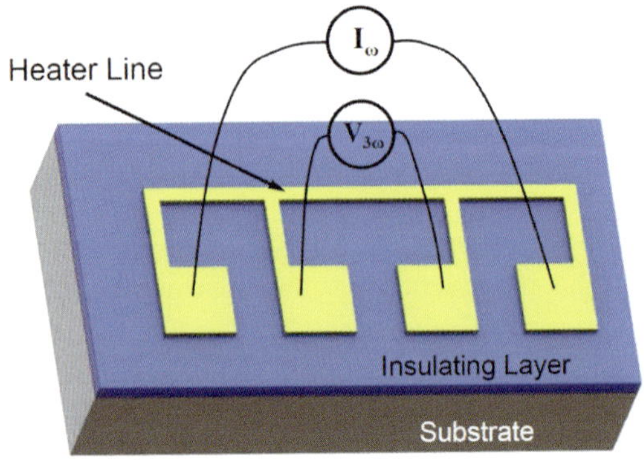

Fig. 5.7: The schematic diagram for the 3ω method.

5.2.2.1. *Three omega method*

The three omega (3ω) method invented by Cahill in 1990 [81] has been widely used for thermal conductivity measurements of both thin films and nanocomposite materials without knowing the specific heat or density. Since it is very sensitive to small temperature fluctuations, it has high accuracy and low thermal radiation loss. The 3ω technique measures the surface temperature oscillation induced by periodic heating of a patterned metal line by monitoring the voltage oscillation. The thermal conductivity of the sample is determined by comparing the temperature oscillation in a film/substrate structure with the corresponding value in the substrate only. The benefits for this method are high measurement sensitivity and well-defined thermal boundary conditions, but at the cost of a sample preparation effort such as patterning a thin film using lithography.

The 3ω method utilizes a micrometal line as both heater and thermometer [94, 95]. A typical implementation of the 3ω method is shown in Fig. 5.7. The 3ω pattern consists of a thin Au metal line with four contact pads for four-probe electrical resistance measurement, which is fabricated on the sample surface by standard photolithography and sputtering or evaporation processes. During a measurement, the strip metal film is driven by a periodically oscillating heat source (an AC current source) at a frequency of 'ω'.

$$I = I_0 \sin(\omega t) \tag{5.9}$$

The heat dissipated in the line heater self-heats itself and the underlying sample and causes temperature oscillations at '2ω' frequency. The Joule heating generated in the heater is

$$Q = I_0^2 \sin^2(\omega t) R_{h,Q} \approx I_Q^2 \sin^2(\omega t) R_{h,0} \tag{5.10}$$

$$Q = \frac{1}{2} I_0^2 R_{h,0}[1 - \cos(2\omega t)] = Q_{DC} + Q_{2\omega}\cos(2\omega t) \tag{5.11}$$

Since the resistance change caused by the Joule heating is small, we can assume $R_{h,Q} \approx R_{h,0}$ in Eq. (10). Since the electrical resistance of the metal line is dependent on the temperature, the temperature rise on the sample surface can be determined by measuring the resistance change of the metal line. The temperature oscillations make the resistance of line heater vary at '2ω' frequency and consequently leading to voltage oscillations at '3ω' frequency. The resistance of the heater will change with its temperature as

$$R_h = R_{h,0} + \frac{dR}{dT}\Delta T = R_{h,0} + \frac{dR}{dT}(\Delta T_{DC} + \Delta T_{2\omega}\cos(2\omega t + \theta)) \tag{5.12}$$

In the first-order analysis, the voltage drop across the heater is

$$\begin{aligned} V_h = IR_h &= I_0\left(R_{h,0} + \frac{dR}{dT}\Delta T_{DC}\right)\sin(\omega t) \\ &\quad + I_0\frac{dR}{dT}\Delta T_{2\omega}\sin(\omega t)\cos(2\omega t + \theta) \\ &= I_0\left(R_{h,0} + \frac{dR}{dT}\Delta T_{DC}\right)\sin(\omega t) \\ &\quad + \frac{1}{2}I_0\frac{dR}{dT}\Delta T_{2\omega}[\sin(3\omega t + \theta) - \sin(\omega t - \theta)] \\ &= V_{1\omega}\sin(\omega t + \theta') + V_{3\omega}\sin(3\omega t + \theta) \end{aligned} \tag{5.13}$$

As seen from Eq. (13), the voltage drop on the heater contains components of two frequencies, $V_{1\omega}$ and $V_{3\omega}$. The temperature rise on the sample surface can be determined by using the voltage drop at the frequency of 3ω,

$$\Delta T_{2\omega} = 2\frac{dT}{dR}R_{h,0}\frac{V_{3\omega}}{V_0} \approx 2\frac{dT}{dR}R_{h,0}\frac{V_{3\omega}}{V_{1\omega}} \tag{5.14}$$

In practice, the temperature change due to the joule heating is quite small, and hence we can assume $V_0 \approx V_{1\omega}$. In order to determine $\Delta T_{2\omega}$, the temperature coefficient of resistance or dT/dR of the metal heater should also be characterized. Borca-Tascuic *et al.* [96] reported an analytic solution of the transient heat conduction equation for an anisotropic system with n layers to quantify the relationship between the heat flux and temperature gradient, assuming a uniform heat flux and very thin heater on an isotropic semi-infinite substrate. Cahill [81] further simplified the formula to

$$\Delta T = \frac{P}{\pi L k}\left(0.5\ln\left(\frac{k}{\rho C b^2}\right) - 0.5\ln(2\omega) + \eta - i\frac{\pi}{4}\right) \tag{5.15}$$

where ΔT is the temperature fluctuation of the substrate, P/L is the amplitude of the heating power per unit length generated at frequency 2ω, k is thermal conductivity of the substrate, ρ is density, C is specific heat, b is the half of the heater line width, η is a constant which has been numerically evaluated to be ~0.923 [81]. Although we can obtain k either by the in-phase part or the out-of-phase part, usually a more reliable result can be obtained by fitting the slope of ΔT versus $\ln(\omega)$ curve,

$$k = -\frac{P}{2\pi L}\left(\frac{d(\Delta T)}{d(\ln(\omega))}\right)^{-1} \tag{5.16}$$

For thermal conductivity measurement of thin films, the film should be prepared on a substrate with high thermal conductivity. The temperature drop on the thin film can be expressed as

$$\Delta T_{film} = \frac{P t_{film}}{k_{film} L w} \tag{5.17}$$

where ΔT_{film} is the temperature drop across the film, t_{film} is the film thickness, and $w = 2b$ is the heater line width. During the 3ω measurement, the total temperature fluctuation ΔT_{total} can be obtained by using Eq. (14). Then, by calculating the slope of ΔT_{total} vs $\ln(\omega)$ curve, thermal conductivity of the substrate (k_{sub}) can be determined by Eq. (5.16). Once k_{sub} is determined, the temperature rise of the substrate ΔT_{sub} can be calculated by Eq. (5.15). Since thermal conductivity of the thin film can be extracted by $\Delta T_{film} = \Delta T_{total} - \Delta T_{sub}$, thermal conductivity of the thin film can be extracted by

$$k_{film} = \frac{P t_{film}}{L w \Delta T_{film}} \tag{5.18}$$

This method is called the slope method. It works well for a single layer on a substrate. But for the sample with multiple layers, or when the properties of the substrate, such as the density and heat capacity, are unknown, the slope method is not applicable.

In order to overcome these issues, a differential 3ω method [43] is adopted to determine thermal conductivity of thin films. For the differential method, two samples need to be measured: one with the film-of-interest on the substrate, the other is a reference sample which contains all other layers

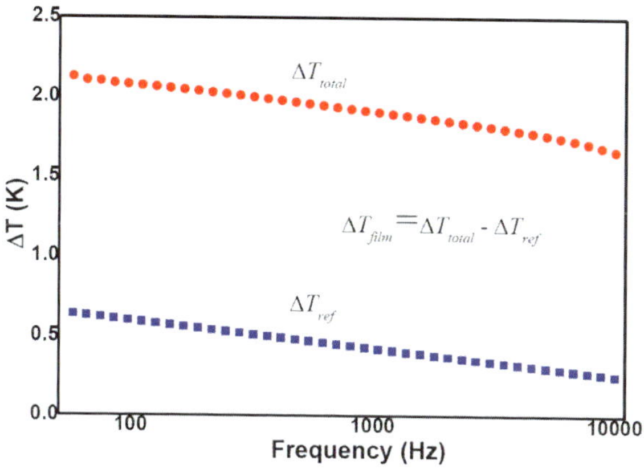

Fig. 5.8: Temperature rise of the heater on the sample with the thin film (ΔT_{total}) and temperature rise of the heater on the reference sample (ΔT_{film}).

except the film to be measured. These two samples should be fabricated at the same time to make sure that they are identical except the thin film layer. The difference between temperature drop on the sample with thin film and the counterpart on the reference sample is purely due to the presence of the film as shown in Fig. 5.8. Although it is difficult to guarantee that the same heat flux is applied in two experiments, we can normalize the temperature drop by the heating power and obtain

$$\frac{\Delta T_{film}}{P_{sam}} = \frac{\Delta T_{total}}{P_{sam}} - \frac{\Delta T_{ref}}{P_{ref}} \tag{5.19}$$

where P_{sam} and P_{ref} are the heating power applied on the sample with thin film and the reference sample, respectively.

Thermal conductivity of the thin film can be determined directly by using Eq. (5.18) without the knowledge of exact geometries and material properties of each layer, or interfaces in the system. A drawback of this method is the difficulty and lack of accuracy in measuring the thermal boundary resistance between a metal and dielectric. The thermal boundary resistance, rather than the individual thermal properties of a thin film layer, can dominate the overall heat transfer process in a multilayer structure.

The 3ω method is limited to dielectric materials. When the specimen is capable of self-heating (metals and semiconductors), an insulating layer needs to be fabricated on top of the sample in order to do 3ω measurements. It requires additional processing steps and careful consideration on not affecting the specimen itself, which is generally difficult to execute on nanoscale specimens. Any current leakage between the metal line and substrate will corrupt the measurement.

5.2.2.2. *Transient thermoreflectance method*

Non-contact optical measurement technique drew attention to be applicable to measure thermal properties using pump-probe thermoreflectance. A schematic of the experimental setup for the TTR is shown in Fig. 5.9. In thermoreflectance measurements, a sample is first coated with a thin metal film, which acts as a temperature transducer layer. A short pulse Laser beam is applied to the film to avoid heat loss with the inclusion of the interfacial conductance and extract thermal diffusivity which is coupled to the thermal conductivity and the heat capacity of the film. During each pulse, a given volume on the sample surface heats up to a temperature level above ambient due to the laser light energy absorbed into the sample. The heating area is specified by adjusting the pulsing laser aperture and the optics of the system. The depth of the volumetric heating, on the other hand, is determined by the optical penetration depth, which is a function of

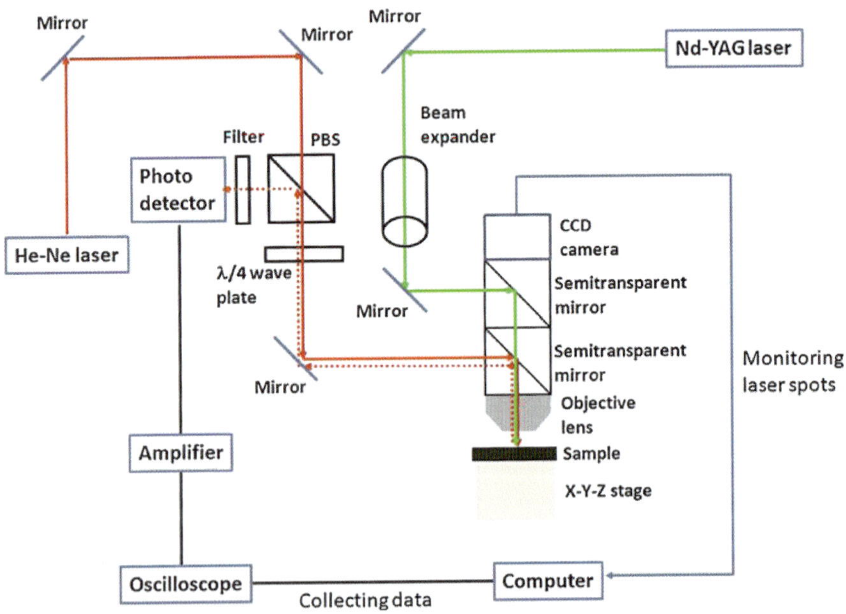

Fig. 5.9: Schematic of the transient thermoreflectance thermometry for thermal-conductivity measurement [101].

laser wavelength and surface material properties. The heating level through the light penetration depth obeys an exponential decay law after each laser pulse is completed, the sample begins to cool down to the initial ambient temperature. During this process, a second laser is used to monitor the change in reflectivity of the metal layer as a function of time [97, 98] or modulation frequency [99, 100]. Because the changes in surface reflectivity are linearly proportional to the changes in surface temperature, within a wide but finite temperature range, this second beam allows contactless assessment of changes in the temperature of the metal surface layer. Following the transient heating pulse, the temperature of the surface layer decreases at a rate determined by the thermal properties of the sample. Applying heat diffusion equation with appropriate initial and boundary condition, the thermal conductivity can be extracted with more accuracy and avoiding experimental tolerance. The influence of a pulsed laser irradiation on a given material depends both on the optical properties of that material as well as on the wavelength and pulse duration of the laser itself. Thus, the wavelength and pulse width are important parameters in the determination of the effectiveness of the TTR method for different materials [88].

Specifically, the heat transfer within the metal film and the substrate material heated by a short-pulsed laser on a timescale greater than 100ps is governed by the time dependent heat conduction equation, which describes the thermal diffusion in the metal film and substrate, respectively:

$$C_f \frac{\partial \theta_f}{\partial t} = k_f \frac{\partial^2 \theta_f}{\partial x^2} \tag{5.20}$$

$$C_s \frac{\partial \theta_s}{\partial t} = k_s \frac{\partial^2 \theta_s}{\partial x^2} \tag{5.21}$$

where θ is the temperature above ambient, C is the thermal capacitance, and k is thermal conductivity. Upon absorption of a laser pulse, there is an immediate temperature gradient in the metal film and a negligible temperature rise in the substrate. Therefore, initial conditions for the metal film and the substrate are as follows:

$$\partial \theta_{f(x)} = \frac{F(1-R)}{C_f \delta} e^{-x/\delta} \tag{5.22}$$

$$\theta_s = 0 \tag{5.23}$$

The time constant, τ, associated with diffusion of heat in the film can be approximated using

$$\tau \approx \frac{d^2}{\alpha} \tag{5.24}$$

where α is thermal diffusivity of the metal film and d is the film thickness. By measuring a temporal transient thermoreflectance signal using an optical pump-probe method, the thermal conductivity of a thin film can be deduced.

Transient Thermoremectance was first introduced by Eesley *et al.* [88] to measure thermal properties of metal film using an intense ultrashort laser pulse to generate a transient thermal heat flux at the metal surface and another laser (probe) was used to monitor the change in reflectance at the surface. Since then, the transient thermoreflectance method became very popular due to its nature of measurement having no contact with the surface. Later, transient thermoreflectance method (TTR) is used to determine the thermal conductivity of thin-film and multilayered materials. The main advantage of the TTR method is that it is a non-contact and nondestructive optical approach both for heating a sample under test and for probing the variations of its surface temperature. Because the method is noninvasive, it is attractive for the measurement of thermal properties of thin-layer materials whose investigation by invasive methods would present the difficulties. Another important feature of this technique is the ability to measure the thermal boundary resistance [101].

5.2.2.3. *Suspended thermal bridge method*

However, conventional techniques for thin film and nanocomposite thermal property measurement cannot be readily extended to 1D nanostructures due to the small sample size. Shi and coworkers developed a suspended microdevice for measuring thermal conductivity and thermoelectric properties of individual 1D nanostructures [28, 102]. The microdevice was designed to be suspended to eliminate the parasitic heat loss to the substrate and it is fabricated through the standard microfabrication processes. Figure 5.10 shows a SEM image of the microdevice. The device consists of two adjacent low-stress silicon nitride (SiN_x) membranes suspended with six long SiN_x beams. There is a long and thin platinum (Pt) coil designed on each membrane, which can act as both the heater and resistance thermometer in the measurement. The platinum resistance thermometer (PRT) is connected to platinum contact pads on the substrate via platinum leads on the long SiN_x beams. In addition, two wide Pt electrodes were patterned on two membranes for thermal and electrical contacts with the sample.

Thermal conductivity measurement technique was first designed by Kim *et al.* in 2001 and then widely used and further developed to characterize

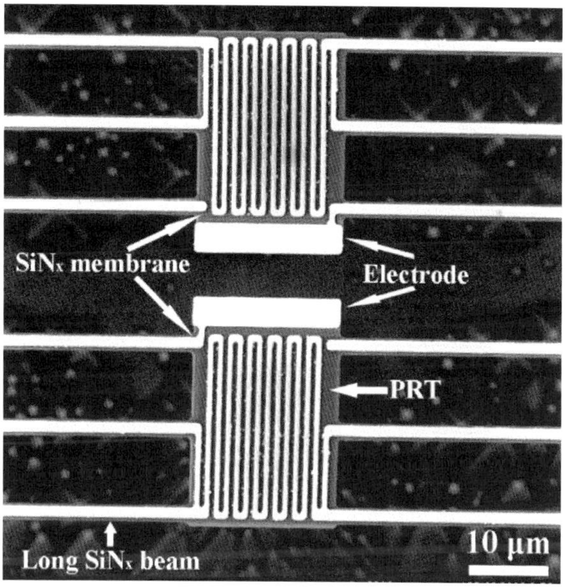

Fig. 5.10: SEM micrograph of the suspended microdevice with two electrodes.

thermal transport property of various 1D nanostructured materials [103]. Figure 5.11 shows the schematic diagram of the heat transfer model under the steady state. A 1D or 2D nanostructure is placed just like a bridge across two suspended membranes. R_h and R_s represent the electrical resistance of Pt coils on heating and sensing membranes, respectively. When a dc current (I) passes through the PRT on the heating membrane, a Joule heat $Q_h = I^2R_h$ is generated and the temperature of the heating membrane is raised to a uniform temperature T_h. A certain amount of heat Q_s is conducted through the sample from the heating membrane to the sensing membrane and raises the temperature of the sensing side to T_s. The PRT on each membrane connects to the contact pads on the substrate at temperature T_0 by four long Pt beams, which enables the four-probe $I - V$ resistance measurement using a small ac current (v_{ac}) applied by using a lock-in amplifier. All the 12 beams in the same device are designed to be identical. Therefore, the resistance R_b of each beam is supposed to be the same. Two Pt leads on two suspended beams are used to supply the dc current I to the heating RTP. The Joule heating in these two beams $2Q_b = 2I^2R_b$ is also dissipated into the substrate. The part of heat transferred through the nanostructure, Q_s, is further conducted to the substrate

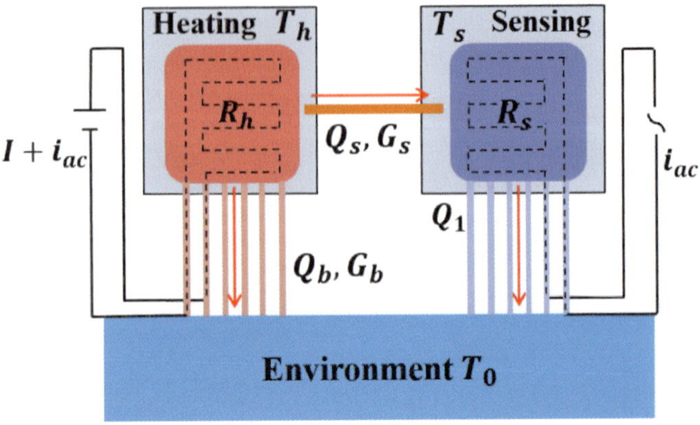

Fig. 5.11: Schematic diagram of the heat transfer model under the steady state.

through the six beams connecting the sensing membrane. The rest heat, i.e., $Q_1 = Q_h + 2Q_b - Q_s$, is conducted to the environment through the other six beams supporting the heating membrane.

Based on the heat transfer model, the thermal conductance of the nanostructure can be analyzed. We can obtain the following equation from the equivalent thermal circuit shown in Fig. 5.11:

$$Q_s = G_b(T_s - T_0) = G_s(T_h - T_s) \tag{5.25}$$

where G_s is the thermal conductance of the nanostructure, which is related to both the intrinsic thermal resistance of the nanostructure and the contact thermal resistance between the sample and the two membranes, i.e.,

$$G_s = \frac{1}{R_s} = \left(\frac{1}{G_n} + \frac{1}{G_c}\right)^{-1} \tag{5.26}$$

Here, $G_n = k_n A_n / L_n$ is the intrinsic thermal conductance of the sample, where k_n, A_n and L_n are the intrinsic thermal conductivity, cross-sectional area, and length of the sample segment between two membranes, respectively, and G_c is the contact thermal conductance between the sample and two membranes. Since all the beams in one device are assumed to be identical and share the same electrical resistance, the Joule heating generated by the two conducting beams is uniform and can be simplified as Q_b. Therefore, the heat conduction to the environment from the two Joule-heated beams can be derived as

$$Q_{h,2} = 2(G_b \Delta T_h / 6 + Q_b / 2) \tag{5.27}$$

and that from the remaining four beams connected to the heating membrane is

$$Q_{h,4} = 4G_b\Delta T_h/6 \tag{5.28}$$

and that from the six beams connected to the sensing membrane is

$$Q_{h,6} = G_b\Delta T_s \tag{5.29}$$

where $\Delta T_s \equiv T_s - T_0$. Considering heat conservation from Fig. 5.11,

$$Q_{h,2} + Q_{h,4} + Q_{h,6} = Q_h + 2Q_b \tag{5.30}$$

Based on the definition of thermal conductance, the thermal conductance of each beam G_b and the 1D nanostructure G_s can be obtained as

$$G_b = \frac{Q_h + Q_b}{\Delta T_h + \Delta T_s} \tag{5.31}$$

$$G_s = G_b\frac{\Delta T_s}{\Delta T_h - \Delta T_s} \tag{5.32}$$

With the dc current I and the voltage drops across the heating PRT and two supporting Pt beams in heating side, Q_h and Q_b can be calculated readily. ΔT_h and ΔT_s can be calculated from the measured electrical resistance of the two PRTs and their temperature coefficients of resistance (TCR $\equiv (dR/dT)/R$).

As the cross-sectional area A and the suspended length L_s of the 1D nanostructure can be determined from the SEM images, we can finally calculate the effective thermal conductivity k_s based on

$$k_s = \frac{G_s A}{L_s} \tag{5.33}$$

The sensitivity of the aforementioned traditional suspended method is approximately 1 nW/K [102], therefore it is not applicable for measuring samples with thermal conductance lower than 1 nW/K. In order to measure thermal conductivity of very thin 1D nanostructures with low thermal conductance ($G < 1$ nW/K), an on-chip Wheatstone bridge circuit is introduced to the traditional method by Wingert *et al.* [104] and the sensitivity is enhanced to be $\sim 10^{-11}$ W/K. The differential measurement method is able to determine much smaller temperature difference compared with the traditional measurement method. As shown in Fig. 5.12, the resistance of the PRT on the sensing membrane is measured using the Wheatstone bridge circuit instead of the four-probe $I-V$ measurement. The bridge circuit consists of the sensing resistor R_s and a pair resistor R_1, both of which are the

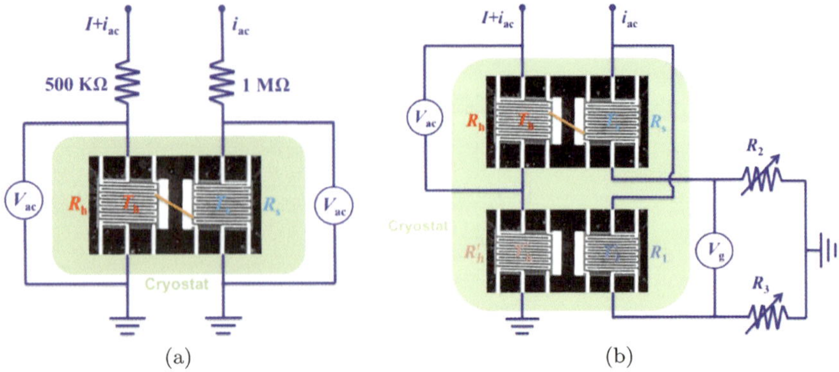

Fig. 5.12: Schematic diagrams of (a) the traditional method and (b) the differential bridge method for thermal conductivity measurement of 1D nanostructured materials.

PRTs patterned on two microdevices located inside the cryogenic chamber, and two high-precision variable resistors (R_2 and R_3) located outside the cryogenic chamber. As shown in Fig. 5.12(b), Joule heating generated on two heating membranes is almost the same and radiation heat losses from the heating membrane to the sensing membrane on two devices are almost the same. Then, the thermal conductance of the 1D nanostructure without the effect of the background conductance signal can be determined from the offset voltage V_g of the Wheatstone bridge circuit on the sensing side. In addition, because of the symmetry of the bridge setup, the fluctuation in the ambient temperature of the cryostat will affect both R_s and R_1; as a result, it will not cause much fluctuation in the measured gate voltage v_g. In other words, its effect on thermal conductivity measurement can be effectively eliminated. The use of a Wheatstone bridge circuit in conjunction with the microfabricated measurement devices not only eliminates the effect of the ambient temperature fluctuation on the measurement but also improves the measurement sensitivity of 1D nanostructures. The differential measurement method provides an effective method to characterize thermal properties of less conductive 1D nanostructures.

In the suspended microdevice method, it is important to ensure a good thermal contact between the sample and the membrane; otherwise, the contact thermal resistance will result in an underestimation of the thermal conductivity of the nanostrucuture. Multiple-measurement approach was adopted to estimate the contact thermal resistance. Each sample was measured on devices with different separations between two membranes and contact thermal resistance can be determined by linearly fitting the

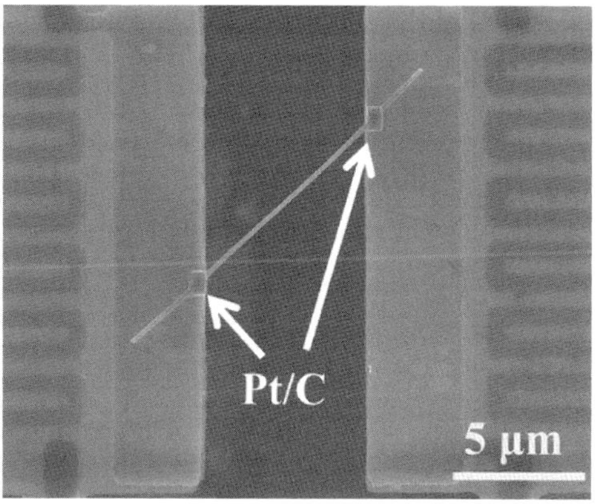

Fig. 5.13: The SEM images of a Bi_2Se_3 nanoribbon with Pt/C composites thermal anchor bonding at the contacts [105].

measured thermal resistance data. In addition, thermal anchor bonding by electron beam-induced deposition (EBID) technique is proved to be another effective way to reduce contact thermal resistance between the one-dimensional nanostructure and suspended membranes, as shown in Fig. 5.13.

5.3. Application

In the regime where quantum effects are dominant, the energy spectra of electrons and phonons can be controlled through altering the size of the structures, leading to new ways to manipulate the properties of these materials, especially their thermal transport properties for selected applications. Each low-dimensional structure can be considered to give rise to a new material, even though the material may be made of the same atomic structure as its parent material. Such small nanostructures bring us numerous opportunities and challenges. For example, the decreasing device size enables the significant increase of integration degree in integrated circuits, but on the other hand, the enhancement of integration requires better thermal management, including both heat dissipation, heat energy conversion, etc. Two interesting limits of the thermal conductivity in nanosystems can be considered.

1. In the low thermal conductivity limit, effort is going into developing very low thermal conductivity nanostructures which might be used for thermoelectric devices, thermal insulation applications, etc.

2. In the opposite limit, one might consider carbon allotropes and their derivatives, which possess high thermal conductivity, for thermal management in nanoelectrics.

5.3.1. *Thermoelectrics*

5.3.1.1. *Background of thermoelectric*

Energy and the environment have become some of the most critical current issues. The need for sources of energy other than fossil fuels, as well as the efficient use of our current fossil fuel supply, has sparked significant research into alternative energy sources and different types of energy conversion technologies. Currently, only 30% of the engine power is converted into useful work, like driving a vehicle and its accessories load [106]. About 70% of the energy consumption of fossil fuels has been wasted in the form of heat by engines or factory systems, which was emitted with exhausted gas (500–800 K) or taken by the cooling systems. If the waste heat can be recovered, a considerable amount of fuel can be saved [107]. A serious and concrete effort should be made towards conserving waste energy through waste heat recovery techniques to increase the energy efficiency of fossil fuels [108]. Thermoelectrics have long been recognized as a potentially transformative energy conversion technology due to their ability to convert heat directly into electricity or create a temperature difference by consuming electricity for cooling or heating. They can be used in many specialty all-solid-state applications, such as coolers, heat pumps, and electrical power generators. Thus, thermoelectric devices have the potential to make significant contribution to reducing greenhouse gas emission and providing clean energy.

State-of-the-art thermoelectric devices contain up to several thousand individual thermoelectric modules. Thermoelectric modules consist of alternating n-type and p-type semiconductors, which are connected electrically in series and thermally in parallel, as shown in Fig. 5.14. Electrical current flows through thermoelectric devices instead of gases or liquids in conventional power generation or heat pump systems. So, there is no moving fluids or moving parts in thermoelectric devices. In this way, thermoelectric devices can provide good performance because of their reliability, silent and vibration-free operation. All of these advantages make them promising candidates to recover electrical power from the waste heat, such as that in automotive exhaust systems to increase fuel efficiency. In the past decades, several commercial applications have been realized by using the available

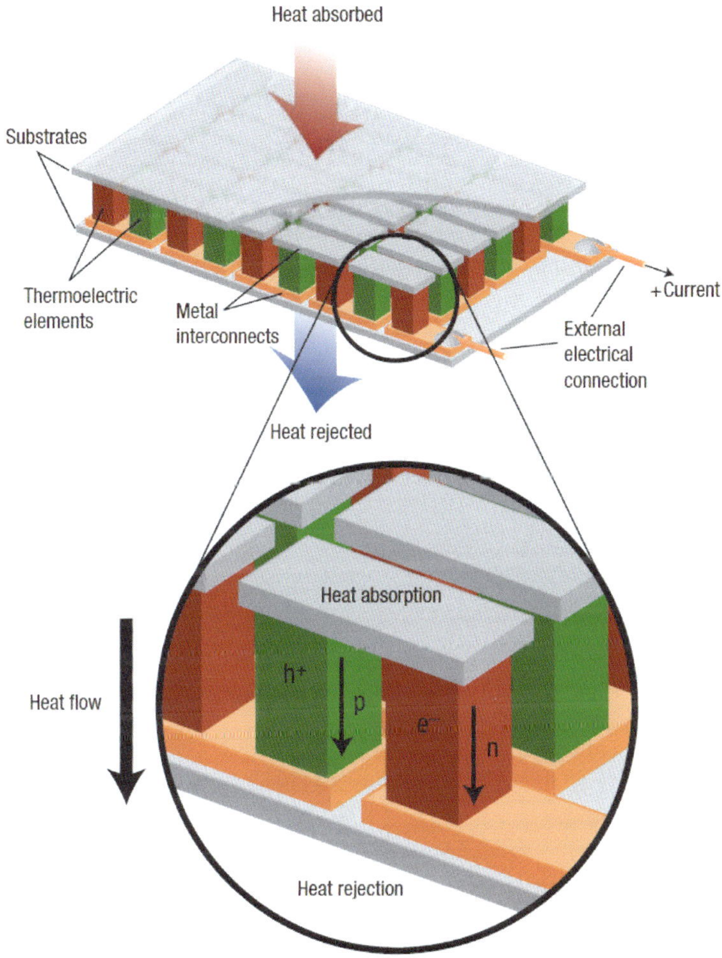

Fig. 5.14: Thermoelectric module showing the direction of charge flow on both cooling and power generation [112].

thermoelectric materials [109–111]. But thermoelectric devices are not in common use because of their low efficiency.

5.3.1.2. *Thermoelectric efficiency*

The link between the intrinsic properties of a thermoelectric device and its performance usually is given by the so-called figure of merit. The figure of merit (ZT), which is directly related to the device efficiency, has remained near 1 or even lower for commercial thermoelectric materials. Indeed, even at the macroscale, the best energy conversion efficiency of thermoelectric devices typically is of the order of 10% of the efficiency of the ideal Carnot thermodynamic cycle. A ZT value of 3–4 [113] is required to compete effectively with conventional refrigeration and power generation units. The efficiency of a thermoelectric device is defined as the conversion efficiency from heat to electrical energy, which depends on temperatures, T_{high} and T_{low}, at hot and cold junctions, respectively:

$$\eta = \frac{T_{high} - T_{low}}{T_{high}} \frac{M - 1}{M + T_{high}/T_{low}} \tag{5.34}$$

$$M = \left[1 + \frac{Z}{2}(T_{high} - T_{low})\right]^{1/2} \tag{5.35}$$

$$Z = \frac{S^2\sigma}{\kappa_l + \kappa_e} \tag{5.36}$$

Z is called the figure of merit, which is a material parameter and is determined by the Seebeck coefficient (S), electrical conductivity (σ) and thermal conductivity (κ), which is shown in Eq. (36). It is obvious that only increasing the figure of merit (Z) and the hot-junction temperature (T_{high}) can enhance the conversion efficiency. Hence, a material with high Seebeck coefficient and electrical conductivity but low thermal conductivity is desirable. The challenge of enhancing ZT is the interdependence of three thermoelectric properties S, σ, and κ [112]. Specifically, electrical conductivity increases with increasing the carrier concentration of the material, while Seebeck coefficient shows a decreasing trend, as shown in Fig. 5.15. In addition, both electrical and thermal conductivities increase with carrier concentration due to the contribution of electrons, which carry both charge and heat. Since these properties are not independent, it is extremely difficult to maximize ZT.

As we discussed before, if the dimensionality of the material is decreased to nanoscale, the introduction of phonon confinement effect and many interfaces, which scatter phonons more effectively than electrons, allows the thermal conductivity reduction of nanostructured materials with enhanced ZT, suitable for thermoelectric applications. So, nanostructuration is one of the approaches to increase the figure of merit in thermoelectric materials [114–120]. For example, Z. F. Ren's group reported a peak ZT of 1.4 in p-type Bi_2Te_3 nanocrystalline [121] (30% higher than the commercial ingot) and 0.95 in p-type SiGe nanocrystalline [52] (50% higher than the previous

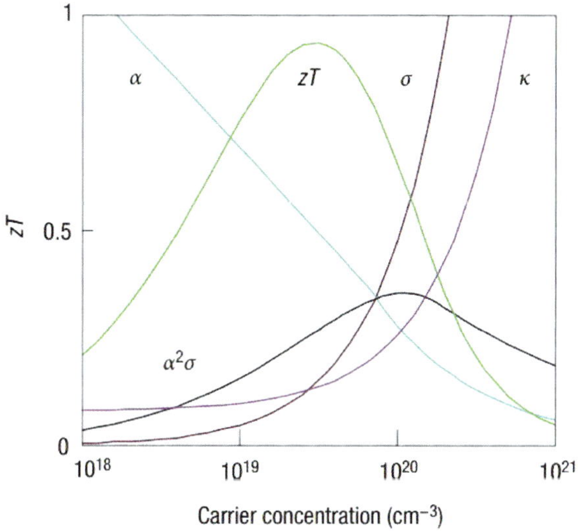

Fig. 5.15: Maximizing ZT of a thermoelectric material involves a compromise of thermal conductivity (k; plotted on the y axis from 0 to a top value of $10\,\mathrm{W/m \cdot K}$), Seebeck coefficient (α; 0 to $500\,\mu\mathrm{V/K}$), and electrical conductivity (σ; 0 to $5,000/\Omega\,\mathrm{cm}$) through tuning carrier concentration [112].

reported record in p-type SiGe alloy). It has been reported that $ZT = 2.6$ has been achieved in SnSe single crystals with layered nanostructure due to exceptionally low lattice thermal conductivity of $0.23\,\mathrm{W\,m^{-1}\,K^{-1}}$ [122]. Therefore, the ability to create nanostructured thermoelectric materials has led to remarkable progress in enhancing thermoelectric properties, making it plausible that thermoelectrics could start being used in new settings in the near future.

5.3.1.3. *Thermoelectric applications*

Applications of thermoelectric generators range from the portable energy sources powered by the human body heat to electrical power generation from the vehicle exhaust heat [111, 112, 123–125]. Solar energy is also an alternative heat source via a photo-thermal conversion, and the solar thermoelectric generator (STEG) exhibits much potential to convert solar energy to electricity [126–128]. In particular, flexible thin-film STEG is expected to be used as power supply for the wireless sensor and the microscale devices because of its portability and high integration [129, 130].

For instance, in a STEG, solar radiation is used as the heat source for a TEG. Sunlight consists of three main components in terms of wavelength: ultraviolet (UV) light ($k < 400\,\text{nm}$), visible light ($400\,\text{nm} < k < 800\,\text{nm}$), and infrared light ($k > 800\,\text{nm}$), accounting for 4%, 53%, and 43% of the total solar energy, respectively. While photovoltaic devices may efficiently convert visible and near-infrared light into electrical power, thermoelectrics could efficiently convert longer infrared wavelengths into electrical power [131–137]. Therefore, the use of thermoelectric devices in tandem with photovoltaics could convert almost the entire solar spectrum into electrical power. To date, the best experimental result for a solar thermoelectric generator has a maximum efficiency of 4.6% for a device fabricated by Kraemer *et al.* [131] using nanostructured thermoelectric materials of Bi_2Te_3 with $ZT \sim 1.03$. To achieve high efficiency in a STEG, high optical concentration can be used. A STEG under a solar concentration of 100, dual-axis tracking, and a hot side temperature of 1000°C could theoretically achieve 14.3% total efficiency using materials with $ZT = 1$ [131–137]. The total efficiency under the same conditions could improve to 21.1% with $ZT = 2$ [130].

5.3.2. *Nanoelectronics*

Microelectronic devices have been continuously scaled down in the drive for higher yield, lower cost, and better performance. As the electronic industry moves toward nanometer designs, one of the most important challenges is growing chip power consumption. Heat removal has become a crucial issue for continuing progress in the electronic industry, and thermal conduction in low-dimensional structures has revealed truly intriguing features. Thus, thermal management in electronic circuits is becoming an integral part of the design [138]. As the performance of ultralarge-scale integrated circuits depends on temperature T, even a small increase in T results in reduction of the device lifetime. A possible approach for solving the thermal problem is finding non-conventional materials with extremely high thermal conductivity k, which can be integrated with Si complementary metal-oxide semiconductor (CMOS) technology. The superb thermal conduction property of carbon allotropes and their derivatives is beneficial for the proposed electronic applications and establishes excellent materials for thermal management.

For instance, an important implication of extremely high thermal conductivity of graphene is its possible use for thermal management in future

ULSI circuits. Graphene layers can be naturally attached to heat sinks, thus, avoiding the problem of thermal contact resistance. Furthermore, it increases the range of graphene applications as the thermal management material in optoelectronics, photonics, and bioengineering.

5.3.2.1. *Heat pipe*

The boosting development of electronic technologies has enabled smart devices that are smaller but more powerful. A growing challenge under such a trend is to develop matching cooling technologies with higher heat transfer capability but in more compact size. In order to address this challenge, some attempts have been made to enhance the traditional materials and cooling devices such as heat pipes and vapor chambers by introducing carbon nanotubes (CNTs) [139, 140], microscale or nanoscale pillar array [141, 142] and other nanoscale enhancing structures [143].

Heat pipes are heat transfer mediums with significantly high effective thermal conductivity [144]. Various types of heat pipes including wick heat pipes, pulsating heat pipes and thermosyphons are applied in several energy systems and cooling devices [145–150]. The main parts of the heat pipes are evaporator and condenser. Heat pipes are made of tubes, in most of the cases, the metal tubes achieve the highest thermal conductivity, and are partially charged with an operating fluid [151, 152]. Fluid evaporation in heat source and its condensation in heat sink result in two-phase heat transfer [153]. Various types of heat pipes utilize different driving forces to return condensed liquid from condenser to evaporator. Figure 5.16 shows the schematic of the pulsating heat pipe model. The most influential factors on heat transfer capability of the heat pipes can be listed as structure, heat load, inclination angle, filling ratio and the applied working fluid [154–157]. Among the mentioned effective parameters, working fluid and its thermophysical specifications significantly affect thermal performance [158, 159]. Fluids with high thermal conductivity and low dynamic viscosity are more appropriate for heat transfer augment of the heat pipes [160]. The most conventional method to enhance thermal conductivity of the fluids is adding nanoparticles to them [161–163]. Dispersion of nanosized solid materials in the base fluid increases the thermal conductivity due to higher thermal conductivity of the solid phase compared with the liquid. Moreover, other properties of fluid such as dynamic viscosity can be changed because of adding the solid structures. A common feature of these approaches is the adoption of micro- or nanoscale structures for the purpose of taking

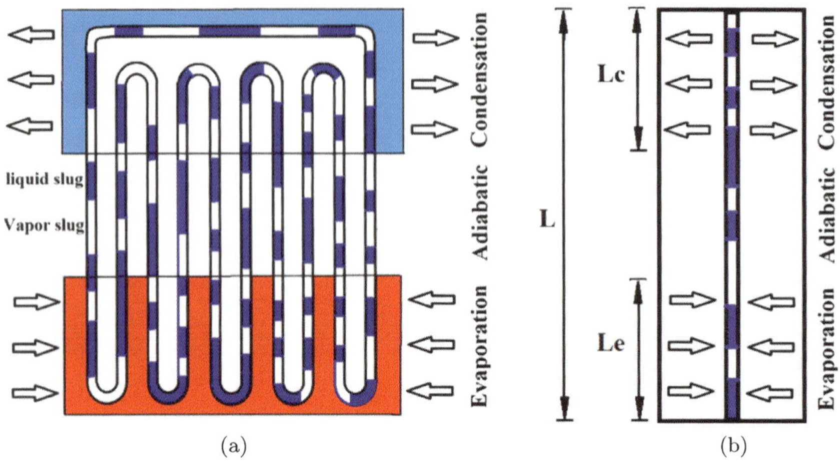

Fig. 5.16: Schematic of the pulsating heat pipe model.

advantage of their superior interfacial and thermal properties over traditional materials.

Tharayil *et al.* [164] utilized graphene–water nanofluid with tested concentrations being 0 vol%, 0.003 vol% and 0.006 vol%, respectively, as working fluid in a miniature loop heat pipe. The obtained results revealed that utilizing the nanofluid led to reduction in evaporator wall temperature and enhancement in heat transfer coefficient. The effect of concentration and heat load on heat transfer coefficient of the heat pipe is shown in Fig. 5.17. Graphene nanofluids can also be used in other types of heat pipes such as pulsating heat pipes [165] (Fig. 5.16), wick heat pipe [166] and grooved heat pipe [167]. The obtained results revealed that using the nanofluid enhanced the heat transfer in these different heat pipes. Based on the literature, graphene/CNT nanofluids are influential in heat transfer behavior in heat pipes.

5.3.2.2. *Thermal interface materials (TIMs)*

Development of the next generations of integrated circuits (ICs), three-dimensional (3D) integration and ultra-fast high-power density transistors makes the thermal management requirements extremely severe [168–173]. Efficient heat removal became a critical issue for the performance and reliability of modern electronic, optoelectronic, photonic devices and systems. The problem of thermal management of the chip includes two tasks:

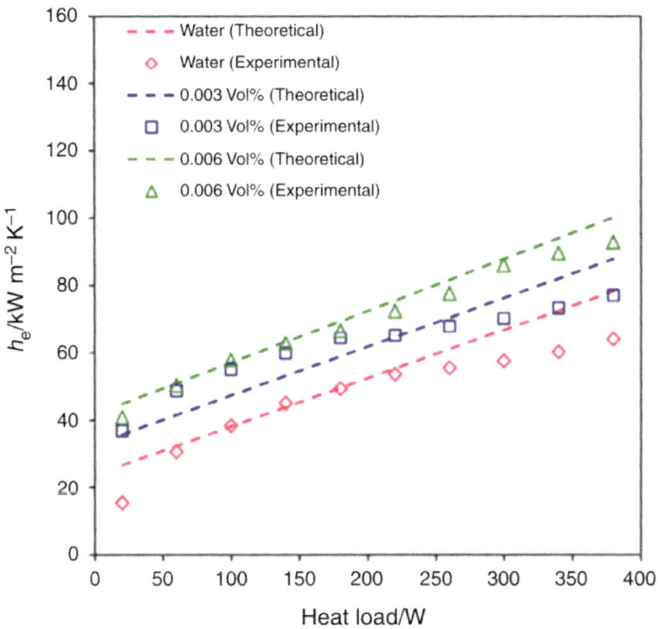

Fig. 5.17: Effect of graphene concentration on heat transfer coefficient [164].

maintaining the average temperature of the chip below a certain design point and keeping the temperature of the hottest spot below a certain level. Improved thermal interface materials (TIMs) can help to accomplish both tasks. The TIMs applied between heat sources and heat sinks are essential ingredients of thermal management [169–173]. The function of TIMs is to fill the voids and grooves created by imperfect surface finish of two mating surfaces (Fig. 5.18), thus forcing air out and improving surface contact and the conduction pathway across the interface (Fig. 5.18(b)). An ideal TIM consists of a material combining low bond line thickness with high thermal conductivity and low thermal contact resistance.

TIMs' performance is characterized by $R_{TIM} = \frac{BLT}{k} + R_{C1} + R_{C1}$, where BLT is the bond line thickness, i.e., the thickness of the TIM, and $R_{C1,2}$ are the TIM's contact resistance with the two bounding surfaces. Depending on the type of TIM and application, the different terms of the equation will have more or less impact on the overall performance, and when choosing or designing a TIM, it is important to identify which parameters to optimize. TIMs have to be mechanically stable, reliable, non-toxic, low cost and easy

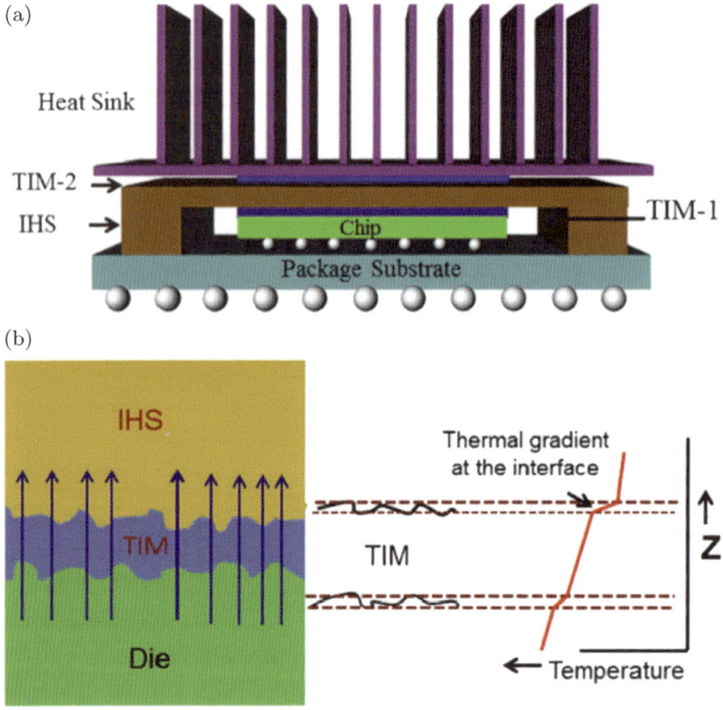

Fig. 5.18: (a) Schematic of the action of the thermal interface material, which fills the gaps between two contacting surfaces. (b) The heat removal improves with the higher thermal conductivity, smaller bond line thickness and smaller contact resistance of TIMs with the surfaces [174].

to apply [169–173]. They should possess as high k as possible, as well as low coefficient of thermal expansion.

1. CNT: One promising TIM candidate is the CNT array, which has been shown with excellent thermal conductivity. The emerging challenge is how to minimize the thermal contact resistance between CNT arrays and the mating surface, and how to fabricate well-aligned high-quality CNT arrays.
2. Graphene: In 2008, it was discovered that graphene has extremely high intrinsic k, which exceeds that of CNTs [10, 57, 175, 176]. Multilayer graphene (MLG) retains excellent thermal properties [57, 176]. It has been demonstrated that addition of an optimized mixture of graphene and multilayer graphene to the composites with different

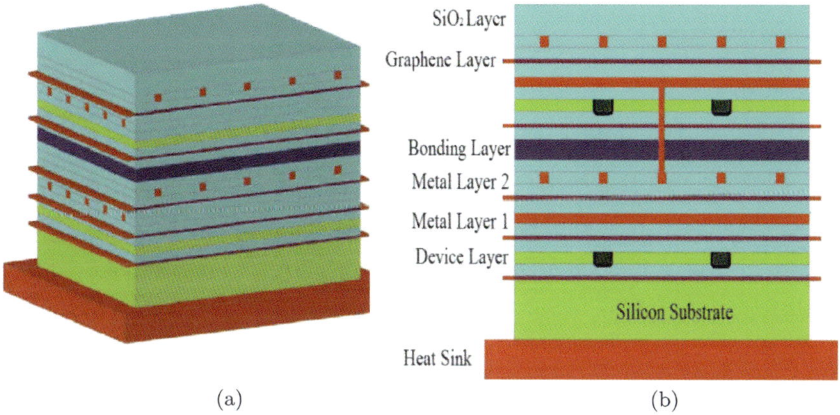

Fig. 5.19: Schematic of the 3D chip with Graphene lateral heat spreaders or TIM (a) 3D view and (b) cross sectional view. Graphene heat spreaders are attached to side heat sinks and embedded underneath the device and interconnect layers [177].

matrix materials produces the record-high enhancement of the effective thermal conductivity at the small filler loading fraction ($f \leq 10\,\text{vol}\%$). The thermal conductivity enhancement due to the presence of graphene in the composites has been observed for a range of matrix materials used by industry. Theoretical considerations indicate that the graphene-based thermal interface materials can outperform those with carbon nanotubes, metal nanoparticles and other fillers owing to graphene's geometry, mechanical flexibility and lower Kapitza resistance at the graphene-based material interface [174].

For instance, Fig. 5.19 shows a three-dimensional model of 3D chip, composed of two strata face-to-face stacking [177]. Device and interconnect layers generate heat due to Joule heating and cause localized heat generation. Graphene heat spreaders with two-side heat sinks are placed underneath the interconnect layers and the device layers.

Using the software COMSOL with the help of finite element analysis, 3D thermal modeling of 3D integrated circuit (IC) was conducted [177]. As seen in Fig. 5.20(a), the equivalent thermal resistance for 3D IC with the incorporation of graphene heat spreaders is only half of that without graphene layer. And the relation between maximum temperature and the variation of different thermal conductivity of single-layer graphene (SLG) and multiple layer graphene (MLG) was explored. As shown in Fig. 5.20(b),

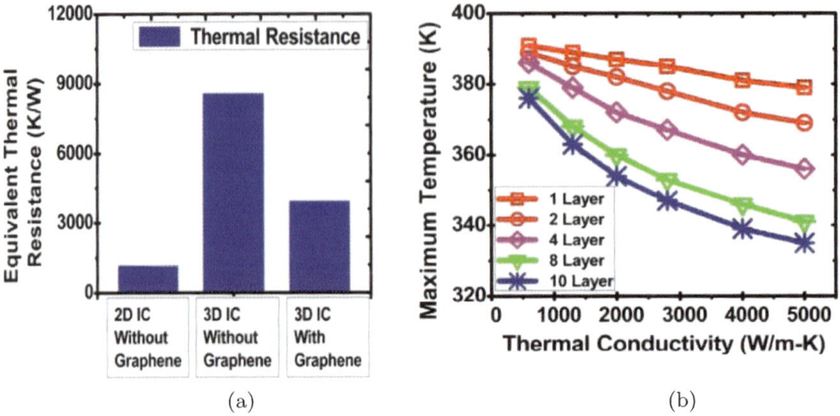

Fig. 5.20: (a) Equivalent thermal resistance for different design schemes. (b) Maximum temperature variation with increasing numbers of Graphene layers and varying thermal conductivity for 3D IC [177].

the results illustrate that when both the atomic layer and thermal conductivity are larger, the maximum temperature is lower. Compared with SLG, graphene multilayers are much cheaper and are expected to retain the heat conducting property.

In addition, graphene layers can be naturally attached to heat sinks, thus, avoiding the problem of thermal contact resistance, which is a major issue for CNTs. Using MLG as TIM might be more technologically feasible, which is important for the heat management of the growing industry of nanoelectronics.

5.4. Future Development

Nanoscale structures allow for the tailoring of thermal transport properties of semiconductors to obtain materials with thermal features suited for targeted applications [178–180]. Depending on the applications, different nanostructures can be prepared or designed to either achieve a low thermal conductivity for thermal insulation and thermoelectric devices or a high thermal conductivity for thermal interface materials used in the continuing miniaturization of electronics. A broad range of nanostructures has been discovered, investigated and discussed, ranging from array structures, holey structures, hierarchical structures, to nanofilms/layers, fibers, nanowires and nanotubes. Different factors that impact the thermal conductivity of different materials with nanostructures make empirical searching for optimal thermal materials a big challenge and exhausting work due to the tremendous degrees of freedom in the composition and structure of nanomaterials. Exploration and designing of materials and structures with desired thermal transport properties have large potential for application. However, two bottlenecks limit the designing efficiency: materials selection and structure designing. The application of materials informatics is a potential tool to help solving the above two challenges in the heat transfer field.

5.4.1. *Materials informatics*

Materials informatics (MI) is integration of material–property calculations and measurements with informatics method to accelerate the material discovery and design [181–183]. MI is mainly used in two directions: materials selection and structure designing.

5.4.1.1. *Materials selection*

Selecting the most appropriate material from a large number of candidates is the first key during the designing of thermal devices and it remains an experimental challenge due to high-cost and time-consuming synthesis procedures due to a wide range of thermal property of materials. High-throughput screening (HTS) for ultimate high/low lattice thermal conductivity materials is a combination of machine-learning algorithms, physical insights, and automatic *ab-initio* calculations. Application of machine learning requires data, descriptors, and machine-learning models. Currently, databases including tens of thousands of crystal compounds have been constructed, including Materials Project [185], AFLOW [186], ICSD [187], OQMD [188, 189], and AtomWork [190], as shown in Fig. 5.21(a). The general strategy in HTS is to incorporate every information of the crystal that may have correlation with the objective thermal properties, and the actual implementation depends on the availability of the data.

The anharmonic lattice dynamics (ALD) method using interatomic force constants (IFCs) obtained by first-principles has been developed to accurately calculate thermal conductivity [191–193]. As shown in Fig. 5.22, three types of data are involved in the ALD calculation: (I) general properties associated with the crystals, the data of which include lattice parameters, atom coordinates, number of elements, number of sites, density, volume, volume/atom, spacegroup number (symmetry), atom types, atom numbers, etc.; (II) harmonic properties associated with harmonic IFCs, the data of which include phonon dispersion relations, phonon density of states, and

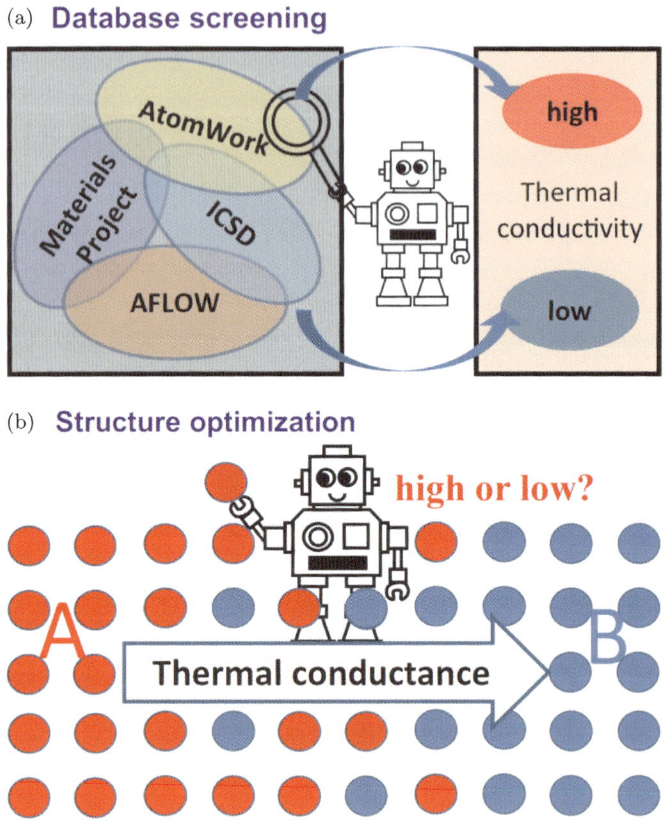

Fig. 5.21: Application of materials informatics on thermal transport: (a) database screening and (b) structure optimization [184].

group velocity; (III) anharmonic properties associated with anharmonic IFCs, the data of which include phonon relaxation time, mean free path, and Grüneisen parameter. When screening materials using machine learning, all the three types of data involved in ALD calculation can be used as descriptors, and the structure descriptors or/and elemental descriptors are often adopted as they are largely accessible. However, it was suggested that combinations of different types of descriptors can be useful for machine learning [194]. Wang *et al.* [195] screened several thousands of compounds from the ICSD database [187] and gave guiding rules for searching for better thermoelectric materials according to the correlations between the power factor and different physical properties. For example, sintered thermoelectric

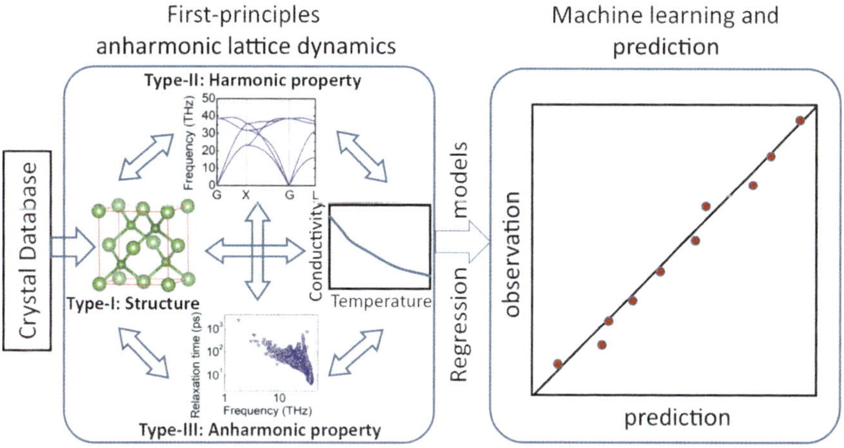

Fig. 5.22: Schematics of the high-throughput screening by combining the first-principles anharmonic lattice dynamics and machine learning. The descriptors currently widely used for machine learning are structure and chemical information [184].

compounds with large band gaps, heavy carrier effective masses, and more atoms per primitive cell are expected to have large power factors.

5.4.1.2. *Structure designing*

As the length scale of materials decreases to nanoscale, the transport of main heat carriers (phonons) becomes more ballistic and heat conduction becomes more controllable through manipulating the nanostructures, as shown in Fig. 5.21(b). However, it is rather difficult to identify the detailed optimal structure for phonon transport due to various and coupled parameters including roughness [196, 197], vacancy defects [198], lattice orientation [199, 200], nanoinclusions [201], and interfacial adhesion or bonding [202, 203]. Besides, the coupling of constructive/deconstructive phonon interference and resonance effects in different types of nanostructures makes the structure designing and optimization more complicated. Developing an effective optimization method for designing nanostructures with desired thermal property is necessary and has great potential for application. Two informatics methods were widely used: Bayesian optimization (BO) [204] and MCTS [205].

BO is an experimental design algorithm based on machine learning. As shown in Fig. 5.23, supposing that the transport property of n candidate structures is initially calculated, the Bayesian linear regression model

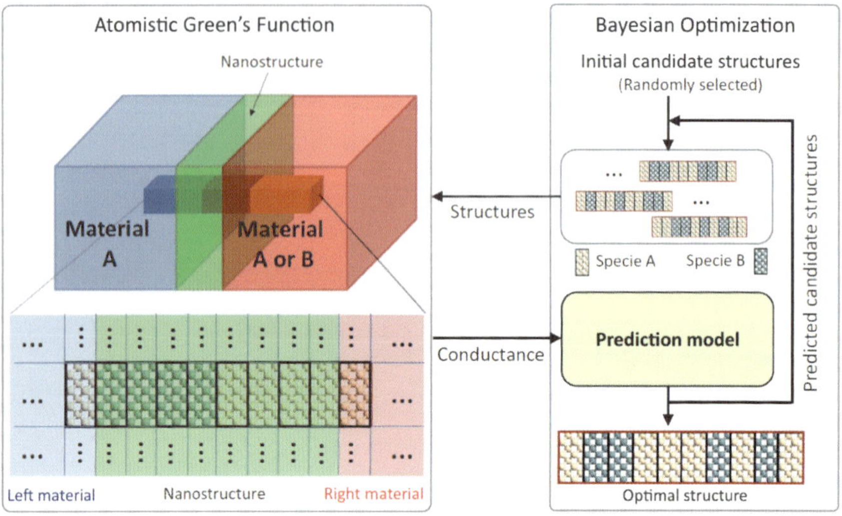

Fig. 5.23: Schematics of the Bayesian optimization combing with the atomistic Green's function for transport property calculation [184].

with random feature map is used to build the prediction model based on n pairs of structures and calculated transport properties. The open-source BO library COMBO [206] was developed to perform the optimization process automatically. After the prediction model is constructed, a predictive distribution of transport property is estimated for each remaining candidate. The best candidate is chosen based on the criterion of expected improvement. Finally, the exact transport property is calculated for the chosen candidate, and it is added to the training examples. By repeating this procedure, the calculation of transport property is scheduled optimally, and the best candidate can be found quickly [184]. The BO is very effective and accurate when the total number of candidates is around several hundred thousand. However, the total number of candidates is relatively larger, BO becomes more memory and time consuming.

Another effective method, MCTS, combines the generality of random simulation with precision tree search. The MCTS algorithm is based on a search tree that is built node by node according to the evaluation of each simulated case, as shown in Fig. 5.24. Each node contains two important information: an estimated value based on simulation results and the number of times it has been visited. In the process of MCTS, there are four typical steps: selection, expansion, simulation, and backpropagation [184].

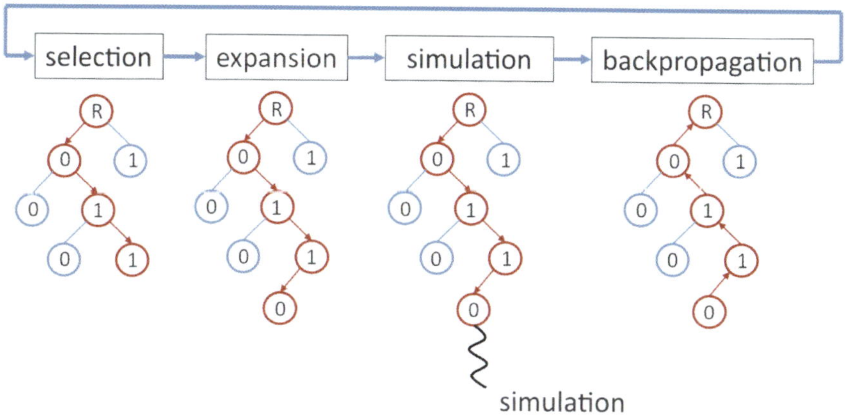

Fig. 5.24: Schematics of the Monte Carlo tree search (MCTS) algorithm, which is composed of four processes: selection, expansion, simulation, and backpropagation. The R, L, and C are nodes in the tree [184].

1. Selection: Starting at root node R, recursively select optimal child nodes according to larger or small upper confidence bound (UCB) score until a leaf node L is reached. The UCB score is calculated by

$$u_i = \frac{V_i}{n_i} \pm b\sqrt{\frac{2\ln N_{parent}}{n_i}} \tag{5.37}$$

   where b is a tunable bias parameter to balance the tree exploration and exploitation, V_i is the accumulated simulation values of all structures that was played out from this node through all visits, n_i is the number of the times the node has been visited, N_{parent} is the total number of times that its parent has been visited. + and − are for maximum and minimum conductance optimization case, respectively.
2. Expansion: If the leaf node L is a not a terminal node, then create one or more child nodes and select one node C.
3. Simulation: Randomly select one playout from node C and do the conductance calculation.
4. Backpropagation: Use the calculated thermal conductance value to update the n_i, N_{parent}, and V_i values of the nodes on the path back from node C to node R.

MCTS is a popular method for making optimal decisions in artificial intelligence problems. It has to be mentioned that MCTS just offers structure close to the global optimal one with high efficiency.

5.4.1.3. *Challenges*

HTS can considerably speed up the selection of best materials from databases for a given objective and has been successfully applied in many fields including catalysis [207, 208], battery technologies [209], thermoelectric materials [195, 210], chemical probes [211], polymers [212, 213], and magnetic materials [214]. MI not only discovers optimal structures efficiently, but also helps us understand new physics behind the found/designed novel materials and structures. It is also worth mentioning that the MI-based design algorithm can be easily extended to transport of other quasi particles (e.g., electron, photon, magnon).

In the future, researchers need to overcome the following technical challenges [184]:

1. There is large gap between the "big data" required for credible machine learning and the "small data" we can collect. We expect more data to come in near future.
2. Since thermal transport consists of multiscale structures ranging from nanometer to micrometers, the current structure design is still limited to small scales and systems. It's necessary to balance the accuracy and speed to an acceptable level to save the calculation time and improve the designing efficiency.
3. To further improve the efficiency of the optimization, new effective descriptors should be developed.
4. There is often some gap between the optimal design obtained and the actual realization of the material. Some structural deviations, such as the non-sharp interface in actual fabrication due to inter-diffusion at the interface and the error in each layer thickness due to uncertainty in the growth/deposition speed, can spoil the optimization.

There have been successful applications of MI on heat transfer problems during the past decade. In the near future, the application of MI to heat transfer is expected to expand in various forms of heat transfer (conduction, convection, and radiation), from nanoscales to macroscales, and from simulations to experiments.

5.4.2. *Application development*

Depending on the nanomaterials with either a high thermal conductivity or a strongly suppressed thermal conductivity, different applications were developed, as shown in Fig. 5.25 [215].

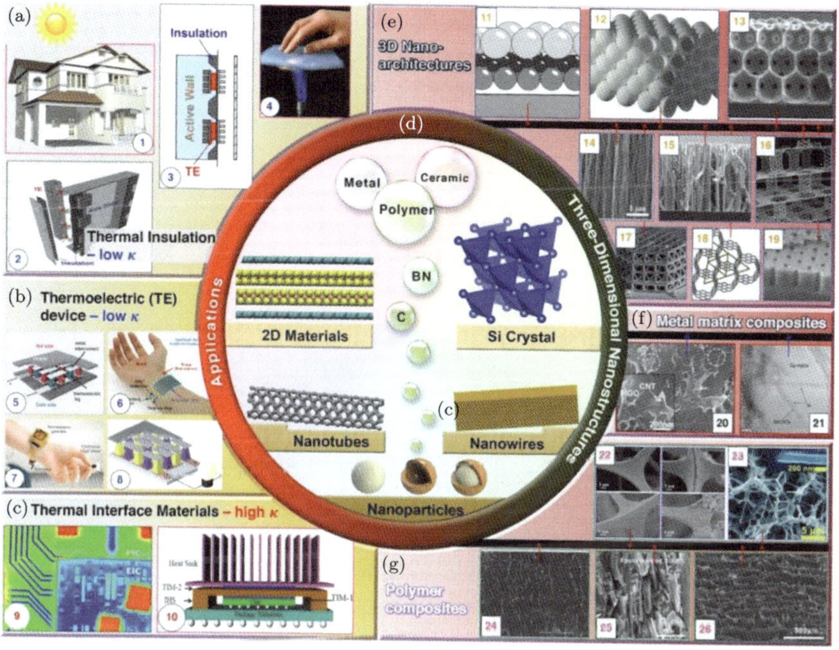

Fig. 5.25: The palette of 3D nanostructures for thermal management. (a) Example applications for thermal insulation, including the house (1 [238]), the active window (2 [239]), the active wall (3 [239]), and the aerogel above a blow torch (4 [240]). (b) Example application of thermoelectric devices, including a thermoelectric module sketch (5 [241]), a wearable thermoelectric generator (6 [242]), a body-powered wireless pulse oximeter (7 [243]), a thermoelectric generator powered flat bulk (8 [244]). (c) Electronics, including a printed circuit board (9 [245]), a schematic view of the electronic packing (10 [174]). (d) Different low-dimensional nanomaterials. (e) Different 3D nanoarchitectures, including colloidal assemblies (11 [246]), silica hollow sphere colloidal crystals (12 [247]), Cu inverse opals (13 [248]), vertically aligned carbon nanotube array (14 [249]), aligned Cu nanowire arrays (15 [250]), the atomic model of a pillared graphene (16 [251]), a carbon nanolattice (17 [252]), the atomic model of a carbon honeycomb structure (18 [253]), and a holey silicon (19 [254]). (f) Different metal matrix composites, including Cu matrix with carbon nanotube and reduced graphene oxide network (20 [255]), Cu matrix with embedded carbon nanotubes (21 [256]). (g) Polymer composites with 3D nanostructures, including melamine foam with BN nanosheets assemblies (22 [257]), 3D BN cellular architecture (23 [258]), polystyrene composite with 3D segregated double networks (24 [259]), epoxy composite with 3D BN network (25 [260]), and 3D BN nanosheets aerogels (26 [261]) [215].

5.4.2.1. *Thermal insulation*

Energy savings for both residential and commercial buildings result in a lower carbon footprint, as promoted by the U.S. Department of Energy (Fig. 5.25(a)) [216]. Nanotechnology-based materials have previously been used by space research, pharmaceuticals and electronics, but in the last decade several nanotechnology-based thermal insulation materials have appeared in the building industry. For energy saving in both residential and commercial buildings, there has been a continuing search for high performance, light weight, and mechanically strong, thermally insulating materials (low thermal conductivity). For example, the addition of expanded polystyrene products, including graphite powder additive (hereafter, "EPS Graphite"), to a wide range of materials in building construction can cause 20% lower thermal conductivity [217]. They have a much lower thermal conductivity because of the reduced interior thermal radiation blocked by the fine graphite powder. EPS Graphite becomes a widely-used thermal insulation material. The new EPS graphite products with their higher compressive strength are useable for insulating flat and high roofs, floors and slab structures [218]. Nanoceramic thermal insulation coatings are considered to be the most critical nanotechnology-based thermal insulation material [217]. They are capable of producing an adequate thermal insulation value of building structures in an extremely thin layer.

Good thermal insulation is also required for the electrical, optical, and space applications in order to tightly regulate heat transfer during operation. Typical thermally insulating materials, including wool, styrofoam, and wood cork, often exhibit a thermal conductivity close to air ($\sim$0.03 W/m $\cdot$ K) [219–221], which is isotropic in nature but not ideal for effective thermal management. However, the anisotropic materials usually require complex designs and energy-consuming manufacturing processes, which thus prohibit their widespread application toward large-scale systems. Nanocellulose is one of the anisotropic nanomaterials for thermal insulation application. It is an earth-abundant biomass resource that exhibits great potential for generating sustainable products that have low environmental, human health, and safety risks [222–227]. Nanocellulose-based products are super thermal insulators with a low thermal conductivity that can potentially find applications in energy-efficient buildings, thermal insulation for space applications, and electrical devices insulation.

Nowadays, nanotechnology-based thermal insulation materials only feature in a narrow range of practice, but they offer many potential applications. Further improvement in mechanical strength and fabrication process [228] of potential nanomaterials for thermal insulation is necessary for large-scale and realistic applications.

5.4.2.2. *Thermoelectrics*

On the one hand, thermoelectric (TE) materials harvesting electricity from waste heat offer us an appealing option to increase the energy efficiency of fossil fuels and help us to overcome the energy crisis and environmental deterioration. Thermoelectric devices possess a vast potential for power generation from solar, automobile, industrial heat sources, and even human body heat (Fig. 5.25(b)) [229]. Nanostructuring has been shown as an effect avenue to suppress phonon transport by enhancing phonon scattering or introducing phonon confinement effect. In recent years, there has been great interest on the manipulation of the so-called wave-like coherent phonon transport, which can be realized through the presence of a secondary periodicity in the nanostructure. The phonon coherence has already been demonstrated to result in remarkably reduced thermal conductivity in phononic crystals, which opens up a new and promising avenue to construct high performance TE devices [230]. To achieve industrial application of TE devices, researchers should focus on TE materials that are composed from low-toxic, earth-abundant elements, and the mechanical strength, processability, durability, and stability of materials when developing high performance TE materials for industrial applications. The future development of TE materials should lie in the synergetic optimization and in-depth explorations (such as using novel technologies to provides a potential opportunity to decouple the k_e and σ).

On the other hand, another type of thermoelectric applications is to convert electrical power to thermal energy for cooling applications, which is based on the Peltier effect. As electronic devices become smaller in size, the power dissipated by modern electronic packages is non-uniform, leading to areas of elevated temperature, known as hotspots. The value of the heat flux at the hotspot location 2 to 5 times higher than the average heat flux [172]. The hotspot temperature limits performance and compromises reliability [231]. The new method for providing high flux, localized cooling for the removal of hotspots involves Thermoelectric Coolers (TECs). Charge coupled device (CCD), microelectronic devices, laser diodes, portable picnic coolers and blood analyzers are some applications that use TECs [232, 233]. The self-cooling framework using thermoelectric has been designed such that the computer chip hotspot at a given operating condition can successfully be cooled at an acceptable temperature without any need for additional power requirements [234]. Since nanoscale multilayer structure can significantly enhance thermoelectric response, and increase the voltage and power produced, nanoscale structure-based thermoelectric device is the trend for low-power and fast-response on-chip refrigeration solutions. Two points need more attention: the fabrication techniques of the integrated thermoelectric devices should be compatible with the standard fabrication of integrated circuits (IC); and the TE device can be scaled for producing higher voltage and power, or achieving solid-state cooling for applying on-chip temperature monitoring and for stabilizing surface temperature.

As thermometric devices are static (i.e., no moving parts and therefore silent, reliable and scalable), they are quite attractive in many applications as they operate for long time while requiring less maintenance. Although challenges still exist in further enhancing ZT and developing highly efficient TE devices for commercial applications, which require multidisciplinary and intense effort, the TE field will have a bright future.

5.4.2.3. *TIM*

The issues of controlling of heat flow, heat dissipation, and heat energy conversion in nanoscale systems have become a specific science field nowadays [235]. High thermal conductivity of the packaging substrates and thermal interface materials (TIMs) play a key role in the continuing miniaturization of electronic devices/systems (such as light-emitting diodes, integrated circuits and microprocessors) to remove heat dissipated during the high-power density operations to ensure their reliable performance and lifetime (Fig. 5.25(c)). The growing field of nanotechnology is providing new opportunities for advancement in the development of novel TIMs through control of the structure at all length scales. For instance, graphene-based TIMs achieved thermal resistances as low as $1.5 \times 10^{-2}\,\text{mm}^2\,\text{K/W}$ at the nanoscale [57, 236]. And metal-organic–inorganic nanocomposites also demonstrate high bulk thermal conductivities ranging from 211–277 $\text{W/m} \cdot \text{K}$ [237]. Metal nanowire filler is a potential breakthrough technology, despite a very high thermal conductivity, the thermal interface resistance is still very high due to high contact resistance and large BLTs. Finally, polymer nanowire array with a thermal interface resistance comparable to conventional grease is a completely new concept, which could spawn novel research, such as chain alignment of the polymer in a polymer

CNT/NW composite. All in all, the development of novel methods and technics in future can provide a promising thermal management solution that can significantly reduce the thermal resistance in applications with high heat dissipation.

5.5. Conclusions

The advancement of nanotechnology enables the construction of novel materials from the bottom up, based on low-dimensional nanomaterials (Fig. 5.25(d)), including 0D (e.g., fullerene and nanoparticles), 1D (e.g., nanotubes, nanowires, and nanothreads), [262] 2D (e.g., nanoribbons or nanosheets) nanomaterials [263], and 3D nanostructures [215], ranging from 3D nanoarchitectures (Fig. 5.25(e)), metal–matrix composites (MMCs) with nanostructured fillers (Fig. 5.25(f)), and the polymer composites with nanostructured fillers (Fig. 5.25(g)). The use of nanostructuring of materials is an effective strategy to modify the thermal conductivity of materials (either by phonon or electron transport). To facilitate engineering applications, it is vital to explore and develop cost-effective and efficient approaches to prepare high-quality, eco-friendly and naturally abundant nanostructures. With the help of materials informatics (MI) technique, significant research efforts will be devoted to either screening the materials with desired thermal transport properties or design novel materials/structures that satisfy different thermal management requirements. In addition, it is still an open question whether these nanomaterials/nanodevices can retain their efficient heat transfer capability under repetitive thermal or mechanical stress during service, which needs to be explored and verified in future research.

Acknowledgments

The authors gratefully acknowledge Qaing Fu for his fruitful collaboration in the study of thermal transport properties of 2D nanomaterials and 3ω measurement method. The work is also funded by Fundamental Research Project of Shenzhen Sci. & Tech. Project Fund (JCYJ20170412153845293, JCYJ20180227175343343, JCYJ20180507182843072), and also by Technical Project of Shenzhen Sci. & Tech. Fund (JSGG20180508151812903), and Guangdong Province SNSFC Project Fund (2018A030313973, 2020A1515011179).

References

1. J. M. Ziman, *Electrons and Phonons: The Theory of Transport Phenomena in Solids.* 1960, Oxford U6 — ctx_ver=Z39.88-2004&ctx_enc=info%3Aofi%2Fenc%3AUTF-8&rfr_id=info%3Asid%2Fsummon.serialssolutions.com&rft_val_fmt=info%3Aofi%2Ffmt%3Akev%3Amtx%3Abook&rft.genre=book&rft.title=Electrons+and+phonons&rft.au=Ziman%2C+J.+M&rft.series=The+International+series+of+monographs+on+physics&rft.date=1960-01-01&rft.pub=Clarendon+Press&rft.externalDBID=5A&rft.externalDocID=1753587&paramdict=en-US U7 — Book: Clarendon Press.
2. W. Kim and A. Majumdar, Phonon scattering cross section of polydispersed spherical nanoparticles. *J. Appl. Phys.* 99(8), 084306–084306-7 (**2006**).
3. D. Jou, V. A. Cimmelli, and A. Sellitto, Nonlocal heat transport with phonons and electrons: Application to metallic nanowires. *In. J. Heat Mass Transf.* 55(9–10), 2338–2344 (**2012**).
4. J. S. Lee and J. Jang, Hetero-structured semiconductor nanomaterials for photocatalytic applications. *J. Ind. Eng. Chem.* 20(2), 363–371 (**2014**).
5. Z. Zhang. and W. Guo, Intrinsic metallic and semiconducting cubic boron nitride nanofilms. *Nano Lett.* 12(7), 3650–3655 (**2012**).
6. B. Feng, Z. Li, and X. Zhang, Role of phonon in the thermal and electrical transports in metallic nanofilms. *J. Appl. Phys.* 105(10), 104315–104315-7 (**2009**).
7. B. Saha, T. D. Sands, and U. V. Waghmare, Thermoelectric properties of HfN/ScN metal/semiconductor superlattices: A first-principles study. *J. Phys. Condens. Matter.* 24(41), 415303 (**2012**).
8. E. S. Landry and A. J. H. McGaughey, Effect of interfacial species mixing on phonon transport in semiconductor superlattices. *Phys. Rev. B — Condens. Matter Mater. Phys.* 79(7) (**2009**).
9. Y. Shao, *et al.*, Graphene-based materials for flexible supercapacitors. *Chem. Soc. Rev.* 44(11), 3639–3665 (**2015**).
10. A. A. Balandin, *et al.*, Superior thermal conductivity of single-layer Graphene. *Nano Letters* 8(3), 902–907 (**2008**).
11. A. Lekawa-Raus, *et al.*, Electrical properties of carbon nanotube based fibers and their future use in electrical wiring. *Adv. Funct. Mater.* 24(24), 3661–3682 (**2014**).
12. L. Qiu, *et al.*, Functionalization and densification of inter-bundle interfaces for improvement in electrical and thermal transport of carbon nanotube fibers. *Carbon* 105, 248–259 (**2016**).
13. H. Li, *et al.*, Experimental investigation of thermal conductivity and viscosity of ethylene glycol based ZnO nanofluids. *Appl. Therm. Eng.* 88, 363–368 (**2015**).
14. K. Zhao, *et al.*, High thermoelectric performance and low thermal conductivity in Cu2-yS1/3Se1/3Te1/3 liquid-like materials with nanoscale mosaic structures. *Nano Energy* 42, 43–50 (**2017**).
15. W.-X. Zhou and K.-Q. Chen, Enhancement of thermoelectric performance by reducing phonon thermal conductance in multiple core-shell nanowires. *Sci. Rep.* 4(1), 7150 (**2014**).

16. L. M. Jiji, *Microscale Conduction* (**2009**).
17. G. P. Chen, *Nanoscale Energy Transport and Conversion: A Parallel Treatment of Electrons, Molecules, Phonons, and Photons* (New York, Oxford: Oxford University Press, **2005**).
18. Z. M. Zhang, *Nano/Microscale Heat Transfer* (New York, NY: McGraw-Hill, **2007**).
19. G. Chen, Phonon heat conduction in nanostructures. *Int. J. Therm. Sci.* 39(4), 471–480 (**2012**).
20. D. G. Cahill, *et al.*, Nanoscale thermal transport. *J. Appl. Phys.* 93(2), 793–818 (**2003**).
21. D. G. Cahill, K. Goodson, and A. Majumdar, Thermometry and thermal transport in micro/nanoscale solid-state devices and structures. *J. Heat Transf.* 124(2), 223–241 (**2002**).
22. Y. Cui and C. Lieber, Functional nanoscale electronic devices assembled using silicon nanowire building blocks. *Science* 291, 851–853 (**2001**).
23. M. Law, L. E. Greene, J. C. Johnson, R. Saykally, and P. Yang, Nanowire dye-sensitized solar cells. *Nat. Mater.* 4, 455 (**2005**).
24. X. Sun, Z. Zhang, and M. Dresselhaus, Theoretical modeling of thermoelectricity in bi nanowires. *Appl. Phys. Lett.* 74 (**1999**).
25. S. G. Walkauskas, *et al.*, Lattice thermal conductivity of wires. *J. Appl. Phys.* 85(5), 2579–2582 (**1999**).
26. Y. Chen, *et al.*, Monte Carlo simulation of silicon nanowire thermal conductivity. *J. Heat Transf.* 127(10), 1129–1137 (**2005**).
27. T. S. Tighe, J. M. Worlock, and M. L. Roukes, Direct thermal conductance measurements on suspended monocrystalline nanostructures. *Appl. Phys. Lett.* 70(20), 2687–2689 (**1997**).
28. D. Li, *et al.*, Thermal conductivity of individual silicon nanowires. *Appl. Phys. Lett.* 83(14), 2934–2936 (**2003**).
29. A. I. Boukai, *et al.*, Silicon nanowires as efficient thermoelectric materials. *Nature* 451(7175), 168–171 (**2008**).
30. S. M. Lee and D. G. Cahill, Heat transport in thin dielectric films. *J. Appl. Phys.* 81(6), 2590–2595 (**1997**).
31. W. Liu and M. Asheghi, Phonon-boundary scattering in ultrathin single-crystal silicon layers. *Appl. Phys. Lett.* 84(19), 3819–3821 (**2004**).
32. Y. S. Ju and K. E. Goodson, Phonon scattering in silicon films with thickness of order 100 nm. *Appl. Phys. Lett.* 74(20), 3005–3007 (**1999**).
33. M. Asheghi, *et al.*, Phonon-boundary scattering in thin silicon layers. *Appl. Phys. Lett.* 71(13), 1798–1800 (**1997**).
34. W. Dai and R. Nassar, A compact finite difference scheme for solving a three-dimensional heat transport equation in a thin film. *Numer. Methods Partial. Differ. Equ.* 16(5), 441–458 (**2000**).
35. J. R. Lukes, *et al.*, Molecular dynamics study of solid thin-film thermal conductivity. *J. Heat Transf.* 122(3), 536–543 (**2000**).
36. F. X. Alvarez and D. Jou, Memory and nonlocal effects in heat transport: From diffusive to ballistic regimes. *Appl. Phys. Lett.* 90(8), 83109 (**2007**).

37. C. Bi, *et al.*, Phonon boundary scattering effect on thermal conductivity of thin films. *J. Appl. Phys.* 110(4), 046102–046102-3 (**2011**).
38. A. J. Minnich, Thermal phonon boundary scattering in anisotropic thin films. *Appl. Phys. Lett.* 107(18), 183106 (**2015**).
39. F. Yang and C. Dames, Mean free path spectra as a tool to understand thermal conductivity in bulk and nanostructures. *Phys. Rev. B Condens. Matter Mater. Phys.* 87(3) (**2013**).
40. R. Taylor, The thermal conductivity of pyrolytic graphite. *Philos. Mag.* 13(121), 157–166 (**1966**).
41. K. Fuchs, The conductivity of thin metallic films according to the electron theory of metals. *Math. Proc. Cambridge Philos. Soc.* 34(1), 100–108 (**1938**).
42. E. H. Sondheimer, The mean free path of electrons in metals. *Adv. Phys.* 50(6) 499–537 (**2001**).
43. A. J. Griffin Jr, F. R. Brotzen, and P. J. Loos, Effect of thickness on the transverse thermal conductivity of thin dielectric films. *J. Appl. Phys.* 75(8), 3761–3764 (**1994**).
44. J. H. Kim,, A. Feldman, and D. Novotny, Application of the three omega thermal conductivity measurement method to a film on a substrate of finite thickness. *J. Appl. Phys.* 86(7), 3959–3963 (**1999**).
45. T. Yamane, *et al.*, Measurement of thermal conductivity of silicon dioxide thin films using a 3 omega method. *J. Appl. Phys.* 91(12), 9772–9776 (**2002**).
46. Q. Fu, *et al.*, Experimental evidence of very long intrinsic phonon mean free path along the c-axis of graphite. *Appl. Phys. Lett.* 106(3), 31905 (**2015**).
47. Y. Lan, *et al.*, Structure study of bulk nanograined thermoelectric bismuth antimony telluride. *Nano Lett.* 9, 1419–1422 (**2009**).
48. J. M. Worlock, Thermal conductivity in sodium chloride crystals containing silver colloids. *Phys. Rev.* 147(2), 636–643 (**1966**).
49. Kim, W., *et al.*, Reducing thermal conductivity of crystalline solids at high temperature using embedded nanostructures. *Nano Lett.* 8(7), 2097–2099 (**2008**).
50. Yang, J., H.-L. Yip, and A. Jen, Rational design of advanced thermoelectric materials. *Adv. Energy Mater.* 3 (**2013**).
51. Kim, W., *et al.*, Thermal conductivity reduction and thermoelectric figure of merit increase by embedding nanoparticles in crystalline semiconductors. *Phys. Rev. Lett.* **96**(4) 045901 (**2006**).
52. Joshi, G., *et al.*, Enhanced thermoelectric figure-of-merit in nanostructured p-type silicon germanium bulk alloys. *Nano Lett.* 8(12), 4670–4674 (**2008**).
53. Ma, Y., *et al.*, Enhanced thermoelectric figure-of-merit in p-type nanostructured bismuth antimony tellurium alloys made from elemental chunks. *Nano Lett.* 8(8), 2580–2584 (**2008**).
54. Lan, Y., *et al.*, Enhancement of thermoelectric figure-of-merit by a bulk nanostructuring approach. *Adv. Funct. Mater.* 20(3), 357–376 (**2010**).
55. A. Minnich and G. Chen, Modified effective medium formulation for the thermal conductivity of nanocomposites. *Appl. Phys. Lett.* 91(7), 73105 (**2007**).

56. A. J. Minnich, *et al.*, Modeling study of thermoelectric SiGe nanocomposites. *Phys. Rev. B Condens. Matter Mater. Phys.* 80(10) (**2009**).
57. A. A. Balandin, Thermal properties of graphene and nanostructured carbon materials. *Nat. Mater.* 10(8) 569–581 (**2011**).
58. S. Jin and H. Mavoori, Processing and properties of CVD diamond for thermal management. *J. Electron. Mater.* 27(11), 1148–1153 (**1998**).
59. M. J. Biercuk, *et al.*, Carbon nanotube composites for thermal management. *Appl. Phys. Lett.* 80(15), 2767–2769 (**2002**).
60. S. Ghosh, *et al.*, Extremely high thermal conductivity of graphene: Prospects for thermal management applications in nanoelectronic circuits. *Appl. Phys. Lett.* 92(15), 151911–151911-3 (**2008**).
61. S. Iijima, Helical microtubules of graphitic carbon. *Nature* 354(6348), 56–58 (**2001**).
62. M. P. Anantram and F. Léonard, Physics of carbon nanotube electronic devices. *Rep. Prog. Phys.* 69(3), 507–561 (**2006**).
63. W. Zhou, *et al.*, Synthesis, structure, and properties of single — walled carbon nanotubes. *Adv. Mater.* 21(45), 4565–4583 (**2009**).
64. R. S. Ruoff and D. C. Lorents, Mechanical and thermal properties of carbon nanotubes. *Carbon* 33(7), 925–930 (**1995**).
65. H. Y. Chiu, *et al.*, Ballistic phonon thermal transport in multiwalled carbon nanotubes. *Phys. Rev. Lett.* 95(22), 226101 (**2005**).
66. C. Yu, *et al.*, Thermal conductance and thermopower of an individual single-wall carbon nanotube. *Nano Lett.* 5(9), 1842–1846 (**2005**).
67. E. Pop, *et al.*, Thermal conductance of an individual single-wall carbon nanotube above room temperature. *Nano Lett.* 6(1), 96–100 (**2006**).
68. J. Hone, *et al.*, Thermal conductivity of single-walled carbon nanotubes. *Phys. Rev. B* 59(4), R2514–R2516 (**1999**).
69. A. V. Sukhadolau, *et al.*, Thermal conductivity of CVD diamond at elevated temperatures. *Diam. Relat. Mater.* 14(3), 589–593 (**2005**).
70. M. M. Sarafraz, F. Hormozi, and V. Nikkhah, Thermal performance of a counter-current double pipe heat exchanger working with COOH-CNT/water nanofluids. *Exp. Therm. Fluid Sci.* 78, 41–49 (**2016**).
71. S. J. Tans, *et al.*, Individual single-wall carbon nanotubes as quantum wires. *Nature* 386(6624), 474–477 (**1997**).
72. E. Katz and I. Willner, Biomolecule — functionalized carbon nanotubes: Applications in nanobioelectronics. *ChemPhysChem* 5(8), 1084–1104 (**2004**).
73. A. Merkoçi and A. Merkoçi, Carbon nanotubes in analytical sciences. *Microchimica Acta*, 152(3), 157–174 (**2006**).
74. A. K. Geim and K. S. Novoselov, The rise of graphene. *Nat. Mater.* 6(3), 183–191 (**2007**).
75. M. A. Worsley, *et al.*, Synthesis of graphene aerogel with high electrical conductivity. *J. Am. Chem. Soc.* 132(40), 14067–14069 (**2010**).
76. M. S. Dresselhaus and P. C. Eklund, Phonons in carbon nanotubes. *Adv. Phys.* 49(6), 705–814 (**2000**).

77. N. Mounet and N. Marzari, First-principles determination of the structural, vibrational and thermodynamic properties of diamond, graphite, and derivatives. *Phys. Rev. B Condens. Matter Mater. Phys.* 71(20) (**2005**).
78. T. M. Tritt and SpringerLink, *Thermal Conductivity: Theory, Properties, and Applications.* (Boston: Springer, **2006**).
79. G. F. Hundy, *Refrigeration, Air Conditioning and Heat Pumps.* Fifth edn. (San Diego, CA, USA: Elsevier Science, **2016**).
80. J. R. Drabble and H. J. Goldsmid, *Thermal Conduction in Semiconductors* (**1963**).
81. D. Cahill, Thermal conductivity measurement from 30 to 750 K: the 3omega method. *Rev. Sci. Instrum.* 61, 802–808 (**1990**).
82. F. Völklein, Thermal conductivity and diffusivity of a thin film SiO2□Si3N4 sandwich system. *Thin Solid Films* 188(1), 27–33 (**1990**).
83. M. Okuda and S. Ohkubo, A novel method for measuring the thermal conductivity of submicrometre thick dielectric films. *Thin Solid Films* 213(2) 176–181 (**1992**).
84. R. Garrelts, A. Marconnet, and X. Xu, Assessment of thermal properties via nanosecond thermoreflectance method. *Nanosc. Microsc. Therm. Eng.* 19(4) 245–257 (**2015**).
85. S. Périchon, *et al.*, Measurement of porous silicon thermal conductivity by micro-Raman scattering. *J. Appl. Phys.* 86(8) 4700–4702 (**1999**).
86. Z. Luo, *et al.*, Measurement of in-plane thermal conductivity of ultrathin films using micro-raman spectroscopy. *Nanosc. Microsc. Therm. Eng.* 18(2) 183–193 (**2014**).
87. X. Wang, H. Hu, and X. Xu, Photo-acoustic measurement of thermal conductivity of thin films and bulk materials. *J. Heat Transf.* 123(1) 138–144 (**2001**).
88. C. A. Paddock and G. L. Eesley, Transient thermoreflectance from thin metal films. *J. Appl. Phys.* 60(1) 285–290 (**1986**).
89. M. Omini, A. Sparavigna, and A. Strigazzi, Dilatometric determination of thermal diffusivity in low conducting materials. *Meas. Sci. Technol.* 1(2) 166–171 (**1990**).
90. W. J. Parker, *et al.*, Flash method of determining thermal diffusivity, heat capacity, and thermal conductivity. *J. Appl. Phys.* 32(9) 1679–1684 (**1990**).
91. D. Zhao, *et al.*, Measurement techniques for thermal conductivity and interfacial thermal conductance of bulk and thin film materials. *J. Electron. Packag. Trans. ASME.* 138(4) (**2016**).
92. L. F. C. Harald Mehling, *Heat and Cold Storage with PCM* (Springer Berlin Heidelberg, **2008**).
93. Y. Li, *et al.*, Improving the accuracy of the transient plane source method by correcting probe heat capacity and resistance influences. *Meas. Sci. Technol* 25(1), 15006 (**2014**).
94. J. Healy, J. DeGroot, and J. Kestin, The theory of the transient hot-wire method for measuring thermal conductivity. *Physica. Grav.* 82, 392–408 (**1975**).

95. S. Gustafsson, E. Karawacki, and M. Khan, Transient hot-strip method for simultaneously measuring thermal conductivity and thermal diffusivity of solids and fluids. *J. Phys. D Appl. Phys.* 12, 1411 (**2001**).
96. T. Borca-Tasciuc, A. R. Kumar, and G. Chen, Data reduction in 3ω method for thin-film thermal conductivity determination. *Rev. Scientif. Instrum.* 72, 2139–2147 (**2001**).
97. B. C. Daly, *et al.*, Optical pump and probe measurement of the thermal conductivity of low-k dielectric thin films. *J. Appl. Phys.* 92(10), 6005–6009 (**2002**).
98. C. J. Morath, *et al.*, Picosecond optical studies of amorphous diamond and diamondlike carbon: Thermal conductivity and longitudinal sound velocity. *J. Appl. Phys.* 76(5), 2636–2640 (**1994**).
99. A. J. Schmidt, R. Cheaito, and M. Chiesa, A frequency-domain thermoreflectance method for the characterization of thermal properties. *Rev. Scientif. Instrum.* 80(9), 094901 (**2009**).
100. J. A. Malen, *et al.*, Optical measurement of thermal conductivity using fiber aligned frequency domain thermoreflectance. *J. Heat Transf.* 133(8) (**2011**).
101. T. Jeong, *et al.*, Thermal characterization of SiC amorphous thin films. *Intl. J. Thermophysi.* 33(6), 1000–1012 (**2012**).
102. Li, *et al.*, *Measuring Thermal and Thermoelectric Properties of One-Dimensional Nanostructures Using a Microfabricated Device.*
103. P. Kim, *et al.*, Thermal transport measurements of individual multiwalled nano-tubes. *Phys.Rev. Lett.* 87, 215502 (**2001**).
104. M. C. Wingert, *et al.*, Ultra-sensitive thermal conductance measurement of one-dimensional nanostructures enhanced by differential bridge. *Rev. Sci. Instrum.* 83(2), 024901 (**2012**).
105. H. Tang, *et al.*, Thermoelectric characterization of individual bismuth selenide topological insulator nanoribbons. *Nanoscale* 7(15), 6683–6669 (**2015**).
106. E. H. Wang, *et al.*, Study of working fluid selection of organic Rankine cycle (ORC) for engine waste heat recovery. *Energy* 36(5), 3406–3418 (**2011**).
107. S. Mavridou, *et al.*, *C*omparative design study of a diesel exhaust gas heat exchanger for truck applications with conventional and state of the art heat transfer enhancements. *Appl. Therm. Eng.* 30(8), 935–947 (**2010**).
108. V. Pandiyarajan, *et al.*, Experimental investigation on heat recovery from diesel engine exhaust using finned shell and tube heat exchanger and thermal storage system. *Appl. Energy* 88(1), 77–87 (**2011**).
109. J. P. Heremans, *et al.*, Enhancement of thermoelectric efficiency in PbTe by distortion of the electronic density of states. *Science* 321(5888), 554–557 (**2008**).
110. F. J. DiSalvo, Thermoelectric cooling and power generation. *Science* 285(5428), 703–706 (**1999**).
111. L. E. Bell, Cooling, heating, generating power, and recovering waste heat with thermoelectric systems. *Science* 321(5895), 1457–1461 (**2008**).
112. G. J. Snyder and E. S. Toberer, Complex thermoelectric materials. *Nat. Mater.* 7(2), 105–114 (**2008**).

113. C. B. Vining, An inconvenient truth about thermoelectrics. *Nat. Mater.* 8(2), 83–85 (**2009**).
114. M. S. Dresselhaus, *et al.*, New directions for low-dimensional thermoelectric materials. *Adv. Mater.* 19(8), 1043–1053 (**2007**).
115. M. Martín-González, O. Caballero-Calero, and P. Díaz-Chao, Nanoengineering thermoelectrics for 21st century: Energy harvesting and other trends in the field. *Renew. Sust. Energy Rev.* 24, 288–305 (**2013**).
116. A. Ali, *et al.*, Nanowire-based thermoelectrics. *Nanotechnology* 28(28), 282001–282001 (**2017**).
117. I. J. Chen, *et al.*, Thermoelectric power factor limit of a 1D nanowire. *Phys. Rev. Lett.* 120(17), 177703 (**2018**).
118. N. I. Goktas, *et al.*, Nanowires for energy: A review. *Am. Inst. Phys. Melville.*, 41305 (**2018**).
119. M. Y. Swinkels and I. Zardo, Nanowires for heat conversion. *J. Phys. D Appl. Phys.* 51(35), 353001 (**2018**).
120. K. V. Selvan, M. N. Hasan, and M. S. Mohamed Ali, Methodological reviews and analyses on the emerging research trends and progresses of thermoelectric generators. *Int. J. Energy Res.* 43(1), 113–140 (**2019**).
121. B. Poudel, *et al.*, High-thermoelectric performance of nanostructured bismuth antimony telluride bulk alloys. *Science* 320(5876), 634–638 (**2008**).
122. L.-D. Zhao, *et al.*, Ultralow thermal conductivity and high thermoelectric figure of merit in SnSe crystals. *Nature* 508(7496), 373–377 (**2014**).
123. L. Francioso, *et al.*, Flexible thermoelectric generator for ambient assisted living wearable biometric sensors. *J. Power Sources* 196(6), 3239–3243 (**2011**).
124. W. Zhu, *et al.*, Thin-film solar thermoelectric generator with enhanced power output: Integrated optimization design to obtain directional heat flow. *Energy* 89, 106–117 (**2015**).
125. X. Liu, *et al.*, Performance analysis of a waste heat recovery thermoelectric generation system for automotive application. *Energy Convers. Manag.* 90, 121–127 (**2015**).
126. W. He, *et al.*, Parametrical analysis of the design and performance of a solar heat pipe thermoelectric generator unit. *Appl. Energy* 88(12), 5083–5089 (**2011**).
127. F. S. Moraes, *et al.*, Solar thermoelectric generator performance relative to air speed. *Energy Convers. Manag.* 99, 326–333 (**2015**).
128. J. Xiao, *et al.*, Thermal design and management for performance optimization of solar thermoelectric generator. *Appl. Energy* 93, 33–38 (**2012**).
129. M. Mizoshiri, *et al.*, Thin-film thermoelectric modules for power generation using focused solar light. *J. Electron. Mater.* 41(6), 1713–1719 (**2012**).
130. W. Zhu, *et al.*, Hierarchical Bi–Te based flexible thin-film solar thermoelectric generator with light sensing feature. *Energy Convers. Manag.* 106, 1192–1200 (**2015**).
131. D. Kraemer, *et al.*, High-performance flat-panel solar thermoelectric generators with high thermal concentration. *Nat. Mater.* 10(7), 532–538 (**2011**).

132. L. L. Baranowski, G. J. Snyder, and E. S. Toberer, Concentrated solar thermoelectric generators. *Energy Environ. Sci.* 5(10), 9055–9067 (**2012**).
133. S. Su, *et al.*, Performance optimization analyses and parametric design criteria of a dye-sensitized solar cell thermoelectric hybrid device. *Appl. Energy* 120, 16–22 (**2014**).
134. M. Fisac, F. X. Villasevil, and A. M. López, High-efficiency photovoltaic technology including thermoelectric generation. *J. Power Sources* 252, 264–269 (**2014**).
135. P. Sundarraj, *et al.*, Recent advances in thermoelectric materials and solar thermoelectric generators-a critical review. *RSC Adv.* 4(87), 46860–46874 (**2014**).
136. T. M. Tritt, H. Böttner, and L. Chen, Thermoelectrics: Direct solar thermal energy conversion. *MRS Bull.* 33(4), 366–368 (**2008**).
137. S. A. Kalogirou, Solar thermal collectors and applications. *Prog. Energy Combust. Sci.* 30(3), 231–295 (**2004**).
138. A. Vassighi, M. Sachdev, and I. Books24x, *Thermal and Power Management of Integrated Circuits* (US: Springer, **2006**).
139. Q. Cai and C.-L. Chen, Design and test of carbon nanotube biwick structure for high-heat-flux phase change heat transfer. *J. Heat Transf.* 132(5), 1–8 (**2010**).
140. E. Terrado, *et al.*, Modifying the heat transfer and capillary pressure of loop heat pipe wicks with carbon nanotubes. *J. Phys. Chem. C* 115(19), 9312–9319 (**2011**).
141. S. Tej, D. Horner, and S. Moghaddam, *Parametric Study of the Transport Properties of Wicking Micro-Pillar Arrays.*
142. R. S. Hale, R. T. Bonnecaze, and C. H. Hidrovo, Optimization of capillary flow through square micropillar arrays. *Int. J. Multiph. Flow* 58, 39–51 (**2014**).
143. Y. Nam, S. Sharratt, and Y. S. Ju, *Heat Transfer Performance of Superhydrophilic Nanostructured Cu Micropost Wicks for Micro Heat Pipes.*
144. M. A. Nazari, *et al.*, Experimental Investigation of triton X-100 solution on pulsating heat pipe thermal performance. *J. Thermophy. Heat Transf.* 32(3), 806–812 (**2018**).
145. Kargar Sharif Abad, H., *et al.*, A novel integrated solar desalination system with a pulsating heat pipe. *Desalination* 311, 206–210 (**2013**).
146. M. Arab, M. Soltanieh, and M. B. Shafii, Experimental investigation of extra-long pulsating heat pipe application in solar water heaters. *Exp.Therm. Fluid Sci.* 42, 6–15 (**2012**).
147. M. Alhuyi Nazari, *et al.*, A review on pulsating heat pipes: From solar to cryogenic applications. *Appl. Energy* 222, 475–484 (**2018**).
148. M. Faegh and M. B. Shafii, Experimental investigation of a solar still equipped with an external heat storage system using phase change materials and heat pipes. *Desalination* 409, 128–135 (**2017**).
149. G. V. Kuznetsov, M. A. Al-Ani, and M. A. Sheremet, Numerical analysis of convective heat transfer in a closed two-phase thermosyphon. *J. Eng. Thermophys.* 20(2), 201–210 (**2011**).

150. G. Kumaresan, *et al.*, Experimental study on effect of wick structures on thermal performance enhancement of cylindrical heat pipes. *J. Therm. Anal. Calorim.* 136(1), 389–400 (**2019**).
151. M. L. Rahman, *et al.*, Effect of using ethanol and methanol on thermal performance of a closed loop pulsating heat pipe (CLPHP) with different filling ratios. *AIP Conf. Proc.* 1754(1) (**2016**).
152. F. Y. Lv, *et al.*, Heat transfer performance of a lubricant-infused thermosyphon at various filling ratios. *Int. J. Heat Mass Transf.* 115, 725–736 (**2017**).
153. S. Wannapakhe, *et al.*, Heat transfer rate of a closed-loop oscillating heat pipe with check valves using silver nanofluid as working fluid. *J. Mech. Sci. Technol.* 23(6), 1576–1582 (**2009**).
154. Q. Wu, *et al.*, Effect of C60 nanofluid on the thermal performance of a flat-plate pulsating heat pipe. *Int. J. Heat Mass Transf.* 100, 892–898 (**2016**).
155. M. B. Shafii, *et al.*, Experimental investigation of pulsating heat pipes and a proposed correlation. *Heat Transf. Eng.* 31(10), 854–861 (**2010**).
156. M. B. Shafii, A. Faghri, and Y. Zhang, Analysis of heat transfer in unlooped and looped pulsating heat pipes. *Int. J. Numer. Method. Heat Fluid Flow* 12(5), 585–609 (**2002**).
157. J. Abolfazli Esfahani, *et al.*, Heat transfer in an eight-pass oscillating loop heat pipe equipped with cooling tower: An experimental study. *J. Therm. Analy. Calorim.* 136(4), 1869–1877 (**2019**).
158. S. Wang, *et al.*, Experimental study on pulsating heat pipe with functional thermal fluids. *Int. J. Heat Mass Transf.* 52(21), 5276–5279 (**2009**).
159. M. Gonzalez and Y. J. Kim, *Experimental study of a pulsating heat pipe using nanofluid as a working fluid.*
160. M. Alhuyi Nazari, *et al.*, How to improve the thermal performance of pulsating heat pipes: A review on working fluid. *Renew. Sust. Energy Rev.* 91, 630–638 (**2018**).
161. M. H. Ahmadi, *et al.*, A proposed model to predict thermal conductivity ratio of Al2O3/EG nanofluid by applying least squares support vector machine (LSSVM) and genetic algorithm as a connectionist approach. *J. Therm. Anal. Calorim.* 135(1), 271–281 (**2019**).
162. M. H. Ahmadi, *et al.*, A review of thermal conductivity of various nanofluids. *J. Mol. Liq.* 265, 181–188 (**2018**).
163. M. H. Ahmadi, *et al.*, Thermal conductivity ratio prediction of Al2O3/water nanofluid by applying connectionist methods. *Colloids Surf. A Physicochem. Eng. Asp.* 541, 154–164 (**2018**).
164. T. Tharayil, *et al.*, Entropy generation analysis of a miniature loop heat pipe with graphene–water nanofluid: Thermodynamics model and experimental study. *Int. J. Heat Mass Transf.* 106, 407–421 (**2017**).
165. Y. Zhou, *et al.*, Experimental investigation of the heat transfer performance of an oscillating heat pipe with graphene nanofluids. *Powder Technol.* 332, 371–380 (**2018**).
166. E. Sadeghinezhad, *et al.*, Experimental investigation of the effect of graphene nanofluids on heat pipe thermal performance. *Appl. Therm. Eng.* 100, 775–787 (**2016**).

167. M. Mehrali, *et al.*, Effect of nitrogen-doped graphene nanofluid on the thermal performance of the grooved copper heat pipe. *Energy Convers. Manag.* 118, 459–473 (**2016**).
168. A. A. Balandin Chill out: New materials and designs can keep chips cool. 46, 34 (**2009**).
169. S. V. Garimella, *et al.*, Thermal challenges in next-generation electronic systems. *IEEE Trans. Compon. Packaging Technol.* 31(4), 801–815 (**2008**).
170. R. Prasher, Thermal interface materials: Historical perspective, status, and future directions. *Proc.e IEEE*, 2006. 94(8): p. 1571–1586 (**2005**).
171. F. Sarvar, D. C. Whalley, and P. P. Conway, *Thermal Interface Materials — A Review of the State of the Art.* IEEE.
172. R. Prasher, Nano and micro technology-based next-generation package-level cooling solutions. *Intel Technol. J.* 9(4) (**2005**).
173. J. Felba, Thermally conductive nanocomposites, in *Nano-Bio- Electronic, Photonic and MEMS Packaging*, C. P. Wong, K.-S. Moon, and Y. Li (Eds.) (Springer US: Boston, MA, **2010**), pp. 277–314.
174. K. M. F. Shahil, and A. A. Balandin, Thermal properties of graphene and multilayer graphene: Applications in thermal interface materials. *Solid State Commun.* 152(15), 1331–1340 (**2012**).
175. D. L. Nika, *et al.*, Phonon thermal conduction in graphene: Role of Umklapp and edge roughness scattering. *Phys. Rev. B Condens. Matter Mater. Phys.* **79**(10) (**2009**).
176. E. P. Pokatilov, *et al.*, Dimensional crossover of thermal transport in few-layer graphene. *Nat. Mater.* 9(7), 555–558 (**2010**).
177. A. Barua, *et al.*, Thermal management in 3-D integrated circuits with graphene heat spreaders. *Phys. Procedia* 25, 311–316 (**2012**).
178. M. Law, J. Goldberger, and P. Yang, Semiconductor nanowires and nanotubes, *Annu. Rev. Mater. Res.* 83–122 (**2004**).
179. L. Shi, *et al.*, Evaluating broader impacts of nanoscale thermal transport research. *Nanoscale Microscale Thermophys. Eng.* 19(2), 127–165 (**2015**).
180. D. Cahill, *et al.*, Nanoscale thermal transport. II. 2003–2012. *Appl. Phys. Rev.* 1 (**2014**).
181. K. Rajan, Materials informatics: The materials "Gene" and Big Data. *Annu. Rev. Mater. Res.* 45(1), 153–169 (**2015**).
182. K. Rajan, Materials informatics. *Mater. Today* 15(11), 470 (**2012**).
183. A. Agrawal and A. Choudhary, Perspective: Materials informatics and big data: Realization of the "fourth paradigm" of science in materials science. *APL Mater.* 4(5), 53208–053208–10 (**2016**).
184. S. Ju and J. Shiomi, Materials informatics for heat transfer: Recent progresses and perspectives. *Nanoscale Microscale Thermophys. Eng.* 23(2), 157–172 (**2019**).
185. A. Jain, *et al.*, Commentary: The materials project: A materials genome approach to accelerating materials innovation. *APL Mater.* 1(1), 11002–011002–11 (**2013**).
186. S. Curtarolo, *et al.*, AFLOWLIB.ORG: A distributed materials properties repository from high-throughput ab initio calculations. *Comput. Mater. Sci.* 58(C), 227–235 (**2012**).

187. A. Belsky, *et al.*, New developments in the Inorganic Crystal Structure Database (ICSD): accessibility in support of materials research and design. *Acta Crystallogr. Sect. B* 58(3–1): p. 364–369 (**2002**).
188. J. E. Saal, *et al.*, Materials design and discovery with high-throughput density functional theory: The open quantum materials database (OQMD). *JOM* 65(11), 1501–1509 (**2013**).
189. S. Kirklin, *et al.*, The open quantum materials database (OQMD): Assessing the accuracy of DFT formation energies. *NPJ Comput.Mater.* 1(1), 15010 (**2015**).
190. Y. Xu, M. Yamazaki, and P. Villars, Inorganic materials database for exploring the nature of material. *Jpn. J. Appl. Phys.* 50(11), 11RH02–11RH02-5 (**2011**).
191. K. Esfarjani, *et al.*, Heat transport in silicon from first-principles calculations. *Phys. Rev. B Condens. Matter Mater. Phys.* 84(8) (**2011**).
192. K. Esfarjani and H. T. Stokes, Method to extract anharmonic force constants from first principles calculations. *Phys. Rev. B Condens. Matter Mater. Phys.* 77(14) (**2008**).
193. J. Shiomi, *et al.*, Thermal conductivity of half-Heusler compounds from first-principles calculations. *Phys. Rev. B Condens. Matter Mater. Phys.* 84(10) (**2011**).
194. A. Seko, A. Togo, and I. Tanaka, *Descriptors for Machine Learning of Materials Data* (**2017**).
195. S. Wang, *et al.*, Assessing the Thermoelectric properties of sintered compounds via high-throughput ab-initio calculations. *Phys. Rev. X* 1(2), 1–8 (**2011**).
196. Z. Tian, K. Esfarjani, and G. Chen, Enhancing phonon transmission across a Si/Ge interface by atomic roughness: First-principles study with the Green's function method. *Phys. Rev. B Condens. Matter Mater. Phys.* 86(23) (**2012**).
197. S. Merabia and K. Termentzidis, Thermal boundary conductance across rough interfaces probed by molecular dynamics. *Phys. Rev. B Condens. Matter Mater. Phys.* 89(5) (**2014**).
198. Y. Liu, *et al.*, Tuning interfacial thermal conductance of graphene embedded in soft materials by vacancy defects. *J. Chem. Phys.* 142(24), 244703 (**2015**).
199. P. K. Schelling, S. R. Phillpot, and P. Keblinski, Kapitza conductance and phonon scattering at grain boundaries by simulation. *J. Appl. Phys.* 95(11 I), 6082–6091 (**2004**).
200. S.-H. Ju, and X.-G. Liang, Investigation on interfacial thermal resistance and phonon scattering at twist boundary of silicon. *J. App. Phys.* 113(5), 53513 (**2013**).
201. M. Sakata, *et al.*, Tuning thermal conductance across sintered silicon interface by local nanostructures. *Nano Energy* 13, 601–608 (**2015**).
202. M. Sakata, *et al.*, Thermal conductance of silicon interfaces directly bonded by room-temperature surface activation. *Appl. Phys. Lett.* 106(8), 81603 (**2015**).

203. K. Zheng, *et al.*, Tuning the interfacial thermal conductance between polystyrene and sapphire by controlling the interfacial adhesion. *ACS Appl. Mater. Interf.* 7(42), 23644–23649 (**2015**).
204. S. Ju, *et al.*, Designing nanostructures for phonon transport via Bayesian optimization. *Phys. Rev. X* 7(2), 021024 (**2017**).
205. M. Dieb, T., *et al.*, MDTS: Automatic complex materials design using Monte Carlo tree search. *Sci. Technol. Adv. Mater.* 18(1), 498–503 (**2017**).
206. T. Ueno, *et al.*, COMBO: An efficient Bayesian optimization library for materials science. *Mater. Discov.* 4, 18–21 (**2016**).
207. J. K. Nørskov, *et al.*, Computational high-throughput screening of electrocatalytic materials for hydrogen evolution. *Nat. Mater.* 5(11), 909–913 (**2006**).
208. Y. Wu, *et al.*, First principles high throughput screening of oxynitrides for water-splitting photocatalysts. *Energy Environ. Sci.* 6(1), 157–168 (**2012**).
209. T. Mueller, *et al.*, Evaluation of tavorite-structured cathode materials for lithium-ion batteries using high-throughput computing. *Chem. Mater.* 23(17), 3854–3862 (**2011**).
210. A. Seko, *et al.*, Prediction of low-thermal-conductivity compounds with first-principles anharmonic lattice-dynamics calculations and Bayesian optimization. *Phys. Rev. Lett.* 115(20), 205901 (**2015**).
211. J. Inglese, *et al.*, High-throughput screening assays for the identification of chemical probes. *Nat. Chem. Biol.* 3(8), 466–479 (**2007**).
212. R. L. Martin, *et al.*, In silico design of porous polymer networks: High-throughput screening for methane storage materials. *J. Am. Chem. Soc.* 136(13), 5006–5022 (**2014**).
213. J. J. Haven, *et al.*, High-throughput polymer screening in microreactors boosting the Passerini three component reaction. *Polym. Chem.* 8(19), 2972–2978 (**2017**).
214. D. Goll, *et al.*, High-throughput search for new permanent magnet materials. *J. Phys. Condens. Matter* 26(6), 064208 (**2014**).
215. H. Zhan, *et al.*, Thermal transport in 3D nanostructures. *Adv. Funct.Mater.* 30(8), 1903841–n/a (**2020**).
216. T. Li, *et al.*, Anisotropic, lightweight, strong, and super thermally insulating nanowood with naturally aligned nanocellulose. *Sci. Adv.* 4(3), eaar3724 (**2018**).
217. R. G. Gellert, *et al.*, *Dämmstoffe: Grundlagen, Materialien, Anwendungen* (München, **2008**).
218. D. Bozsaky, Application of nanotechnology-based thermal insulation materials in building construction. *SJCE* 24(1), 17–23 (**2016**).
219. B. P. Jelle, Traditional, state-of-the-art and future thermal building insulation materials and solutions — Properties, requirements and possibilities. *Energy Build.* 43(10), 2549–2563 (**2011**).
220. B. Wicklein, *et al.*, Thermally insulating and fire-retardant lightweight anisotropic foams based on nanocellulose and graphene oxide. *Nat. Nanotechnol.* 10(3), 277–283 (**2015**).

221. A. M. Papadopoulos, State of the art in thermal insulation materials and aims for future developments. *Energy Build.* 37(1), 77–86 (**2005**).
222. Y. H. Jung, *et al.*, High-performance green flexible electronics based on biodegradable cellulose nanofibril paper. *Nat. Commun.* 6(1), 7170–7170 (**2015**).
223. I. A. Sacui, *et al.*, Comparison of the properties of cellulose nanocrystals and cellulose nanofibrils isolated from bacteria, tunicate, and wood processed using acid, enzymatic, mechanical, and oxidative methods. *ACS Appl. Mater. Interf.* 6(9), 6127–6138 (**2014**).
224. J. Majoinen, *et al.*, Chiral plasmonics using twisting along cellulose nanocrystals as a template for gold nanoparticles. *Adv. Mater.* 28(26), 5262–5267 (**2016**).
225. J. Y. Zhu, R. Sabo, and X. Luo, Integrated production of nano-fibrillated cellulose and cellulosic biofuel (ethanol) by enzymatic fractionation of wood fibers. *Green Chem.* 13(5), 1339–1344 (**2011**).
226. Y. Habibi, L. A. Lucia, and O. J. Rojas, Cellulose nanocrystals: Chemistry, self-assembly, and applications. *Chem. Rev.* 110(6), 3479–3500 (**2010**).
227. J. Simonsen, Utilizing straw as a filler in thermoplastic building materials. *Constr. Build. Mater.* 10(6), 435–440 (**1996**).
228. A. J. Svagan, *et al.*, Biomimetic foams of high mechanical performance based on nanostructured cell walls reinforced by native cellulose nanofibrils. *Adv. Mater.* 20(7), 1263–1269 (**2008**).
229. L. Yang, *et al.*, High performance thermoelectric materials: Progress and their applications. *Adv. Energy Mater.* 8(6), 1701797 (**2018**).
230. G. Xie, D. Ding, and G. Zhang, Phonon coherence and its effect on thermal conductivity of nanostructures. *Adv. Phys. X* 3(1), 1480417 (**2018**).
231. H. F. Hamann, *et al.*, Hotspot-limited microprocessors: Direct temperature and power distribution measurements. *IEEE J. Solid-State Circ.* 42(1), 56–65 (**2007**).
232. J. U. Knickerbocker, *et al.*, Three-dimensional silicon integration. *IBM J. Res. Develop.* 52(6), 553–569 (**2008**).
233. J. U. Knickerbocker, *et al.*, An advanced multichip module (MCM) for high-performance UNIX servers. *IBM J. Res. Develop.* 46(6), 779–804 (**2002**).
234. S. Al-Shehri and H. H. Saber, Experimental investigation of using thermoelectric cooling for computer chips. *J. King Saud Univ. Eng. Sci.* (**2019**).
235. N. Li, *et al.*, Colloquium: Phononics: Manipulating heat flow with electronic analogs and beyond. *Rev. Mod. Phys.* 84 (**2011**).
236. Z. Yan, *et al.*, Graphene quilts for thermal management of high-power GaN transistors. *Nat. Commun.* 3(1), 827 (**2012**).
237. C. Yegin, *et al.*, Metal–organic–inorganic nanocomposite thermal interface materials with ultralow thermal resistances. *ACS Appl. Mater. Interf.* 9(11), 10120–10127 (**2017**).
238. S. Maneewan, *et al.*, Heat gain reduction by means of thermoelectric roof solar collector. *Solar Energy*, 78(4), 495–503 (**2005**).
239. Z. Liu, *et al.*, Review of solar thermoelectric cooling technologies for use in zero energy buildings. *Energy Build.* 102, 207–216 (**2015**).

240. L. Aditya, *et al.*, A review on insulation materials for energy conservation in buildings. *Renew. Sust. Energy Rev.* 73, 1352–1365 (**2017**).
241. O. Bubnova, *et al.*, Towards polymer-based organic thermoelectric generators. *Energy Environ. Sci.* 5(11), 9345–9362 (**2012**).
242. Y. Wang, *et al.*, Wearable thermoelectric generator to harvest body heat for powering a miniaturized accelerometer. *Appl. Energy* 215, 690–698 (**2018**).
243. V. Leonov and R. J. M. Vullers, Wearable electronics self-powered by using human body heat: The state of the art and the perspective *AIP: MELVILLE*, 062701–062701–14 (**2009**).
244. R. He, G. Schierning, and K. Nielsch, Thermoelectric devices: A review of devices, architectures, and contact optimization. *Adv. Mater. Technol.* 3(4), 1870016–n/a (**2018**).
245. L. Carroll, *et al.*, Photonic packaging: Transforming silicon photonic integrated circuits into photonic devices. *Appl. Sci.* (Switzerland) 6(12), 426 (**2016**).
246. N. Vogel, *et al.*, Advances in Colloidal Assembly: The Design of Structure and Hierarchy in Two and Three Dimensions. Chemical Reviews 115(13), 6265–6311 (**2015**).
247. P. Ruckdeschel, A. Philipp, and M. Retsch, Understanding thermal insulation in porous, particulate materials. *Adv. Funct. Mater.* 27(38), 1702256 (**2017**).
248. M. T. Barako, *et al.*, Quasi-ballistic electronic thermal conduction in metal inverse opals. *Nano Lett.* 16(4), 2754–2761 (**2016**).
249. L. Ping, *et al.*, Vertically aligned carbon nanotube arrays as a thermal interface material. *APL Mater.* 7(2), 20902–020902–9 (**2019**).
250. M. T. Barako, *et al.*, Thermal conduction in vertically aligned copper nanowire arrays and composites. *ACS Appl. Mater. Interf.* 7(34), 19251–19259 (**2015**).
251. G. K. Dimitrakakis, E. Tylianakis, and G. E. Froudakis, Pillared graphene: A new 3-D network nanostructure for enhanced hydrogen storage. *Nano Lett.* 8(10), 3166–3170 (**2008**).
252. J. Bauer, *et al.*, Approaching theoretical strength in glassy carbon nanolattices. *Nat. Mater.* 15(4), 438–443 (**2016**).
253. N. V. Krainyukova and E. N. Zubarev, Carbon honeycomb high capacity storage for gaseous and liquid species. *Phys. Rev. Lett.* 116(5), 055501 (**2016**).
254. J. Lim, *et al.*, Simultaneous Thermoelectric property measurement and incoherent phonon transport in holey silicon. *ACS Nano* 10(1), 124–132 (**2016**).
255. X. Chen, *et al.*, Interface interaction and synergistic strengthening behavior in pure copper matrix composites reinforced with functionalized carbon nanotube-graphene hybrids. *Carbon* 146, 736–755 (**2019**).
256. S. Zhao, *et al.*, Cu matrix composites reinforced with aligned carbon nanotubes: Mechanical, electrical and thermal properties. *Mater Sci Eng: A* 675, 82–91 (**2016**).

257. X. Wang, and P. Wu, Melamine foam-supported 3D interconnected boron nitride nanosheets network encapsulated in epoxy to achieve significant thermal conductivity enhancement at an ultralow filler loading. *Chem. Eng. J.* 348, 723–731 (**2018**).
258. Y. Xue, *et al.*, Multifunctional superelastic foam-like boron nitride nanotubular cellular-network architectures. *ACS Nano* 11(1), 558–568 (**2017**).
259. K. Wu, *et al.*, Design and preparation of a unique segregated double network with excellent thermal conductive property. *ACS Appl. Mater. Interf.* 9(8), 7637–7647 (**2017**).
260. J. Hu, *et al.*, Polymer composite with improved thermal conductivity by constructing a hierarchically ordered three-dimensional interconnected network of BN. *ACS Appl. Mater. Interf.* 9(15), 13544–13553 (**2017**).
261. X. Zeng, *et al.*, Ice-templated assembly strategy to construct 3D boron nitride nanosheet networks in polymer composites for thermal conductivity improvement. *Small* 11(46), 6205–6213 (**2015**).
262. H. F. Zhan, *et al.*, Thermal conductivity of Si nanowires with faulted stacking layers. *J. Phys. D: Appl. Phys.* 47(1), 015303 (**2013**).
263. G. Zhang and Y.-W. Zhang, Thermal properties of two-dimensional materials. *Chinese Phys. B* 26(3), 034401 (**2017**).

Chapter 6

Thermoelectric Phenomena at Nanoscale Level

Moumita Dey[*,‡] **and Santanu K. Maiti**[†,§]

[*]Department of Physics, Adamas University,
Adamas Knowledge City, Barasat-Barrackpore Road,
24 Parganas North, Kolkata 700 126, India
[†]Physics and Applied Mathematics Unit,
Indian Statistical Institute,
203 Barrackpore Trunk Road,
Kolkata 700 108, India
[§]santanu.maiti@isical.ac.in

6.1. Introduction

Recovering waste heat and converting it into usable electrical energy has been considered as one of the major challenges both in science and technology over the past few decades while addressing the energy crisis across the globe [1]. Though several prescriptions are available to recycle waste heat, much emphasis is given in *thermoelectric* (TE) energy conversion because of the 'green' nature of conversion process, i.e., harvesting power from the waste heat and easy device maintenance due to the absence of moving mechanical parts. This mechanism is no doubt technologically intriguing.

First thermoelectric effect, known as the Seebeck effect [2], was discovered in 1822, where a voltage drop is induced across the junctions of two distinct metals kept at two different temperatures (see Fig. 6.1). Thus, a heat energy is being converted into an electrical one through this process. If ΔV is the amount of voltage generated due to temperature difference ΔT, then the Seebeck coefficient becomes [3]

$$S = \frac{\Delta V}{\Delta T} \tag{6.1}$$

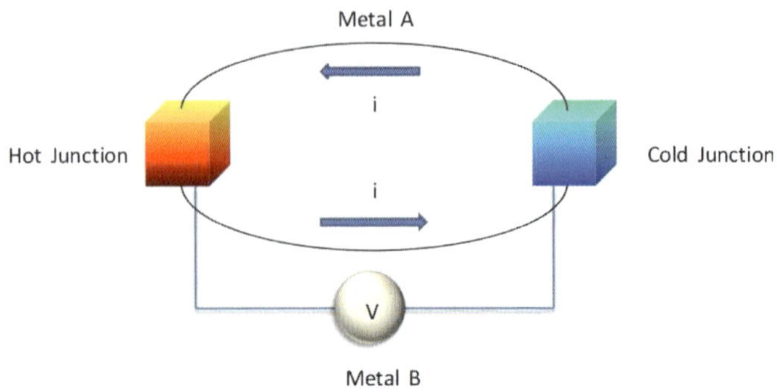

Fig. 6.1: Schematic diagram of the Seebeck effect where a thermal gradient induces a voltage drop at two junctions.

Twelve years later, in 1834, Peltier [4] noted that the temperature changes at the junctions of two different materials when an electric current is passed through it. If Q amount of heat is generated or absorbed due to current I flowing through a metallic branch, then these two factors are related as $Q = \Pi I$, where Π is the Peltier coefficient.

Within few years of this discovery, a famous scientist Lenz performed an experiment in 1838 which had a great impact on technological applications of thermoelectric effects [3]. He put a water droplet at each of the two junctions of Bi–Sb closed loop and passed an electric current through the system. One water droplet gets frozen into ice, while the other remains unchanged in the form of water. By altering the direction of the current flow through the loop, the ice at one junction melts into water and the water droplet at the other junction freezes into ice. This simple experiment suggested that the thermoelectric effect can probably be utilized for both power generation and refrigeration.

Now, if a thermoelectric material is coupled to two heat baths (Fig. 6.2) having the temperatures T_h (temperature of the hot junction) and T_c (temperature of the cold junction) ($T_h > T_c$), then from the argument of thermodynamics, it is shown that the *efficiency* of thermoelectric power generation can be written as

$$\eta_{\max} = \left(1 - \frac{T_c}{T_h}\right)\left(\frac{r-1}{r+\frac{T_c}{T_h}}\right) \tag{6.2}$$

where

$$r = \sqrt{1+ZT}$$

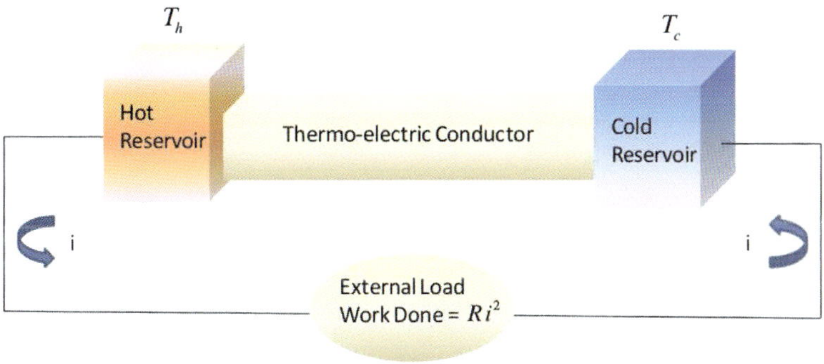

Fig. 6.2: Schematic diagram of a thermoelectric power generator.

Here, T denotes the equilibrium temperature and it is defined as

$$T = \frac{T_c + T_h}{2} \tag{6.3}$$

The parameter ZT, a dimensionless quantity, is called the *figure of merit* (FOM), which plays the crucial role in determining the quality of the thermoelectric material used. If the value of ZT is increased, η_{max} approaches the ideal Carnot cycle efficiency.

The FOM is defined as

$$ZT = \frac{GS^2T}{\kappa} \tag{6.4}$$

Here, G represents the electrical conductance of the material used. It is a physical measure of how easily charges can cross the junction to generate the voltage drop. The parameter S describes the thermopower which is commonly referred to as the Seebeck coefficient. It gives a measure of how large can a voltage drop be developed for a given temperature bias across the junction. The other physical quantity κ corresponds to thermal conductance of the functional material and it includes both electronic and phononic contributions, i.e., it is written as a sum $\kappa = \kappa_{\text{el}} + \kappa_{\text{ph}}$. It gives an estimate of how hard it is to maintain the temperature gradient. Thermoelectric materials with low value of κ is undoubtedly desirable to get higher ZT.

Similarly, for thermoelectric refrigeration [5] (see Fig. 6.3 for the schematic illustration), the *coefficient of performance* is given by

$$K_{max} = \left(\frac{T_c}{T_h - T_c}\right)\left(\frac{r - \frac{T_h}{T_c}}{r + 1}\right) \tag{6.5}$$

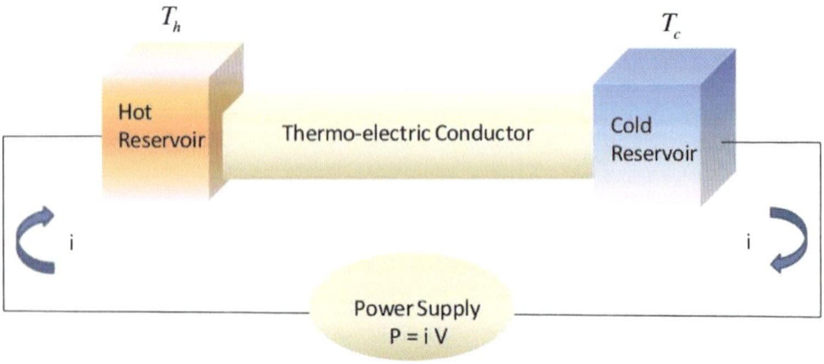

Fig. 6.3: Schematic diagram of a thermoelectric refrigerator.

Like the earlier case, here also a higher ZT will lead to a favorable value of K.

Now, before the widespread usage of semiconductors, metals were the obvious first choice for implementing these ideas in technology. But, for metals, the value of ZT was found to be much less than unity for all temperatures, which are thus not suitable enough in the context of efficiency perspective.

It was shown by Ioffe [6] that value of Z can have a maximum in Z–n characteristic curve when the concentration n approximately reaches $10^{19}\,\text{cm}^3$. This value of n can easily be reached in semiconductors doped with impurities [3]. So, naturally the attention got shifted to bulk semiconductors and it was studied from both theoretical and experimental perspectives. Theoretically, thermoelectricity was studied with the help of Boltzmann transport equation within the framework of relaxation time approximation for analytic solutions. Otherwise, analytic solution is impossible to obtain. But, in this prescription, theoretical prediction was that due to scattering process, the resulting transport mechanism is diffusive and slow, and therefore, it cannot result for a high value of ZT. It was also verified experimentally. So far, lots of thermoelectric materials have been taken into account, such as Bi_2Te_3, Mn–Si, Bi–Sb–Te–Se and to name a few. But, in most of the cases, ZT is quite closed to unity at room temperature, and for bulk samples, it has not been possible to enhance ZT very much, which was also the theoretical estimate.

The fact is that for bulk thermoelectric materials ZT can be increased by changing the carrier concentration by doping, though it has some limitations as electrical conductance G and Seebeck coefficient S change in opposite senses as carrier concentration gets tuned in bulk materials. Therefore, to enhance ZT lower thermal conductance (κ) and at the same time higher electrical conductance (G) and Seebeck coefficient (S) are preferred. To obtain low thermal conductance, quantum confinement was an appropriate option. So, people started to think about using quantum wires and molecular junctions, which are even smaller in size than regular quantum systems, as the prospective thermoelectric materials [7, 8].

In this review, while discussing the concept of thermoelectricity at nanoscale system, our main focus is on molecular systems which are sandwiched between two macroscopic electrodes, but one can also consider quantum dots (QD), quantum wires (QW), nanotubes, different non-trivial topologically intriguing systems, etc., in place of the molecular ones. Such low-dimensional nanostructured systems can have crucial significance for efficient energy conversion even in the room temperature limit [9–23].

The availability, low dimensionality, and low thermal conductivity make the molecular systems a natural choice for next generation thermoelectric materials with enhanced ZT. As molecules are automatically nanostructured and their electrical conductance (G) and Seebeck coefficient (S) can be monitored externally by applying a suitable gate voltage and/or some other possible ways, they can be chosen as potential candidates for efficient and high-power thermoelectric devices.

Till date, quite a number of calculations were carried out, where thermoelectric properties of molecular junctions were studied extensively, mostly within the framework of ballistic transport theory. In 2015, Dubi *et al.* proposed [24] a new way to tune and enhance ZT of a molecular junction by introducing semiconducting nanoparticles within the junction, thereby making a hybrid single-molecule–nanoparticles junction. The main essence of this work is that by selectively tuning the energy levels of the single molecule and aligning one of the molecular energy levels across the band gap that is generated due to the presence of the semiconducting nanoparticles, much higher thermopower efficiency can be achieved, which yields higher ZT.

The transport of electrons in a nanojunction can be affected by phonons. When an electron interacts with phonons, the tunneling becomes inelastic i.e., it may gain or lose energy [25]. This inelastic scattering can substantially affect the transmission probability and may lead to several inelastic processes like heat dissipation. In a work, Galperin *et al.* proposed [26] a simple phenomenological model to investigate inelastic scattering effects at molecular scale level that can influence the thermopower.

From the recent experimental findings of thermopower measurements in nanowires [21, 27–29], it has been established that there is a possibility of enhancement of S and G and reduction of κ independently by designing a system which is smaller than phonon mean free path and larger than electronic mean free path (or holes in case), and thus, there exists a finite possibility to get enhanced and favorable ZT. Very few experiments have been performed so far on molecular junctions for measuring thermopower. Among them, in 2007, Reddy *et al.* carried out an experiment [30] which demonstrates that the Seebeck coefficient (S) and the electrical conductance (G) can be increased simultaneously by applying a gate voltage, i.e., by shifting equilibrium Fermi level closer to the resonant peak. It strongly suggests that much higher ZT can be achieved in molecular systems.

The next challenge is to find which molecule and/or the material junction leads to the best and experimentally achievable thermoelectric properties. Keeping this in mind, in this chapter, we focus on two different molecular systems and discuss the thermoelectric effects in detail. For one setup, we consider a molecular junction where the molecule contains loop sub-structures. The interferometric geometries always provide some non-trivial signatures that we want to expose here. In the other setup, we choose a helical molecule that exhibits several interesting features. Our analysis can easily be extended to any other simple and/or complex molecular systems as well as tailor-made geometries to investigate thermoelectric phenomena.

The arrangement of the remaining parts of this chapter is as follows. In Section 6.2, we give a comprehensive theoretical description for the calculation of all the thermoelectric quantities. This part is extremely helpful, especially for beginners, as all the mathematical steps are derived. Then, in Section 6.3, we highlight some important results of thermoelectric effects considering different molecular wires. The analysis can be easily generalized further to investigate thermoelectric phenomena in other similar kind of molecular wires and tailor-made geometries as well. Finally, we summarize in Section 6.4.

6.2. Theoretical Framework

In this section, we present elaborately the theoretical framework, for the benefit of general readers, to calculate all the required thermoelectric quantities which include electrical conductance, thermopower, and thermal conductance to estimate the figure of merit ZT.

6.2.1. *Figure of merit ZT*

It has already been mentioned that the efficiency of thermoelectric energy conversion depends on a dimensionless quantity which is referred to as 'figure of merit' and it is defined as [3, 7, 12, 24, 31–35]

$$ZT = \frac{GS^2T}{\kappa_{\text{el}} + \kappa_{\text{ph}}} \tag{6.6}$$

where G is the electrical conductance, S is the thermopower or the Seebeck coefficient, κ_{el} and κ_{ph} are the thermal conductances due to electronic and phononic contributions, respectively.

In the following, we present the detailed derivations for calculating the thermopower S, electrical conductance G and thermal conductance due to electrons κ_{el} of a two-terminal nanojunction. Here, we have made a valid approximation that as the phononic contribution is too small in molecular systems, especially for smaller molecules, we have ignored this contribution towards thermal conductance κ.

6.2.2. *Calculation of Seebeck coefficient S*

Let us consider a setup where a conductor (a molecular system in our case) is attached to two semi-infinite leads with electrochemical potentials μ_1 and μ_2 and temperatures T_1 and T_2, respectively (see Fig. 6.4). Under this non-equilibrium condition, both charge current and heat current will flow through the system. The equilibrium electrochemical potential and temperature are defined as $\mu = (\mu_1 + \mu_2)/2$ and $T = (T_1 + T_2)/2$. The electrochemical potentials again related to the bias drop V through the relation $\mu_1 - \mu_2 = eV$.

In a two-terminal bridge system if a charge current I_c is passed through the conductor where the drop across the conductor becomes ΔV, then the

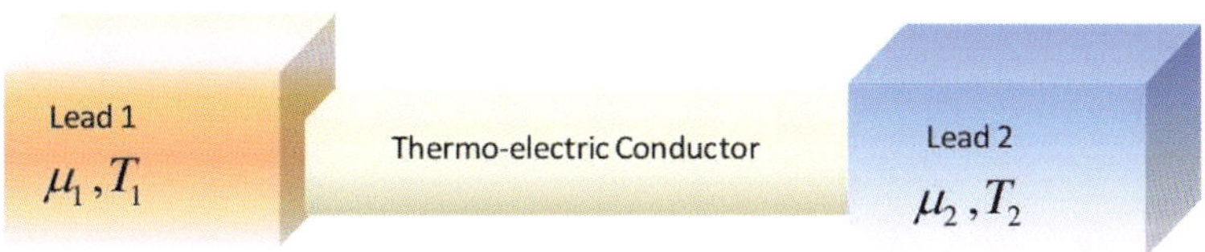

Fig. 6.4: Junction setup of a thermoelectric device where a conductor is coupled to two semi-infinite leads having temperatures T_1 and T_2, and electrochemical potentials mu_1 and μ_2.

conductance is defined as [3]

$$G = \frac{I_c}{\Delta V} \tag{6.7}$$

Following the Landauer–Büttiker formalism [36–40] in linear transport regime, the expression of charge current in a two-terminal system can be written as

$$I_c = \frac{2e}{h} \int dE\, \mathcal{T}_{12}(E)(f_1 - f_2) \tag{6.8}$$

where the Fermi functions f_1 and f_2 are defined as

$$f_1(E) = \frac{1}{1 + e^{\frac{E-\mu_1}{k_B T}}} \quad \text{and} \quad f_2(E) = \frac{1}{1 + e^{\frac{E-\mu_2}{k_B T}}}$$

$\mathcal{T}_{12}$ is the electronic transmission probability through the junction.

In the low bias regime, we can expand the Fermi functions ($f_1(E)$ and $f_2(E)$) about the equilibrium Fermi energy of the system (μ) and temperature (T). Following the Taylor expansion of the Fermi functions, keeping the first-order terms only, we get

$$f_1(E) = f_0(E) + \left.\frac{\partial f}{\partial \mu}\right|_{T=T} + (\mu - \mu_1) + \left.\frac{\partial f}{\partial T}\right|_{\mu=\mu} + (T - T_1) \tag{6.9}$$

and

$$f_2(E) = f_0(E) + \left.\frac{\partial f}{\partial \mu}\right|_{T=T} + (\mu - \mu_2) + \left.\frac{\partial f}{\partial T}\right|_{\mu=\mu} + (T - T_2) \tag{6.10}$$

From these two equations, we have

$$\begin{aligned}(f_1 - f_2) &= \frac{\partial f}{\partial \mu}(\mu_1 - \mu_2) + \frac{\partial f}{\partial T}(T_2 - T_1) \\ &= e\frac{\partial f}{\partial \mu}\Delta V + \frac{\partial f}{\partial T}\Delta T \end{aligned} \tag{6.11}$$

Substituting Eq. (6.11) in Eq. (6.8), we get

$$I_c = \frac{2e}{h} \int dE\, \mathcal{T}_{12}(E) \left[e\frac{\partial f}{\partial \mu}\Delta V + \frac{\partial f}{\partial T}\Delta T \right] \tag{6.12}$$

It can easily be shown that

$$\frac{\partial f}{\partial \mu} = -\frac{\partial f}{\partial E} \quad \text{and} \quad \frac{\partial f}{\partial T} = -\frac{(E-\mu)}{T}\frac{\partial f}{\partial E}$$

Now, the Seebeck effect is the thermal gradient-induced bias in a two-probe system (or current under closed boundary condition). The Seebeck coefficient is defined as the ratio of thermal gradient-induced voltage to the applied thermal gradient, i.e.,

$$S = -\frac{\Delta V}{\Delta T} \tag{6.13}$$

Setting $I_c = 0$ (open circuit condition) in Eq. (6.12) and substituting the values of $\frac{\partial f}{\partial \mu}$ and $\frac{\partial f}{\partial T}$, we get

$$e\Delta V \int dE\, \mathcal{T}_{12}(E)\frac{\partial f}{\partial E} = \frac{\Delta T}{T}\int \mathcal{T}_{12}(E)\frac{\partial f}{\partial E}(E-\mu) \tag{6.14}$$

Therefore, the expression of Seebeck coefficient in terms of transmission probability becomes

$$\boxed{S = -\frac{1}{eT}\frac{\displaystyle\int \mathcal{T}_{12}(E)(E-\mu)\frac{\partial f}{\partial E}dE}{\displaystyle\int \mathcal{T}_{12}(E)\frac{\partial f}{\partial E}dE}} \tag{6.15}$$

6.2.3. *Calculation of electrical conductance G*

Two-terminal electrical conductance in the linear response regime is defined by the ratio of charge current passing through the conductor under zero thermal bias, i.e.,

$$G = \left.\frac{I_c}{\Delta V}\right|_{\Delta T=0}$$

So, from Eq. (6.12), the expression of G gets the form

$$\boxed{G = -\frac{2e^2}{h}\int \mathcal{T}_{12}(E)\frac{\partial f}{\partial E}dE} \tag{6.16}$$

6.2.4. *Calculation of thermal conductance κ_{el}*

Under non-equilibrium condition, the heat current flowing through the system can be written as

$$I_Q = \frac{2}{h}\int \mathcal{T}_{12}(E)(E-\mu)(f_1 - f_2)\,dE \tag{6.17}$$

In the linear response regime, thermal conductance of a two-probe system is defined as the heat current flowing through the system per unit temperature gradient, i.e.,

$$\kappa_{\text{el}} = \left.\frac{I_Q}{\Delta T}\right|_{I_c=0}$$

Using the condition $I_c = 0$ in Eq. (6.12), we determine ΔV which is

$$\Delta V = \frac{\displaystyle\int \mathcal{T}_{12}(E)(E-\mu)\frac{\partial f}{\partial E}dE}{\displaystyle\int \mathcal{T}_{12}(E)\frac{\partial f}{\partial E}dE}\Delta T \tag{6.18}$$

Substituting the above expression of ΔV and the form of $(f_1 - f_2)$ into Eq. (6.17), we finally reach the relation of thermal conductance as

$$\boxed{\kappa_{\text{el}} = \frac{2}{hT}\left[L_2 - \frac{L_1^2}{L_0}\right]} \tag{6.19}$$

where

$$L_n = \int_{\infty}^{\infty} (E-\mu)^n \mathcal{T}_{12}(E)\frac{\partial f}{\partial E}\, dE \tag{6.20}$$

L_n's are known as Landauer integrals [24, 41–43].

6.2.5. *Evaluation of transmission probability*

What we find from our above description is that in order to find any of the thermoelectric quantities (G, S, κ_{el}), we need to evaluate transmission probability function $\mathcal{T}$. The transmission function or probability $\mathcal{T}_{pq}(E)$ is involved in describing current flow through a conductor sandwiched between the contact leads. $\mathcal{T}_{pq}(E)$ measures the transmission probability of an electron from terminal q to terminal p at a particular energy E. Our main target is to evaluate this T_{pq} for any arbitrarily shaped conductors.

In this chapter, we use Green's function technique [36–40] to obtain the transmission probability for a two and/or multi-terminal system. It is highly robust and elegant technique compared to other available approaches like wave-guide theory and transfer matrix method, and most importantly, it can be utilized to any arbitrarily shaped conductor even in the presence of impurities. Here, we are confined ourselves within the coherent transport regime where the system size is smaller than the phase coherent length. In this case, the transmission function can be evaluated, in principle, by using Schrödinger equation within the framework of S matrix theory [36, 39]. Now, before getting into the details of the relation between S matrix and Green's function, first we introduce two basic concepts about (i) S matrix and (ii) Green's function operator.

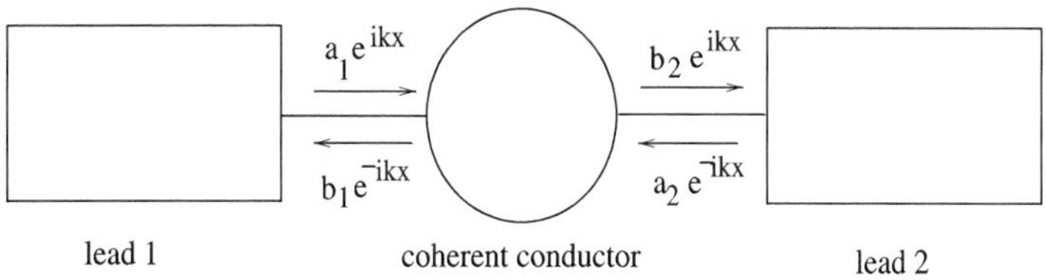

Fig. 6.5: A conductor attached to two leads (lead 1 and lead 2). Different arrows indicate incoming and outgoing waves.

6.2.5.1. *Introduction to S matrix*

Definition

Suppose we have a coherent conductor on which an incoming wave is incident, a part of which is being transmitted and the rest is reflected. There is also an incident wave from the opposite side. The S matrix relates the outgoing wave amplitudes to the incoming ones at different leads [36, 39].

Following Fig. 6.5, we can write

$$\begin{pmatrix} b_1 \\ b_2 \end{pmatrix} = \begin{pmatrix} s_{11} & s_{12} \\ s_{21} & s_{22} \end{pmatrix} \begin{pmatrix} a_1 \\ a_2 \end{pmatrix} \tag{6.21}$$

which is expressed in compact form as $\boldsymbol{b} = \boldsymbol{S}\boldsymbol{a}$.

For a simple one-dimensional problem, the S matrix takes the form

$$\boldsymbol{S} = \begin{pmatrix} r & t' \\ t & r' \end{pmatrix} \tag{6.22}$$

where r and t represent the reflection and transmission amplitudes, respectively. r' and t' are the time-reversed components. The dimension of S matrix becomes $M \times M$, where $M = \sum_p M_p$. M_p is the number of channels in pth lead.

Using the matrix elements of S, transmission probability $\mathcal{T}_{mn}$ for an individual mode can be evaluated through the operation

$$T_{mn} = |S_{mn}|^2 \tag{6.23}$$

and thus, summing over all the available modes, we get the net transmission probability between the leads which is

$$T_{pq} = \sum_m \sum_n T_{m \leftarrow n} \tag{6.24}$$

Here, m and n are associated with the total number of available modes for the lead q and lead p, respectively.

Properties of S matrix

1. S matrix is unitary, viz.,

$$\boldsymbol{S}^{\dagger}\boldsymbol{S} = \boldsymbol{S}\boldsymbol{S}^{\dagger} = \mathbf{1} \tag{6.25}$$

   It means

$$\sum_{m} |S_{mn}|^2 = 1 \quad \text{and} \quad \sum_{n} |S_{mn}|^2 = 1$$

   The first equation represents the sum of the transmission probabilities for a given input channel n over all possible output channels m and it should be equal to 1 because of the current conservation. The second equation is not that obvious, but it follows from the unitarity.

2. From the unitarity of the S matrix, the sum rules of transmission functions and conductances are described as

$$\sum_{q} T_{pq} = \sum_{q} T_{qp} = M_p \tag{6.26}$$

$$\sum_{q} G_{pq} = \sum_{q} G_{qp} = \frac{2e^2}{h} M_p(E_F) \tag{6.27}$$

3. S matrix follows the reciprocity relation with respect to the magnetic field, and thus,

$$[\boldsymbol{S}]_{+B} = [\boldsymbol{S}^{T}]_{-B} \tag{6.28}$$

   From the above equation, it is reflected that the transmission function and the conductances also obey the reciprocity relation. They are

$$[T_{pq}]_{+B} = [T_{qp}]_{-B} \quad \text{and} \quad [G_{pq}]_{+B} = [G_{qp}]_{-B}$$

Combination of S matrices

Coherent Combination: In Fig. 6.6, a coherent conductor is shown, which is coupled to two semi-infinite leads. The conductor is divided into two parts which are described by two individual S matrices, S_1 and S_2. The overall S matrix for the entire conductor is obtained by combining these individual S matrices (S_1 and S_2) coherently.

The scattering matrices for the individual scatterers are described as follows:

For scatterer 1,

$$\begin{pmatrix} b_1 \\ c_3 \end{pmatrix} = \begin{pmatrix} r_1 & t_1' \\ t_1 & r_1' \end{pmatrix} \begin{pmatrix} a_1 \\ d_2 \end{pmatrix} \tag{6.29}$$

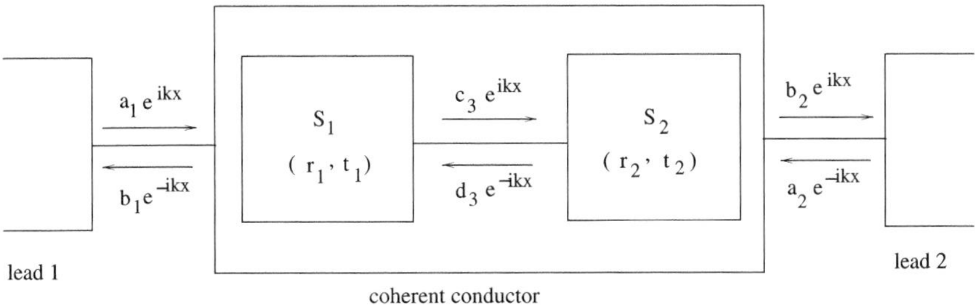

Fig. 6.6: A coherent conductor is divided into two parts, which are described by two individual S matrices S_1 and S_2.

Similarly, for scatterer 2,

$$\begin{pmatrix} d_3 \\ b_2 \end{pmatrix} = \begin{pmatrix} r_2 & t_2' \\ t_2 & r_2' \end{pmatrix} \begin{pmatrix} c_3 \\ a_2 \end{pmatrix} \tag{6.30}$$

Solving Eqs. (6.29) and (6.30), we eliminate c_3 and d_3, and express b_1 and b_2 in terms of a_1 and a_2. After doing a little algebra, we get

$$\begin{pmatrix} b_1 \\ b_2 \end{pmatrix} = \begin{pmatrix} r_1 + \dfrac{t_1' r_2 t_1}{1 - r_1' r_2} & \dfrac{t_1' t_2'}{1 - r_1' r_2} \\ \dfrac{t_2 t_1}{1 - r_1' r_2} & r_2' + \frac{t_2 r_1' t_2'}{1 - r_1' r_2} \end{pmatrix} \begin{pmatrix} a_1 \\ a_2 \end{pmatrix} \tag{6.31}$$

Comparing Eqs. (6.21) and (6.31), we get

$$\boxed{\begin{aligned} r &= r_1 + \frac{t_1' r_2 t_1}{1 - r_1' r_2} \\ r' &= r_2' + \frac{t_2 r_1' t_2'}{1 - r_1' r_2} \\ t &= \frac{t_2 t_1}{1 - r_1' r_2} \\ t' &= \frac{t_1' t_2'}{1 - r_1' r_2} \end{aligned}} \tag{6.32}$$

The transmission probability of an electron in forward direction (from lead 1 to lead 2) is given by

$$\boxed{T = |t|^2 = \frac{T_1 T_2}{1 - 2\sqrt{R_1 R_2} \cos\theta + R_1 R_2}} \tag{6.33}$$

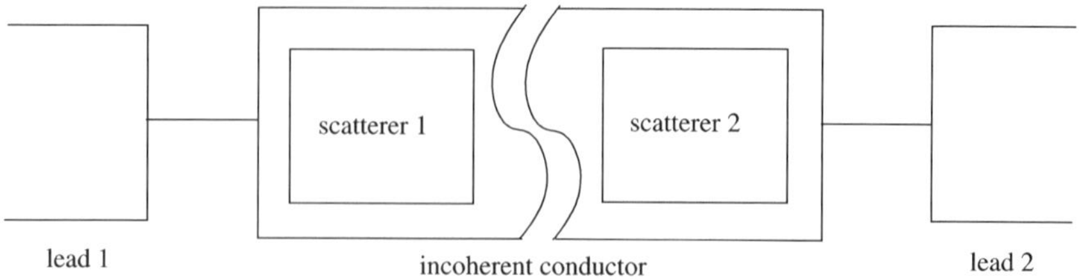

Fig. 6.7: The conductor consists of two scatterers which do not have any coherence between them.

where $T_1 = |t_1|^2$, $T_2 = |t_2|^2$, $R_1 = |r_1|^2$ and $R_2 = |r_2|^2$. θ describes the net phase shifts acquired in one round trip between the scatterers.

Incoherent Combination: Now, we move to the case where the size of the conductor is much greater than the phase coherence length. In Fig. 6.7, we consider one such incoherent conductor which is assumed to be divided into two subsections which are incoherent in nature. It means the phase of an electron after being scattered from scatterer 1 is completely uncorrelated with the phase it acquires after scattering from the second scatterer. We also assume that both the scatterers are described by individual S matrices S_1 and S_2, like the previous case. It is known that the conductance of a single-channel two-terminal conductor is described as

$$G = \frac{2e^2}{h} T$$

Hence, the net resistance becomes

$$R = G^{-1} = \frac{h}{2e^2}\left(\frac{1}{T}\right)$$

This resistance can be rewritten as a sum of two parts, one is associated with contact resistance R_c and the other is related to intrinsic resistance R_i, and it becomes

$$\begin{aligned} R &= \frac{h}{2e^2}\left(1 + \frac{1-T}{T}\right) \\ &= \frac{h}{2e^2} + \frac{h}{2e^2}\left(\frac{1-T}{T}\right) \\ &= R_c + R_i \end{aligned} \tag{6.34}$$

We can see from the above equation that when $T \to 1$, the intrinsic resistance, $R_i \to 0$. Now, when two scatterers are coupled incoherently, the above relations must be true for each individual scatterer. Therefore, the net intrinsic resistance for the whole conductor must be the sum of the contributions from these two parts. Thus,

$$\begin{aligned} R_i &= R_i^1 + R_i^2 \\ &= \frac{h}{2e^2}\left(\frac{1-T_1}{T_1} + \frac{1-T_2}{T_2}\right) \\ &= \left(\frac{R_1}{T_1} + \frac{R_2}{T_2}\right) \end{aligned} \tag{6.35}$$

So, the net resistance becomes

$$\begin{aligned} R &= \frac{h}{2e^2}\left(1 + \frac{R_1}{T_1} + \frac{R_2}{T_2}\right) \\ &= \frac{1-R_1R_2}{T_1T_2} \end{aligned} \tag{6.36}$$

Hence, the conductance is

$$G = \frac{2e^2}{h}\left(\frac{T_1T_2}{1-R_1R_2}\right) \tag{6.37}$$

So, in short, after combining the scatterers incoherently, we can write the transmission probability as

$$\boxed{T = \frac{T_1T_2}{1-R_1R_2}} \tag{6.38}$$

Equation (6.38) is independent of any phase factor θ as expected.

Now, suppose there are N number of incoherent scatterers that are combined. We assume that all the scatterers have the same transmission probability T and reflection probability R. Hence, the net intrinsic resistance is

$$\begin{aligned} R_{\text{in}} &= \left(\frac{h}{2e^2}\right)\sum_{k=1}^{N}\frac{1-T_k}{T_k} \\ &= \left(\frac{h}{2e^2}\right)\left[\frac{1-T_1}{T_1} + \frac{1-T_2}{T_2} + \cdots + \frac{1-T_N}{T_N}\right] \\ &= \left(\frac{h}{2e^2}\right)\frac{N(1-T)}{T} \quad [\text{as all } T_k\text{'s are equal}] \end{aligned} \tag{6.39}$$

Thus, the net resistance becomes

$$
\begin{aligned}
R &= \frac{h}{2e^2}\left(1+\frac{(1-T)N}{T}\right) \\
&= \frac{h}{2e^2}\left(\frac{T+N(1-T)}{T}\right)
\end{aligned}
\tag{6.40}
$$

Hence, the conductance is

$$
G = \frac{2e^2}{h}\left(\frac{T}{T+N(1-T)}\right) \tag{6.41}
$$

Therefore, the total transmission probability after combining N number of scatterers gets the form

$$
\boxed{T(N) = \frac{T}{T+N(1-T)}} \tag{6.42}
$$

6.2.5.2. *Green's function*

Definition

The S matrix provides the information about the response at one lead due to an excitation at the other. Green's function is a more powerful concept in a sense that it tells us about the response at any point (inside or outside the conductor) due to an excitation at any other [36, 39].

Suppose a response R is related to the excitation by the relation

$$
D_{\text{op}}R = S \tag{6.43}
$$

Then the Green's function can be defined as

$$
G = D_{\text{op}}^{-1} \tag{6.44}
$$

In our transport problem, we define the Green's function in the form

$$
G = (E - H_{\text{op}})^{-1} \tag{6.45}
$$

Green's function for a single-channel wire

For a differential operator, the inverse is not properly specified until the boundary conditions are mentioned. Depending on two types of boundary conditions, Green's function is also defined in two different types: (i) retarded Green's function and (ii) advanced Green's function.

First, we consider a simple one-dimensional single-channel wire with constant potential U_0 and zero vector potential. In this case, the Hamiltonian is given by

$$H = \left[-\frac{\hbar^2}{2m}\frac{\partial^2}{\partial x^2} + U_0\right] \tag{6.46}$$

Therefore, Green's function can be defined as

$$G = \left[E + \frac{\hbar^2}{2m}\frac{\partial^2}{\partial x^2} - U_0\right]^{-1} \tag{6.47}$$

i.e.,

$$\boxed{\left(E + \frac{\hbar^2}{2m}\frac{\partial^2}{\partial x^2} - U_0\right) G(x, x') = \delta(x - x')} \tag{6.48}$$

Equation (6.48) looks like the well-known Schrödinger equation apart from the source term at the right. This Green's function can be viewed physically as a wave function at x, resulting due to an excitation at x'.

Such excitations can generate two different types of waves travelling in two opposite directions with amplitudes A^+ and A^-, as presented in Figs. 6.8 and 6.9.

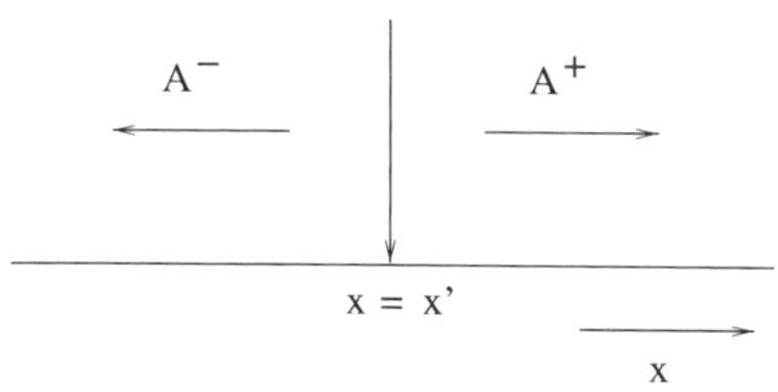

Fig. 6.8: Two waves travelling in opposite directions "away" from the point of excitation ($x = x'$).

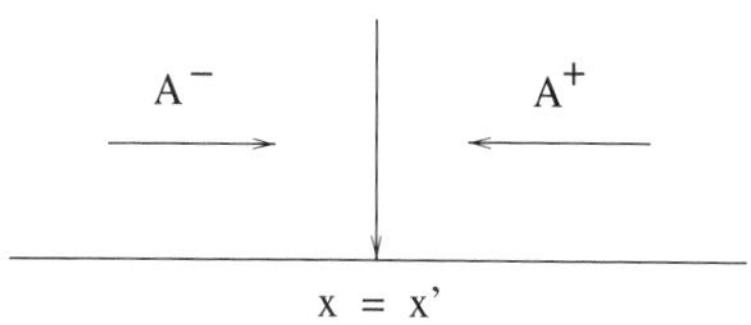

Fig. 6.9: Two waves travelling in opposite directions "towards" the point of excitation ($x = x'$).

First, we concentrate on the case described in Fig. 6.8. Here, the waves are travelling in outward direction from the point of excitation. We define

$$\begin{aligned} G(x, x') &= A^+ e^{ik(x-x')} \quad \text{for } x > x' \\ &= A^- e^{-ik(x-x')} \quad \text{for } x < x' \end{aligned} \tag{6.49}$$

where $k = \sqrt{\frac{2m(E-U_0)}{\hbar^2}}$.

Now, we apply the condition of continuity of Green's function to satisfy Eq. (6.49) at $x = x'$, i.e.,

$$\begin{aligned} [G(x, x')]_{x=x'+\epsilon} &= [G(x, x')]_{x=x'-\epsilon} \\ A^+ &= A^- \end{aligned} \tag{6.50}$$

So, we can write Green's function as

$$G(x, x') = Ae^{ik|x-x'|} \tag{6.51}$$

Let us rewrite Eq. (6.48) in the following form:

$$\frac{d^2G(x, x')}{dx^2} + \frac{2m}{\hbar^2}(E - U_0)G(x, x') = \frac{2m}{\hbar^2}\delta(x - x') \tag{6.52}$$

We integrate the above equation within the interval $x' + \epsilon$ to $x' - \epsilon$ for $\epsilon \to 0$:

$$\begin{aligned} &\left[\frac{dG}{dx}|_{x'+\epsilon} - \frac{dG}{dx}|_{x'-\epsilon}\right] + k^2 \int_{x'-\epsilon}^{x'+\epsilon} G(x, x')\, dx \\ &= \frac{2m}{\hbar^2} \int_{x'-\epsilon}^{x'+\epsilon} \delta(x - x')\, dx \end{aligned} \tag{6.53}$$

In the limit $\epsilon \to 0$, the second term in the L.H.S. of the above equation $\to$ 0 which leads to the equation

$$\frac{dG}{dx}\,|_{x=x'+\epsilon} - \frac{dG}{dx}|_{x=x'-\epsilon} = \frac{2m}{\hbar^2} \tag{6.54}$$

Now, following the definition of Green's function, Eq. (6.51), we get from the above equation under the limiting condition $\epsilon \to 0$

$$A = \frac{m}{ik\hbar^2} = \frac{-i}{\hbar v}$$

where $v = \frac{\hbar k}{m}$. Therefore,

$$\boxed{G(x, x') = -\left(\frac{i}{\hbar v}\right) e^{ik|x-x'|}} \tag{6.55}$$

Equation (6.55) denotes the solutions which are referred to as the retarded Green's function. It represents the outgoing waves moving away from the point of excitation $x = x'$.

There are another set of solutions (see Fig. 6.9) which represent the incoming waves moving towards the point of excitation. These solutions are referred to as the advanced Green's function and expressed as

$$\boxed{G(x, x') = +\left(\frac{i}{\hbar v}\right) e^{-ik|x-x'|}} \tag{6.56}$$

Thus, in brief, the two types of Green's functions (retarded and advanced) refers to two different types of solutions subjected to two different boundary conditions. The retarded function represents the outgoing waves (viz., towards the source), whereas the advanced function represents the incoming waves (viz., away from the source).

Incorporation of a small imaginary term η: The common practice to incorporate boundary conditions in the Green's function is to add or subtract an infinitesimal imaginary number to the equation. We rewrite Eq. (6.48) including η as

$$\frac{\hbar^2}{2m}\frac{d^2G(x, x')}{dx^2} + (E - U_0 + i\eta)G(x, x') = \delta(x - x')$$

$$\Rightarrow \quad \frac{d^2G(x, x')}{dx^2} + \frac{2m(E - U_0 + i\eta)}{\hbar^2}G(x, x') = \frac{2m}{\hbar^2}\delta(x - x')$$

$$\Rightarrow \quad \frac{d^2G(x, x')}{dx^2} + k_1^2 G(x, x') = \frac{2m}{\hbar^2}\delta(x - x') \tag{6.57}$$

Here, the momentum is modified as

$$\begin{aligned} k_1 &= \frac{\sqrt{2m(E - U_0 + i\eta)}}{\hbar} \\ &= \frac{\sqrt{2m(E - U_0)}}{\hbar}\left(1 + \frac{i\eta}{2(E - U_0)}\right) \\ &= k(1 + i\delta) \end{aligned} \tag{6.58}$$

where $\delta = \frac{\eta}{2(E-U_0)}$. As $\eta \to 0$, δ becomes vanishingly small. Therefore, we see that as we incorporate a small imaginary number to Green's function equation, the momentum is also modified, that is, an infinitesimally small quantity is also added to the momentum, which means $k \to k(1 + i\delta)$. From Eq. (6.58), we see that due to this modified momentum, the retarded function becomes the only acceptable solution as the advanced part grows indefinitely as we go away from the point of excitation, i.e., $x - x'$ gets increased.

Similarly, if Eq. (6.48) is modified by subtracting $i\eta$, then we get

$$\begin{aligned} \frac{\hbar^2}{2m}\frac{d^2G(x,x')}{dx^2} + (E - U_0 - i\eta)G(x,x') &= \delta(x - x') \\ \Rightarrow \quad \frac{d^2G(x,x')}{dx^2} + \frac{2m(E - U_0 - i\eta)}{\hbar^2}G(x,x') &= \frac{2m}{\hbar^2}\delta(x - x') \\ \Rightarrow \quad \frac{d^2G(x,x')}{dx^2} + k_2^2 G(x,x') &= \frac{2m}{\hbar^2}\delta(x - x') \end{aligned} \tag{6.59}$$

Here, the momentum is modified as

$$k_2 = k(1 - i\delta) \tag{6.60}$$

In this case, by a similar argument as stated earlier, the advanced Green's function becomes the only acceptable solution:

$$\boxed{\begin{aligned} G^r &= [E - H_{op} + i\eta]^{-1} \\ G^a &= [E - H_{op} - i\eta]^{-1} \\ &\text{where } \eta \longrightarrow 0 \end{aligned}} \tag{6.61}$$

Green's function for a multi-channel wire

Now, we are going to derive an expression of Green's function for a multi-channel conducting wire. The Hamiltonian is written as

$$H = -\frac{\hbar^2}{2m}\left(\frac{\partial^2}{\partial x^2} + \frac{\partial^2}{\partial y^2}\right) + U(y) \tag{6.62}$$

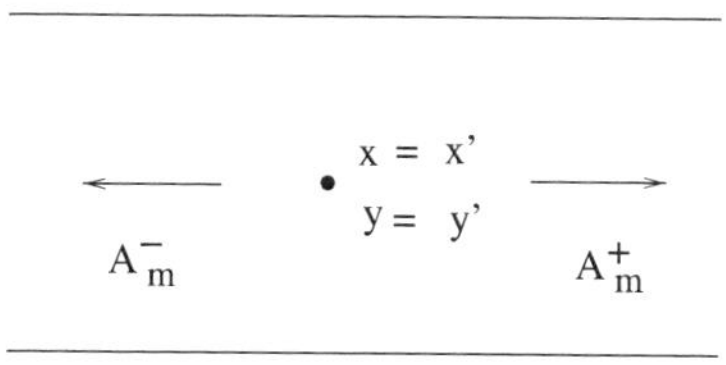

Fig. 6.10: Waves are travelling in two opposite directions from the point of excitation ($x = x'$, $y = y'$). The wire has a finite width along Y direction:

where $U(y)$ is the transverse confining potential along Y direction. Here, Green's function can be written as

$$\begin{aligned} G^r(x, y; x', y') &= \left(E + i\eta + \frac{\hbar^2}{2m}\frac{\partial^2}{\partial x^2} + \frac{\hbar^2}{2m}\frac{\partial^2}{\partial y^2} - U(y)\right)^{-1} \\ &\Rightarrow \left(E + i\eta + \frac{\hbar^2}{2m}\frac{\partial^2}{\partial x^2} + \frac{\hbar^2}{2m}\frac{\partial^2}{\partial y^2} - U(y)\right) \\ &\quad \times G^r(x, y; x', y') = \delta(x - x')\delta(y - y') \end{aligned} \tag{6.63}$$

Physically, $G^r(x, y; x', y')$ represents a wave function at (x, y) due to an excitation at (x', y'). As we have already discussed, it can propagate in two directions. So, we can write Green's function as

$$G^r(x, x') = \sum_m A_m^{\pm}\, \chi_m(y)\, e^{ik_m|x-x'|} \tag{6.64}$$

Here, A_m^+ and A_m^- are the amplitudes of two oppositely propagating modes through channel m. $\chi_m(y)$'s are the transverse mode wave functions, which satisfy the following relation:

$$\left[-\frac{\hbar^2}{2m}\frac{\partial^2}{\partial y^2} + U(y)\right]\chi_m(y) = \epsilon_m \chi_m(y) \tag{6.65}$$

$\chi_m(y)$'s are orthogonal to each other, which means

$$\int \chi_m(y)\chi_n(y)\, dy = \delta_{mn} \tag{6.66}$$

Now, proceeding as earlier, we get two conditions. First one is

$$G^r(x, x')|_{x=x'+\epsilon} = G^r(x, x')|_{x=x'-\epsilon} \tag{6.67}$$

which, in the limit $\epsilon \to 0$, reduces to the following:

$$\sum_m A_m^+ \chi_m(y) = \sum_m A_m^- \chi_m(y) \tag{6.68}$$

and the second condition is

$$\left.\frac{\partial G^r}{\partial x}\right|_{x'+\epsilon} - \left.\frac{\partial G^r}{\partial x}\right|_{x'-\epsilon} = \frac{2m}{\hbar^2}\delta(y-y') \tag{6.69}$$

It gives

$$\sum_m ik_m(A_m^+ + A_m^-)\chi_m(y) = \frac{2m}{\hbar^2}\delta(y-y') \tag{6.70}$$

Now, multiplying Eq. (6.68) by $\chi_n(y)$ and integrating over y, we get

$$\begin{aligned}
\int \sum_m A_m^+ \chi_m(y)\chi_n(y)\,dy &= \int \sum_m A_m^- \chi_m(y)\chi_n(y)\,dy \\
\Rightarrow \quad \sum_m A_m^+ \delta_{mn} &= \sum_m A_m^- \delta_{mn} \\
\Rightarrow \quad A_m^+ = A_m^- &= A_m \quad \text{(say)}
\end{aligned} \tag{6.71}$$

Similarly, from Eq. (6.70), we have

$$\begin{aligned}
\int \sum_m ik_m\,(A_m^+ + A_m^-)\chi_m(y)\chi_n(y)\,dy &= \frac{2m}{\hbar^2}\int \delta(y-y')\chi_n(y)\,dy \\
\Rightarrow \quad \sum_m ik_m\,(A_m^+ + A_m^-)\delta_{mn} &= \frac{2m}{\hbar^2}\chi_n(y') \\
\Rightarrow \quad ik_m(A_m^+ + A_m^-) &= \frac{2m}{\hbar^2}\chi_m(y')
\end{aligned} \tag{6.72}$$

Using Eq. (6.71) in Eq. (6.72), we find

$$A_m = -\frac{i}{\hbar v_m}\chi_m(y')$$

Therefore, the expression of Green's function becomes

$$\boxed{G^r(x,y;x',y') = \sum_m -\frac{i}{\hbar v_m}\chi_m(y')\,\chi_m(y)\,e^{ik_m|x-x'|}} \tag{6.73}$$

where

$$k_m = \frac{\sqrt{2m(E-\epsilon_m)}}{\hbar} \quad \text{and} \quad v_m = \frac{\hbar k_m}{m}$$

Eigenfunction expansion of Green's function

Now, we derive an expression of $G(\vec{r}, \vec{r'})$ in terms of energy eigenfunctions. Suppose our system is described by the Hamiltonian H_{op} so that we can write

$$H_{op}\psi_\alpha(\vec{r}) = \epsilon_\alpha \psi_\alpha(\vec{r}) \tag{6.74}$$

where $\psi_\alpha(\vec{r})$'s are the eigenstates of H with eigenvalue ϵ_α.

In general, we expand Green's function in terms of $\vec{r}$ and $\vec{r'}$ as

$$G^r(\vec{r}, \vec{r'}) = \sum_\alpha C_\alpha(\vec{r'})\psi_\alpha(\vec{r}) \tag{6.75}$$

where $G^r(\vec{r}, \vec{r'})$ follows the relation

$$\begin{aligned}
(E - H_{op} + i\eta)\, G^r(\vec{r}, \vec{r'}) &= \delta(\vec{r} - \vec{r'}) \\
\Rightarrow \quad \sum_\alpha (E - H_{op} + i\eta)\, C_\alpha(\vec{r'})\psi_\alpha(\vec{r}) &= \delta(\vec{r} - \vec{r'}) \\
\Rightarrow \quad \sum_\alpha (E - \epsilon_\alpha + i\eta)\, C_\alpha(\vec{r'})\psi_\alpha(\vec{r}) &= \delta(\vec{r} - \vec{r'})
\end{aligned} \tag{6.76}$$

Now, multiplying the above equation by $\psi^*_\beta(\vec{r})$ and integrating over $\vec{r}$, we get

$$\sum_\alpha C_\alpha(\vec{r'})\,(E - \epsilon_\alpha + i\eta) \int \psi_\alpha(\vec{r})\psi^*_\beta(\vec{r})\, d\vec{r} = \int \psi^*_\beta(\vec{r})\delta(\vec{r} - \vec{r'})\, d\vec{r}$$

$$\Rightarrow \quad C_\beta(\vec{r'})\,(E - \epsilon_\beta + i\eta) = \psi^*_\beta(\vec{r'})$$

$$\Rightarrow \quad C_\alpha(\vec{r'}) = \frac{\psi^*_\alpha(\vec{r'})}{(E - \epsilon_\alpha + i\eta)} \tag{6.77}$$

Therefore, we can write Green's function in terms of the eigenfunctions as

$$\boxed{G^r(\vec{r}, \vec{r'}) = \sum_\alpha \frac{\psi^*_\alpha(\vec{r'})\psi_\alpha(\vec{r})}{E - \epsilon_\alpha + i\eta}} \tag{6.78}$$

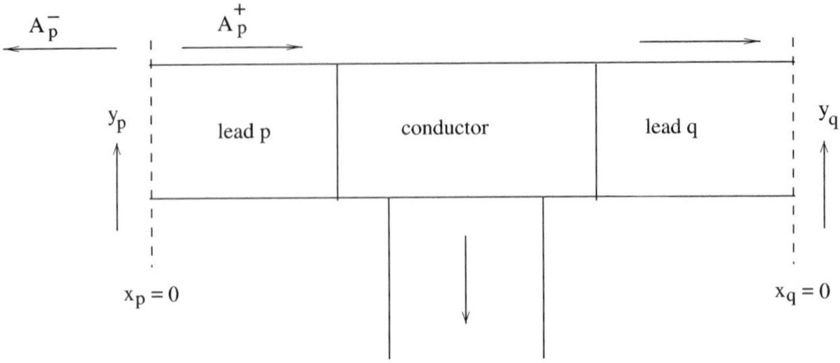

Fig. 6.11: An incident wave is generated due to excitation in lead p, which is partially transmitted to other leads.

In a similar way, the advanced Green's function can also be evaluated as

$$\boxed{G^a(\vec{r},\vec{r'}) = \sum_\alpha \frac{\psi^*_\alpha(\vec{r'})\psi_\alpha(\vec{r})}{E - \epsilon_\alpha - i\eta}} \tag{6.79}$$

6.2.5.3. *Relation between S matrix and Green's function*

For single-channel lead

We consider a conductor which is attached to a set of multiple leads, as shown in Fig. 6.11. The leads are assumed to have single channel. We want to relate the S matrix and Green's function between the two planes placed at $x_p = 0$ and $x_q = 0$. We assume G^r_{qp} is Green's function which relates the excitation to the response between a point lying on $x_p = 0$ and the other lying at $x_q = 0$.

It is already known that a unit excitation at any point on $x_p = 0$ gives rise to two waves with amplitudes A_p^+ and A_p^-. A_p^+ moves in $+X$ direction, while A_p^- goes along the opposite direction (not shown here in the figure). This A_p^+ is then scattered by the conductor into different leads. So, keeping this picture in mind, we can write Green's function as

$$G^r_{qp} = \delta_{qp}A_p^- + s'_{qp}A_p^+ \tag{6.80}$$

where

$$s'_{qp} = \sqrt{\frac{v_p}{v_q}}s_{qp}$$

We have already shown in our earlier calculations that $A_p^+ = A_p^- = -\frac{i}{\hbar v_p}$.

In the same way, if we now consider the same system, but with multi-channel leads, then following the previous arguments, we can write the Green's function as

$$G^r_{qp}(y_q; y_p) = \sum_{m\in p}\sum_{n\in q}\left[\delta_{nm}A^-_m + s'_{nm}A^+_m\right]\chi_n(y_q) \tag{6.81}$$

Now, $A^+_m = A^-_m = -\frac{i}{\hbar v_m}\chi_m(y_p)$; $s'_{nm} = s_{nm}\sqrt{\frac{v_m}{v_n}}$, and thus,

$$G^r_{qp}(y_q; y_p) = \sum_{m\in p}\sum_{n\in q} -\frac{i}{\hbar v_m}\chi_m(y_p)\left(\delta_{nm} + \sqrt{\frac{v_m}{v_n}}s_{nm}\right)\chi_n(y_q)$$

$$\Rightarrow \quad i\hbar v_m G^r_{qp}(y_q; y_p) = \sum_{m\in p}\sum_{n\in q}\chi_m(y_p)\left(\delta_{nm} + \sqrt{\frac{v_m}{v_n}}s_{nm}\right)\chi_n(y_q) \tag{6.82}$$

Multiplying both sides of Eq. (6.82) by $\chi_m(y_p)$ and $\chi_n(y_q)$, and integrating over y_p and y_q by using the orthogonality relation,

$$\int \chi_m(y_p)\chi_n(y_p)\,dy_p = \delta_{mn}$$

we get

$$\begin{aligned}
&i\hbar v_m \int\int G^r_{qp}(y_q; y_p)\chi_n(y_q)\chi_m(y_p)\,dy_p\,dy_q \\
&\quad = \sum_{\substack{m\in p\\ n\in q}}\int\int \chi_m(y_p)\chi_m(y_p)\,dy_p \\
&\quad \times \left(\delta_{nm} + \sqrt{\frac{v_m}{v_n}}s_{nm}\right)\chi_n(y_q)\chi_n(y_q)\,dy_q \\
&\quad \Rightarrow \left(\delta_{nm} + \sqrt{\frac{v_m}{v_n}}s_{nm}\right) = i\hbar v_m \\
&\quad \times \int\int \chi_n(y_q)G^r_{qp}(y_q; y_p)\chi_m(y_p)\,dy_p\,dy_q
\end{aligned} \tag{6.83}$$

Finally, we reach the relation

$$\boxed{s_{nm} = -\sqrt{\frac{v_n}{v_m}}\delta_{nm} + i\hbar\sqrt{v_m v_n}\int\int \chi_n(y_q)G^r_{qp}(y_q; y_p)\chi_m(y_p)\,dy_p\,dy_q} \tag{6.84}$$

Equation (6.84) describes the relation between S matrix elements and Green's function for a conductor attached to multi-channel leads.

6.2.5.4. *Partitioning the Hamiltonian matrix*

Usually in our problems, we work in a single-electron tight-binding (TB) framework, so that we can express the Hamiltonians in a matrix form very easily. Probably this is the simplest way of simulating Hamiltonians of low-dimensional quantum systems, especially in the absence of electron–electron correlation. Following the definition of Green's function, we can write

$$G^r = [(E + i\eta)\mathbb{I} - H]^{-1} \tag{6.85}$$

But, there is a problem with Eq. (6.85) when we work on an open quantum system i.e., when a conductor is connected with two or even more semi-infinite contact leads. Therefore, H becomes infinite dimensional and so also G^r. It is therefore not possible to do any calculation with a matrix whose dimension is infinity. So, to get rid of this situation, we apply the partitioning technique which maps the Green's function matrix in the reduced Hilbert space of the conductor itself which is sandwiched between the contact leads, and the effects of the leads are incorporated there.

Suppose we have a conductor which is attached to lead p. Hence, the total Hamiltonian of the system can be written in a matrix form as

$$H = \begin{pmatrix} \boldsymbol{H_p} & \boldsymbol{\tau_p} \\ \boldsymbol{\tau_p^\dagger} & \boldsymbol{H_c} \end{pmatrix} \tag{6.86}$$

Here, $\boldsymbol{H_c}$ and $\boldsymbol{H_p}$ are the Hamiltonians describing the conductor and the side-attached lead. $\boldsymbol{\tau_p}$ and $\boldsymbol{\tau_p^\dagger}$ correspond to the coupling of the conductor to the side-attached lead, i.e., $\langle p_i|\boldsymbol{\tau_p}|i\rangle = t$ (see Fig. 6.12).

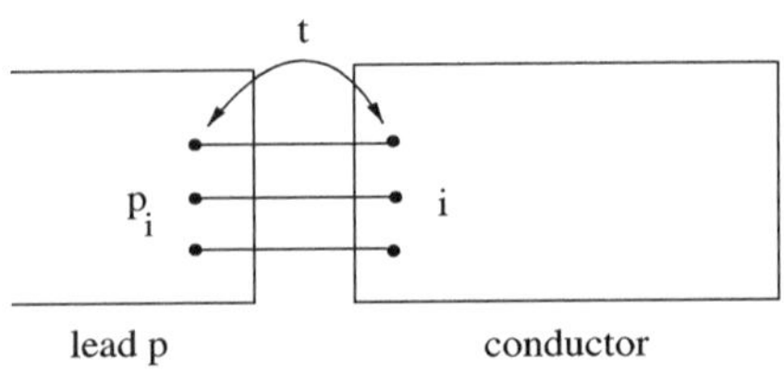

Fig. 6.12: Coupling between the conductor and the semi-infinite lead p is depicted here.

Hence, Green's function becomes

$$G^r = ((E + i\eta)\mathbb{I} - \boldsymbol{H})^{-1}$$
$$= \begin{pmatrix} (E + i\eta)\mathbb{I} - \boldsymbol{H_p} & -\boldsymbol{\tau_p} \\ -\boldsymbol{\tau_p^\dagger} & E\mathbb{I} - \boldsymbol{H_{pc}} \end{pmatrix}^{-1} \tag{6.87}$$

Now, partitioning the Green's function matrix in the same way as we do for the Hamiltonian matrix, we can rewrite the above equation as

$$\begin{pmatrix} \boldsymbol{G_p} & \boldsymbol{G_{pc}} \\ \boldsymbol{G_{cp}} & \boldsymbol{G_c} \end{pmatrix} = \begin{pmatrix} (E + i\eta)\mathbb{I} - \boldsymbol{H_p} & -\boldsymbol{\tau_p} \\ -\boldsymbol{\tau_p^\dagger} & E\mathbb{I} - \boldsymbol{H_c} \end{pmatrix}^{-1}$$
$$\Rightarrow \begin{pmatrix} \boldsymbol{G_p} & \boldsymbol{G_{pc}} \\ \boldsymbol{G_{cp}} & \boldsymbol{G_c} \end{pmatrix} \begin{pmatrix} (E + i\eta)\mathbb{I} - \boldsymbol{H_p} & -\boldsymbol{\tau_p} \\ -\boldsymbol{\tau_p^\dagger} & E\mathbb{I} - \boldsymbol{H_c} \end{pmatrix}$$
$$= \begin{pmatrix} \mathbf{1} & \mathbf{0} \\ \mathbf{0} & \mathbf{1} \end{pmatrix} \tag{6.88}$$

The above equation yields two decoupled equations which are

$$\boldsymbol{G_{cp}}\left[(E + i\eta)\mathbb{I} - \boldsymbol{H_p}\right] - \boldsymbol{G_c}\boldsymbol{\tau_p^\dagger} = \mathbf{0} \tag{6.89}$$

and

$$-\boldsymbol{G_{cp}}\boldsymbol{\tau_p} + \boldsymbol{G_c}(E\mathbb{I} - \boldsymbol{H_c}) = \mathbf{1} \tag{6.90}$$

Defining $(E + i\eta)\mathbb{I} - \boldsymbol{H_p} = (\boldsymbol{g_p^r})^{-1}$, we get from Eq. (6.89)

$$\boldsymbol{G_{cp}}(\boldsymbol{g_p^r})^{-1} = \boldsymbol{G_c}\boldsymbol{\tau_p^\dagger}$$
$$\Rightarrow \boldsymbol{G_{cp}} = \boldsymbol{G_c}\boldsymbol{\tau_p^\dagger}\boldsymbol{g_p^r} \tag{6.91}$$

Using the expression of $\boldsymbol{G_{cp}}$ from Eq. (6.91) into Eq. (6.90), we get

$$-\boldsymbol{G_c}\boldsymbol{\tau_p^\dagger}\boldsymbol{g_p^r}\boldsymbol{\tau_p} + \boldsymbol{G_c}(E\mathbb{I} - \boldsymbol{H_c}) = \mathbf{1}$$
$$\Rightarrow \boldsymbol{G_c}\left[E\mathbb{I} - \boldsymbol{H_c} - \boldsymbol{\tau_p^\dagger}\boldsymbol{g_p^r}\boldsymbol{\tau_p}\right] = \mathbf{1}$$
$$\Rightarrow \boldsymbol{G_c}\left[E\mathbb{I} - \boldsymbol{H_c} - \boldsymbol{\Sigma_p}\right] = \mathbf{1}$$
$$\Rightarrow \boldsymbol{G_c} = \left[E\mathbb{I} - \boldsymbol{H_c} - \boldsymbol{\Sigma_p}\right]^{-1} \tag{6.92}$$

where

$$\boxed{\boldsymbol{\Sigma_p} = \boldsymbol{\tau_p^\dagger g_p^r \tau_p}} \tag{6.93}$$

Here, $\boldsymbol{G_c}$ ($= \boldsymbol{\mathcal{G}}$) is the effective Green's function which incorporates the effect of the infinite lead coupled to the conductor.

In Eq. (6.92), all matrices have identical dimension ($C \times C$) where C is the dimension of the conductor. But, at first sight, it might appear that the problem of inverting an infinite-dimensional matrix still remains, while evaluating $\boldsymbol{g_p^r}$ to obtain the expression for $\boldsymbol{\Sigma_p}$. Fortunately, for an isolated infinite lead, $\boldsymbol{g_p^r}$ can be calculated analytically.

The ijth element of the self-energy matrix

$$\begin{aligned} \boldsymbol{\Sigma}_{ij} &= \langle i|\boldsymbol{\tau_p^\dagger g_p^r \tau_p}|j\rangle \\ &= \sum_{m,n} \langle i|\boldsymbol{\tau_p^\dagger}|m\rangle\langle m|\boldsymbol{g_p^r}|n\rangle\langle n|\boldsymbol{\tau_p}|j\rangle \\ &= t^2 \boldsymbol{g_p^r}(p_i, p_j) \end{aligned} \tag{6.94}$$

Here, $\boldsymbol{g_p^r}(p_i, p_j)$ denotes the (p_i, p_j)th element of the matrix $\boldsymbol{g_p^r}$. To work with Eq. (6.92), we have to reconstruct the $\boldsymbol{\Sigma_p}$ matrix in $C \times C$ dimension. Here, all the elements of $\boldsymbol{\Sigma_p}$ matrix would be zero except at the points (i,j) inside the conductor, which are adjacent to points (p_i,p_j) inside the lead.

For more than one contact leads, we need to add the effects of all individual leads. It means, if we have p number of side-attached leads, then the effective Green's function will be

$$\boldsymbol{\mathcal{G}} = \left[E\mathbb{I} - \boldsymbol{H_c} - \sum_p \boldsymbol{\Sigma_p}\right]^{-1} \tag{6.95}$$

So, from now on, we have to consider only this effective Green's function ($\boldsymbol{\mathcal{G}}$) in which the effects of all the side-attached leads are already included.

6.2.5.5. *Alternate derivation of Green's function for a multi-channel wire*

In this section, we present an alternate derivation of Eq. (6.73) by using the expression described in Eq. (6.78). It will be helpful for understanding the derivation of self-energy calculations which will be provided in the forthcoming sections [36, 39].

For an infinite wire having finite width in Y direction (see Fig. 6.10), the Hamiltonian can be written as

$$H = -\frac{\hbar^2}{2m}\frac{\partial^2}{\partial x^2} - \frac{\hbar^2}{2m}\frac{\partial^2}{\partial y^2} + U(y) \tag{6.96}$$

where $U(y)$ is the confining potential along Y direction.

The solution of Eq. (6.96) is expressed as

$$\psi_{\beta,m}(x,y) = \frac{1}{\sqrt{L}}\chi_m(y)e^{i\beta x} \tag{6.97}$$

where $\chi_m(y)$'s are transverse mode wave functions and they satisfy the following relation:

$$\left[-\frac{\hbar^2}{2m}\frac{\partial^2}{\partial y^2} + U(y)\right]\chi_m(y) = \epsilon_{m,0}\chi_m(y) \tag{6.98}$$

and

$$\epsilon_{m,\beta} = \epsilon_{m,0} + \frac{\hbar^2\beta^2}{2m} \tag{6.99}$$

Here, $\frac{1}{\sqrt{L}}$ is the normalization constant and L is the length over which wave functions are normalized.

Now, we already know that Green's function can be written using the eigenfunction technique (see Eq. (6.78)). Hence, substituting Eq. (6.97) in Eq. (6.78) and assuming that χ_m's are real, Green's function can be written as

$$G^r(x,y;x',y') = \frac{1}{L}\sum_{\beta,m}\frac{\chi_m(y)\chi_m(y')e^{i\beta(x-x')}}{E-\epsilon_{m,0}-\frac{\hbar^2\beta^2}{2m}+i\eta} \tag{6.100}$$

Converting the discrete sum over β into an integral, we get

$$G^r(x,y;x',y') = \frac{1}{2\pi}\sum_m \chi_m(y)\chi_m(y') \times \int_{-\infty}^{+\infty}\frac{e^{i\beta(x-x')}\,d\beta}{E-\epsilon_{m,0}-\frac{\hbar^2\beta^2}{2m}+i\eta} \tag{6.101}$$

Now,

$$
\begin{aligned}
E - \epsilon_{m,0} - \frac{\hbar^2\beta^2}{2m} + i\eta &= -\frac{\hbar^2}{2m}\left[\beta^2 - \frac{2m}{\hbar^2}(E - \epsilon_{m,0} + i\eta)\right] \\
&= -\frac{\hbar^2}{2m}\left[\beta^2 - \frac{2m}{\hbar^2}(E - \epsilon_{m,0})\left(1 + \frac{i\eta}{E - \epsilon_{m,0}}\right)\right] \\
&= -\frac{\hbar^2}{2m}\left[\beta^2 - k_m^2(1 + i\delta)\right] \qquad (6.102)
\end{aligned}
$$

where

$$
k_m^2 = \frac{2m}{\hbar^2}(E - \epsilon_{m,0}) \quad \text{and} \quad \delta = \frac{\eta}{E - \epsilon_{m,0}}
$$

Therefore, Eq. (6.101) can be rewritten as

$$
\boxed{G^r(x, y; x', y') = -\frac{m}{\pi\hbar^2}\sum_m \chi_m(y)\chi_m(y')\int_{-\infty}^{+\infty}\frac{e^{i\beta(x-x')}\,d\beta}{\beta^2 - k_m^2(1 + i\delta)}} \qquad (6.103)
$$

Here, our first job is to evaluate the contour integration

$$
\begin{aligned}
I &= \int_{-\infty}^{+\infty}\frac{e^{i\beta(x-x')}\,d\beta}{\beta^2 - k_m^2(1 + i\delta)} \\
&= \int_{-\infty}^{+\infty}\frac{e^{i\beta(x-x')}\,d\beta}{\left[\beta + k_m\left(1 + i\frac{\delta}{2}\right)\right]\left[\beta - k_m\left(1 + i\frac{\delta}{2}\right)\right]} \qquad (6.104)
\end{aligned}
$$

We can see that there are two simple poles off the real axis by an infinitesimal quantity $\frac{\delta}{2}$ (see Fig. 6.13). Now, for choosing the contour, lets say $\beta = Re^{i\theta}$.

So,

$$
e^{i\beta(x-x')} = e^{iR\cos\theta(x-x')}e^{-R\sin\theta(x-x')} \qquad (6.105)
$$

Therefore, for $x - x' > 0$, the integrand is bounded in the limit $R \to \infty$ if we consider the upper contour. Similarly, for $x - x' < 0$, we choose the lower contour.

Considering Cauchy's residue theorem for $x - x' > 0$, we get

$$
\begin{aligned}
I &= 2\pi i \lim_{\beta \to k_m(1+i\frac{\delta}{2})}\left[\beta - k_m\left(1 + i\frac{\delta}{2}\right)\right]\frac{e^{i\beta(x-x')}}{\left[\beta + k_m(1 + i\frac{\delta}{2})\right]\left[\beta - k_m(1 + i\frac{\delta}{2})\right]} \\
&= \pi i\frac{e^{ik_m(1+i\frac{\delta}{2})(x-x')}}{k_m(1 + i\frac{\delta}{2})} \qquad (6.106)
\end{aligned}
$$

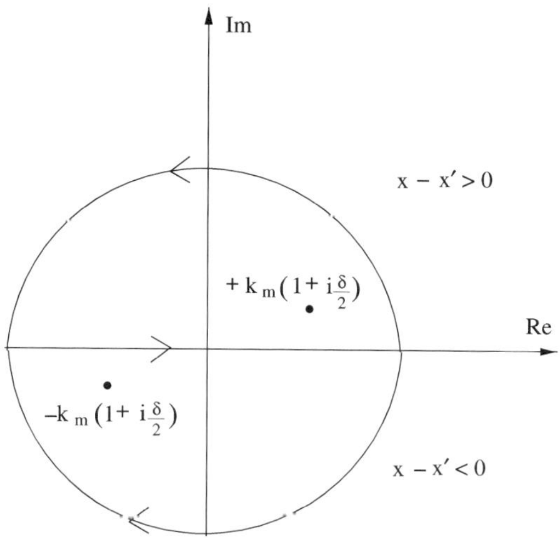

Fig. 6.13: Two separate contours for $x - x' > 0$ and $x - x' < 0$.

In the limit $\delta \to 0$,

$$I = \left(\frac{i\pi}{k_m}\right) e^{ik_m(x-x')} \tag{6.107}$$

Using the relation $\frac{\hbar k_m}{m} = v_m$, we eventually get the expression of Green's function as

$$G^r = \sum_m - \left(\frac{i}{\hbar v_m}\right) \chi_m(y)\chi_m(y') e^{ik(x-x')} \tag{6.108}$$

In the same footing, for $x - x' < 0$,

$$\begin{aligned} I = -2\pi i & \lim_{\beta \to -k_m(1+i\frac{\delta}{2})} \left[\beta + k_m\left(1 + i\frac{\delta}{2}\right)\right] \\ & \times \frac{e^{i\beta(x-x')}}{\left[\beta + k_m\left(1+i\frac{\delta}{2}\right)\right]\left[\beta - k_m(1+i\frac{\delta}{2})\right]} \\ = \pi i & \frac{e^{-ik_m(1+i\frac{\delta}{2})(x-x')}}{k_m(1+i\frac{\delta}{2})} \end{aligned} \tag{6.109}$$

Again, for $\delta \to 0$,

$$I = \frac{i\pi}{k_m} e^{-ik_m(x-x')} \tag{6.110}$$

So, precisely, we can write the expression of Green's function

$$\boxed{G^r = \sum_m -\left(\frac{i}{\hbar v_m}\right)\chi_m(y)\chi_m(y')e^{ik_m|x-x'|}} \tag{6.111}$$

6.2.5.6. *Derivation of self-energy expression for a semi-infinite multi-channel lead*

Here, we will derive the general expression of self-energy term for a semi-infinite, multi-channel lead [36, 37, 39].

To do this, first we calculate the expression for Green's function of a multi-moded semi-infinite lead. We use the same technique following Eq. (6.78), as discussed in the previous section. The Hamiltonian is the same as given in Eq. (6.96), but as the lead is terminated on one side, the wave function must vanish at that side. The eigenfunctions can be written as

$$\psi_{m,\beta}(x) = \sqrt{\frac{2}{L}}\chi_m(y)\sin(\beta x) \tag{6.112}$$

with eigenenergies

$$\epsilon_{m,\beta} = \epsilon_{m,0} + \frac{\hbar^2\beta^2}{2m} \tag{6.113}$$

The factor $\sqrt{\frac{2}{L}}$ is the normalization constant and $\chi_m(y)$'s are the transverse mode eigenfunctions. Substituting the expression of $\psi_{m,\beta}(x)$ in Eq. (6.78), we obtain the expression of Green's function as

$$G^r(x, y; x, y') = \frac{2}{L}\sum_{m,\beta}\frac{\chi_m(y)\chi_m(y')\sin^2(\beta x)}{E - \epsilon_{m,0} - \frac{\hbar^2\beta^2}{2m} + i\eta} \tag{6.114}$$

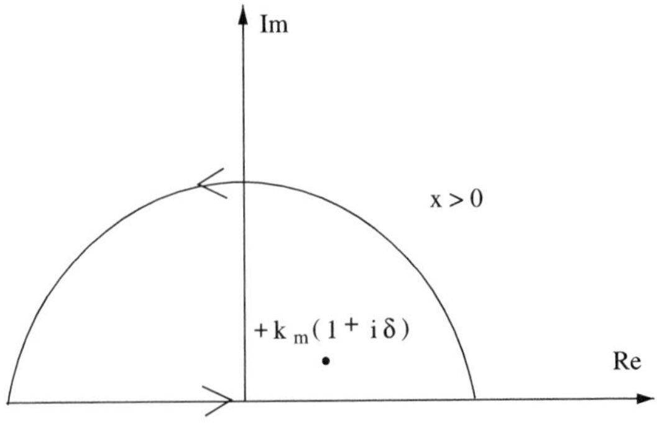

Fig. 6.14: Contour of integration for $x > 0$.

Here, we choose $x' = x$ because we need to evaluate Green's function between two points having same X coordinate. Now, converting the discrete sum into an integral, we get

$$G^r(x, y, x, y') = \frac{2}{L}\sum_m \chi_m(y)\chi_m(y')\int_0^{+\infty}\frac{\sin^2(\beta x)\,d\beta}{E - \epsilon_{m,0} - \frac{\hbar^2\beta^2}{2m} + i\eta} \tag{6.115}$$

We have already got

$$E - \epsilon_{m,0} - \frac{\hbar^2\beta^2}{2m} + i\eta = -\frac{\hbar^2}{2m}\left[\beta^2 - k_m^2(1 + i\delta)\right]$$

where

$$k_m^2 = \frac{2m}{\hbar^2}(E - \epsilon_{m,0}) \quad \text{and} \quad \delta = \frac{\eta}{E - \epsilon_{m,0}}$$

So, we rewrite Eq. (6.115) as

$$G^r(x, y; x, y') = \frac{4m}{\pi\hbar^2}\sum_m \chi_m(y)\chi_m(y')\int_0^{+\infty}\frac{\sin^2(\beta x)\,d\beta}{\beta^2 - k_m^2(1 + i\delta)} \tag{6.116}$$

Again,

$$\sin^2(\beta x) = \frac{(1 - e^{i2\beta x}) + (1 - e^{-i2\beta x})}{4}$$

Thus, finally, the Green's function looks like

$$G^r(x, y; x, y') = \frac{m}{\pi\hbar^2}\sum_m \chi_m(y)\chi_m(y')\int_{-\infty}^{+\infty}\frac{(1 - e^{i2\beta x})\,d\beta}{\beta^2 - k_m^2(1 + i\delta)} \tag{6.117}$$

Our job now boils down to the evaluation of the following integral:

$$\begin{aligned} I &= \int_{-\infty}^{+\infty}\frac{(1 - e^{i2\beta x})\,d\beta}{\beta^2 - k_m^2(1 + i\delta)} \\ &= \int_{-\infty}^{+\infty}\frac{(1 - e^{i2\beta x})\,d\beta}{\left[\beta + k_m(1 + i\frac{\delta}{2})\right]\left[\beta - k_m(1 + i\frac{\delta}{2})\right]} \end{aligned} \tag{6.118}$$

Following the same logic as in the previous calculation, here we choose the upper contour for integration (as $x > 0$), and following the Cauchy's residue theorem, we get

$$
\begin{aligned}
I &= 2\pi i \lim_{\beta\to k_m(1+i\frac{\delta}{2})} \left[\beta - k_m\left(1+i\frac{\delta}{2}\right)\right] \\
&\quad \times \frac{1-e^{i2\beta x}}{\left[\beta + k_m(1+i\frac{\delta}{2})\right]\left[\beta - k_m(1+i\frac{\delta}{2})\right]} \\
&= \frac{\pi i}{k_m}\frac{1-e^{i2k_m(1+i\frac{\delta}{2})x}}{k_m\left(1+i\frac{\delta}{2}\right)} \qquad (6.119)
\end{aligned}
$$

In the limit $\delta \to 0$,

$$
\begin{aligned}
I &= \frac{\pi i}{k_m}(1-e^{i2k_m x}) \\
&= \frac{2\pi}{k_m}\sin(k_m x)e^{ik_m x} \qquad (6.120)
\end{aligned}
$$

Using Eqs. (6.117) and (6.120), we get

$$
G^r(x,y;x,y') = -\sum_m \frac{2m}{\hbar^2 k_m}\chi_m(y)\chi_m(y')\sin(k_m x)e^{ik_m x} \qquad (6.121)
$$

Therefore, using the relation $\hbar k_m = m v_m$, the final expression of Green's function reads as

$$
\boxed{G^r(x,y;x,y') = -\sum_m \frac{2\sin(k_m x)}{\hbar v_m}\chi_m(y)e^{ik_m x}\chi_m(y')} \qquad (6.122)
$$

Equation (6.122) describes Green's function for a multi-channel wire, where both X and Y coordinates are continuous. In our problems, we usually model the entire system on a discrete lattice. So, what we need now is the expression of Green's function for a multi-channel lead on discrete lattice basis. Therefore, discretizing Eq. (6.122), we obtain the Green's function elements $g_p^r(p_i,p_j)$ as

$$
\begin{aligned}
\boldsymbol{g_p^r}(p_i,p_j) &= G^r(x,y,x,y')|_{x=a,y=p_i,y'=p_j} \\
&= -\sum_m \frac{2\sin(k_m a)}{\hbar v_m}\chi_m(p_i)e^{ik_m a}\chi_m(p_j) \qquad (6.123)
\end{aligned}
$$

It can be shown that any general n-tuple chain can be decomposed into n single tight-binding chains with site energies given by

$$\epsilon_m = 2v_l \cos\left(\frac{m\pi a}{N+1}\right) \quad \text{where} \quad m = 1, 2, \ldots, N \tag{6.124}$$

Here, m denotes the mth mode, and the dispersion relation for each chain is

$$E_m = \epsilon_m + 2t_l \cos(k_m a)$$

Here, t_l and v_l stand for the nearest-neighbor hopping (NNH) along the X and Y directions, respectively. The velocity for the mth mode is given by

$$\begin{aligned} v_m &= \frac{1}{\hbar}\left(\frac{\partial E}{\partial k}\right) \\ \Rightarrow \hbar v_m &= -2at_l \sin(k_m a) \end{aligned} \tag{6.125}$$

Using Eqs. (6.125) and (6.123), we get (setting $a = 1$ for simplification)

$$\boldsymbol{g}_{\boldsymbol{p}}^{\boldsymbol{r}}(p_i, p_j) = \sum_m \frac{1}{t_l} \chi_m(p_i) e^{ik_m a} \chi_m(p_j) \tag{6.126}$$

Therefore, from Eq. (6.94), finally we obtain

$$\boxed{\boldsymbol{\Sigma}_{ij} = \frac{t^2}{t_l} \sum_m \chi_m(p_i) e^{ik_m a} \chi_m(p_j)} \tag{6.127}$$

In the following section, we will present the self-energy expression in terms of tight-binding parameters.

6.2.5.7. *Expression of self-energy for single- and multi-channel leads in terms of TB parameters*

As we work mostly using a TB model, it is customary for us to write down the self-energy matrix in discrete Wannier basis in terms of the TB parameters like site energy, nearest-neighbor hopping strength, etc.

Single-channel TB lead

Let us begin with a simple one-dimensional (1D) semi-infinite TB chain as a contact lead. In this case, Eq. (6.127) is simplified as

$$\Sigma = \frac{t^2}{t_l} e^{ika} \tag{6.128}$$

Here, Σ is an element instead of being a matrix. Now, for a 1D chain, the energy dispersion relation becomes

$$\begin{aligned} E &= \epsilon_l + 2t_l \cos(ka) \\ &\epsilon_l + t_l e^{ika} + t_l e^{-ika} \end{aligned} \tag{6.129}$$

Defining $e^{ika} = x$, we rewrite Eq. (6.129) as

$$x^2 - \left(\frac{E - \epsilon_l}{t_l}\right) x + 1 = 0 \tag{6.130}$$

Solving the above equation, we get

$$x = \frac{1}{2t_l}\left[(E - \epsilon_l) \pm ic\sqrt{4t_l^2 - (E - \epsilon_l)^2}\right] \tag{6.131}$$

Now, substituting back into the above expression, we find

$$\boxed{\Sigma = \frac{t^2}{2t_l^2}\left[(E - \epsilon_l) - i\sqrt{4t_l^2 - (E - \epsilon_l)^2}\right]} \tag{6.132}$$

This expression is valid within the range, $|E - \epsilon_l| < 2t_l$. If the energy lies outside this range, the self-energy expression is modified in the form

$$\boxed{\begin{aligned} \Sigma &= \frac{t^2}{2t_l^2}\left[(E - \epsilon_l) - \sqrt{(E - \epsilon_l)^2 - 4t_l^2}\right] \quad \text{for } E > \epsilon_l + 2|t_l| \\ \Sigma &= \frac{t^2}{2t_l^2}\left[(E - \epsilon_l) + \sqrt{(E - \epsilon_l)^2 - 4t_l^2}\right] \quad \text{for } E < \epsilon_l - 2|t_l| \end{aligned}}$$

Multi-channel TB lead

Next, we consider the case where the lead has N number of transverse modes (see Fig. 6.15). Hence, the ijth element of the self-energy matrix, having dimension $N \times N$, is given by

$$\mathbf{\Sigma}_{ij} = \frac{t^2}{t_l}\sum_m \chi_m(p_i) e^{ik_m a} \chi_m(p_j) \tag{6.133}$$

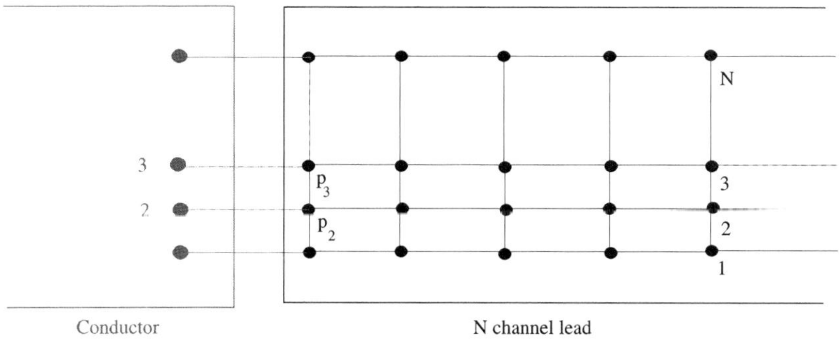

Fig. 6.15: An N-channel lead is attached to a conductor. Here, boundary sites of the conductor and the leads are described by i's and p_i's, respectively.

Here, $\chi_m(p_i)$ is the discretized transverse mode wave function at the point p_i, and it is

$$\chi_m(p_i) = \sqrt{\frac{2}{N+1}}\sin\left(k_y(m)p_i a\right) \tag{6.134}$$

where

$$k_y(m) = \frac{m\pi}{(N+1)a} \quad m = 1, 2, \ldots, N$$

We have already calculated that for an isolated 1D chain with on-site energy ϵ_m,

$$e^{ik_m a} = \frac{1}{2t_l}\left[(E-\epsilon_m) - i\sqrt{4t_l^2 - (E-\epsilon_m^2)^2}\right] \quad \text{for } |E-\epsilon_m| < 2|t_l| \tag{6.135}$$

and

$$e^{ik_m a} = \frac{1}{2t_l}\left[(E-\epsilon_m) \mp \sqrt{(E-\epsilon_m^2)^2 - 4t_l^2}\right] \quad \text{for } |E-\epsilon_m| > 2|t_l| \tag{6.136}$$

with

$$\epsilon_m = 2t_l \cos\left(\frac{m\pi}{N+1}\right)$$

Therefore, substituting the above expression in Eq. (6.127), finally we obtain

$$\boldsymbol{\Sigma}_{ij} = \frac{2}{(N+1)}\left[\frac{t^2}{2t_l^2}\sum_m \sin(k_y(m)p_i)\left((E-\epsilon_m) - i\sqrt{4t_l^2-(E-\epsilon_m)^2}\right)\sin(k_y(m)p_j)\right]$$

for $|E-\epsilon_m| < 2|t_l|$

$$\boldsymbol{\Sigma}_{ij} = \frac{2}{(N+1)}\left[\frac{t^2}{2t_l^2}\sum_m \sin(k_y(m)p_i)\left((E-\epsilon_m) - \sqrt{(E-\epsilon_m)^2-4t_l^2}\right)\sin(k_y(m)p_j)\right]$$

for $E > \epsilon_m + 2|t_l|$

$$\boldsymbol{\Sigma}_{ij} = \frac{2}{(N+1)}\left[\frac{t^2}{2t_l^2}\sum_m \sin(k_y(m)p_i)\left((E-\epsilon_m) + \sqrt{(E-\epsilon_m)^2-4t_l^2}\right)\sin(k_y(m)p_j)\right]$$

for $E < \epsilon_m - 2|t_l|$

6.2.5.8. *Coupling matrix*

The coupling matrix Γ_p for pth lead is defined as

$$\boldsymbol{\Gamma_p} = i\left[\Sigma_p^r - \Sigma_p^a\right] \tag{6.137}$$

i.e.,

$$\boldsymbol{\Gamma_p}(i,j) = i\left[\boldsymbol{\Sigma_p^r}(i,j) - \boldsymbol{\Sigma_p^a}(i,j)\right] \tag{6.138}$$

Now, we already know the expression of Σ_p^r (see Eq. (6.127)) which is

$$\boldsymbol{\Sigma_p^r}(i,j) = \frac{t^2}{t_l}\sum_m \chi_m(p_i)e^{ik_m a}\chi_m(p_j) \tag{6.139}$$

Therefore,

$$\boldsymbol{\Sigma_p^a}(i,j) = \frac{t^2}{t_l}\sum_m \chi_m(p_i)e^{-ik_m a}\chi_m(p_j) \tag{6.140}$$

Hence, we obtain the coupling matrix as

$$\begin{aligned}\boldsymbol{\Gamma_p}(i,j) &= i\left[\boldsymbol{\Sigma_p^r}(i,j) - \boldsymbol{\Sigma_p^a}(i,j)\right] \\ &= \frac{it^2}{t_l}\sum_m \chi_m(p_i)\left(e^{ik_m a} - e^{-ik_m a}\right)\chi_m(p_j) \\ &= \frac{-2t^2}{t_l}\sum_m \chi_m(p_i)\sin(k_m a)\chi_m(p_j)\end{aligned} \tag{6.141}$$

From the energy–momentum dispersion relation, we have

$$\begin{aligned} E &= \epsilon_m + 2t_l \cos(k_m a) \\ \Rightarrow \frac{\partial E}{\partial k} &= -2t_l a \sin(k_m a) \\ \Rightarrow -2\sin(k_m a) &= \frac{\hbar v_m}{t_l a} \end{aligned} \tag{6.142}$$

Substituting the above relation in Eq. (6.141), we get

$$\mathbf{\Gamma}_{\boldsymbol{p}}(i,j) = \frac{t^2}{t_l^2} \sum_m \chi_m(p_i) \left(\frac{\hbar v_m}{a}\right) \chi_m(p_j) \tag{6.143}$$

If we assume $t = t_l$, then the above equation is modified to

$$\boxed{\mathbf{\Gamma}_{\boldsymbol{p}}(i,j) = \sum_m \chi_m(p_i) \left(\frac{\hbar v_m}{a}\right) \chi_m(p_j)} \tag{6.144}$$

The coupling matrix elements are symmetric, i.e., $\Gamma_p(i,j) = \Gamma_p(j,i)$.

6.2.5.9. *Evaluation of transmission function*

In this section, we derive the expression of the transmission function T_{qp}, which describes the transmission probability of an electron from lead p to lead q. $|s_{nm}|^2$ denotes the transmission probability of an electron from channel m ($m \in p$) to channel n ($n \in q$).

The discretized version of Eq. (6.84) (for $n \neq m$) is given by

$$s_{nm} = \frac{i\hbar\sqrt{v_m v_n}}{a} \sum_{i,j} \chi_n(q_j) G^r(j,i) \chi_m(p_i) \tag{6.145}$$

Therefore, using the relation $G^r(j,i)^* = G^a(i,j)$, we can write

$$s_{nm}^* = \frac{-i\hbar\sqrt{v_m v_n}}{a} \sum_{i',j'} \chi_n(q_{j'}) G^a(i',j') \chi_m(p_{i'}) \tag{6.146}$$

So,

$$\begin{aligned} |s_{nm}|^2 = \frac{\hbar^2 v_m v_n}{a^2} &\sum_{i,j,i',j'} \chi_n(q_j) G^r(j,i) \chi_m(p_i) \chi_n(q_{j'}) \\ &\times G^a(i',j') \chi_m(p_{i'}) \end{aligned} \tag{6.147}$$

Now,

$$\begin{aligned}
T_{qp} &= \sum_{n\in q}\sum_{m\in p} |s_{nm}|^2 \\
&= \sum_{i,j,i',j'} \left(\sum_{n\in q} \chi_n(q_j)\frac{\hbar v_n}{a}\chi_n(q_{j'})\right) G^r(j,i) \\
&\quad \times \left(\sum_{m\in p} \chi_m(p_i)\frac{\hbar v_m}{a}\chi_m(p_{i'})\right) G^a(i',j') \\
&= \sum_{i,j,i',j'} \Gamma_q(j',j)G^r(j,i)\Gamma_p(i,i')G^a(i',j') \\
&= \mathrm{Tr}[\boldsymbol{\Gamma_q G^r \Gamma_p G^a}]
\end{aligned}$$

Therefore, the transmission probability of an electron from lead p to lead q is evaluated from the relation

$$\boxed{T_{qp} = \mathrm{Tr}[\boldsymbol{\Gamma_q G^r \Gamma_p G^a}]} \tag{6.148}$$

In the case of spin-dependent transmission, while evaluating the spin-resolved transmission probabilities, the main idea remains the same, but we need to determine each quantity for $\uparrow$ and $\downarrow$ spins separately. In such cases, the spin-resolved transmission probability of an electron injected from lead p with spin σ and collected at lead q with spin σ' can be written as

$$\boxed{T_{qp}^{\sigma'\sigma} = \mathrm{Tr}[\boldsymbol{\Gamma_q^{\sigma'} G^r \Gamma_p^{\sigma} G^a}]} \tag{6.149}$$

6.2.5.10. *Spectral function*

The spectral function is defined by the following relation

$$A = i[G^r - G^a] \tag{6.150}$$

Spectral function in terms of energy eigenfunctions

For an isolated system (i.e., conductor without any external leads), which is described by the Hamiltonian H_c, we can always write

$$H_c\psi_{\alpha 0} = \epsilon_{\alpha 0}\psi_{\alpha 0} \tag{6.151}$$

where $\epsilon_{\alpha 0}$ and $\psi_{\alpha 0}$ are the energy eigenvalues and eigenvectors, respectively. But, when we deal with an open system, the effective Hamiltonian looks like

$$H_{\mathit{eff}} = H_c + \Sigma^r \tag{6.152}$$

Here, $\Sigma = \sum_i \Sigma_i^r$, and Σ_i^r's are the self-energy matrices introduced to incorporate the effects of side-attached leads.

For this effective Hamiltonian. we can write

$$[H_c + \Sigma^r]\psi_\alpha = \epsilon_\alpha \psi_\alpha \tag{6.153}$$

Here, ϵ_α and ψ_α are the eigenvalues and eigenfunctions for this new effective Hamiltonian.

One important point to be noted here is that, in this case, the energy eigenvalues ϵ_α's are in general complex because H_{eff} is not hermitian.

Therefore, we can expand the energy eigenvalues as

$$\epsilon_\alpha = \epsilon_{\alpha 0} - \Delta_\alpha - \frac{i\gamma_\alpha}{2} \tag{6.154}$$

Now, unlike Eqs (6.78) and (6.79), here, in the case of an open system (i.e., conductor + leads), the Green's function cannot be expanded in terms of ψ_α's only because ψ_α's do not form an orthogonal set as the effective Hamiltonian $(H_c + \Sigma^r)$ is not hermitian.

So, we define another set of eigenfunctions (ϕ_α's) for the adjoint Hamiltonian $H_{\text{eff}}^\dagger$ as well.

$$[H_c + \Sigma^a]\phi_\alpha = \epsilon_\alpha^* \phi_\alpha \tag{6.155}$$

Together ϕ's and ψ's form a biorthonormal set which satisfy

$$\int \phi_\alpha(\vec{r})\psi_\beta^*(\vec{r})\, d\vec{r} = \delta_{\alpha\beta} \tag{6.156}$$

Let us write Green's function as

$$G^r(\vec{r}, \vec{r'}) = \sum_\alpha C_\alpha(\vec{r'})\psi_\alpha(\vec{r}) \tag{6.157}$$

Now, following the definition of Green's function, we have

$$[E - H_c - \Sigma^r]G^r(\vec{r}, \vec{r'}) = \delta(\vec{r} - \vec{r'}) \tag{6.158}$$

Using Eq. (6.157), we can rewrite Eq. (6.158) as

$$\begin{aligned} &\sum_\alpha C_\alpha(\vec{r'})(E - H_c - \Sigma^r)\psi_\alpha = \delta(\vec{r} - \vec{r'}) \\ &\Rightarrow \sum_\alpha C_\alpha(\vec{r'})(E - \epsilon_\alpha)\psi_\alpha = \delta(\vec{r} - \vec{r'}) \end{aligned} \tag{6.159}$$

Multiplying the above equation by $\phi_\beta^*(\vec{r})$ and by integrating over $\vec{r}$, we get

$$\sum_\alpha C_\alpha(\vec{r'})(E-\epsilon_\alpha)\int \phi_\beta^*(\vec{r})\psi_\alpha(\vec{r})\,d\vec{r} = \int \phi_\beta^*(\vec{r})\delta(\vec{r}-\vec{r'})\,d\vec{r} \tag{6.160}$$

Now, using the biorthogonality relation, we find

$$C_\alpha(\vec{r'}) = \frac{\phi_\alpha^*(\vec{r'})}{E-\epsilon_\alpha} \tag{6.161}$$

Hence, substituting the value of $C_\alpha(\vec{r'})$ from Eq. (6.161) into Eq. (6.157), we finally obtain

$$G^r(\vec{r},\vec{r'}) = \sum_\alpha \frac{\phi_\alpha^*(\vec{r'})\psi_\alpha(\vec{r})}{E-\epsilon_\alpha} \tag{6.162}$$

Using the form of ϵ_α, we write

$$\boxed{G^r(\vec{r},\vec{r'}) = \sum_\alpha \frac{\psi_\alpha(\vec{r})\phi_\alpha^*(\vec{r'})}{(E-\epsilon_{\alpha 0}+\Delta_\alpha)+\frac{i\gamma_\alpha}{2}}} \tag{6.163}$$

Similarly, it can be shown that

$$\boxed{G^a(\vec{r},\vec{r'}) = \sum_\alpha \frac{\psi_\alpha(\vec{r})\phi_\alpha^*(\vec{r'})}{(E-\epsilon_{\alpha 0}+\Delta_\alpha)-\frac{i\gamma_\alpha}{2}}} \tag{6.164}$$

Therefore, the spectral function is given by

$$\begin{aligned} A &= i[G^r - G^a] \\ &= \sum_\alpha \psi_\alpha(\vec{r})\phi_\alpha^*(\vec{r'})\frac{\gamma_\alpha}{\left[(E-\epsilon_{\alpha 0}+\Delta_\alpha)^2+\frac{\gamma_\alpha^2}{4}\right]} \end{aligned} \tag{6.165}$$

Hence,

$$|A(\vec{r},\vec{r'},E)| = \sum_\alpha \frac{\gamma_\alpha}{\left[(E-\epsilon_{\alpha 0}+\Delta_\alpha)^2+\frac{\gamma_\alpha^2}{4}\right]} \tag{6.166}$$

The quantity $|A(\vec{r},\vec{r'},E)|$ shows some peaks as a function of energy. Here, Δ_α's and γ_α's represent the shift in peak positions and broadening of the peaks, respectively. As the coupling of the conductor to the leads is increased, the quantities Δ_α's and γ_α's are also increased, and therefore, the peaks get shifted and also broadened as compared to the case of unperturbed Hamiltonian, i.e., the uncoupled conductor only.

One important point is there to mention. The peaks associated with the plot of $|A(\vec{r},\vec{r'},E)|$ vs. E need not have the Lorentzian shapes as may be assumed from Eq. (6.166) because Σ's themselves are functions of energy and so also ϵ_α's, Δ_α's and γ_α's. So, the peaks in general can have different non-Lorentzian shapes.

6.2.5.11. *Density of states*

The trace of the spectral function is defined as the density of states (DOS), i.e.,

$$N(E) = \left(\frac{1}{2\pi}\right) \mathrm{Tr}[A(E)] \tag{6.167}$$

Now,

$$\begin{aligned} \mathrm{Tr}[A(E)] &= \int A(\vec{r}, \vec{r'})\, dE \\ &= \sum_\alpha \int \psi_\alpha(\vec{r})\phi^*_\alpha(\vec{r})\, d\vec{r} \frac{\gamma_\alpha}{\left[(E - \epsilon_{\alpha 0} + \Delta_\alpha)^2 + \frac{\gamma_\alpha^2}{4}\right]} \\ &= \sum_\alpha \frac{\gamma_\alpha}{\left[(E - \epsilon_{\alpha 0} + \Delta_\alpha)^2 + \frac{\gamma_\alpha^2}{4}\right]} \end{aligned} \tag{6.168}$$

Therefore,

$$\boxed{N(E) = \frac{1}{2\pi} \sum_\alpha \frac{\gamma_\alpha}{\left[(E - \epsilon_{\alpha 0} + \Delta_\alpha)^2 + \frac{\gamma_\alpha^2}{4}\right]}} \tag{6.169}$$

In the limit $\gamma_\alpha \to 0$, we get

$$N(E) = \delta(E - \epsilon_{\alpha 0} + \Delta_\alpha) \tag{6.170}$$

Again,

$$\begin{aligned} A(E) &= i[G^r - G^a] \\ &= -2\,\mathrm{Im}[G^r] \end{aligned} \tag{6.171}$$

Hence,

$$\boxed{N(E) = -\frac{1}{\pi} \mathrm{Tr}[\mathrm{Im}(G^r)]} \tag{6.172}$$

The average density of states (ADOS) is given by

$$\boxed{\rho(E) = -\frac{1}{N\pi} \mathrm{Tr}[\mathrm{Im}(G^r)]} \tag{6.173}$$

This is all about the theoretical prescriptions that we have wanted to explore here for a clear understanding of all the required mathematical steps for any general reader. Now, in the following section, we will concentrate on the thermoelectric phenomena in two different molecular systems.

6.3. Molecular Systems as Functional Elements

6.3.1. *Molecular wire with multiple loops*

Although a substantial amount of works has already been done [24, 44–48] to investigate thermoelectric phenomena using simple and/or complex molecular structures along with different tailor-made geometries, exploring several important results, still many open questions persist. For instance, can we have a proper functional element that can exhibit required thermoelectric efficiency or can we design a device whose efficiency can be tuned externally? Constructing a device of favorable ZT and its possible tuning will undoubtedly bring significant impact to hit the present demand of energy conversion.

In this section, we put forward an idea of getting higher ZT as well as its proper regulation considering a naphthalene molecule as the functional element [49]. The two loops of this molecular system are subjected to magnetic fluxes, the so-called Aharonov–Bohm (AB) fluxes, which play the central role of our analysis. As AB flux has significant impact on transport phenomena in interferometric geometries [50–55], especially in molecular junctions having single and/or multiple loops which has already been established with several works, we can also expect unusual signature in thermoelectric properties since two fluxes are there in the two sub-rings of the molecular system. The key idea is that, by selectively choosing the AB fluxes, we can have much higher asymmetric transmission function compared to the flux-free molecular system, and this is one of the primary requirements to achieve better ZT. Surprisingly, we see that much higher ZT can be achieved and it can be tuned as well by regulating the AB fluxes threaded by the sub-rings. The interplay between two fluxes in interferometric geometry and their critical roles on thermoelectric properties may bring some important inputs towards thermoelectric study. Employing a tight-binding (TB) framework, we calculate all the required thermoelectric quantities using Landauer prescription together with Green's function formalism, as discussed above. We hope that our results can be generalized to any higher multi-loop systems.

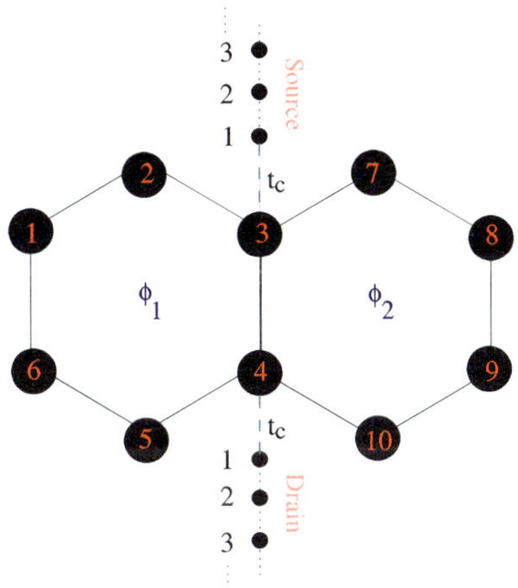

Fig. 6.16: Molecular junction where a naphthalene molecule is coupled to source and drain electrodes. The molecular sub-rings are threaded by the AB fluxes ϕ_1 and ϕ_2, which are the key controlling parameters of our analysis.

6.3.1.1. *Molecular system and the TB Hamiltonian*

The molecular system shown in Fig. 6.16 is taken into account to investigate the thermoelectric properties. The molecule is directly coupled to two semi-infinite one-dimensional perfect electrodes. A finite temperature difference ΔT is made between these electrodes, setting their temperatures as $T + \Delta T/2$ and $T - \Delta T/2$, respectively, where T is the equilibrium temperature. Because of this temperature difference, a net voltage is developed across the molecular junction. Though in Fig. 6.16, it is shown that the electrodes are coupled to sites 3 and 4 of the naphthalene molecule, we will consider different other junction configurations as well to examine thermoelectric behavior, since quantum interference associated with the junction configuration has an important effect on transport properties, which may influence the thermoelectric performance. In addition to junction configuration, the AB fluxes ϕ_1 and ϕ_2, threaded by two phenyl rings will have a significant impact on electron transfer from source to drain that can generate non-trivial thermoelectric features, and here, we will examine all these issues critically.

The Hamiltonian of the full system can be written as

$$H = H_m + H_s + H_d + H_c \tag{6.174}$$

where H_m, H_s, H_d and H_c are associated with different parts of the molecular wire. All these sub-Hamiltonians are expressed within a TB framework, and their explicit forms are described one by one as follows.

In the absence of AB fluxes ϕ_1 and ϕ_2, the sub-Hamiltonian H_m of the naphthalene molecule looks quite simple and it reads as

$$H_m = \sum_i \epsilon c_i^\dagger c_i + t \sum_i \left(c_{i+1}^\dagger c_i + c_i^\dagger c_{i+1} \right) \tag{6.175}$$

where, ϵ and t correspond to the molecular site energy and NNH integral, respectively. Now, when we include magnetic fluxes, the hopping integrals are no longer same for clockwise and anti-clockwise movements of electrons, and they are modified by proper phase factors. For such a connected geometry, two types of phase factors appear: one for the outer NNH bonds and other for the inner ones. Here, the total number of outer bonds is 10 and that of the inner bonds is one. The inner bond is formed by connecting sites 3 and 4 of the molecule (see Fig. 6.16). The phase factors in each outer NNH arm is [56] $\theta_{\text{out}} = 2\pi(\phi_1 + \phi_2)/10$, and in the inner arm, it becomes $\theta_{\text{in}} = 2\pi(\phi_1 - \phi_2)/2$. Thus, for one particular direction if the NNH integrals are $te^{i\theta_{\text{out}}}$ and $te^{i\theta_{\text{in}}}$, then for the reverse direction, the conjugates of these integrals are taken into account.

The terms H_s and H_d are used to describe the sub-Hamiltonians of the source and drain electrodes, which are considered as perfect, 1D and semi-infinite. In TB notation, they look like

$$H_s = \sum_p \epsilon_0 a_p^\dagger a_p + t_0 \sum_p \left(a_{p+1}^\dagger a_p + a_p^\dagger a_{p+1} \right) \tag{6.176}$$

and

$$H_d = \sum_q \epsilon_0 b_q^\dagger b_q + t_0 \sum_q \left(b_{q+1}^\dagger b_q + b_q^\dagger b_{q+1} \right) \tag{6.177}$$

where ϵ_0 and t_0 are the site energy and NNH integrals, respectively. a_p, $a_p^\dagger$, b_p and $b_p^\dagger$ are the fermionic operators in the electrodes.

These electrodes are coupled to the naphthalene molecule via the coupling parameter t_c. For the junction configuration shown in Fig. 6.16, the source and drain are coupled to sites 3 and 4 of the molecule, and thus, the coupling Hamiltonian H_c (mentioned in Eq. (6.174) reads as

$$H_c = t_c \left(a_1 c_3^\dagger + c_3 a_1^\dagger + b_1 c_4^\dagger + c_4 b_1^\dagger \right) \tag{6.178}$$

where the lattice sites of both the source and drain electrodes are numbered as 1, 2, 3,... from the molecular end (see Fig. 6.16). Here, we would like to note that, to explore the effect of quantum interference on thermoelectric performance, we consider different source–molecule–drain configurations. It is done by connecting the drain electrode at different sites of the molecule, keeping the source position fixed, as shown in Fig. 6.16 (viz., source is connected at site 3 for our entire analysis). Depending on the drain-to-molecular configuration, the site indices in the last two terms of the sub-Hamiltonian, Eq. (6.178), will be changed.

6.3.1.2. *Numerical results and discussion*

Here, we present our numerical results which essentially include the interplay between two fluxes threaded by two sub-rings on the thermoelectric performance. Our key focus is to search higher figure of merit for a particular junction set up under realistic parameter values. Before presenting the results, lets us mention the physical parameters that are kept constant for this molecular wire. We set $\epsilon_0 = \epsilon = 0$, $t = 1$, $t_0 = 2$, $t_c = 0.5$ and $T = 300\,\mathrm{K}$. All the energies are measured in the unit of electron volt (eV). The other parameter values that are not common are mentioned at the appropriate places.

Let us begin with Fig. 6.17 where the variations of G, S, k and ZT as a function of Fermi energy are shown, connecting the source and drain electrodes at sites 3 and 4 of the naphthalene molecule. Three different cases are considered, depending on the values of ϕ_1 and ϕ_2 which are presented by three-colored curves in each of these spectra. Several important features emerge, which are analyzed one by one as follows. The first key observation is that the behavior of electrical conductance changes drastically with the inclusion of magnetic flux (Fig. 6.17(a)). When both the fluxes are zero, almost vanishing conductance is obtained for the wide Fermi energy window (red line). Naturally, we cannot expect good thermoelectric performance within this window in the absence of magnetic flux, as higher G is the primary criterion to have better thermoelectric performance. The situation becomes quite interesting with the inclusion of magnetic flux in any one of the two sub-rings (blue curve). Two large conductance peaks appear. Interestingly, we see that these two peaks come closer with increasing heights when the other sub-ring is also subjected to a magnetic flux (black line). All these peaks are associated with the resonant energy channels of the system. Thus, the first impression is that we can selectively tune the position of the

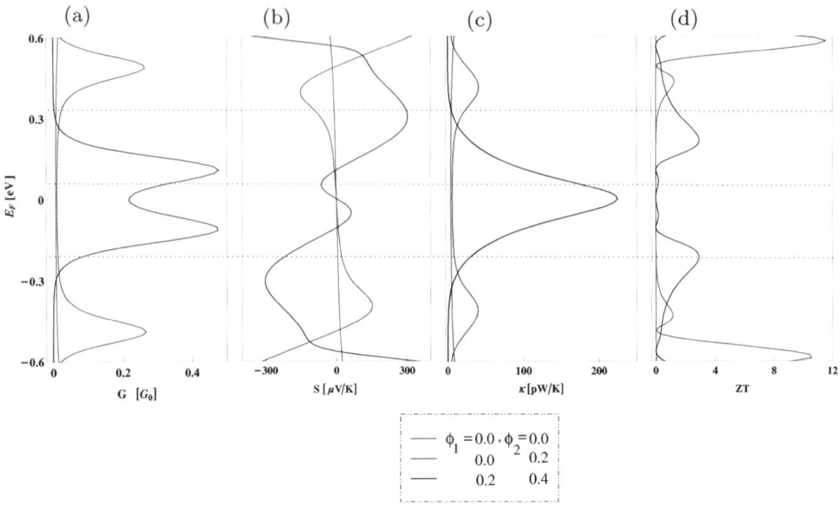

Fig. 6.17: (a) Electrical conductance, (b) Seebeck coefficient, (c) thermal conductance, and (d) ZT as the function of Fermi energy for the naphthalene molecule at some typical values of ϕ_1 and ϕ_2. All these results are computed by connecting the source and drain electrodes at the molecular sites 3 and 4, respectively.

resonant peak as well as its magnitude with the help of AB fluxes ϕ_1 and ϕ_2. It gives us a confidence that we have the finite possibility to tune the device performance externally.

But whenever we talk about the performance, we need to focus on S and κ as well, since these two factors have a significant impact on ZT. The results are presented in Figs. 6.17(b) and (c). In the absence of ϕ_1 and ϕ_2, both S and κ are vanishingly small (red curves) over the Fermi energy window shown in the spectra, whereas reasonably large values of S and κ are obtained with the addition of magnetic fluxes. It is always good to have higher S for better performance, but at the same time, we also need lower κ. Now, carefully examining the spectra in Figs. 6.17(a)–(c), it can be observed that at $E_F \sim \pm 0.55$, G and S have reasonably large values, while κ becomes too small for the case when $\phi_1 = 0$ and $\phi_2 = 0.2$. As a result of this, we get a peak value of $ZT \sim 8$ (blue line of Fig. 6.17(d)), which can be expected to satisfy the present requirement of ZT. Now, for the situation when both the rings are threaded by finite magnetic fluxes, apparently it seems that quite higher κ is obtained for wide Fermi energies, so favorable ZT may not be obtained. But, we can find suitable E_Fs, say for instance, $E_F \sim \pm 0.2$ where ZT reaches ~ 3 (black line of Fig. 6.17(d)) which also

Table 6.1: Interplay between magnetic fluxes and molecule-to-electrode junction configuration on thermoelectric efficiency. ZT_{max} is obtained by taking the maximum of ZT, varying the Fermi energy E_F from −0.6 to 0.6.

	ZT_{max}		
Drain position	$\phi_1 = 0$, $\phi_2 = 0$	$\phi_1 = 0$, $\phi_2 = 0.2$	$\phi_1 = 0.2$, $\phi_2 = 0.4$
1	0.767	2.762	1.495
2	0.029	1.408	4.849
4	0.023	11.525	2.931
5	0.759	2.854	1.609
6	0.034	1.912	5.232
7	0.029	0.840	0.848
8	0.767	1.859	1.239
9	0.034	1.743	1.408
10	0.759	1.318	1.108

satisfies the required demand of efficiency. At these Fermi energies, viz., $E_F \sim \pm 0.2$, on the other hand, ZT is too small for the system where one ring is free from any magnetic flux. These features are solely associated with the nature of conductance, i.e., how the transmission function gets modified with the physical parameters. Thus, depending on ϕ_1, ϕ_2 and E_F, we can have a chance to find a favorable ZT.

As already stated that the thermoelectric performance strongly depends on the transmission function which is quite sensitive to the molecule-to-electrode interface geometry, further scrutiny is required, considering different interface geometries to probe into it in more detail. Keeping the source position fixed at site 3, we connect the drain at all other sites of the molecule and find the maximum of ZT (which we refer to as ZT_{max}) by varying the Fermi energy from −0.6 to 0.6. The results are presented in Table 6.1. Three identical sets of ϕ_1 and ϕ_2 are considered to compare these results with the previous analysis given in Fig. 6.17. From the data presented in Table 6.1, it is clear that when both the rings are free from magnetic flux, favorable thermoelectric performance cannot be obtained. For most of the junction configurations, ZT_{max} is relatively low ($ZT_{\text{max}} < 1$). Now, comparing the results given in last two columns of Table 6.1, we can see that in almost all the source–molecule–drain junction configurations, affirmative response is obtained. Depending on ϕ_1 and ϕ_2, we have several options to select the specific molecule-to-electrode junction configuration to achieve reasonably large ZT, and it is clearly reflected from the data given in Table 6.1. All

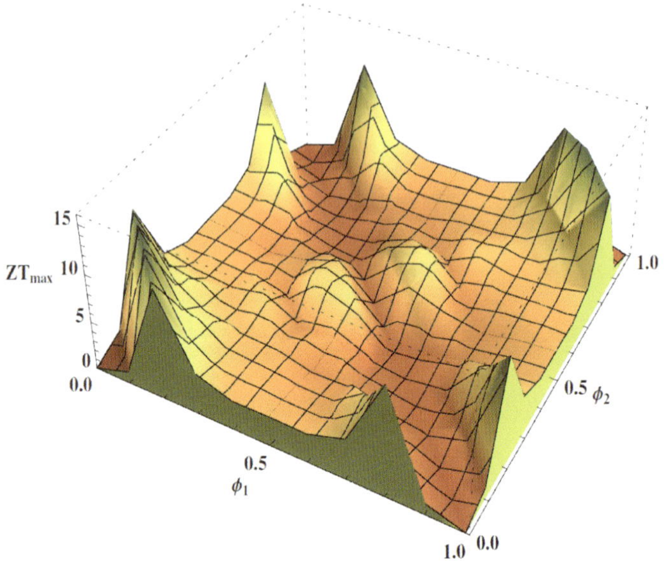

Fig. 6.18: Simultaneous variation of $ZT_{\max}$ as functions of ϕ_1 and ϕ_2, where $ZT_{\max}$ is calculated by varying the Fermi energy from -0.6 to 0.6. The source and drain electrodes are coupled to sites 3 and 4 of the molecule.

these features are associated with the resultant effect of quantum interference among the electronic waves passing through different arms of the molecular rings, which are affected by the junction configuration and the AB fluxes as well.

With a definite signature that the magnetic fluxes have a strong effect on the thermoelectric performance, now we illustrate the simultaneous variation of $ZT_{\max}$ in Fig. 6.18 with the fluxes ϕ_1 and ϕ_2 by changing these fluxes in a wide range. The main aim of this part is to have a clear idea about the choices of ϕ_1 and ϕ_2 for the best thermoelectric performance using our molecular system. Almost for all the values of ϕ_1 and ϕ_2, $ZT_{\max}$ shows a reasonable value, and it reaches surprisingly a large value ($\sim$13) comparatively at low values of these fluxes. On the other hand, the performance is relatively poor when any one of the two fluxes becomes zero, and it becomes more inferior when $\phi_1 = \phi_2 = 0$. Here, it is important to note that the results given in Fig. 6.18 are computed for a specific molecule-to-electrode junction configuration where source and drain are connected at sites 3 and 4, respectively. Similar kind of behavior of $ZT_{\max}$ is also obtained by changing these fluxes for the other junction setups which we

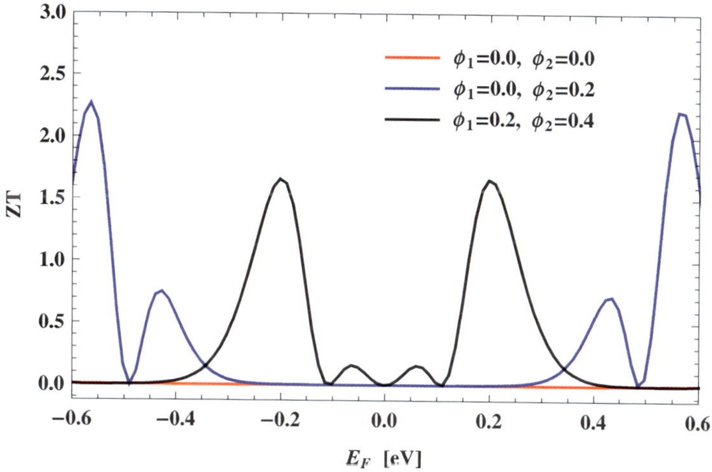

Fig. 6.19: Simultaneous variation of ZT_{max} as functions of ϕ_1 and ϕ_2, where ZT_{max} is calculated by varying the Fermi energy from −0.6 to 0.6. The source and drain electrodes are coupled to sites 3 and 4 of the molecule.

confirm through our detailed numerical calculations, and thus, here we do not present the results for other configurations.

Finally, for the sake of completeness of our analysis, we include the effect of phonon contribution in thermal conductance along with the electronic part. It is well established that phonon thermal conductance is too small for molecular systems, which becomes more weaker in lowering the molecular size. The reported typical value in molecular junctions varies in a wide range [24] from ∼5 to 50 pW/K. Now, for our system considering the thermal conductance due to phonon as 25 pW/K, in Fig. 6.19, we present the variation of ZT as a function of E_F for the same molecular junction configuration, as taken in Fig. 6.17. Three identical cases, depending on the values of ϕ_1 and ϕ_2, are also taken into account to compare the results presented in Fig. 6.19 with that shown in Fig. 6.17(d). Comparing the spectra, it is clear that even in the presence of finite phonon thermal conductance, the thermoelectric efficiency ZT is quite significant upon the inclusion of magnetic fluxes, satisfying the previous analysis, and thus, we hope that the present prescription can be useful for converting waste heat into usable electrical energy.

As an additional point, here we would like to note that the central mechanism of getting higher ZT and its possible tuning rely on the magnetic fluxes ϕ_1 and ϕ_2 threaded by two sub-rings. Apparently, it seems that the inclusion of magnetic flux in a (small) molecular ring is a difficult task as it requires a very strong magnetic field because of the small loop area. One can circumvent this difficulty by considering bigger rings. But, for our case, the added advantage is that, even for very weak magnetic fluxes, a reasonably large ZT can be achieved (see Fig. 6.18), and thus, we hope that this amount of magnetic flux can easily be established, and the results can be examined in a suitable laboratory setup.

6.3.2. *Molecular wire with helical geometry*

In this section, we consider a helical molecule, be it a single stranded DNA (ssDNA) or a α-helical protein one, as a functional element to explore the topological effect on thermoelectric phenomena [57]. The motivations behind this analysis are many-fold. First, though the ssDNA and the α-helical protein exhibit identical physical structure, they provide markedly different band structures and hence electronic transmission spectra [58]. This distinct feature is solely associated with the correlation of one atomic site with its neighboring sites. In ssDNA, atomic sites are less packed and thus electrons usually hop from one site to its immediate neighboring sites, whereas for the α-helical protein, atoms are closely packed as staking distance is much smaller than ssDNA which allows higher-order hopping of electrons. These higher-order hopping integrals play the key role in providing non-trivial signature in thermoelectric performance compared to a regular ssDNA one, which will be clearly understood from our analysis. The critical role of long-range hopping (LRH) integrals in the context of thermoelectric study will be extremely important for understanding and devising suitable thermoelectric converters. Second, due to the diverse characteristic features, DNA and other helical molecules have been the compelling elements for doing research in nanoscience communities and biological sciences [59]. Third, we are also looking for a thermoelectric device where the performance can be tuned externally. This is always beneficial for us since it yields more control, and to substantiate it, we apply an external electric field via gate electrodes [60–62] by placing it suitably in the vicinity of the molecular system. Apart from enhancing the efficiency, the electric field also tunes it, which thus leads to a proposition of externally controlled thermoelectric device.

6.3.2.1. *Helical geometry and the TB Hamiltonian*

Structurally both the α-helical protein and ssDNA look identical, and they can be represented schematically, as shown in Fig. 6.20. Three key

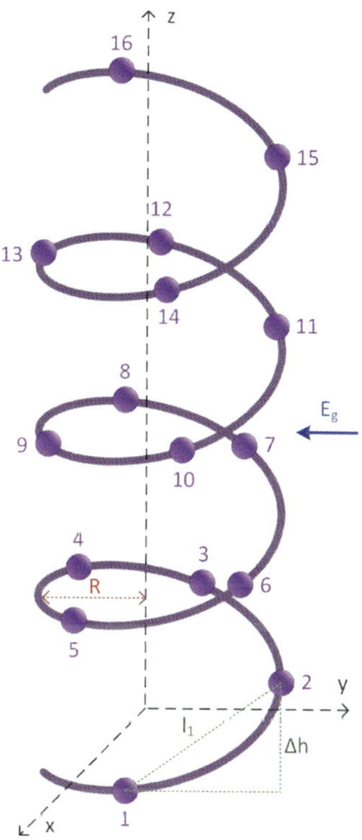

Fig. 6.20: Helical molecule with radius R, stacking distance Δh and nearest-neighbor distance l_1. The molecule is subjected to an external electric field of strength E_g.

parameters that essentially separate these two molecular models are the stacking distance Δh, radius R, and the distance of separation among the sites. Associated with these parameters, primarily the stacking distance, LRH integrals are significant in α-helical protein, whereas these integrals are less significant in the ssDNA molecule. Both these molecular systems can be modeled by a general Hamiltonian in a tight-binding (TB) framework, and for an N-site molecule, it reads as [57, 60]

$$H_{\text{mol}} = \sum_{n=1}^{N} \epsilon_n c_n^\dagger c_n + \sum_{n=1}^{N-1} \sum_{m=1}^{N-n} \left(t_{nm} c_n^\dagger c_{n+m} + h.c. \right) \tag{6.179}$$

where $c_n^\dagger$, c_n are the Fermionic creation and annihilation operators, respectively, and ϵ_n gives the on-site energy which can be set to zero, unless otherwise specified, for the ordered case. t_{nm} denotes the hopping strength between nth and $(n+m)$th sites, and assuming its exponential decay with respect to nearest-neighbor hopping (NNH) integral, labeled as t, following the Slater–Koster scheme, we can write

$$t_{nm} = te^{-(l_{nm}-l_1)/\xi} \tag{6.180}$$

where ξ is the decay parameter that is usually determined from first-principles calculations [63]. l_{nm} measures the distance between n and $(n+m)$ sites, which shrinks to the nearest-neighbor distance for $m=1$. In terms of the coordinates R, ϕ_n and h_n (described in cylindrical system) of site n, l_{nm} can be expressed as

$$l_{nm} = \sqrt{(2R\sin\phi_{nm})^2 + (h_{n+m} - h_n)^2} \tag{6.181}$$

with $\phi_{nm} = (\phi_{n+m} - \phi_n)/2$. Thus, for $m=1$, the nearest-neighbor distance becomes

$$l_1 = \sqrt{2R\sin(\Delta\phi/2)^2 + (\Delta h)^2} \tag{6.182}$$

where $\Delta\phi$ and Δh represent the twisting angle and stacking distance between nearest-neighbor sites.

When the bridging molecule is subjected to a transverse electric field E_g, its site energies are modified following the relation [61]

$$\epsilon_n = \epsilon_n(0) + eE_gR\cos(n\Delta\phi + \beta) = \epsilon_n(0) + eV_g\cos(n\Delta\phi + \beta) \tag{6.183}$$

where $\epsilon_n(0)$ is the site energy in the absence of any electric field, V_g $(= E_gR)$ is the gate voltage associated with E_g, and β describes the angle between the direction of electric field and X-axis (+ve).

This is all about the Hamiltonian of the helical molecule which is directly coupled to two one-dimensional semi-infinite reflectionless contacts, (say) lead 1 and lead 2, and we describe them using conventional nearest-neighbor TB Hamiltonians which are parameterized by the on-site energy ϵ_0 and NNH integrals t_0. The molecule-to-lead couplings are described by τ_1 and τ_2, associated with lead 1 and lead 2, respectively.

6.3.2.2. *Numerical results and discussion*

Before going to analyze the results, let us look into Table 6.2 where physical parameters of α-helical protein and DNA are given [58, 63]. From now on, we call ssDNA as DNA for simplification. The common parameter values that are used throughout the numerical calculations in this part are as follows. For the bridging molecular systems, we choose the field-independent site energies $\epsilon_n(0) = 0\ \forall n$, NNH integral $t = 1\,\text{eV}$, and the decay exponent

Table 6.2: Physical parameters associated with DNA and protein molecules.

Molecule	R (in Å)	$\Delta\phi$ (in rad)	Δh (in Å)
DNA	7	$\frac{\pi}{5}$	3.4
Protein	2.5	$\frac{5\pi}{9}$	1.5

$\xi = 0.9$ Å [63]. In the side-attached leads, we set $\epsilon_0 = 0$ and $t_0 = 2\,\text{eV}$, and the molecule-to-lead coupling strengths $\tau_1 = \tau_2 = 0.5\,\text{eV}$. Unless otherwise stated, we compute all the results considering the molecular systems with $N = 20$.

Usually for molecular systems, especially for too smaller molecules, the phononic contribution towards thermal conductance is quite small (see [24] and the references therein), which normally allows us to ignore this effect. However, it is not a general statement as the phononic contribution strongly depends on the type of a molecule, and sometimes, it is reasonably large, which cannot be avoided [64]. Although here we have simplified our analysis by considering only the electronic contribution on thermal conductance, but the incorporation of phononic part is also required which might reduce the value of ZT compared to the present discussion. But as far as the basic phenomenology is concerned for such helical topology, even electronic contribution alone provides an insight into the deeper understanding for efficient thermoelectric energy conversion at molecular length scale.

Molecular junction without any electric field

This part deals with the detailed description of electrical conductance G, thermopower S, thermal conductance due to electrons κ_{el} which is simply referred to as κ without loss of any generality and ZT to understand the critical roles of structural effects on thermoelectric efficiency.

Figure 6.21 shows the variation of electrical conductance G (measured in units of G_0) as a function of equilibrium Fermi energy E_F at two typical temperatures for the two molecular systems. Several noteworthy features are visible. The first observation is that for the protein molecule, the allowed energy window for which finite conduction takes place, associated with energy eigenvalues of the molecule, is markedly different compared to the DNA one. DNA molecule exhibits sharp peaks within the energy region $-2\,\text{eV}$ to $2\,\text{eV}$, symmetric about $E_F = 0$, similar to what we get in a perfect 1D chain with short-range hopping. The widening of accessible energy zone in the case of protein molecule is solely associated with the longer-range

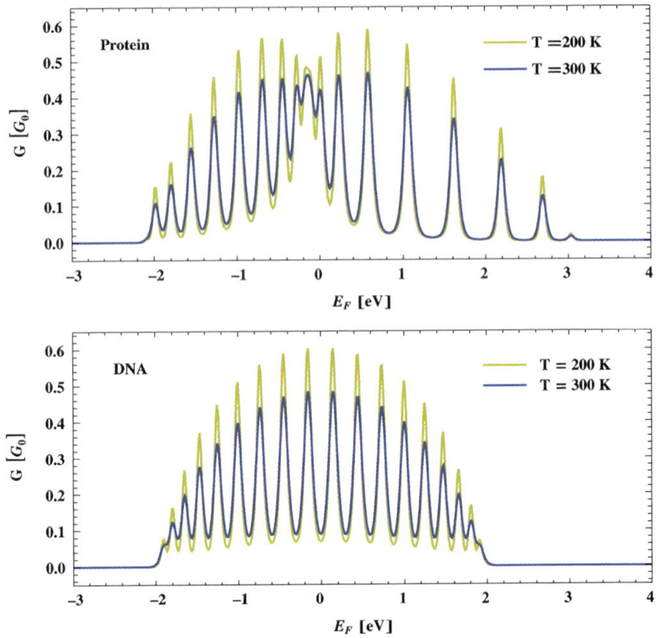

Fig. 6.21: G-E_F characteristics for the two molecular systems at two typical temperatures.

hopping of electrons. This is the generic feature of LRH model where the particle–hole symmetry is broken, and accordingly, the conductance envelop is no longer symmetric across $E_F = 0$ (Fig. 6.21(a)). Among these two molecular systems, why in one case (protein molecule) longer-range hopping integrals are contributing and for the other case (DNA), short-range interaction (equivalent to NNH model) is mostly available can be understood by noting the hopping integrals up to some range of hoppings associated with the lengths $l_1, l_2, l_3, \ldots$, etc., (these are the distances between first, second, third neighbors, and so on). For DNA molecule, $l_1 = 5.502\,\text{Å}$, $l_2 = 10.675\,\text{Å}$, $l_3 = 15.242\,\text{Å}, \ldots$, whereas for the protein one, these lengths are $4.113\,\text{Å}$, $5.766\,\text{Å}$, $5.148\,\text{Å}, \ldots$, and thus, longer-range interactions are naturally quite reasonable in the protein molecule. In both the cases, the peak heights in conductance G get reduced with temperature which are clearly noticed from Fig. 6.21, though the other effect, i.e., the thermal broadening is not so transparent as it is too low compared to the broadening caused by the molecule-to-lead coupling.

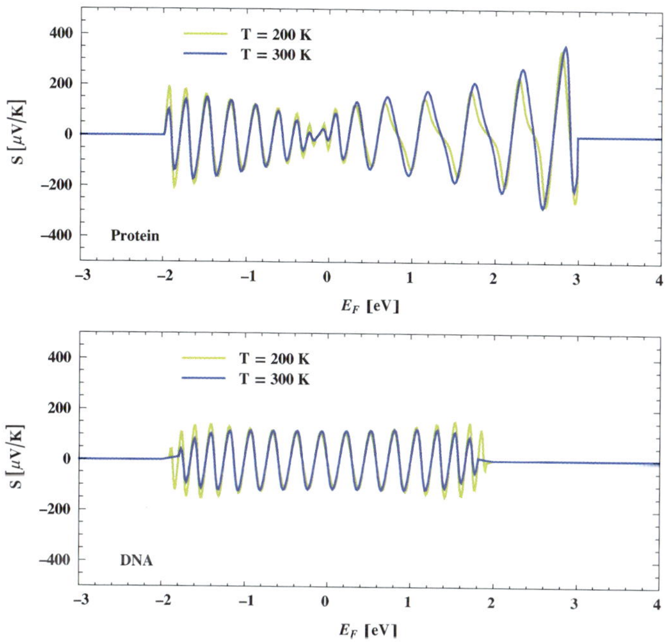

Fig. 6.22: Variation of S with E_F for the two molecular systems at two typical temperatures.

The other notable feature observed from Fig. 6.21 is that, unlike the case of DNA molecule where the resonant peaks are uniformly spaced across $E_F = 0$ (Fig. 6.21(b)), for the protein molecule, these peaks in the region $E_F < 0$ are densely packed than the other side of $E_F = 0$, revealing the fact of breaking particle–hole symmetry.

The spacings as well as the nature of resonant peaks at different Fermi energies have a significant effect on thermopower and thermal conductance, and thus on ZT. To reveal this fact, let us start with the spectra given in Fig. 6.22. Across $E_F = 0$, the thermopower S behaves differently with temperature for the protein molecule (Fig. 6.22(a)). Far below from $E_F = 0$, S shows higher magnitudes at lower temperature, while the scenario becomes opposite in the positive higher Fermi energies. The essential mechanism of getting higher or lower values of S at a particular E_F relies on the asymmetric nature of transmission function $\mathcal{T}(E)$ across that E_F and the thermal broadening function $(\partial f/\partial E)$. The expression of Seebeck coefficient (S) contains the factor $(E - E_F)(\partial f/\partial E)$ in L_1 which is anti-symmetric across $E = E_F$, and its width depends on $(\partial f/\partial E)$. Thus, the area under this

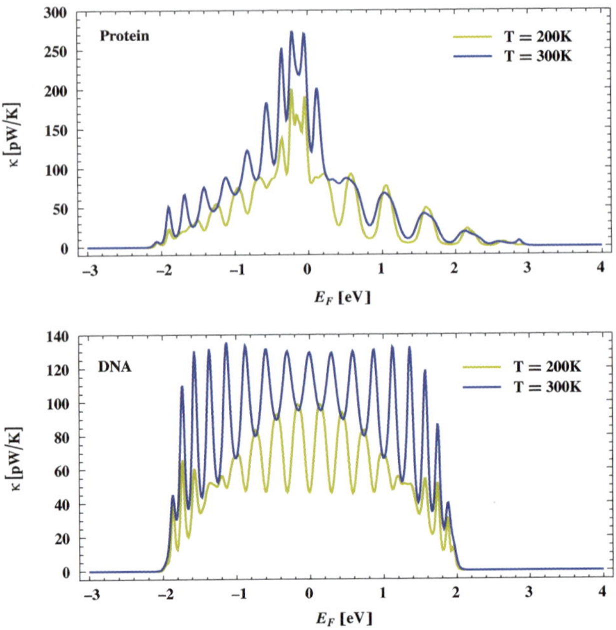

Fig. 6.23: Variation of thermal conductance with E_F for the two molecular systems at two typical temperatures.

curve will always be zero. So, to get non-vanishing contribution, one needs to break the anti-symmetric nature and that can be done by incorporating asymmetric transmission function (see the expression of L_1). Thus, more asymmetric transmission leads to more thermopower, and undoubtedly, it is one of the important requirements to achieve better performance.

For the case of DNA, the pattern of S is identical in both sides of $E_F = 0$ (Fig. 6.22(b)), exhibiting regular oscillation consistent with the G–E_F spectrum (Fig. 6.21(b)).

Now, let us move to discuss the behavior of thermal conductance κ. The results computed for two different temperatures, as taken in the earlier cases, are presented in Fig. 6.23. Several anomalous features are observed. A sharp contrast is noticed between the two molecular systems. For the protein molecule κ is quite large across $E_F = 0$, while it falls rapidly away from this energy (Fig. 6.23(a)) in contrast to the DNA molecule where almost uniform oscillatory pattern is obtained. Reduction of κ is another key requirement of achieving efficient thermoelectric device. Now, scrutinizing the spectra given in Figs. 6.22 and 6.23, one can conclude that for the

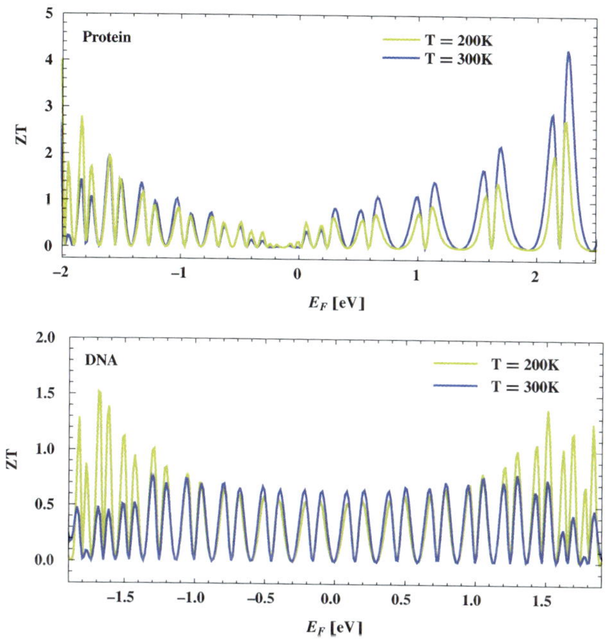

Fig. 6.24: ZT as a function of E_F at two different temperatures for the two molecular systems.

Fermi energies where S is maximum, κ exhibits minimum. This feature can be explained from the expression of κ ($\propto L_2 - L_1^2/L_0$), as higher S means higher L_1 and thus (L_1^2/L_0). Since L_0, L_1^2 and L_2 are of the same sign, the negative sign in the relation of κ overall reduces the magnitude.

The temperature dependence is also interesting. In DNA, overall κ gets increased with temperature, whereas protein molecule shows different nature across $E_F = 0$. For $E_F > 0$, there are some typical Fermi energies where higher k is noticed at lower temperature.

From the understanding of all the characteristic features of G, S and κ (shown in Figs. 6.21–6.23), now we can easily describe the behavior of ZT. For the energies where κ is vanishingly small ($\kappa \to 0+$), the efficiency diverges as this factor comes in the denominator of the efficiency expression, and therefore, we select the energy zone accordingly. Figure 6.24 shows the variation of ZT as a function of E_F at two typical temperatures. Very close to zero Fermi energy, the protein molecule exhibits vanishingly small ZT (Fig. 6.24(a)), as on one hand S is too small within this window, and on the other hand, κ is too high. But, as we move away from this zone, the overall

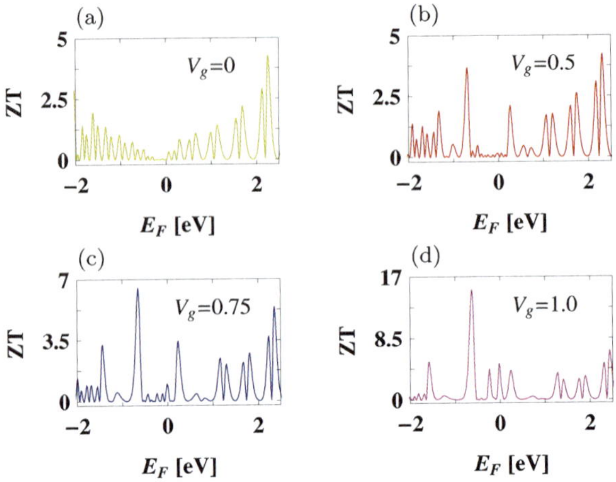

Fig. 6.25: Figure of merit of α-helical protein as a function of Fermi energy for different gate voltages considering $\beta = \pi/4$ and $T = 300\,\mathrm{K}$.

ZT increases and reaches a reasonably large value $\sim$4.5 at $E_F \sim 2.3\,\mathrm{eV}$. Apart from this higher ZT, its temperature dependence is also very much interesting. For two different energy zones (positive and negative), it shows completely opposite signature with temperature, satisfying the natures of electrical and thermal conductances. This scenario is no longer observed in the case of short-range hopping model which is clearly reflected from Fig. 6.24(b). Comparing the spectra given in Figs. 6.24(a) and (b), we can clearly emphasize that the DNA molecule is less efficient than the protein one.

Molecular junction with transverse electric field

This part deals with the possible tuning mechanism of ZT with the help of an external electric field. In the presence of this field as site energies get modified in the form of deterministic disordered lattices, more precisely in the form of well-known Aubry–André or Harper (AAH) model [65–70], and they can be further modulated by the phase factor β, we may expect suitable control over thermoelectric performance.

Let us begin with the spectra given in Figs. 6.25 and 6.26, where the results are shown for the two molecular systems at some typical values of gate voltage V_g. Here, we choose $\beta = \pi/4$ and fix the temperature at $T = 300\,\mathrm{K}$. Several interesting features are observed. It is seen that for the α-helical protein and DNA molecule, high degree of enhancement in ZT takes place in the presence of electric field. It is important to note that,

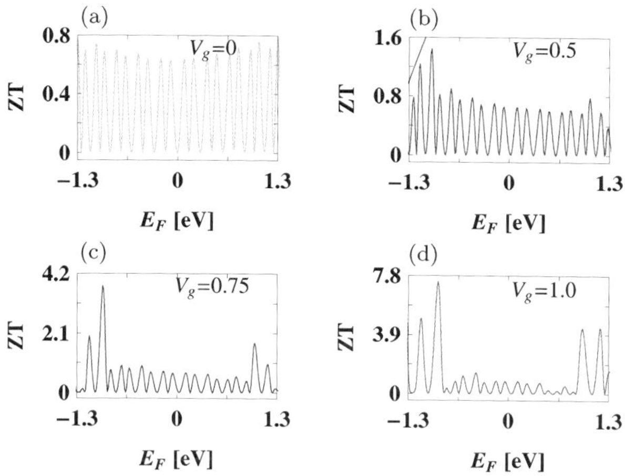

Fig. 6.26: Figure of merit of DNA molecule as a function of Fermi energy for different gate voltages, setting $T = 300\,\text{K}$ and $\beta = \pi/4$.

especially for the protein molecule, higher ZT is obtained at some typical Fermi energies which are not so far from $E_F = 0$ (i.e., not restricted at the edges of the allowed energy zone), which thus can be adjusted properly. The underlying physics of getting such enhanced ZT is that the transmission function becomes asymmetric in the presence of electric field, and the asymmetry gets increased more with the rising field strength, yielding much higher efficiency. Care should be taken on the field strength E_g because for higher E_g all energy eigenstates will be localized, and hence, electronic transfer through the molecule is no longer possible. In this limiting condition, no such phenomena will be noticed.

To inspect the dependence of phase factor β which can be tuned quite easily, as it is associated with the orientation of electric field, in Figs. 6.27 and 6.28, we present the results of ZT for some typical values of β, considering $V_g = 0.5\,\text{V}$ and $T = 300\,\text{K}$. What we see is that ZT can be regulated quite well by rotating the field direction. For this case also, the same argument is valid for changing the efficiency, i.e., the transmission function gets modified with β as it directly modulates molecular site energy.

Comparing the results given in Figs. 6.25–6.28, we can argue that the protein molecule is much more sensitive to the electric field than the DNA one.

The above analysis raises an obvious question about how ZT gets modified with gradual change of electric field strength and its direction and how

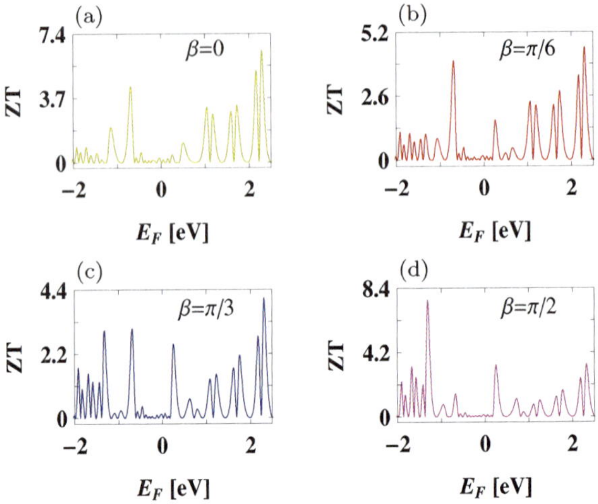

Fig. 6.27: ZT as a function of E_F for the protein molecule at some typical phase factors β. Here, we set $V_g = 0.5\,\mathrm{V}$ and $T = 300\,\mathrm{K}$.

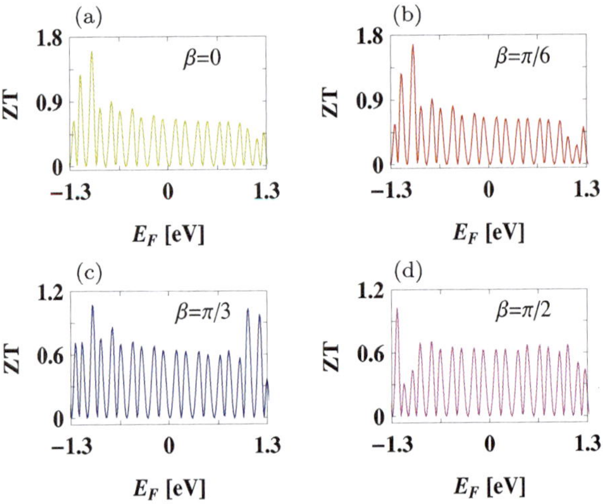

Fig. 6.28: ZT as a function of E_F for the DNA molecule at some typical phase factors β. Here, we set $V_g = 0.5\,\mathrm{V}$ and $T = 300\,\mathrm{K}$.

much sensitive it is with temperature. These features are explored in two separate figures (Figs. 6.29 and 6.30), where in one case, we vary the field strength and temperature, keeping the field direction unchanged, whereas in the other case, we fix the field strength and vary the other two. As here

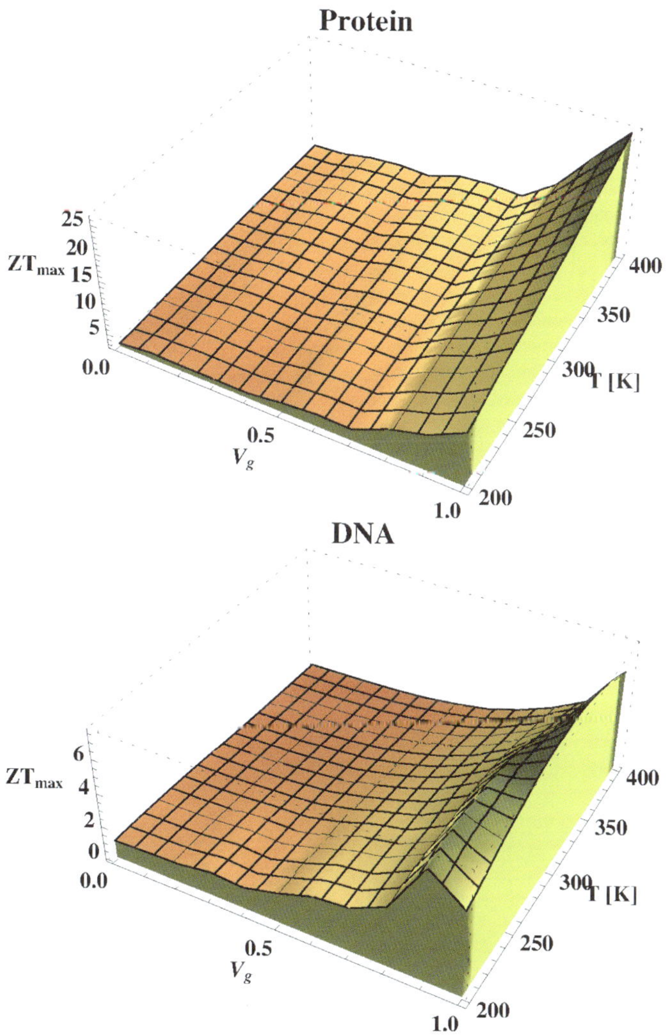

Fig. 6.29: Simultaneous variation of ZT_{max} with gate voltage and temperature for the two molecular systems. Here, we choose $\beta = 0$.

we are trying to find the maximum efficiency, we compute the maximum of ZT by scanning it over the full energy region. From these spectra, we can see that for both the molecular systems, a wide range of physical parameters is available, for which sufficiently large ZT can be obtained. Thus, one can safely select the suitable parameters to achieve desired thermoelectric

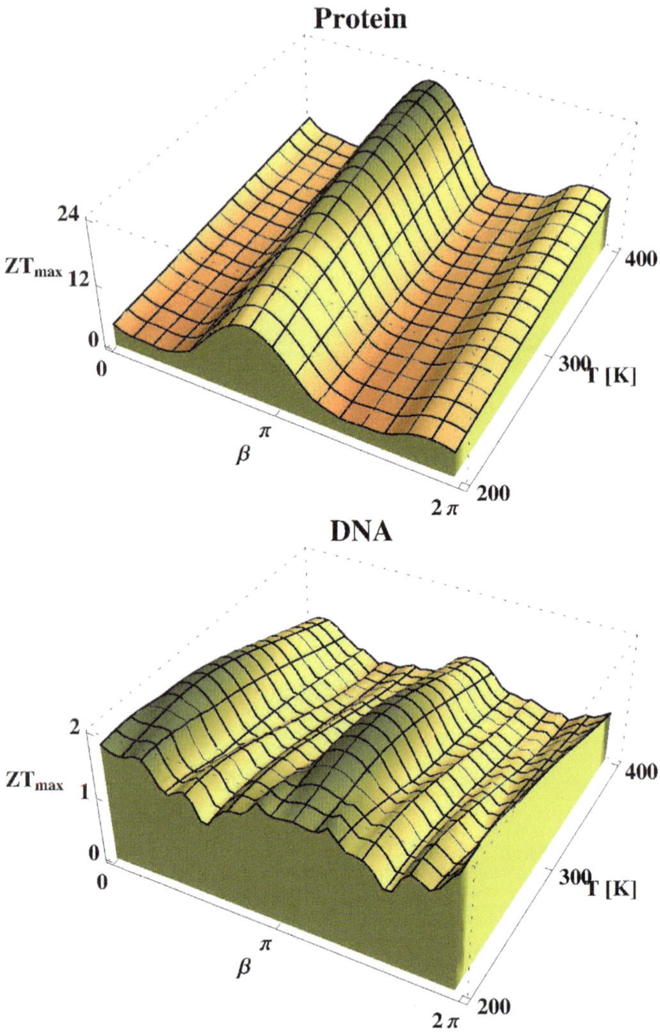

Fig. 6.30: Simultaneous variation of $ZT_{\max}$ with temperature and phase factor for the two molecular systems. Here, we choose $V_g = 0.5$V.

performance. Finally, it is again verified that the protein molecule is much superior than the DNA one, as the previous one exhibits much higher ZT.

Effect of system size

The results studied so far are worked out for a fixed system size $N = 20$. Now, in order to test the size dependence on ZT and also to examine

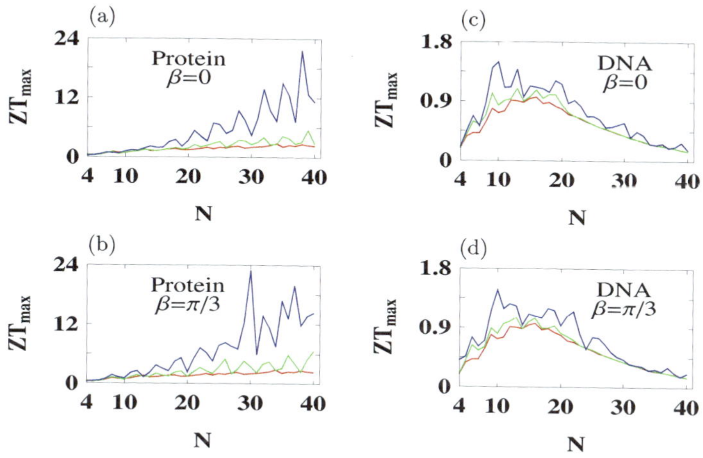

Fig. 6.31: System size dependence of $ZT_{\max}$ for the two molecular systems under different phase factors and gate voltages, where the red, green and blue lines correspond to $V_g = 0$, 0.2 and 0.4 V, respectively.

whether any favorable response is obtained in other system sizes, let us look into the spectra given in Fig. 6.31. The results are computed for the three different gate voltages $V_g = 0$, 0.2 and 0.4 V. Both for the protein and DNA molecules, oscillation takes place with N, though the oscillation in the absence of gate voltage (red curve) is vanishingly small and not so transparent due to large amplitude oscillations in the presence of finite gate voltages. This oscillation is caused because of the quantum interference effects. Apart from this, we find another interesting feature in these two molecular systems. For DNA, $ZT_{\max}$ reaches a maximum quite quickly and then it gradually decreases with N. This is the usual scenario for short-range hopping model, and from this behavior, we can argue that longer DNA chain will not be a good choice for effective thermoelectric energy conversion. On the other hand, due to the longer-range hopping integrals in protein, favorable response is obtained for larger N, and the efficiency gets triggered more in the presence of transverse electric field.

Here, we want to put a caution that, although a favorable response is obtained for longer protein molecule in the presence of electric field, care should be taken on selecting N and V_g, as for very large N all states will be localized in the presence of large V_g and no such phenomenon will be noticed, which we confirm through our exhaustive numerical calculations.

6.4. Closing Remarks and Outlook

In this chapter, we have discussed thermoelectric phenomena, which is one of the most promising research areas in condensed matter physics. The main aim of this review is to provide the basic understanding of energy conversion, from waste heat to usable electrical energy, using functional element at nanoscale region. A considerable amount of this chapter has been devoted to illustrate the theoretical prescription for the calculation of all the required thermoelectric quantities starting from the very fundamental level. We have illustrated each and every mathematical step, to the base of our concern, for the determination of electrical conductance, Seebeck co-efficient, thermal conductance and figure of merit. A common factor, known as transmission function, is involved in all these quantities, and we have given an elaborate theoretical description for it, considering both single- and multi-channel contacts.

Following the theoretical methods, we have highlighted thermoelectric properties considering some molecular wires which can be generalized further in any other simple and/or complex molecular systems as well as in tailor-made geometries. We have most concentrated on devising a thermoelectric setup where a reasonably large figure of merit ($ZT > 1$) can be obtained to hit the present energy market, and at the same time, the energy conversion efficiency along with other physical phenomena can be tuned selectively with the help of some simple prescription. For one molecular wire containing naphthalene molecule, it is done by adjusting the AB flux, and the interplay between the fluxes in two neighboring molecular ring sometimes may yield non-trivial signatures, whereas for the other molecular wire having a helical system, the tuning is made by applying a transverse electric field. The correlation between helical topology, higher-order hopping and the transverse field has an important effect in thermoelectric phenomena. Due to unique and diverse characteristic features, DNA and protein molecules could be the interesting candidates for future thermoelectric as well as cooling devices.

Though a significant progress has been made in energy conversion at nanoscale level due to enormous development in science and technology, a long-standing discrepancy still persists between the theoretical estimates and the experimental evaluation in molecular junctions, and thus, further probing is undoubtedly required.

Acknowledgments

This work is supported by the DST-SERB, India under the Project Grant No. EMR/2017/000504.

References

1. USDOE, *Advanced Thermoelectric Materials for Efficient Waste Heat Recovery in Process Industries* (**2009**).
2. T. J. Seebeck, Magnetic polarization of metals and minerals by temperature differences. *Abh. Dtsch. Akad. Wiss. Berl.* 265–373 (**1825**).
3. K. A. Chao and M. Larsson, *Physics of Zero- and One-Dimensional Nanoscopic System.* S. N. Karmakar, S. K. Maiti, and J. Chowdhury (eds.), Vol. 156, Solid State Sciences (Springer, **2007**).
4. J. Peltier, *Ann. Chim.* LVI, 371 (**1834**).
5. H. J. Goldsmid, *Thermoelectric Refrigeration* (Plenum: New York, **1964**).
6. A. F. Ioffe, *Semiconductor Thermoelements and Thermoelectric Cooling.* (Infosearch limited: London, **1957**).
7. Y. Dubi and M. Di Ventra, *Rev. Mod. Phys.* 83, 131 (**2011**).
8. G. Chen, *Nanoscale Energy Transport and Conversion.* (Oxford University Press: New York, **2005**).
9. A. Majumdar, *Science* 303, 777 (**2004**).
10. P. Rodgers, *Nat. Nanotech.* 3, 76 (**2008**).
11. L. E. Bell, *Science* 321, 1457 (**2008**).
12. N. A Zimbovskaya, *J. Phys. Condens. Matter* 28, 183002 (**2016**).
13. L. D. Hicks and M. S. Dresselhaus, *Phys. Rev. B* 47, 12727 (**1993**).
14. L. D. Hicks and M. S. Dresselhaus, *Phys. Rev. B* 47, 16631 (**1993**).
15. D. Segal and A. Nitzan, *Phys. Rev. Lett.* 94, 034301 (**2005**).
16. N. Yang, G. Zhang, and B. Li, *Appl. Phys. Lett.* 95, 033107 (**2009**).
17. L. Wang and B. Li, *Phys. Rev. Lett.* 101, 267203 (**2008**).
18. L. Wang and B. Li, *Phys. Rev. Lett.* 99, 177208 (**2007**).
19. B. Li, L. Wang, and G. Casati, *Appl. Phys. Lett.* 88, 143501 (**2006**).
20. A. Shakouri, *Proc. IEEE* 94, 1613 (**2006**).
21. A. I. Hochbaum, R. Chen, R. D. Delgado, W. Liang, E. C. Garnett, M. Najarian, A. Majumdar, and P. Yang, *Nature* 451, 163 (**2008**).
22. A. I. Hochbaum and P. Yang, *Chem. Rev.* 110, 527 (**2010**).
23. C. Yu, Y. S. Kim, D. Kim, and J, C. Grunlan, *Nano Lett.* 8, 12, 4428 (**2008**)
24. E. Zerah-Harush and Y. Dubi, *Phys. Rev. Appl.* 3, 064017 (**2015**).
25. N. Sergueev, S. Shin, M. Kaviany, and B. Dunietz, *Phys. Rev. B* 83, 195415 (**2011**).
26. M. Galperin, A. Nitzan, and M. A. Ratner, *Mol. Phys.* 106, 397 (**2008**).
27. A. I. Boukai, Y. Bunimovich, J. T.-Kheli, J.-K. Yu, W. A. Goddard III, and J. R. Heath, *Nature* 451, 168 (**2008**).
28. J. P. Small, K. M. Perez, and P. Kim, *Phys. Rev. Lett.* 91, 256801 (**2003**).

29. W. J. Kong, L. Lu, H. W. Zhu, B. Q. Wei, and D. H. Wu, *J. Phys. Condens. Matter* 17, 1923 (**2005**).
30. P. Reddy, S.-Y. Jang, R. A. Segalman, and A. Majumdar, *Science* 315, 1568 (**2007**).
31. S. Chakraborty and S. K. Maiti, *ChemPhysChem* 20, 848 (**2019**).
32. J. H. O. Silva and S. K. Maiti, *ChemPhysChem* 20, 3346 (**2019**).
33. S. Ganguly and S. K. Maiti, *Superlattices Microstruct.* 136, 106264 (**2019**).
34. S. Chakraborty and S. K. Maiti, *J. Appl. Phys.* 127, 024302 (**2020**).
35. S. Chakraborty and S. K. Maiti, *AIP Conf. Proc.* 2072, 020018 (**2019**).
36. S. Datta, *Electronic Transport in Mesoscopic Systems* (Cambridge University Press: Cambridge, **1997**).
37. S. Datta, *Quantum Transport: Atom to Transistor* (Cambridge University Press: Cambridge (**2005**).
38. M. D. Ventra, *Electrical Transport in Nanoscale Systems* (Cambridge University Press, Cambridge, **2008**).
39. P. A. Mello and N. Kumar, *Quantum Transport in Mesoscopic Systems: Complexity and Statistical Fluctuations* (Oxford University Press: New York, **2004**).
40. Y. V. Nazarov and Y. M. Blanter, *Quantum Transport: Introduction to Nanoscience* (Cambridge University Press: Cambridge, **2009**).
41. N. R. Claughton and C. J. Lambert, *Phys. Rev. B* 53, 6605 (**1996**).
42. C. M. Finch, V. M. Garcia-Suarez, and C. J. Lambert, *Phys. Rev. B* 79, 033405 (**2009**).
43. U. Peskin, *J. Phys. B* 43, 153001 (**2010**).
44. O. Entin-Wohlman, Y. Imry, and A. Aharony, *Phys. Rev. B* 82, 115314 (**2010**).
45. R.-Q. Wang, L. Sheng, R. Shen, B. Wang, and D. Y. Xing, *Phys. Rev. Lett.* 105, 057202 (**2010**).
46. S.-H. Ke, W. Yang, S. Curtarolo, and H. U. Baranger, *Nano Lett.* 9, 1011 (**2009**).
47. D. Nozaki, H. Sevinli, W. Li, R. Gutiérrez, and G. Cuniberti, *Phys. Rev. B* 81, 235406 (**2010**).
48. L. Cui, R. Miao, C. Jiang, E. Meyhofer, and P. Reddy, *J. Chem. Phys.* 146, 092201 (**2017**).
49. S. K. Maiti and M. Dey, *Chem. Phys. Lett.* 731, 136601 (**2019**).
50. J. H. Ojeda, P. A. Orellana, and D. Laroze, *J. Chem. Phys.* 140, 104308 (**2014**).
51. M. Dey, S. K. Maiti, S. N. Karmakar, *J. Appl. Phys.* 109, 024304 (**2011**).
52. S. Sil, S. K. Maiti, and A. Chakrabarti, *Phys. Rev. B* 79, 193309 (**2009**).
53. S. Bala Kumar, M. B. A. Jalil, S. G. Tan, and G. Liang, *J. Phys. Condens. Matter* 22, 375303 (**2010**).
54. D. Rai, O. Hod, and A. Nitzan, *Phys. Rev. B* 85, 155440 (**2012**).
55. M. Dey, S. K. Maiti, and S. N. Karmakar, *Eur. Phys. J. B* 80, 105 (**2011**).
56. S. Saha, S. K. Maiti, and S. N. Karmakar, *Eur. Phys. J. B* 85, 283 (**2012**).
57. M. Dey, S. F. Aman, and S. K. Maiti, *Europhys. Lett.* 126, 27003 (**2019**).
58. A.-M. Guo and Q.-F. Sun, *Proc. Natl. Acad. Sci. U.S.A.* 111, 11658 (**2014**).

59. D. Klotsa, R. A. Römer, and M. S. Turner, *Biophys. J.* 89, 2187 (**2005**).
60. T.-R. Pan, A.-M. Guo, and Q.-F. Sun, *Phys. Rev. B* 92, 115418 (**2015**).
61. A.-M. Guo and Q.-F. Sun, *Phys. Rev. B* 95, 155411 (**2017**).
62. S. Sarkar and S. K. Maiti, *Phys. Rev. B* 100, 205402 (**2019**).
63. R. G. Endres, D. L. Cox, and R. R. P. Singh, Colloquium: The quest for high conductance DNA. *Rev. Mod. Phys.* 76, 195 (**2004**).
64. J. C. Klöckner, J. C. Cuevas, and F. Pauly, *Phys. Rev. B* 96, 245419 (**2017**).
65. S. Aubry and G. André, *Ann. Isr. Phys. Soc.* 3, 133 (**1980**).
66. S. Sil, S. K. Maiti, and A. Chakrabarti, *Phys. Rev. Lett.* 101, 076803 (**2008**).
67. Y. E. Kraus, Y. Lahini, Z. Ringel, M. Verbin, and O. Zilberberg, *Phys. Rev. Lett.* 109, 106402 (**2012**).
68. S. Ganeshan, J. H. Pixley, and S. Das Sarma, *Phys. Rev. Lett.* 114, 146601 (**2015**).
69. M. Rossignolo and L. Dell'Anna, *Phys. Rev. B* 99, 054211 (**2019**).
70. M. Patra and S. K. Maiti, *Sci. Rep.* 7, 14313 (**2017**).

Chapter 7

Application of Green's Functions and Difference Equations in Theoretical Analyses of Nanostructures

Vjekoslav Sajfert[*,‡] **and Jovan P. Šetrajčić**[†,§]

[*] *University of Novi Sad, Technical Faculty "M. Pupin", Djure Djakovica Street bb, Zrenjanin, Vojvodina – Serbia*

[†] *University "Union – Nikola Tesla", Faculty of Sport, Narodnih heroja 30/I, Novi Beograd, Vojvodina – Serbia;*

Academy of Sciences and Arts of the Republic of Srpska, Bana Lazarevića 1, Banja Luka, Republic of Srpska – Bosnia and Herzegovina

[‡] *sajfert.vjekoslav@gmail.com*

sajfertv@tfzr.uns.ac.rs

[§] *jovan.setrajcic@gmail.com*

7.1. Introduction[1]

The standard theory of the solid state physics of the 20th century is based on methods that are "borrowed" from the physics of the continuum, i.e., from field theory [1–5]. What enabled the incorporation of these methods is the fact that in the solid body elementary cells are reflected translationally in all spatial directions, i.e., there is a translational invariance in the space. When the space is unlimited, the differential displacements of elemental excitations can be considered infinitesimal, which allows for switching to a continuous representation of the methods. Thus, the Schrödinger equation,

[1] In this section, parts of the text from two papers were used with permission: [6] *J. Comput. Theor. Nanosci.* **7**/1, 15–84 (2010) and [10] *Encyclopedia of Nanoscience and Nanotechnology* **50**, 281–350 (2011); Copyright @ASP.

the Heisenberg equations of motion, Green's functions, etc. get their full meaning. Based on such an approach, the physics of condensed systems has emerged, many phenomena and laws have been explained, and numerous applications have been found.

This type of solid state physics was the theory of ideal (spatial translation invariant) crystalline structures, since they could not be produced in practice, as is still the case today. On the contrary, defects, impurities, boundaries, and structural irregularities have been known to cause many exotic and significant phenomena. Basically, the main problem was skillfully masked, but also always noticeably present. The beginning of solving these natural puzzles had to wait for a more favorable moment — the appropriate advances in engineering and technology, as well as measuring and computing equipment.

7.1.1. *Broken symmetry problem*

As we have pointed out in Ref. [6], the research of broken symmetry structures has a relatively long history. In the first stage, the analysis of random symmetry breaks (impurities, vacancies, etc.) was performed. It is understandable since synthesis of absolutely ideal structures was not achieved. These analyses were mathematically complex and, more importantly, the approach to the problem was as chaotic as the researched structure itself.

The described example was the reason for the slow development of the theory of broken symmetry structures. This research started from more complex problems to less complicated ones. At first glance, it can be concluded that the analysis of random broken symmetry structures is nothing new, but rather only more complex mathematically. However, that presumption would be wrong. By using precise mechanical formulas, it was possible to solve the problem of one and maximally two impurities or vacancies. The problem of three impurities is an unsolvable three-body problem. Although it is well known that a similar problem in classical mechanics was statistically solved by Boltzmann, for a very long time, nobody tried to translate Boltzmann's ideas into the problematic of chaotically broken symmetry structures. Naturally, the statistical approach can give only the most probable solution, but not the exact solution. Boltzmann's genial ideas had become a formidable adversary. Albert Einstein, being one of the authors of statistical formulas (Bose–Einstein statistics), insisted that the complete theory must give improbable, but also accurate laws (actually, he was against quantum mechanics). Nevertheless, Boltzmann's ideas obtained a wide field of application and in the practical sense advanced development and application of physics.

The theoretical analysis of planned nanostructures (in geometrical sense) is less complicated than in the case of random disturbed structures. In the basis of those analyses is one of the fundaments of quantum mechanics — Heisenberg uncertainty principle. This drastically changes the usual habits taken from the analysis of ideal structures concerned, primarily with the existence (or non-existence) of the energy gap and with configuration dependence of essential physical characteristics. The equidistance of energetic levels, which are found in ideal structures, is also discarded in corresponding nanostructures. The mentioned characteristics of nanostructures forced some authors to declare that the occurrence of nanostructures with respect to the ideal structures is in the same position as quantum mechanics is with respect to classical mechanics [7]. We are of the opinion that such ideas cannot be considered correct, since the increase of nanocharacteristics (thickness, cross section, etc.) constantly draws nearer to the behavior of ideal structures. Mathematical difficulties are primarily connected to the geometry of structure. The most convenient for analysis are the systems of difference equations obtained for simple cubic geometry. So, a new idea emerged: when it is possible, we should determine statistically and geometrically equivalent simple structures for more complicated geometries (triclinic, monoclinic, etc.) [8]. Non-isomorphism of transition configuration space–momentum space (autoreduction) is a very common characteristic of nanostructures and this also requires some additional research [9]. Analysis of anharmonic (nonlinear) effects in nanostructures is mathematically complex. It is the reason for the very common use of self-consistent Tyablikov's approximation in the analysis of thermodynamics of nanostructures.

7.1.2. *Progress of nanotechnology*

As we have noted in Ref. [10], it seems that in the 21st century we enter the nanoworld. By observing expressions in physics, it can be concluded that an enormous number of terms has the prefix "nano" (nanoparticles, nanowires, nanostructures, nanocomposites materials, nanoprobe microscopy, nanoelectronics, nanotechnology, etc.).

It is not quite clear when this "nanorevolution" began. Most probably, it was in the mid-1980s, when scanning tunneling microscopy was invented [11]. Colloid chemists are of the opinion that in the 1960s and 1970s they made Langmuir–Blodgett (LB) films with extremely high periodicity in nanometer scale. From this point of view, the nanorevolution originated from the works of Irving Langmuir and Katherine Blodgett [12, 13] in 1930s or from later works of Mann and Kühn [14–16], Aviram and Ratner [17], and Carter [18], which published ideas of molecular electronics in the 1970s. What is the point of these arguments? It seems that all of them were right. Modern nanotechnology cannot be imagined without any of the above-mentioned contributions. The fact is that we are in the nanoworld now, and the words with prefix "nano" have become everyday reality. Also, Moore's Law, which determines the growth of nanotechnology, is used for the progress of nanotechnology.

Defining nanotechnology is not a simple problem. As it was said, it arose and has developed progressively during several decades using results of development of series of physical and chemical researches. Independently of the definition of nanotechnology, nanoresearch is becoming more and more intensive everywhere. It is to be expected that achievements of nanotechnology will have an important role in medicine, biology, technology, energetics, etc.

According to the views expressed in Ref. [6], the general opinion is that the first nanorevolution steps were made in 1980. Then scanning instruments microscopy was invented. Molecular beam epitaxy, the revolutionary technology of the 1980s, which allows production of layered structures with the thickness of each layer in the nanometer range, can be added to this. In the colloid chemistry, on the other hand, Langmuir–Blodgett films with extremely high periodicity in nanometer scale were made. It was done in 1970s. This is given in Refs. [19–25]. In our opinion, the question who was first is irrelevant. The nanoscience building was built in different sciences and it was caused by different needs.

Research of nanostructures represents a new stage in the development of physics and biology since they have the midplace between molecules and macroobjects. Nanoobjects are characterized with small dimensions, complex internal organization, and intention for dense packing and strong side interactions. They also have a very high surface-area-to-volume ratio [10]. On the basis of nanostructures, one can make materials with new electric, magnetic and optical properties. Besides, it can be expected that nanostructure research will cause advancing of computer technology since analyses of nanostructures require, in principle, very complicated numerical evaluations.

A relevant role in the development of nanotechnology belongs to the increase of computing speed. The starting point was the choice of different materials causing the idea of molecular electronics to appear. Many interesting and fascinating ideas of logical devices based on one molecule or group of molecules, and revolutionary novel principles of molecular computing systems were suggested at the time. Although most of these ideas have not yet been fulfilled, the efforts were not wasted. The research in molecular electronics and thin organic films forced the technology and instrumentation into the nanometer zone. The molecular electronics moved towards sensors and biosystems, while solid state electronics prevailed with several brilliant ideas of quantum electron devices. Three major breakthrough developments should be mentioned here: resonance tunneling devices, single-electron devices, quantum dots. Quantum phenomena begin to affect the properties of materials when the dimensions are smaller than 10 nm, which is currently not the case [6]. The energy spectra and possible states of elementary excitations in low-dimensional solid states (e.g., two-dimensional — ultrathin films, one-dimensional — quantum wires or nanotubes and nanorods — and zero-dimensional — quantum dots) dramatically change in comparison to those in three-dimensional bulk materials [6, 8–10]. On the other hand, quantum phenomena may have a rather positive effect on the progress of solid state microelectronics.

Reducing the thickness of the insulating layers seems to increase the electric field, and the probability of field-related effects, such as electron tunneling and avalanche breakdown. Scaling of the dimensions may have the following consequences:

- Dispersion of bulk properties of materials;
- Quantum phenomena in low-dimensional systems;
- Problems of heat dissipation;
- Limitation of computing speed;
- Thermo-isolation and acoustic isolation;
- Synthesis of completely new superconductors;
- Appearance of Bose–Einstein condensation of quasi-particle systems together with superfluidity of quasi-particles and its applications.

7.1.3. *Some relevant experimental data*

In the review paper [6], we have already emphasized that the rapid development of nanotechnology was followed by intensive testing of physical characteristics and practical implications of nanomaterials. The extensive experimental material is concerned with semiconductor nanocrystals and their biological application.

In recent years, a number of reviews on the synthesis [26], size- and shape-dependent properties [27], characterization [28], and electron–phonon interactions [29] of these nanocrystals have been published; indeed, there are also several books available that emphasize the theory and experimental characterization of semiconductor nanocrystals [30–33]. The basic physical properties of semiconductor nanocrystals have been treated in detail: treatments include techniques for Raman scattering from arrays of semiconductor nanocrystals [34], optical lattice vibrations of polar semiconductor quantum dots [35] the effective-phonon approximation of polarons in ternary mixed crystals [36] and models of optical line-widths of individual quantum dots [37]. Moreover, the basic electrical, optical, and mechanical properties of GaAs-based semiconductor nanocrystals have been studied extensively [38–48]. The many applications of semiconductor nanocrystals include the single-photon detection in the far-infrared portion of the electromagnetic spectrum [49].

Furthermore, electrical, optical, and mechanical properties of InAs-based semiconductor nanocrystals have been studied extensively [50–59]. Applications included the following [6]:

- mid-infrared second-harmonic generation in p-type InAs/GaAs self-assembled quantum dots [60],
- self-assembled InAs/GaAs quantum dot intersubband detectors [61],
- InAs-based infrared photodetectors [62, 63],
- mid-infrared absorption and photocurrent spectroscopy of InAs/GaAs self-assembled quantum dots [64],
- InAs quantum dot field effect transistors [65] and
- InGaAs/GaAs quantum dot lasers [66].

In the last few years, a large number of groups have succeeded in making nanorods. Various nanomaterials such as silver [67, 68], gold [69–73], platinum [74, 75], copper [76] and classical-type semiconductors [77, 78] were produced.

The relatively very fresh experiments of testing are concerned with superfluidity of magnons in magnetic films YIG [88] and with polarons [89]. It turned out that a high gap in spectrum of mentioned elementary excitations leads to short-time statistical equilibrium having in this time short interval magnons and polarons have superfluid properties. It is interesting that the high gap as necessary condition for arising of superfluidity of quasiparticles was predicted in [90]. The experimental justification of this idea after 40 years was achieved owing to the development of nanotechnology.

7.1.4. *Expectations of nanoscience research*

As we have already pointed out in the review paper [6] the main reason for intensive development of nanotechnology and nanoscience is the fact that the nature did not produce all materials necessary for further civilization and therefore we have decided to produce them artificially.

Nanotechnology holds the promise of advances that exceed those achieved in recent decades in computers and biotechnology. Its applications will have dramatic infrastructural impacts, such as building tremendously faster computers, constructing lighter aircraft, finding cancerous tumors still invisible to the human eye or generating vast amounts of energy from highly efficient solar cells. Nanotechnology will manifest in innovations both large and small in diverse industries, but the real benefit will accumulate in small cascades over decades rather than in a sudden, engulfing wave of change. It is not the "Next Big Thing" but rather will be any number of "next large things". Nanotechnology may not yield a result as dramatic as Edison's light bulb but rather numerous gains as pervasive as the integrated-circuit-controlled light bulbs in the traffic lights that are ubiquitous in modern life.

Although the light bulb breakthroughs will be few, there will be numerous benefits taken for granted, such as the advantages that the automated intelligence of traffic grids provides to major cities. This should not be a surprise, because nanotechnology is not an invention but rather a range of fields of study and applications, defined by size, that use tools, ideas, and intuitions available to innumerable scientific disciplines. Thus nanotechnology offers tremendous potential for several key reasons. Materials and processes at that size have unique properties not seen at a larger scale, offer proportionately greater reactive surface area than their larger counterparts, and can be used in or with living organisms for medical applications. As a result, familiar materials can have completely different properties at the nanoscale. For example, carbon atoms form both coal and diamonds, but with different molecular arrangements. Scientists now know that carbon molecules at the nanoscale can form cylindrical tubes, called carbon nanotubes, that are much stronger than steel and conduct electricity, neither of which is possible with the carbon found in coal or diamonds. Carbon nanotubes may one day provide key breakthroughs in medicine and electronics. Likewise, nanotechnology can provide breakthroughs in industrial uses [6]. The electrical current produced in solar cells or batteries reflects the flow of electrons from one surface to another. Nanotechnology has already enabled the demonstration of a vastly increased surface area of electrodes that allows electrons to flow much more freely, along with corresponding improvements in battery performance. Safer, cheaper, and cleaner electricity and electrical storage would obviously have a dramatic impact on our society.

Another reason nanotechnology holds so much promise is that it enables solutions at the same size scale as biological organisms, such as the individual cells in our bodies. Engineered materials are possible, such as ultra-small particles made in the exact size to perform like a "smart bomb" in delivering drugs in the blood stream. Other applications might detect cancer when it is only a few cells in size [79–83]. Future convergence of nanotechnology and biotechnology may combine biological and man-made devices in a variety of applications, such as batteries for implanted heart pacemakers that draw electrical current from the wearer's glucose rather than from surgically implanted batteries. Yet another important facet of nanotechnology that underpins both its promise and the challenges is that it embraces and attracts so many different disciplines that researchers and business leaders are working in, among them, chemistry, biology, materials science, physics, and computer science [84–87]. Although each field has tremendously talented people, each also has its own somewhat unique training and terminology. Almost like the parable of the blind men and the elephant, each group approaches the molecular level with unique skills, training, and language. Communication and research between academic disciplines and between researchers and their business counterparts is critical to the advancement of nanotechnology.

With the diversity of professional cultures in mind, a central goal of this book-chapter is to promote communication and cooperation between researchers and industry by including similarly diverse articles written by experts but accessible to everyone. The depth of scientific talent and the substantial resources being devoted to nanotechnology are a tremendous cause for optimism for both near-term and long-term gains. Ultimately nanotechnology will yield greater impact than information technology or biotechnology has. However, the tempo of technology is not set by the velocity of novel discoveries, but rather by the pace of what the market will embrace and pay for. The medium term in nanotechnology will be difficult and delayed by issues far beyond scientific research or product prototyping namely, by the long, difficult process of new products gaining traction in the marketplace. To reach the stage of a viable product, the innovations will have to overcome issues such as how they are integrated, how much power they consume, and how they are controlled.

In the last decade, a commonly mentioned idea has been that nanoscience is a fundamentally new domain of physics, something like "new solid state physics". We are of the opinion that this idea is wrong. The main reason for such an opinion is the fact that the increase of nanostructure characteristics (thickness, crossection, etc.) leads to the disappearance of nanoscience. It is better to think that nanoscience is a domain of physics which requires new efforts in the domain of technology, methodology and field of applications. The general characteristic of nanostructures is appearance of energy gaps of some excitations where this effect was not expected. This discovery is of a great practical interest since the high gap causes and amplifies superconductive and superfluid effects, whose perspective is boundless.

7.1.5. *Necessity of theoretical research*

As we have already pointed out in the review paper [6] the rapid development of nanotechnology was followed by the intensive testing of physical characteristics and practical implications of nanomaterials. The extensive experimental material is concerned with semiconductor nanocrystals and their biological application.

Over the last decade, we have witnessed the rapid development of nanotechnology, i.e., production technologies of structures with nanodimensions [91–93]. The trend of rapid development of nanotechnologies is not accidental since nanostructures possess physical characteristics that are substantially different from the physical characteristics of larger structures. Nanostructures have a number of qualitatively new effects. Superconductive, thermal insulating, acoustic and other features that characterize nanomaterials are better than in bulk structures. It is also expected that the usage of nanostructured elements may increase the measuring sensitivity of instruments, and that is leading to new experimental discoveries. All of the abovementioned leads to the conclusion that nanomaterials can be a good basis for further development of Physics [94–96]. In some way, nanostructure effects are connected to the effects in bulk structures, similarly to the relationship between quantum and classical effects. Therefore, further and even faster development of the science of nanostructure materials and nanotechnology is to be expected.

Application of nanostructures requires knowledge of their fundamental physical (mechanical, electro-magnetic, optical, etc.) characteristics. Thermodynamic properties associated with phonon displacements through the nanosamples are particularly interesting. Independent of the type of lattices, the thermodynamics of their subsystems (electrons, excitons, spin waves, etc.) is determined when the subsystem is in thermodynamic equilibrium with phonons. Phonons are quasiparticles that represent collective mechanical oscillations of molecules or atoms and they are at the most important system of excitations [97–99]. Besides, the acoustical characteristics, as well as conductive and superconductive properties, etc., could not realistically be explained without knowing phonon behavior. The main fact concerning phonon properties in nanostructures is the absence of the so-called acoustic phonons, i.e., phonons whose energy tends to zero when phonon momentum tends to zero. For the exciting of phonons in nanostructures activation energy different from zero is necessary. What is more, the appearance of additional energy gaps in the spectra of other elemental excitations is shown. These unexpected characteristics require a revision of all conclusions obtained by bulk theories of condensed matter [100–105]. This work describes a major aspect of the effort to understand nanostructures, namely the study of elementary excitations and mediated effects with phonons in structures with nanoscale-dimensional confinement in one or more spatial dimensions.

Since quantum low-dimensional structures are of the great interest in science, we have researched the basic microscopic behavior of different elementary subsystems in following models of nanostructures: ultra-thin films (UTF), nanorods (NR) and quantum dots (QD). The influence of changed microscopic properties of elementary excitations in these structures is decisive on macroscopic properties, especially from the aspect of their possible usage in nanotechnology. This work is based on research from our group of theoretical researchers gathered around the University of Novi Sad in the last decades [106–110]. In the opposite of the films which are quasi-2D structures, limited by two surfaces parallel to XY planes, nanorods and quantum dots are quasi 1D and 0D crystal systems limited in two or three directions: y and z, or x, y and z respectively.

At the beginning, we must redefine theoretical methods and mathematical apparatus and adapt them to discretely changeable spatial variables.

7.2. Solving of Difference Equations of Second Order By Translation Operators[2]

A large part of this work is devoted to the methodology of analysis of nanostructures and, generally, of broken symmetry structures. This is due

[2]In this section, parts of the text from the cited paper were used with permission: Ref. [10] *Encyclopedia of Nanoscience and Nanotechnology* **50**, 281–350 (2011); Copyright @ASP.

to the fact that nowadays there is no unique method of analysis of nano- and broken symmetry structures. Besides, we are of the opinion, that the development of nanotechnology is more rapid than the development of corresponding theories and therefore the efforts in the theoretical field must be intensified.

7.2.1. *Difference equations with constant and valuable coefficients*

In the review paper [10], we exposed one convenient method for solving difference equations with constant coefficients by translation operators [111–120]. This type of equations is found in solid state analyses of one-dimensional ideal[3] crystalline structures, but it can easily be applied to multi-dimensional systems. Presently we are focused on some nonlinear difference equations which play an important role in the condensed matter physics, especially for the problems related with crystal structures with broken translational symmetry [121–123]. Probability amplitudes in one-particle wave functions are the solutions of systems of precisely such difference equations. General type of difference equations defining mentioned quantities are the following:

$$Y_{n+1} + Y_{n-1} + FY_n = 0 \tag{7.1}$$

where function $Y_n \equiv Y_n(x)$, integer n denotes location of atom or molecule, while F is the parameter related with the physical characteristic of the system. If the crystal structure is transitionally invariant [1–5], parameter x does not depend on integer n (see, for example, [124–126]). Since $F = \text{const}$, this equation can be solved easily by the substitution:

$$Y_n = Y_0 \mathrm{e}^{kn} \tag{7.2}$$

By putting (7.2) into (7.1), we obtain the following equation:

$$(\mathrm{e}^k)_{1,2} = \frac{F}{2}\left(-1 \pm \sqrt{1 - \frac{4}{F^2}}\right) \tag{7.3}$$

[3]Ideal crystalline structures are those with a simple cubic lattice and with one motive (atom, molecule or ion) in the elementary cell. We will explain in the appendix how this idealized problem can be applied to more complex crystalline systems with a different symmetry and a complex elementary cell.

wherefrom two solutions of (7.1) follow:

$$Y_n^{(1)} = Y_0 \left[\frac{F}{2}\left(-1+\sqrt{1-\frac{4}{F^2}}\right)\right]^n; \quad Y_n^{(2)} = Y_0 \left[\frac{F}{2}\left(-1-\sqrt{1-\frac{4}{F^2}}\right)\right]^n \tag{7.4}$$

In broken symmetry structures (the structures with impurities, vacancies, ideal structures with finite boundaries, etc.), the constant F goes over to the function F_n depending on the position. Thus, for broken symmetry structures, Eq. (7.1) becomes

$$Y_{n+1} + Y_{n-1} + F_n Y_n = 0 \tag{7.5}$$

This equation will be solved by means of operator method [127, 128]. The first step of solving is the transformation of (7.6) into the form

$$(\hat{A} + \hat{B}_n) Y_n = 0 \tag{7.6}$$

where

$$\hat{B}_n = F_n + 2 \tag{7.7}$$

is the multiplicative operator, while $\hat{A}$ is the operator of translatory type given by

$$\hat{A} = \hat{T}_1 + \hat{T}_{-1} - 2 \tag{7.8}$$

Operators $\hat{T}$ in (7.9) are translational operators defined in the following way:

$$\hat{T}_{\pm k} f_n = f_{n\pm k}; \quad \hat{T}_{\pm k}\mathrm{const} = \mathrm{const}; \quad \hat{T}_{\pm k} = \hat{T}_1^{\pm k}; \quad \hat{T}_k \hat{T}_{-k} = \hat{T}_{-k}\hat{T}_k = 1 \tag{7.9}$$

The next step is defining of inverse operator $(\hat{A} + \hat{B}_n)^{-1}$. Operator $\hat{A} + \hat{B}_n$ can be rewritten in two forms:

$$\hat{A} + \hat{B}_n = \hat{B}_n (1 + \hat{B}_n^{-1} \hat{A}) \tag{7.10a}$$

$$\hat{A} + \hat{B}_n = \hat{A}(1 + \hat{\hat{A}}^{-1} \hat{B}_n) \tag{7.10b}$$

Consequently, inverse operators have two different forms, also.

$$(\hat{A} + \hat{B}_n)^{-1} = (1 + \hat{B}_n^{-1} \hat{A}) \hat{B}_n = \sum_{k=0}^{\infty} (-1)^k (\hat{B}_n^{-1} \hat{A})^k \hat{B}_n^{-1} \tag{7.11a}$$

$$(\hat{A} + \hat{B}_n)^{-1} = (1 + \hat{A}^{-1} \hat{B}_n)^{-1} \hat{A}^{-1} = \sum_{k=0}^{\infty} (-1)^k (\hat{A}^{-1} \hat{B}_n)^k \hat{A}^{-1} \tag{7.11b}$$

In connection to the formulas (7.13) and (7.14), it is important to note the following:

- Infinite series in (7.13) and (7.14) are operator geometry progressions. Whether series converge or not cannot be answered until we figure out operator terms in the series. The convergence can be examined only after application of operator series to some function.
- Since operators $\hat{A}$ and $\hat{B}_n$ do not commute, the terms $(\hat{B}_n^{-1}\hat{A})^k$ and $(\hat{A}^{-1}\hat{B}_n)^k$ must be written as follows:

$$(\hat{B}_n^{-1}\hat{A})^k = \underbrace{\hat{B}_n^{-1}\hat{A}\hat{B}_n^{-1}\hat{A}\dots\hat{B}_n^{-1}\hat{A}}_{k-\text{times}} \tag{7.12a}$$

$$(\hat{A}^{-1}\hat{B}_n)^k = \underbrace{\hat{A}^{-1}\hat{B}_n\hat{A}^{-1}\hat{B}_n\dots\hat{A}^{-1}\hat{B}_n}_{k-\text{times}} \tag{7.12b}$$

The solving of Eq. (7.6) requires introduction of the auxiliary non-homogeneous difference equation

$$(\hat{A} + \hat{B}_n)y_n = \Psi_n \tag{7.13}$$

Taking into account that the inverse operator $(\hat{A} + \hat{B}_n)^{-1}$ has two different forms (7.12a,b) it can be concluded that Eq. (7.14) has two solutions $y_n^{(1)}$ and $y_n^{(2)}$. So, we write

$$(\hat{A} + \hat{B}_n)y_n^{(1)} = \psi_n \tag{7.14a}$$

$$(\hat{A} + \hat{B}_n)y_n^{(2)} = \psi_n \tag{7.14b}$$

By subtracting (7.18b) and (7.19b), we obtain $(\hat{A} + \hat{B}_n)(y_n^{(1)} - y_n^{(2)}) = 0$, wherefrom it is clear that the solution of (7.6) is given by

$$Y_n^{(I)} \equiv y_n^{(1)} - y_n^{(2)} = \sum_{k=0}^{\infty}(-1)^k[(\hat{B}_n^{-1}\hat{A})^k\hat{B}_n^{-1} - (\hat{A}^{-1}\hat{B}_n)^k\hat{A}^{-1}]\Psi_n \tag{7.15}$$

The second solution of (7.6) can be found by means of artificial non-homogenous equation

$$(\hat{A} + \hat{B}_n)z_n = \Phi_n \tag{7.16}$$

and this solution is given by

$$Y_n^{(II)} = \sum_{k=0}^{\infty}(-1)^k[(\hat{B}_n^{-1}\hat{A})^k\hat{B}_n^{-1} - (\hat{A}^{-1}\hat{B}_n)^k\hat{A}^{-1}]\Phi_n \tag{7.17}$$

It is convenient that infinite series for $y_n^{(1)}$ and $z_n^{(1)}$ are as simple as possible. It is easily seen that this will be achieved for the following choice of functions Ψ_n and Φ_n:

$$\Psi_n = B_n \tag{7.18a}$$

$$\Phi_n = nB_n \tag{7.18b}$$

For $\Psi_n = B_n$ we obtain $y_n^{(1)} = 1$, while for $\Phi_n = nB_n$ we have $z_n^{(1)} = n$.

The solutions $y_n^{(2)}$ and $z_n^{(2)}$ are infinite series whose convergence has to be examined. The operator series (7.12b), which gives solutions $y_n^{(2)}$ and $z_n^{(2)}$, contains operator $\hat{A}^{-1}$ whose properties have to be analyzed. Operator $\hat{A}$ given by (7.9) can be written in two different ways:

$$\hat{A}_I = \hat{T}_1\hat{T}_{-1}(\hat{T}_1 + \hat{T}_{-1} - 2) = \hat{T}_1(\hat{T}_{-2} - 2\hat{T}_{-1} + 1) = \hat{T}_1(1 - \hat{T}_{-1})^2 \tag{7.19a}$$

$$\hat{A}_{II} = \hat{T}_{-1}\hat{T}_1(\hat{T}_1 + \hat{T}_{-1} - 2) = \hat{T}_{-1}(\hat{T}_2 - 2\hat{T}_1 + 1) = \hat{T}_{-1}(1 - \hat{T}_1)^2 \tag{7.19b}$$

In order to find inverse operators $\hat{A}_I^{-1}$ and $\hat{A}_{II}^{-1}$ we shall use analogy with the known formula:

$$\frac{1}{(1-x)^2} = \frac{d}{dx}\frac{1}{1-x} = \frac{d}{dx}\sum_{s=0}^{\infty} x^s = \sum_{s=1}^{\infty} sx^{s-1}; \quad |x| < 1$$

Combining this with (2.19a) and keeping in mind that it is senseless to take care of convergence of operator series, we can write the following:

$$\hat{A}_I^{-1} = (1 - \hat{T}_{-1})^{-2}\hat{T}_{-1} = \left[\frac{d}{d\hat{T}_{-1}}(1 - \hat{T}_{-1})^{-1}\right]\hat{T}_{-1}$$

$$= \left[\frac{d}{d\hat{T}_{-1}}\sum_{s=0}^{\infty}\hat{T}_{-1}^s\right]\hat{T}_{-1} = \sum_{s=0}^{\infty} s\hat{T}_{-1}^{s-1}\hat{T}_{-1} = \sum_{s=1}^{\infty} s\hat{T}_{-s}$$

Similarly, it can be written as $\hat{A}_{II}^{-1} = \sum_{s=1}^{\infty} s\hat{T}_s$. It means that operator $\hat{A}^{-1}$ has two different forms:

$$\hat{A}_I^{-1} = \sum_{s=1}^{\infty} s\hat{T}_{-s} \tag{7.20a}$$

$$\hat{A}_{II}^{-1} = \sum_{s=1}^{\infty} s\hat{T}_s \tag{7.20b}$$

The choice of the form for $\hat{A}^{-1}$ depends on the function to which operator $\hat{A}^{-1}$ is applied. Obtained functional series must be convergent.

7.2.2. *A more general nonlinear problem*

Now, we will consider the nonlinear difference equation in the following form:

$$Y_{n+1} + Y_{n-1} + (\alpha \mathrm{e}^{\beta n} - 2)\, Y_n = 0; \quad \alpha > 0; \quad \beta > 0 \tag{7.21}$$

where, in this special case, the multiplicative operator is of the form $\hat{B}_n = \alpha \mathrm{e}^{\beta n}$. The auxiliary equation (7.14) is the following:

$$(\hat{A} + \alpha \mathrm{e}^{\beta n}) y_n = \alpha \mathrm{e}^{\beta n} \tag{7.22}$$

while the auxiliary equation (7.17) is

$$(\hat{A} + \alpha \mathrm{e}^{\beta n}) z_n = n \alpha \mathrm{e}^{\beta n} \tag{7.23}$$

We shall examine Eq. (7.23), first. Taking into account (7.21), we can write

$$y_n^{(1)} = \sum_{k=0}^{\infty} (-1)^k \left[\left(\frac{1}{\alpha \mathrm{e}^{\beta n}} \hat{A} \right)^k \frac{1}{\alpha \mathrm{e}^{\beta n}} \alpha \mathrm{e}^{\beta n} \right] = 1 \tag{7.24}$$

Since $\beta > 0$, the evaluation of $y_n^{(2)}$ requires the use of the operator A_I^{-1} because this leads to convergent series. In accordance with (7.21), the following can be written:

$$y_n^{(2)} = \sum_{k=0}^{\infty} (-1)^k [(\alpha \mathrm{e}^{\beta n} \hat{A}_I^{-1})^k \hat{A}_I^{-1} \alpha \mathrm{e}^{\beta n}] \tag{7.25}$$

Now, we shall find first $\hat{A}\, \alpha\, \mathrm{e}^{\beta n}$. In accordance with (2.20a), we can write

$$\begin{aligned} A_I^{-1} \alpha \mathrm{e}^{\beta n} &= \sum_{s=1}^{\infty} s A_{-s} \alpha \mathrm{e}^{\beta n} = \alpha \mathrm{e}^{\beta n} (\mathrm{e}^{-\beta} + 2\mathrm{e}^{-2\beta} + 3\mathrm{e}^{-3\beta} + \cdots) \\ &= -\alpha \mathrm{e}^{\beta n} \frac{\mathrm{d}}{\mathrm{d}\beta} (\mathrm{e}^{-\beta} + \mathrm{e}^{-2\beta} + \mathrm{e}^{-3\beta} + \cdots) \\ &= -\alpha \mathrm{e}^{\beta n} \frac{d}{d\beta} \frac{1}{\mathrm{e}^{\beta} - 1} = \frac{\alpha\, \mathrm{e}^{\beta n}}{4} \frac{1}{\sinh^2 \frac{\beta}{2}} \end{aligned}$$

and now follows

$$y_n^{(2)} = (1 - \alpha \mathrm{e}^{\beta n} \hat{A}_I^{-1} + \alpha \mathrm{e}^{\beta n} \hat{A}_I^{-1} \alpha \mathrm{e}^{\beta n} \hat{A}_I^{-1} + \cdots) \frac{\alpha \mathrm{e}^{\beta n}}{4} \frac{1}{\sinh^2 \frac{\beta}{2}} \tag{7.26}$$

Using the approach which gave the result (7.26), we can write

$$y_n^{(2)} = \frac{\alpha e^{\beta n}}{4}\frac{1}{\sinh^2 1\frac{\beta}{2}} - \left(\frac{\alpha e^{\beta n}}{4}\right)^2 \frac{1}{\sinh^2 1\frac{\beta}{2}\sinh^2 2\frac{\beta}{2}}$$

$$+\left(\frac{\alpha e^{\beta n}}{4}\right)^3 \frac{1}{\sinh^2 1\frac{\beta}{2}\sinh^2 2\frac{\beta}{2}\sinh^2 3\frac{\beta}{2}} - \cdots, \tag{7.27}$$

and substituting (7.25) and (7.28) in (7.21), we obtain the first particular solution of Eq. (7.22):

$$Y_n^{(I)} = \sum_{k=0}^{\infty}(-1)^k\left(\frac{\alpha e^{\beta n}}{4}\right)^k \frac{1}{\prod_{\mu=1}^{k}\sinh^2 \mu\frac{\beta}{2}}; \quad \prod_{\mu=1}^{0}\sinh^2 \mu\frac{\beta}{2} = 1 \tag{7.28}$$

Evaluation of the second particular solution of Eq. (7.21) requires solving of auxiliary equation

$$(\hat{A} + \alpha e^{\beta n})z_n = n\alpha e^{\beta n}$$

Now, using operator A_I^{-1} and the procedure analogous to that from the previous case, we obtain the second particular solution of Eq. (7.22) in the following form:

$$Y_n^{(II)} = n\sum_{k=0}^{\infty}(-1)^k\left(\frac{\alpha e^{\beta n}}{4}\right)^k \frac{1}{\prod_{\mu=1}^{k}\sinh^2 \mu\frac{\beta}{2}}$$

$$-\sum_{k=0}^{\infty}(-1)^k\left(\frac{\alpha e^{\beta n}}{4}\right)^k \frac{\sum_{\mu=1}^{k}\coth \mu\frac{\beta}{2}}{\prod_{\mu=1}^{k}\sinh^2 \mu\frac{\beta}{2}} \tag{7.29}$$

By the exposed procedure we found the algorithm for solving of difference equations appearing in the analysis of broken symmetry structure. The illustrative example is related to the function of the type $\alpha\, e^{\beta n}$. Similarly, to the quoted illustrative example, the solutions of the equation,

$$Y_{n+1} + Y_{n-1} + (\alpha a^{\beta n} - 2)Y_n = 0 \tag{7.30}$$

where $a > 1$, can be found.

In all other cases, unfortunately, the quoted algorithm gives very cumbersome and therefore intractable solutions. Independently, we decided to include this algorithm. The reason is simple: the present computer technique is able to give the solution of Eq. (7.6) in tabular form for arbitrary function without any more serious problems.

7.2.3. *Appropriate example*

For clarity, we shall demonstrate exposed procedure of operator solving on an illustrative example — we will observe the phonons in crystalline chain.

One-dimensional propagation of mechanical waves in condensed matter physics is described, in the nearest-neighbor approximation [1–5], by consideration of the difference equation,

$$u_{n+1} + u_{n-1} + \rho u_n = 0 \tag{7.31}$$

valid for small displacements u_n of atom/molecule[4] from its equilibrium site $n \in [0, N]$. For ultra-short ($L = Na$, $N \leq 10$, a is lattice constant) crystalline chain [127–130], the quantity $\rho \equiv \rho_n = 2(\cos n\xi - 1)$, $\xi \equiv \xi(\omega)$. The expression (7.32) is equivalent to (7.6) and we can apply the described operator method. It is seen from (7.7) to (7.9) that $B_n = e^{in} + e^{-in}$.

At the beginning, we shall look for the function $\hat{A}^{-1}\mathrm{e}^{in\xi}$ representing the operator $\hat{A}^{-1}$ by the formula (2.20a). Since

$$(\hat{T}_{+1} + \hat{T}_{-1})^{-1}\mathrm{e}^{in\xi} = 2\mathrm{e}^{in\xi}\cos\xi$$

it follows that

$$\begin{aligned}
\hat{A}^{-1}\mathrm{e}^{in\xi} &= -\frac{1}{2}\,\mathrm{e}^{in\xi} - \frac{1}{2^2}(\hat{T}_{+1} + \hat{T}_{-1})^{+1}\mathrm{e}^{in\xi} - \frac{1}{2^3}(\hat{T}_{+1} + \hat{T}_{-1})^{+2}\mathrm{e}^{in\xi} - \cdots \\
&= -\left(\frac{\mathrm{e}^{in\xi}}{2} + \frac{\mathrm{e}^{in\xi}}{2}\cos\xi + \frac{\mathrm{e}^{in\xi}}{2}\cos^2\xi + \cdots\right) \\
&= -\frac{\mathrm{e}^{in\xi}}{2}\sum_{k=0}^{\infty}\cos^k\xi = -\frac{\mathrm{e}^{in\xi}}{2(1-\cos\xi)}
\end{aligned}$$

Further, we obtained the following:

$$(\hat{A}^{-1}\mathrm{e}^{in\xi})^2 = \hat{A}^{-1}\mathrm{e}^{in\xi}\left[-\frac{\mathrm{e}^{in\xi}}{2(1-\cos\xi)}\right] = \frac{\mathrm{e}^{2in\xi}}{2^2(1-\cos\xi)\,(1-\cos 2\xi)}$$

wherefrom it can be easily concluded that

$$(\hat{A}^{-1}\mathrm{e}^{in\xi})^k = \frac{(-1)^k\,\mathrm{e}^{k\,in\xi}}{2^k\prod_{q=0}^{k}(1-\cos q\xi)}; \qquad k = 1, 2, 3, \ldots \tag{7.32}$$

[4] Atom displacements $u_n \equiv u_n(\omega)$ are the Fourier's transform ($t \to \omega$) the space displacements $u_n(t)$.

On the basis of the general facts presented in the previous sections, we must look for the solution $u_n(\xi)$ by (7.33), which satisfies the difference equation:

$$(\hat{A} + \mathrm{e}^{in\xi})^{-1} u_n(\xi) = 0 \tag{7.33}$$

If we multiply this equation by $\mathrm{e}^{-2i\,n\xi}$ from left, and to the obtained equation add (7.34), we obtain

$$(\hat{\alpha}_n + B_n)^{-1} u_n(\xi) = 0 \tag{7.34}$$

where: $\hat{\alpha}_n \equiv (1 + \mathrm{e}^{-2in\xi})$ and $B_n \equiv 2\cos n\xi$. From (7.16), the solution of the upper equation is

$$u_n(\xi) = \sum_{k=0}^{\infty} (-1)^k (\hat{\alpha}_n^{-1} B_n)^k \tag{7.35}$$

It is easy to prove that $\hat{\alpha}_n^{-1} B_n \equiv \hat{A}^{-1} \mathrm{e}^{in\xi}$, so – from (7.33) and (7.35), the wanted solution is

$$u_n(\omega) \equiv u_n(\xi) = \sum_{k=0}^{\infty} \frac{\mathrm{e}^{k\, in\xi(\omega)}}{2^k \prod_{q=0}^{k} [1 - \cos q\xi(\omega)]} \tag{7.36}$$

It can be seen that atom displacement represents the specific superposition of harmonic functions which depend on space position. Physically, it is represented by the superposition of the progressive plane and standing waves.

7.3. Green's Functions in Solid State Physics

Green's Functions (GFs) in quantum statistical physics represent the statistical mean value of the time-ordered products of two complexes of Heisenberg's operators [1–5, 131–134]. One of the complexes operates at the moment t, and the other at another time t'. The timing is that the product of the two complexes is multiplied by the unit, if the operator complex acting at an earlier point in time stands in the product to the right of the complex that operates at a later point in time. In case complexes are multiplied in reverse order, the product multiplies by zero, i.e., it is assumed that such order of operation of the operating complexes is inadmissible. Since they are described — the time editing and statistical aggregation operations — independent, the order of these operations is arbitrary.

This definition of GF is not artificially introduced. In the described form, they appear when calculating the mean values of the physical characteristics of the system on which the external, time-dependent fields act. Since the systems (materials) are not and cannot be isolated from the environment, the calculation of their characteristics in the external fields is the part of the theoretical analysis whose results immediately have immediate application. External fields can exist permanently and independently of human desire and will, but it is even more common to place materials in different artificially created fields in which their properties can be better used. A typical example of this is the placing of metals in the external electric fields that direct the chaotic movement of their free electrons, i.e., turn it into an electrical current. In the process of analyzing the influence of external fields on the material environment, a number of essential phenomenological characteristics of the material, such as transport coefficients (coefficient of electrical and thermal conductivity, diffusion coefficient, etc.), dielectric permittivity, magnetic permeability, electrical resistance and many others appear. An important part of the explanation for the phenomenological phenomenon and behavior of these phenomenological features is the function of Green, so it is understandable that special attention is paid to their study.

Operational complexes that are part of GFs are made up of the creation and annihilation of elemental excitations that arise in the middle. Therefore, GF depends on the energy of the elemental excitations of the environment, and beyond it from the microcircuits of the environment such as electronic or some other states of atoms or molecules, intermolecular forces, local fields of elemental cells — in the case of a crystalline environment — and so on. By entering as such in the composition of the phenomenological characteristics of the environment, GF represents a bridge that connects the macroscopic (phenomenological) characteristics of the environment with its micro-characteristics. When this link is known, it is possible to approach the improvement of materials (changing its characteristics, for example, by doping or synthesizing new meters) in order to make their macroscopic characteristics better and more effective in immediate (practical) application.

All that is stated is the reason why the characteristics of GFs and the ways of their calculation for decades are studied and perfected. Methods of calculating GFs have been developed in two directions. One direction was designed by the American-Japanese school (Feynman, Dyson, Swinger, Tomonaga, Matsubara, Kubo), and the second Soviet physics school (Bogolyubov, Bonch-Bruevich, Tyablikov, Zubarev). The method of the US-Japanese school consists in the application of Wick's theorem to operators' time-regulated products, with a higher degree of accuracy of the calculation requiring the inclusion of operator products with an increasing number of factors. This approach has made it possible for each analytic expression that is part of Green's function to correspond to a corresponding diagram representing the inter-process process responsible for the occurrence of the GF repair. The fact that the drawings of particle or quasi-particle processes (scattering, fusion, decay, etc.) can, according to certain rules, correspond to a certain analytic expression that is part of the Green function has brought widespread popularity to this technique of the account. The method of the Soviet school, which is called the method of two-time temperature Grin functions, consists in calculating the time-ordered statistical mean values of the operator's products using the Heisenberg equations of motion. This method gives an infinite chain of interrelated equations that determine the desired GF. In order to obtain an analytic expression for the required GF, the equation chain must be cut out based on some more or less justifiable criteria. The method of cutting the chain of equations, the criteria that are used on that occasion and the results that, depending on the way of cutting, are obtained will be subject to further exposure. The generally established criteria for closing the equation system that determines Green's function (the cutting of the equation chain) do not exist today, which is the part of the theory of two-time temperature Green's functions, which requires further, intensive research.

Each of these two approaches must give the same result, at the same degree of accuracy of the budget. Of course, you need to select one that in fewer moves gives the desired result. This result is, if the cases of square (harmonic) Hamiltonians of the system are excluded, approximate, but it is desirable to maximally adequately reproduce the properties of the examined system. The problem of adequate reproduction, in the method of two-time, temperature GF, is reduced to the problem of precisely closing the chain of equations. In the method using particle process diagrams, the adequacy of the reproduction of the system's properties depends on whether the processes for determining the characteristics of the system are selected for the calculation of GF. All processes cannot be taken into account (as the infinite chain of equations cannot be solved), and the choice of the most relevant processes is still today in the domain of intuition, but in the domain of some general, strict rules.

As already mentioned, the further exposition will refer to the method of quantum two-dimensional, temperature-dependent Green's functions (Q2TT GFs). Before moving to mathematical formalism, it should be said that the popularity of this method is on the increase, among other things, because of its self-consistency. This means that within the framework of the method there is everything necessary to describe the system — and the energy of elementary excitations and the statistical mean value of the physical characteristics of the system — so there is no need for "borrowing" from other areas or from other methods. Such a closed budget has a number of advantages, of which, most probably, the most important thing, is a better possibility of error estimation. When the method combines methods from different areas, the estimation of the accuracy of the account is far more difficult and more complicated.

7.3.1. *Quantum two-time temperature Green's functions*

Here, the properties of the commutative GF will be examined, which represents the time-ordered statistical mean value of commutators of two operator complexes operating at two different times [5, 131–134]. In this commutator form, GF appears when calculating the mean values of the physical characteristics of the system, when an external, time-dependent field is applied to it. Such accounts are of a perturbation type and in the vast majority of practical examples they are carried out in an approximation that is linear by the external perturbation. This linear approximation is called the linear response of the system to an external perturbation. It should be emphasized that the results of this, linear approximation, very faithfully reproduce most of the phenomenological characteristics of the system, and that it is resorting to higher approximations in extremely rare cases. In addition to the commutator GF, there are also anticommutator functions (mean values of two anticommutator complexes) that are suitable for analysis of fermion systems and causal GF (mean values of products of two operator complexes). The results of the analysis of the commutator functions are very easily transferred to the anticommutator and causal functions, so it is not necessary to separately examine these.

The two-time, temperature commutator GF is defined as follows:

$$G(x,x',t,t') = \theta(t-t')\mathrm{Sp}\{[\hat{A}(x,t),\hat{B}(x',t')]\hat{\rho}\} \tag{7.37}$$

where $\Theta(t-t')$ is the Heaviside's step-function, and $\hat{\rho}$ is statistical operator of the canonical ensemble:

$$\hat{\rho} = \mathrm{e}^{(\hat{F}-\hat{H})/\theta} \tag{7.38}$$

while the operator complexes $\hat{A}(x,t)$ and $\hat{B}(x',t')$ are given in Heisenberg's picture:

$$\hat{A}(x,t) = \mathrm{e}^{i\hat{H}t/\hbar}\hat{A}(x)\mathrm{e}^{-i\hat{H}t/\hbar}; \qquad \hat{B}(x',t') = \mathrm{e}^{i\hat{H}t'/\hbar}\hat{B}(x)\mathrm{e}^{-i\hat{H}t'/\hbar} \quad (7.39)$$

In the expression (7.38), the free energy $\hat{F}$ of the system is denoted with the Hamiltonian system $\hat{H}$. The size is characterized by the exchange of energy between the system and the thermostat. Spatial coordinates x and x' in general, can be components of the radius of a vector in some N-dimensional hyper-space: $x = (x_1, x_2, \ldots, x_N)$ and $x' = (x'_1, x'_2, \ldots, x'_N)$.

Definition (7.37) introduced Tyablikov, guided, probably, by the fact that it is exactly this form that appears in the expression for the linear response of the system. Zubarev defines Green's function in the form (7.37), but multiplied by the factor i. This factor also appears in the Feynman–Matsubara theory of Green's functions, and with the introduction of factors in the Q2TT GF technique, the possibility of the results of two different approaches is easier to compare. It is obvious, however, that the multiplicative constant (by which the two definitions differ) cannot significantly change anything in the results and in the efficiency of the calculation. This Tyablikov definition will still be used here.

By Definition (7.37), which is the most general type, GF depends on four variable arguments: x, x', t, t'. These GFs appear in analyses of non-spatially homogeneous spaces and by operator complexes A and B which are not Schrödinger's, but explicitly depend on time. The calculation of these GFs is very complicated, and some concrete conclusions can be made only after a numerical analysis. Therefore, in the GF theory, a certain idealization of the real situation is resorted: the environment is considered to be spatially homogeneous and that the operator complexes $\hat{A}$ and $\hat{B}$ are independent of time. Then GF depends on two "pooled" variables $x - x'$ and $t - t'$.

In a spatially homogeneous environment (all points of the medium are identical, for example, in crystals, which consist of identical atoms and there is no defect in the structure, nor ingredient), the measurable physical characteristic of the system, such as the mean value, which is represented in (7.37), cannot depend on each point separately, but only on their distance $x - x'$. Otherwise, a rule would be to distinguish between two identical points, which is both logical and physically intolerable. This reasoning, of course, is always accompanied by the assumption that the observed environment can be treated as infinite. This second idealization, strictly speaking, at first glance seems to be too rough, because systems, which are taught by physics, are of finite dimensions. Typical physical reasoning here is used: for the middle, say a crystal, where there are characteristic distances between the atoms of a tenth-mile radius of a meter, the linear dimension of the crystal, which is a hundredth order of a meter, can be treated as infinite [1–5]. The practice confirmed the validity of this reasoning, because the results of theoretical analyses, based on idealizations such as spatial homogeneity and infinity of the structure, fit very well into the results of the corresponding experiments, although the experiments use patterns that are neither ideally spatially homogeneous nor infinite in literally meaning.

In addition to all the mentioned restrictions about how real systems can be considered spatially homogeneous and infinite, GF will be further analyzed of spatially homogeneous and infinite environments. So, these GFs do not depend on the coordinates x and x' separately, but on their differences.

If the operator complexes do not depend explicitly on time, the GF will not depend on the time variables t and t' separately, but on their differences $t - t'$. This will be demonstrated in the case of a medium value $\mathrm{Sp}\{\hat{A}(x,t), \hat{B}(x',t')\hat{\rho}\}$, which is a part of GF (7.37). The proof is based on the fact that, under the sign of the spur (trace), cyclic permutation of the operator is permitted and that the operators $\mathrm{e}^{+i\hat{H}t/\hbar}\mathrm{e}^{-\hat{H}/\theta}$: mutually commute because they have functions of the same operator structure $\hat{H}$. So,

$$\begin{aligned}
\mathrm{Sp}\{\hat{A}(x,t), \hat{B}(x',t')\hat{\rho}\} &= \mathrm{Sp}\{\mathrm{e}^{i\hat{H}t/\hbar}\hat{A}(x)\mathrm{e}^{-i\hat{H}t/\hbar}\mathrm{e}^{i\hat{H}t'/\hbar}\hat{B}(x')\mathrm{e}^{-i\hat{H}t'/\hbar}\} \\
&= \mathrm{Sp}\{\mathrm{e}^{(\hat{F}-\hat{H})/\theta}\mathrm{e}^{i\hat{H}t/\hbar}\hat{A}(x)\mathrm{e}^{-i\hat{H}(t-t')/\hbar}\hat{B}(x')\mathrm{e}^{-i\hat{H}t'/\hbar}\} \\
&= \mathrm{Sp}\{\mathrm{e}^{-i\hat{H}t'/\hbar}\mathrm{e}^{(\hat{F}-\hat{H})/\theta}\mathrm{e}^{i\hat{H}t/\hbar}\hat{A}(x)\mathrm{e}^{-i\hat{H}(t-t')/\hbar}\hat{B}(x')\} \\
&= \mathrm{Sp}\{\mathrm{e}^{(\hat{F}-\hat{H})/\theta}\mathrm{e}^{-i\hat{H}(t-t')/\hbar}\hat{A}(x)\mathrm{e}^{-i\hat{H}(t-t')/\hbar}\hat{B}(x')\} \\
&= \phi(t-t') \qquad (7.40)
\end{aligned}$$

Proof that this also applies to the mean value: $\mathrm{Sp}\{[\hat{A}(x,t), \hat{B}(x',t')]\hat{\rho}\}$, which is represented in (7.37), is completely identical to (7.40) and will not be stated.

The summary of the above considerations and proofs would be the following: GF constructed from operator complexes $\hat{A}$ and $\hat{B}$ (which explicitly does not depend on time) for spatially homogeneous and infinite environments depends on the difference between time variables $t - t'$ and the difference in spatial variables $x - x'$.

In the future, such GFs will be analyzed, and instead of the general expression (7.37), the expression

$$G(x, x', t, t') = \theta(t - t')\mathrm{Sp}\{[\hat{A}(x, t), \hat{B}(x', t')]\hat{\rho}\} \tag{7.41}$$

will be used.

Correlation Functions

In the expression (7.41) for the Q2TT GF, we figure out the statistical values of the operator products:

$$\begin{aligned} J_{\hat{A}\hat{B}}(x - x', t - t') &= \mathrm{Sp}\{\hat{A}(x, t), \hat{B}(x', t')\hat{\rho}\}; \\ J_{\hat{B}\hat{A}}(x - x', t - t') &= \mathrm{Sp}\{\hat{B}(x', t'), \hat{A}(x, t)\hat{\rho}\} \end{aligned} \tag{7.42}$$

which is called correlation functions (CFs). It should be noted that CF, constructed from an operator operating at the same point in space and at the same time, does not depend on spatial or time variables. On the basis of (7.42), the expression (7.41) for GF can be written in the form

$$G(x, x', t, t') = \theta(t - t')\mathrm{Sp}\{[\hat{A}(x, t), \hat{B}(x', t')]\hat{\rho}\} \tag{7.43}$$

so some important conclusions can be made by analyzing CFs, which are related to both CF and GF. Spur will, in expression (7.41), be taken by the functions of the Hamiltonian of the system H:

$$\hat{H}|K\rangle = E_k|K\rangle \tag{7.44}$$

where E_k are the values of the energy of the system are in the quantum mechanical state $|K\rangle$. Considering (7.44), the relations are valid

$$\begin{aligned} \mathrm{e}^{i\hat{H}t/\hbar}|K\rangle &= \mathrm{e}^{iE_k t/\hbar}|K\rangle; \quad |K\rangle\mathrm{e}^{i\hat{H}t/\hbar} = |K\rangle\mathrm{e}^{iE_k t/\hbar} \\ \mathrm{e}^{-\hat{H}/\theta}|K\rangle &= \mathrm{e}^{-E_k/\theta}|K\rangle; \quad |K\rangle\mathrm{e}^{-\hat{H}/\theta} = |K\rangle\mathrm{e}^{-E_k t/\theta} \end{aligned} \tag{7.45}$$

which will be used when the expression (7.42) for the CF $J_{\hat{B}\hat{A}}(x - x', t - t')$ is concretized. If a projection operator $\hat{P} = \sum_l |l\rangle\langle l|$ is used, the CF $J_{\hat{B}\hat{A}}(x - x', t - t')$ expression can be transformed as follows:

$$J_{\hat{B}\hat{A}}(x-x',t-t') = \mathrm{Sp}\{\hat{B}(x',t'),\hat{A}(x,t)\hat{\rho}\}$$
$$= \mathrm{e}^{F/\theta}\mathrm{Sp}\{\hat{B}(x',t'),\hat{A}(x,t)\mathrm{e}^{-\hat{H}/\theta}\}$$
$$= \mathrm{e}^{F/\theta}\sum_k \langle \mathrm{K}|\hat{B}(x',t')\hat{A}(x,t)\mathrm{e}^{-\hat{H}/\theta}|K\rangle$$
$$= \mathrm{e}^{F/\theta}\sum_k \langle \mathrm{K}|\hat{B}(x',t')\hat{P}\hat{A}(x,t)\mathrm{e}^{-\hat{H}/\theta}|K\rangle$$
$$= \mathrm{e}^{F/\theta}\sum_k \langle \mathrm{K}|\hat{B}(x',t')|l\rangle\langle l|\hat{A}(x,t)\mathrm{e}^{-\hat{H}/\theta}|K\rangle$$
$$= \mathrm{e}^{F/\theta}\sum_k \langle K|\mathrm{e}^{i\hat{H}t'/\hbar}\hat{B}(x',t')\mathrm{e}^{-i\hat{H}t'/\hbar}$$
$$\times\, \mathrm{e}^{i\hat{H}t/\hbar}\hat{A}(x,t)\mathrm{e}^{-i\hat{H}t/\hbar}|K\rangle\mathrm{e}^{-E_k/\theta}$$
$$= \mathrm{e}^{F/\theta}\sum_k \langle K|\hat{B}(x')|l\rangle\langle l|\hat{A}(x)|K\rangle\mathrm{e}^{-E_k/\theta}\mathrm{e}^{-i(E_k-E_l)(t-t')/\hbar}$$

which after the replacement $t-t'=\tau$ is reduced to

$$J_{\hat{B}\hat{A}}(x-x',\tau) = \mathrm{e}^{F/\theta}\sum_k \langle K|\hat{B}(x')|l\rangle\langle l|\hat{A}(x)|K\rangle\mathrm{e}^{-E_k/\theta}\mathrm{e}^{-i(E_k-E_l)\tau/\hbar} \tag{7.46}$$

After multiplying with $\mathrm{e}^{i\omega\tau}$ and integrating by $\tau\in(-\infty,+\infty)$, (7.46) becomes

$$\int_{-\infty}^{+\infty} \mathrm{d}\tau \mathrm{e}^{i\omega\tau}J_{\hat{B}\hat{A}}(x-x',\tau)$$
$$= \mathrm{e}^{F/\theta}\sum_k \langle K|\hat{B}(x')|l\rangle\langle l||\hat{A}(x)|K\rangle\mathrm{e}^{-E_k/\theta}\int_{-\infty}^{+\infty}\mathrm{d}\tau\mathrm{e}^{-i(\omega-(E_k-E_l))\tau} \tag{7.47}$$

Fourier's CF transform, in relation to transformation time–frequency, is defined as

$$J_{\hat{B}\hat{A}}(x-x',\omega) = \frac{1}{2\pi}\int_{-\infty}^{+\infty}\mathrm{d}\tau\,\mathrm{e}^{i\omega\tau}J_{\hat{B}\hat{A}}(x-x',\tau) \tag{7.48}$$

whereby the inverse transformation is

$$J_{\hat{B}\hat{A}}(x-x',\tau) = \int_{-\infty}^{+\infty}\mathrm{d}\omega\,\mathrm{e}^{-i\omega\tau}J_{\hat{B}\hat{A}}(x-x',\omega) \tag{7.49}$$

If, in addition to (7.48), it is considered that

$$\int_{-\infty}^{+\infty}\mathrm{d}\tau\,\mathrm{e}^{-i(\omega-\omega_d)\tau} = 2\pi\delta\left(\omega-\frac{E_k-E_l}{\hbar}\right) = 2\pi\hbar\delta(E-E_{kl})$$

where $\omega_{kl} = E_{kl}/\hbar$, $E_{kl} = E_k - E_l$ and $E = \hbar\omega$, the relation (7.48) is reduced to

$$\begin{aligned} J_{\hat{B}\hat{A}}(x - x', \omega) &= \mathrm{e}^{F/\theta} \sum_k \langle K|\hat{B}(x')|l\rangle\langle l|\hat{A}(x)|K\rangle\, \mathrm{e}^{-E_k/\theta}\delta(\omega - \omega_{kl}) \\ &= \hbar \mathrm{e}^{F/\theta} \sum_k \langle K|\hat{B}(x')|l\rangle\langle l|\hat{A}(x)|K\rangle \mathrm{e}^{-E_k/\theta}\delta(E - E_{kl}) \end{aligned} \tag{7.50}$$

The Dirac delta function is a singular function, and based on the results (7.50), it can be concluded that Fourier's image of CF $J_{\hat{B}\hat{A}}(x - x', \omega)$ has a singularity for $\hbar\omega \equiv E = E_k - E_l$. Since $E_k - E_l$ represents a change in the energy of a system when moving from a quantum state $|k\rangle$ to a quantum state $|l\rangle$, this difference can be treated as the energy of the elementary system excitation when moving from one state to another.

In this regard, it is necessary to make some other remarks. First of all, it should be recalled that for each singular function the rule of visiting singularities should be defined [131]. Touring singularities on a real axis requires that a part of the contour is deformed so that it enters the complex plane. Because of this, the size E in expression (7.50) must be treated as complex. On the other hand, the empirical experience shows that the system's excitation occurred during the time of the gas, i.e., that elemental revolutions have the final time of life. The reasons for extinguishing elementary excitations are very different, and most commonly interaction between excitations, interaction with other excitations (usually phonons) and friction effects with the atoms constituting the system. For these reasons and sizes, $E_k - E_l$ can have an imaginary correction, which is usually far smaller than the real part.

Because of these remarks, the conclusion that follows from (7.50) can be formulated by the statement: the singularities of CF $J_{\hat{B}\hat{A}}(x - x', \omega) \equiv J_{\hat{B}\hat{A}}(x - x', E/\hbar)$ in the E complex-surface determine the energies and lifetimes of elementary system excitations. Since CF is part of the GF, the same conclusion applies to GF. The singularities of CF, that is, GF, most often fall in the fourth quadrant of the composite plane. The real part of the singularity represents the energy of elementary excitations, and the imaginary part determines the time of life by dividing the constant $\hbar$ by the intensity of the imaginary part.

Correlation function $J_{\hat{A}\hat{B}}(x - x', t - t')$ can be transformed in a similar way to a function $J_{\hat{B}\hat{A}}$. In order to more easily determine the connection between $J_{\hat{A}\hat{B}}$ and $J_{\hat{B}\hat{A}}$, in the expression (7.42) for $J_{\hat{A}\hat{B}}$ the first two cyclic operator permutations are performed, and then only the spur is calculated.

So,

$$
\begin{aligned}
J_{\hat{A}\hat{B}}(x-x',t-t') &= \mathrm{Sp}\{\hat{A}(x,t)\hat{B}(x',t')\hat{\rho}\} \\
&= \mathrm{Sp}\{\hat{\rho}\hat{A}(x,t)\hat{B}(x',t')\} = \mathrm{Sp}\{\hat{B}(x',t')\hat{\rho}\hat{A}(x,t)\} \\
&= \mathrm{e}^{F/\theta}\mathrm{Sp}\{\hat{B}(x',t')\mathrm{e}^{-\hat{H}/\theta}\hat{A}(x,t)\} = \mathrm{e}^{F/\theta}\sum_k \langle K|\hat{B}(x',t')\mathrm{e}^{-\hat{H}/\theta}\hat{A}(x,t)|K\rangle \\
&= \mathrm{e}^{F/\theta}\sum_k \langle K|\hat{B}(x',t')\mathrm{e}^{-\hat{H}/\theta}\hat{A}(x,t)|K\rangle \\
&= \mathrm{e}^{F/\theta}\sum_k \langle K|\hat{B}(x',t')\mathrm{e}^{-\hat{H}/\theta}\hat{P}\hat{A}(x,t)|K\rangle \\
&= \mathrm{e}^{F/\theta}\sum_k \langle K|\mathrm{e}^{i\hat{H}t'/\hbar}\hat{B}(x',t')\mathrm{e}^{-i\hat{H}t'/\hbar}\mathrm{e}^{-\hat{T}t/\hbar}|l\rangle \\
&\quad\times \langle l|\,\mathrm{e}^{i\hat{H}t/\hbar}\hat{H}(x)\mathrm{e}^{-i\hat{H}t/\hbar}|K\rangle\mathrm{e}^{-E_k/\theta} \\
&= \mathrm{e}^{F/\theta}\sum_k \langle K|\hat{B}(x')|l\rangle\langle l|\hat{A}(x)|K\rangle\,\mathrm{e}^{-E_l/\theta}\mathrm{e}^{-i\omega_{kl}(t-t')/\hbar} \\
&= \mathrm{e}^{F/\theta}\sum_k \langle K|\hat{B}(x')|l\rangle\langle l|\,\hat{A}(x)|K\rangle\,\mathrm{e}^{-E_k/\theta}\mathrm{e}^{-i\omega_{kl}\tau/\hbar}\mathrm{e}^{E_{kl}/\theta}
\end{aligned}
$$

which, after being replaced $t-t'-\tau$, becomes

$$
J_{\hat{A}\hat{B}}(x-x',\tau) = \mathrm{e}^{F/\theta}\sum_k \langle K|\hat{B}(x')|l\rangle\langle l|\hat{A}(x)|K\rangle\,\mathrm{e}^{-E_k/\theta}\mathrm{e}^{-i\omega_{kl}\tau}\mathrm{e}^{iE_{kl}/\theta} \tag{7.51}
$$

After the Fourier transforms, the time–frequency, (7.51), is reduced to

$$
\begin{aligned}
J_{\hat{A}\hat{B}}(x-x',\tau) &= \mathrm{e}^{F/\theta}\sum_{k,l} \langle K|\hat{B}(x')|l\rangle\langle l|\hat{A}(x)|K\rangle\,\mathrm{e}^{-E_k/\theta}\delta(\omega-\omega_{kl})\mathrm{e}^{iE_{kl}/\theta} \\
&= \hbar\,\mathrm{e}^{F/\theta}\sum_{k,l} \langle K|\hat{B}(x')|l\rangle\langle l|\hat{A}(x)|K\rangle\,\mathrm{e}^{-E_k/\theta}\delta(E-E_{kl})\mathrm{e}^{iE_{kl}/\theta} \\
&= \mathrm{e}^{E/\theta}\hbar\,\mathrm{e}^{F/\theta}\sum_{k,l} \langle K|\hat{B}(x')|l\rangle\langle l|\hat{A}(x)|K\rangle\,\mathrm{e}^{-E_k/\theta}\delta(E-E_{kl})
\end{aligned} \tag{7.52}
$$

Fourier image of CF, defined with

$$
J_{\hat{A}\hat{B}}(x-x',\omega) = \frac{1}{2\pi}\int_{-\infty}^{+\infty} \mathrm{d}\tau\,\mathrm{e}^{i\omega\tau}J_{\hat{A}\hat{B}}(x-x',\tau) \tag{7.53}
$$

and the inverse transformation, is

$$J_{\hat{A}\hat{B}}(x - x', \tau) = \int_{-\infty}^{+\infty} \mathrm{d}\omega \, \mathrm{e}^{-i\omega\tau} J_{\hat{A}\hat{B}}(x - x', \omega) \tag{7.54}$$

By comparing the expressions (7.50) and (7.53), the relation between the observed CFs is obtained:

$$J_{\hat{A}\hat{B}}(x - x', \omega) = \mathrm{e}^{E/\theta} J_{\hat{B}\hat{A}}(x - x', \omega); \qquad E = \hbar\omega \tag{7.55}$$

Finally, it should be noted that CF $J_{\hat{A}\hat{B}}(x - x', \omega)$ has the same singularities in the complex E-plane as CF $J_{\hat{B}\hat{A}}(x - x', \omega)$, which means that the energy and lifetime of elementary excitations can be determined by any of these CFs.

Relationship between Green and Correlation Functions

The relationship between these functions that depend on spatial-time variables is given in the definition of Green's function and is explicitly expressed by the relation (7.43). Here, we will examine what kind of connection has the definitive Eq. (7.43) after Fourier transforms time–frequency. After substitution $t - t' = \tau$ in (7.43), the following is obtained:

$$G(x - x', \tau) = \Theta(\tau)[J_{\hat{A}\hat{B}}(x - x', \tau) - J_{\hat{B}\hat{A}}(x - x', \tau)] \tag{7.56}$$

In this form, the relation (7.43) is more suitable for further transformations. Since the Heaviside step-function is an integral of Dirac's delta function, the following can be written:

$$\begin{aligned} \Theta(\tau) &= \int \mathrm{d}\tau \, \delta(\tau) = \frac{1}{2\pi} \int \mathrm{d}\tau \int_{-\infty}^{+\infty} \mathrm{d}\omega'' \mathrm{e}^{-i\omega''\tau} \\ &= \frac{1}{2\pi} \int \mathrm{d}\omega'' \int_{-\infty}^{+\infty} \mathrm{d}\tau \, \mathrm{e}^{-i\omega''\tau} = \frac{i}{2\pi} \int_{-\infty}^{+\infty} \mathrm{d}\omega'' \frac{\mathrm{e}^{-i\omega''\tau}}{\omega'' \pm i\delta} \end{aligned} \tag{7.57}$$

If, apart from this, Fourier transforms of Green's and correlation functions are performed,

$$\begin{aligned} G(x - x', \tau) &= \int_{-\infty}^{+\infty} \mathrm{d}\omega' \mathrm{e}^{-i\omega'\tau} G(x - x', \omega'); \, J_{\hat{A}\hat{B},\hat{B}\hat{A}}(x - x', \tau) \\ &= \int_{-\infty}^{+\infty} \mathrm{d}\omega' \mathrm{e}^{-i\omega'\tau} J_{\hat{A}\hat{B},\hat{B}\hat{A}}(x - x', \omega') \end{aligned} \tag{7.58}$$

so (7.57) and (7.58) are replaced by (7.56), we get

$$\begin{aligned}
&\lim_{\delta\to+\infty}\int_{-\infty}^{+\infty} \mathrm{d}\omega' \mathrm{e}^{-i\omega'\tau} G(x-x',\omega'\pm i\delta) \\
&= \frac{i}{2\pi}\int_{-\infty}^{+\infty} \mathrm{d}\omega'\mathrm{d}\omega'' \mathrm{e}^{-i(\omega'+\omega'')\tau} \frac{J_{\hat{A}\hat{B}}(x-x',\omega') - J_{\hat{B}\hat{A}}(x-x',\omega')}{\omega''\pm i\delta} \\
&= \frac{i}{2\pi}\int_{-\infty}^{+\infty} \mathrm{d}\omega'\mathrm{d}\omega'' \mathrm{e}^{-i(\omega'+\omega'')\tau} \frac{(\mathrm{e}^{\omega'/\theta i}-1) J_{\hat{B}\hat{A}}(x-x',\omega')}{\omega''\pm i\delta}
\end{aligned} \tag{7.59}$$

The final position in (7.59) was obtained on the basis of (7.55). After multiplying (7.59) with $\mathrm{e}^{i\omega\tau}$ and integrating at infinite boundaries by τ, the result is

$$\begin{aligned}
\lim_{\delta\to+\infty} G(x-x',\omega\pm i\delta) &= \frac{i}{2\pi}\int_{-\infty}^{+\infty} \mathrm{d}\omega' \frac{(\mathrm{e}^{-i\hbar\omega'/\theta}-1) J_{\hat{B}\hat{A}}(x-x',\omega')}{\omega'-\omega\pm i\delta} \\
&= \frac{1}{2\pi i}\int_{-\infty}^{+\infty} \mathrm{d}\omega' \frac{(\mathrm{e}^{-i\hbar\omega'/\theta}-1) J_{\hat{B}\hat{A}}(x-x',\omega')}{\omega'-\omega\mp i\delta}
\end{aligned} \tag{7.60}$$

How: $\lim_{\delta\to 0}\frac{1}{\omega'-\omega\mp i\delta} = \mathcal{P}\left\{\frac{1}{\omega'-\omega}\right\} \pm i\pi\delta(\omega'-\omega)$, by its substitution in (7.60), we get the following relations:

$$\begin{aligned}
&\lim_{\delta\to+\infty} G(x-x',\omega'\pm i\delta) \\
&= \frac{1}{2\pi i}\int_{-\infty}^{+\infty} \frac{\mathrm{d}\omega'}{\omega'-\omega\mp i\delta} (\mathrm{e}^{-i\hbar\omega'/\theta}-1) J_{\hat{B}\hat{A}}(x-x',\omega') \\
&= \frac{1}{2\pi i}\int_{-\infty}^{+\infty} \mathrm{d}\omega' \left[\mathcal{P}\left\{\frac{1}{\omega'-\omega}\right\} \pm i\pi\,\delta(\omega'-\omega)\right] \\
&\quad\times (\mathrm{e}^{-i\hbar\omega'/\theta}-1) J_{\hat{B}\hat{A}}(x-x',\omega') \\
&= \frac{1}{2\pi i}\mathcal{P}\left\{\int_{-\infty}^{+\infty} \mathrm{d}\omega' \frac{(\mathrm{e}^{-i\hbar\omega'/\theta}-1) J_{\hat{B}\hat{A}}(x-x',\omega')}{\omega'-\omega}\right\} \\
&\quad\pm \frac{\mathrm{e}^{-i\hbar\omega'/\theta}-1}{2} J_{\hat{B}\hat{A}}(x-x',\omega')
\end{aligned} \tag{7.61}$$

After their subtraction (G−G +), the required connection between Fourier's CF and GF characters is obtained:

$$J_{\hat{B}\hat{A}}(x-x',\omega)=\frac{G(x-x',\omega+i\delta)-G(x-x',\omega-i\delta)}{\mathrm{e}^{-i\hbar\omega/\theta}-1} \tag{7.62}$$

If now (7.62) is replaced by (7.54), a formula that is fundamental for determining the statistical mean values of the products of two operator complexes comes to the fore. Considering (7.54) and replacing $\tau = t - t'$, it follows

$$\begin{aligned} J_{\hat{B}\hat{A}}(x-x',\omega) &\equiv \langle \hat{B}(x',t')\hat{A}(x,t)\rangle \\ &= \int_{-\infty}^{+\infty} \mathrm{d}\omega' \mathrm{e}^{-i\omega(t-t')}\frac{G(x-x',\omega+i\delta)-G(x-x',\omega-i\delta)}{\mathrm{e}^{-i\hbar\omega/\theta}-1} \end{aligned} \tag{7.63}$$

In this term, a "more economical" method of recording statistical mean values was used — instead of the symbol Sp $\{(\ldots)\}$ the symbol $\langle(\ldots)\rangle$ was used. In the following text, this symbolic record will be used mainly.

A correlation function $J_{\hat{A}\hat{B}}$ can also be expressed through Green's functions. For this, it is sufficient to express in (7.59), $J_{\hat{B}\hat{A}}$ and using (7.55) $J_{\hat{A}\hat{B}}$, explicitly: $J_{\hat{B}\hat{A}}(x-x',\omega) = \mathrm{e}^{-\hbar\omega/\theta}J_{\hat{A}\hat{B}}$. The procedure for further calculations is exactly the same as in the previous case, so only the final results will be listed:

$$J_{\hat{B}\hat{A}}(x-x',\omega)=\frac{G(x-x',\omega+i\delta)-G(x-x',\omega-i\delta)}{1-\mathrm{e}^{-i\hbar\omega/\theta}} \tag{7.64}$$

i.e., for an inverse form:

$$\begin{aligned} J_{\hat{A}\hat{B}}(x-x',t-t') &\equiv \langle \hat{A}(x,t)\hat{B}(x',t')\rangle \\ &= \int_{-\infty}^{+\infty} \mathrm{d}\omega\, \mathrm{e}^{-i\omega(t-t')}\frac{G(x-x',\omega+i\delta)-G(x-x',\omega-i\delta)}{1-\mathrm{e}^{-i\hbar\omega/\theta}} \end{aligned} \tag{7.65}$$

The summary of the analysis carried out in this section could be formulated as follows. Knowing the commutative two-way, temperature Green function allows you to find expressions for the statistical mean value of the products of the operator complex from which it is constructed. These mean values are correlation functions and enter the definition of many physical quantities.

A Chain of Equations for Green's Functions

In order to increase the transparency of the chain of equations for Green's function, it is convenient to write (7.39) in the new labels:

$$\langle\langle \hat{A}(x,t)|\hat{B}(x',t')\rangle\rangle = \Theta(t-t')\,\langle[\hat{A}(x,t),\hat{B}(x',t')]\rangle \tag{7.66}$$

where $\ll \ldots | \ldots \gg$ signifies GF, while $< [\ldots, \ldots] >$, as already mentioned, a replacement for the spur of the corresponding sizes. Of the two corner brackets in the expression for Green's function, one denotes statistical aggregation, and the other, in combination with the vertical line, "warns" that the product of the operator complexes is arranged by decreasing times.

If expression (7.66) is differentiated by time, then

$$\begin{aligned} &\frac{d}{dt}\langle\langle \hat{A}(x,t)|\hat{B}(x',t')\rangle\rangle \\ &\quad = \delta(t-t')\langle[\hat{A}(x,t),\hat{B}(x',t')]\rangle + \Theta(t-t')\left\langle\left[\frac{d}{dt}\hat{A}(x,t),\hat{B}(x',t')\right]\right\rangle \end{aligned} \tag{7.67}$$

It should be noticed that, in the first article on the right-hand side of this term, both operator complexes act at the same time, which follows from the definite properties of the δ-function: $\delta(x) = |a|\delta(ax)$, when $a = \text{const}$. Furthermore, based on Heisenberg's equations of motion, it follows

$$\frac{d}{dt}\hat{A}(x,t) = \frac{1}{i\hbar}[\hat{A}(x,t),\hat{B}(x',t')] \tag{7.68}$$

so Eq. (7.67) becomes

$$\begin{aligned} i\hbar\tfrac{d}{dt}\langle\langle \ddot{A}(x,t)|\hat{B}(x',t')\rangle\rangle = {}& i\hbar\delta(t-t')\langle[\hat{A}(x,t),\hat{B}(x',t')]\rangle \\ & + \Theta(t-t')\langle[[\hat{A}(x,t),\hat{H}(t)],\hat{B}(x',t')]\rangle \end{aligned} \tag{7.69}$$

Considering (7.66), it is obvious that the other article on the right side of this term represents a new GF, which is called a higher GF. In practice, the higher GF in itself contains more complex operator complexes than the initial one from which the chain of equations arises. If the term for a higher GF

$$\langle\langle[\hat{A}(x,t),\hat{H}(t)]|\hat{B}(x',t')\rangle\rangle = \Theta(t-t')\langle[[\hat{A}(x,t),\hat{H}(t)],\hat{B}(x',t')]\rangle \tag{7.70}$$

differentiates by time and repeats the described procedure, the following equation is obtained:

$$\begin{aligned} & i\hbar\frac{d}{dt}\langle\langle[\hat{A}(x,t),\hat{H}(t)]|\hat{B}(x',t')\rangle\rangle \\ &\quad = i\hbar\,\delta(t-t')\langle[[\hat{A}(x,t),\hat{H}(t)],\hat{B}(x',t')]\rangle \\ &\quad\quad + \Theta(t-t')\langle[[[\hat{A}(x,t),\hat{H}(t)],\hat{H}(t)],\hat{B}(x',t')]\rangle \end{aligned} \tag{7.71}$$

in which the higher GF is expressed through the following, even more GF

$$\begin{aligned}&\langle\langle[[[\hat{A}(x,t),\hat{H}(t)],\hat{H}(t)],\hat{B}(x',t')]\rangle\rangle\\&\quad=\Theta(t-t')\langle[[[\hat{A}(x,t),\hat{H}(t)],\hat{H}(t)],\hat{B}(x',t')]\rangle\end{aligned}\tag{7.72}$$

and this is expressed by the following — even more GFs, ... In this way, there is an infinite chain of equations that determines the initial GF: $\langle\langle\hat{A}(x,t)|\hat{B}(x',t')\rangle\rangle$.

Depending on the form of the operator complexes, it is sometimes more convenient for the chain of equations to be formed by successive differentiation over time t'. Also, it is sometimes advisable to combine differentiations with respect to t and t' in forming a chain of equations. Regardless of the way in which the chain of equations is closed, the chain of equations closes or, in itself, only intercepts, only in cases where the Hamiltonian system is represented by a square operator form, which can be diagonalized or already diagonalized [5, 131–134]. In all other cases, and such is far more, Hamiltonian system contains in addition to the quadratic an operator higher-order form. Then the chain of equations remains infinite and the initial GF can be determined only approximately by the fact that a chain, on the basis of some sufficiently justified assumption, in some place is cut. Formally, cutting the chain consists in expressing Green's functions through the lower ones. Rules for cutting the chain do not exist, although the idea of a small parameter is usually exploited, if one exists. When analyzing the system of elementary excitations, the concentration of elementary excitations is usually used as a small parameter, because, fortunately, more GF is proportional to higher concentration levels.

In order to further express GF formalism, it will be assumed that the chain of equations for $\langle\langle\hat{A}(x,t)|\hat{B}(x',t')\rangle\rangle$ can be exceeded in the first equation. In view of (7.70), this first equation has the form

$$\begin{aligned}i\hbar\,\langle\langle\hat{A}(x,t)|\hat{B}(x',t')\rangle\rangle&=i\hbar\,\delta(t-t')\langle[\hat{A}(x,t),\hat{B}(x',t')]\rangle\\&\quad+\Theta(t-t')\langle[\tfrac{d}{dt}\,[\hat{A}(x,t),\hat{B}(x',t')]]\rangle\end{aligned}\tag{7.73}$$

This equation is "closed" if, for some physically sufficiently justified reasons, an approximation can be used:

$$\langle\langle[\hat{A}(x,t),\hat{H}(t)]|\hat{B}(x',t')\rangle\rangle=\int_{-\infty}^{+\infty}\mathrm{d}y\,E(x-y)\langle\langle\hat{A}(x,t)|\hat{B}(x',t')\rangle\rangle\tag{7.74}$$

The integral occurs because Hamiltonians always represent sums or, in this case, integrate operator forms by coordinates. In the first section, on the right side of the expression (7.73), operators act at the same time, so this mean value, as shown above, does not depend on time. That's why it can be written as follows:

$$\langle[\hat{A}(x,t),\hat{B}(x',t')]\rangle \equiv Q(x-x') \tag{7.75}$$

where the size: $Q(x-x')$ is called Green's correlator. After the substitutions (7.74) and (7.75) in (7.73), this equation becomes

$$\begin{aligned} i\hbar\langle\langle\hat{A}(x,t)|\hat{B}(x',t')\rangle\rangle &= i\hbar\,\delta(t-t')Q(x-x') \\ &\quad + \int_{-\infty}^{+\infty} \mathrm{d}y\, E(x-y)\langle\langle\hat{A}(x,t)|\hat{B}(x',t')\rangle\rangle \end{aligned} \tag{7.76}$$

and, after Fourier's transforms of time–frequency and space–wave vector,

$$\begin{aligned} \langle\langle\hat{A}(x,t)|\hat{B}(x',t')\rangle\rangle &= \frac{1}{(2\pi)^N}\int_{-\infty}^{+\infty} \mathrm{d}k\,\mathrm{d}\omega\,\mathrm{e}^{ik(x-x')-i\omega(t-t')}\langle\langle\hat{A}|\hat{B}\rangle\rangle_{k,\omega}; \\ Q(x-x') &= \frac{1}{(2\pi)^N}\int_{-\infty}^{+\infty} \mathrm{d}k\,\mathrm{e}^{ik(x-x')}Q(k); \\ E(x-y) &= \frac{1}{(2\pi)^N}\int_{-\infty}^{+\infty} \mathrm{d}k'\,\mathrm{e}^{ik'(x-y)}E(k') \end{aligned} \tag{7.77}$$

it moves to

$$\begin{aligned} &\int_{-\infty}^{+\infty} \mathrm{d}k\,\mathrm{d}\omega\,\mathrm{e}^{ik(x-x')-i\omega(t-t')}\hbar\omega\langle\langle\hat{A}|\hat{B}\rangle\rangle_{k,\omega} \\ &= \frac{i\hbar}{2\pi}\int_{-\infty}^{+\infty} \mathrm{d}k\,\mathrm{d}\omega\,\mathrm{e}^{ik(x-x')-i\omega(t-t')}Q(k) \\ &\quad + \frac{1}{(2\pi)^N}\int_{-\infty}^{+\infty} \mathrm{d}k'\,\mathrm{d}k\,\mathrm{d}\omega\,\mathrm{e}^{ik'x-ikx'-i\omega(t-t')}E(k')\langle\langle\hat{A}|\hat{B}\rangle\rangle_{k,\omega} \\ &\quad \times \int_{-\infty}^{+\infty} \mathrm{d}y\,\mathrm{e}^{iy(k-k')} \\ &= \frac{i\hbar}{2\pi}\int_{-\infty}^{+\infty} \mathrm{d}k\,\mathrm{d}\omega\,\mathrm{e}^{ik(x-x')-i\omega(t-t')}Q(k) \\ &\quad + \int_{-\infty}^{+\infty} \mathrm{d}k\,\mathrm{d}\omega\,\mathrm{e}^{ik(x-x')-i\omega(t-t')}E(k)\langle\langle\hat{A}|\hat{B}\rangle\rangle_{k,\omega} \end{aligned} \tag{7.78}$$

Based on the previously defined Fourier's transforms, the following expression for the Fourier's GF is given by

$$G(k,\omega \pm i\delta) \equiv \langle\langle \hat{A}|\hat{B}\rangle\rangle_{k,\omega\pm i\delta} = \lim_{\delta\to+0} \frac{i}{2\pi}\frac{Q(k)}{\omega - \Omega_k \pm i\delta} \tag{7.79}$$

where it is marked $\Omega_k = E(k)/\hbar$.

Before proceeding to further analysis, it should be noted that, by cutting the chain of Eq. (7.74), an integral equation (7.76) was obtained for determining GF and that, only after Fourier's transforms, this integral equation is reduced to an algebraic equation by unknown Fourier's transform of $\langle\langle \hat{A}|\hat{B}\rangle\rangle_{k,\omega}$. Regardless of the fact that the method outlined here is presented schematically, the resolution procedure remains the same in practical problems: in order to obtain a "usable" result of a budget, it must go over to the Fourier images in the above transformations.

The expression for Green's function (7.61) is singular and has poles in the ω-plane: $\omega_\pm = \Omega_k \mp i\delta$. The real part, on the basis of the previously presented, represents the frequency of elemental excitations Ω_k, and $\hbar\Omega_k = E(k)$ the energy of these excitations. As already emphasized, the pole should lie in the fourth quadrant of the complex plane, the expression (7.79) is used with the $+\delta$ sign in the denominator.

By finding the energy of elementary excitations by the GF method, the first, and not small, problem is solved in the analysis of particle or quasi-systemic systems. Another, but not less important problem — determining the statistical mean values of the products of the operator complexes — can also be solved based on the expression (7.79) for GF. Namely, based on this equation, we can write

$$G(k,\omega \pm i\delta) = \frac{i}{2\pi}\mathcal{P}\left\{\frac{Q(k)}{\omega - \Omega_k}\right\} \pm \frac{1}{2}Q(k)\delta(\omega - \Omega_k)$$

from which follows

$$\lim_{\delta\to+0}[G(k,\omega + i\delta) - G(k,\omega - i\delta)] = Q(k)\,\delta(\omega - \Omega_k) \tag{7.80}$$

If in relation

$$G(x - x', t - t') = \frac{1}{(2\pi)^N}\int_{-\infty}^{+\infty} \mathrm{d}k\,\mathrm{d}\omega\, \mathrm{e}^{ik(x-x') - i\omega(t-t')} \tag{7.81}$$

performs inverse Fourier's transforms of frequency–time (by multiplying with $\mathrm{e}^{i\omega(t-t')}$ and integrating by $t-t'$), it gets

$$\begin{aligned} G(x-x',\omega) &\equiv \frac{1}{2\pi}\int_{-\infty}^{+\infty} d(t-t')\,\mathrm{e}^{i\omega(t-t')}\,G(x-x',t-t') \\ &= \frac{1}{(2\pi)^N}\int_{-\infty}^{+\infty} \mathrm{d}k\,\mathrm{e}^{ik(x-x')-i\omega(t-t')}\,G(k,\omega) \end{aligned} \tag{7.82}$$

In the expression (7.63) for the CF $J_{\hat{B}\hat{A}}(x-x',t-t')$, GF $G(x-x',\omega)$ is expressed through (7.82). If this is included in (7.63) and takes into account (7.80), the expression for CF is concretized in the sense that there are already sizes found: the correlator $Q(k)$ and the energy of elementary excitations: $E(k)=\hbar\Omega_k$. So,

$$\begin{aligned} &J_{\hat{B}\hat{A}}(x-x',\omega) \equiv \left\langle \hat{B}(x',t')\hat{A}(x,t)\right\rangle \\ &= \frac{1}{(2\pi)^N}\int_{-\infty}^{+\infty} \mathrm{d}k\,d\omega\,\mathrm{e}^{ik(x-x')-i\omega(t-t')}\,\frac{G(k,\omega+i\delta)-G(k,\omega-i\delta)}{\mathrm{e}^{-i\hbar\omega/\theta}-1} \\ &= \frac{1}{(2\pi)^N}\int_{-\infty}^{+\infty} \mathrm{d}k\,\mathrm{e}^{ik(x-x')-i\Omega_k(t-t')}\,\frac{Q(k)\delta(\omega-\Omega_k)}{\mathrm{e}^{-i\hbar\Omega_k/\theta}-1} \end{aligned} \tag{7.83}$$

In the same way, formula (7.65) becomes

$$\begin{aligned} J_{\hat{A}\hat{B}}(x-x',t-t') &\equiv \langle \hat{A}(x',t')\hat{B}(x,t)\rangle \\ &= \frac{1}{(2\pi)^N}\int_{-\infty}^{+\infty} \mathrm{d}k\,\mathrm{e}^{ik(x-x')-i\Omega_k(t-t')}\,\frac{Q(k)}{1-\mathrm{e}^{-i\hbar\Omega_k/\theta}} \end{aligned} \tag{7.84}$$

As it can be seen, the problem of determining statistical mean values is solved. If the operator complexes $\hat{A}(x,t)$ and $\hat{B}(x',t')$ act at the same time of time $t'=t$ and at the same point of space $x'=x$, the corresponding mean values, based on (7.83) and (7.84), are, respectively,

$$\begin{aligned} \langle \hat{B}(x,t)\hat{A}(x',t')\rangle &= \frac{1}{(2\pi)^N}\int_{-\infty}^{+\infty} \mathrm{d}k\frac{Q(k)}{\mathrm{e}^{-i\hbar\Omega_k/\theta}-1} \\ \langle \hat{A}(x',t')\hat{B}(x,t)\rangle &= \frac{1}{(2\pi)^N}\int_{-\infty}^{+\infty} \mathrm{d}k\frac{Q(k)}{1-\mathrm{e}^{-i\hbar\Omega_k/\theta}} \end{aligned} \tag{7.85}$$

It seems that these mean values do not depend on spatial or temporal coordinates, as has been emphasized earlier. Derived formulas (7.79), (7.83), and (7.84) represent a schematic demonstration of what was said at the beginning of the presentation.

The Q2TT GF method represents a completely closed formalism in which the problem of particle and quasi-particle systems is completely solved, i.e., there are also energies of elemental excitations and a statistical mean value of the corresponding physical quantities. This makes it one of the most fixed instruments of theoretical analysis of phenomena in condensed environments.

Anticommutator and Causal Green's Functions

At the end of this presentation, one should look at the aforementioned anticommutator and causal GFs. Anticommutator GFs are defined [5, 131–134] as

$$G_A(x - x'; t - t') = \Theta(t - t')\mathrm{Sp}\left\{\{\hat{A}(x,t), \hat{B}(x',t')\}\,\hat{\rho}\right\} \tag{7.86}$$

where the anticommutator is $\{\hat{A}(x,t), \hat{B}(x',t')\} \equiv \hat{A}(x,t)\,\hat{B}(x',t') + \hat{B}(x',t')\hat{A}(x,t)$, while the definition of a causal GF is

$$G_C(x - x'; t - t') = \Theta(t - t')\mathrm{Sp}\{\hat{A}(x,t)\hat{B}(x',t')\hat{\rho}\} \tag{7.87}$$

Based on the given definitions, it is clear that the chain of equations for the G_A and G_C functions is formed in the same way as for the commutator GF. For this reason, it can be expected that, at the same degree of budget accuracy, all three functions give the same result for the energy of elementary incentives. Correlators of the G_A and G_C functions, based on their definitions, are different, which means that the mean values are determined by different expressions for all three types of GFs. Regardless of the formal variation of the terms in which they are calculated, the mean values must be equal, of course — at the same degree of accuracy of the applied account. The type of function is chosen according to the operators from which it is constructed. For Bose operators, the mean-value calculation is faster with a commutator, and for Fermi operators it is faster with an anticommuting. In any case, the choice of GF is only a matter of computational commodity and cannot in any way affect the final results.

7.3.2. *Applying the GF method — Examples of bulk cases*

In this part, typical examples of the application of the GF method in the theory of condensed state of matter will be shown.

Magnetization at Low Temperatures

It will be shown here that, using the GF method to investigate the behavior of ferromagnetic with spin S = 1/2, it is very easy to obtain the well-known Dyson magnetization result [5, 135–137], i.e., Pauli operators, who satisfy the commutation relations:

$$\begin{aligned} &[P_{\vec{n}}, P^+_{\vec{m}}] = (1 - 2P^+_{\vec{n}} P_{\vec{n}},)\delta_{\vec{n}\vec{m}}; \quad &\langle P^+_{\vec{n}} P_{\vec{n}}\rangle \simeq 0 \text{ or } 1; \\ &[P_{\vec{n}}, P_{\vec{m}}] = [P^+_{\vec{n}}, P^+_{\vec{m}}] = 0; \quad &P^2_{\vec{n}} = P^{+2}_{\vec{n}} = 0 \end{aligned} \tag{7.88}$$

the performances are exemplary via the Bose operator, according to the formula [89, 90]:

$$\begin{gathered} P = \left[\sum_{\nu=0}^{\infty} \frac{(-2)^\nu}{(1+\nu)!} B^{+\nu} B^\nu\right]^{1/2} B; \quad P^+ = B^+ \left[\sum_{\nu=0}^{\infty} \frac{(-2)^\nu}{(1+\nu)!} B^{+\nu} B^\nu\right]^{1/2}; \\ P^+ P = \left[\sum_{\nu=0}^{\infty} \frac{(-2)^\nu}{(1+\nu)!} B^{+(\nu+1)} B^{\nu+1}\right]^{1/2} \end{gathered} \tag{7.89}$$

In the 50s of the last century, and before that, the mechanism of low temperature magnetization was intensively worked out. The problem itself is more methodological than of practical significance, and has spawned a series of discussions on which theoretical approach gives the correct expression for low-temperature magnetization. In 1956, Dyson found the correct expression for magnetization, but with very complex calculations of spin statistical statistics. He proposed the representation (Pauli's expression, through the Bose operator) in the following form:

$$P_{\vec{n}} = B_{\vec{n}} - B^+_{\vec{n}} B_{\vec{n}} B_{\vec{n}}; \quad P^+_{\vec{n}} = B^+_{\vec{n}}; \quad P^+_{\vec{n}} P_{\vec{n}} = B^+_{\vec{n}} B_{\vec{n}} \tag{7.90}$$

because it faithfully reproduces the results of its statistical calculations, not because it represents a good replacement for Pauli operators.

In order to obtain the Dyson result in a shorter way, Tyablikov used the Q2TT GF method for the ferromagnetic model with spin $S = 1/2$[1–5, 134]. This method was tempting, because —besides the energies of elementary excitations — he gave terms for the mean value of physical characteristics. One of these mean values is magnetization itself. With this approach Tyablikov failed to reproduce Dyson's result — in the expression for magnetization, a member appeared, commensurate with the third degree of temperature, which he entered in the literature called "mist in the slurry." In his works, Dyson proved that such a member cannot appear in the expression for magnetization. Here, we show that the GF method gives the correct result for magnetization, but only when the paulion GF is correctly replaced by the boson.

In the case of ferromagnetic with spin $S = 1/2$, spin operators can be represented by the Pauli operator: $S^+ = P$, $S^- = P^+$ and $S_z = 1/2 - P^+P$, and the Hamiltonian of this ferromagnetic has the form

$$H = \frac{J_0}{2}\sum_{\vec{n}} P^+_{\vec{n}} P_{\vec{n}} - \frac{1}{2}\sum_{\vec{n},\vec{m}} I_{\vec{n}\vec{m}} P^+_{\vec{n}} P_{\vec{m}} - \frac{1}{2}\sum_{\vec{n},\vec{m}} I_{\vec{n}\vec{m}} P^+_{\vec{n}} P_{\vec{n}} P^+_{\vec{m}} P_{\vec{m}};$$
$$J_0 = \sum_{\vec{m}} I_{0\vec{m}},; \quad I_{\vec{n}\vec{m}} = I_{\vec{m}\vec{n}}; \quad I_{\vec{n}\vec{n}} = 0; \quad I_{\vec{n}\vec{m}} \equiv I_{\vec{n}-\vec{m}} \tag{7.91}$$

Tyablikov [1–5, 134] analyzed the ferromagnetic system with such a Hamiltonian using the paulian GFs:

$$\Gamma_{\vec{a}\vec{b}}(t) = \Theta(t)\langle[P_{\vec{a}}(t), P_{\vec{b}}(0)]\rangle \equiv \langle\langle P_{\vec{a}}(t) \mid P_{\vec{b}}(0)\rangle\rangle \tag{7.92}$$

where $\vec{n}$ and $\vec{m}$ are node vectors of a crystal lattice, and $\Theta(t)$ — Heaviside step function. If this term differs in time and uses Heisenberg's equations of motion, it comes to

$$\begin{aligned} i\hbar\frac{d}{dt}\Gamma_{\vec{a}\vec{b}}(t) &= i\hbar\delta(t)\,\delta_{\vec{a}\vec{b}}(1 - 2\langle P^+_{\vec{n}} P_{\vec{n}}\rangle) - \frac{J_0}{2}\Gamma_{\vec{a}\vec{b}}(t) - \frac{1}{2}\sum_{\vec{m}} I_{\vec{n}\vec{m}}\Gamma_{\vec{a}\vec{b}}(t) \\ &\quad + \sum_{\vec{m}} I_{\vec{n}\vec{m}}\,\langle\langle P^+_{\vec{a}}(t) P_{\vec{a}}(t) P_{\vec{m}}(t) \mid P^+_{\vec{m}}(t)\rangle\rangle \\ &\quad - \sum_{\vec{m}} I_{\vec{n}\vec{m}}\left\langle\left\langle P^+_{\vec{m}}(t) P_{\vec{m}}(t) P_{\vec{a}}(t) \mid P^+_{\vec{b}}(0)\right\rangle\right\rangle \end{aligned} \tag{7.93}$$

Several paulian GFs are decoupled using Wick's theorem for the Bose operators (for Pauli operators, it is valid only approximately) [5, 131–133]:

$$\left\langle\left\langle P^+_{\vec{a}}(t) P_{\vec{b}}(t) P_{\vec{c}}(t) | P^+_{\vec{d}}(0)\right\rangle\right\rangle \approx \langle P^+_{\vec{a}}(t) P_{\vec{b}}(t)\rangle \Gamma_{\vec{c}\vec{d}}(t) + \langle P^+_{\vec{a}}(t) P_{\vec{c}}(t)\rangle \Gamma_{\vec{b}\vec{d}}(t) \tag{7.94}$$

After this decoupling and Fourier's-type transformation,

$$F_{\vec{a}-\vec{b}}(t) = \frac{1}{N}\sum_{\vec{k}} \int_{-\infty}^{+\infty} d\omega\, e^{i\vec{k}(\vec{a}-\vec{b}) - i\omega t}\, F_{\vec{k}}(\omega);$$
$$\langle P^+_{\vec{a}} P_{\vec{b}}\rangle = \frac{1}{N}\sum_{\vec{k}} \langle P^+_{\vec{k}} P_{\vec{k}}\rangle\, e^{i\vec{k}(\vec{a}-\vec{b})} \tag{7.95}$$

the following expression is obtained for the Fourier-image of paulion GF $\langle P^+_{\vec{a}} P_{\vec{b}} \rangle$:

$$\Gamma_{\vec{k}}(\omega) = \frac{i}{2\pi} \frac{1 - 2\langle P^+_{\vec{a}} P_{\vec{b}} \rangle}{\omega - \omega^{(1)}_{\vec{k}}};$$

$$\hbar\omega^{(1)}_{\vec{k}} \equiv E^{(1)}_{\vec{k}} = \frac{J_0 - J_{\vec{k}}}{2} + \frac{1}{N} \sum_{\vec{q}} (J_{\vec{k}} + J_{\vec{q}} - J_0 - J_{\vec{k}-\vec{q}}) \langle P^+_{\vec{q}} P_{\vec{q}} \rangle;$$

$$J_{\vec{k}} = \sum_{\vec{m}} I_{0\vec{m}}\, \mathrm{e}^{\mathrm{i}\vec{k}\vec{m}}; \qquad \hbar\omega \equiv E \tag{7.96}$$

On the basis of the theorem on the spectral intensity of the GF, from (3.60), it follows that

$$\langle P^+_{\vec{k}} P_{\vec{k}} \rangle = \frac{\sigma}{\mathrm{e}^{E^{(1)}_{\vec{\mathrm{k}}}/\theta} - 1} \tag{7.97}$$

where $\theta = k_B T$, and $\sigma = 1 - 2\langle P^+ P \rangle$ is the relative magnetization, which should be calculated. Since $\langle P^+ P \rangle = N^{-1} \sum_{\vec{k}} \langle P^+_{\vec{k}} P_{\vec{k}} \rangle$, based on (3.58), the magnetization σ can be determined. If the approximation of the nearest neighbors and the approximation of small wave vectors are used (the size $J_{\vec{k}}$ is broken up by degrees $\vec{k}$, up to the sixth degree — concluding), the following expression is obtained for magnetization:

$$\sigma = 1 - 2\zeta_{3/2}\tau^{3/2} - \frac{3\pi}{2}\zeta_{5/2}\tau^{5/2} - \frac{33\pi^2}{16}\zeta_{7/2}\tau^{7/2} + 4\zeta^2_{3/2}\tau^3 + O(\tau^{9/2}) \tag{7.98}$$

where $\tau = \theta/(2\pi I)$, and I is the integral of the change for the nearest neighbors, while ζ represents the Riemann's zeta-function: $\zeta_p = \sum_{n=1}^{\infty} n^{-p}$.

The term for magnetization given here was given by Tyablikov, which was described by the procedure [5, 131–133]. A member of proportionality τ^3 is a "spur error" that has already been spoken. Formally, this article appears because the size $\langle P^+ P \rangle$ is calculated by definition:

$$\langle P^+ P \rangle = \frac{\sigma}{N} \sum_{\vec{\mathrm{k}}} \left[\mathrm{e}^{E^{(1)}_{\vec{\mathrm{k}}}/\theta} - 1 \right]^{-1} \tag{7.99}$$

The correct magnetization result is obtained if, in (3.56), the Pauli operator is replaced by the Bose operators — according to the formula

$$P_{\vec{n}} = B_{\vec{n}} - B^+_{\vec{n}} B_{\vec{n}} B_{\vec{n}}; \qquad P^+_{\vec{n}} = B^+_{\vec{n}} - B^+_{\vec{n}} B^+_{\vec{n}} B_{\vec{n}} \tag{7.100}$$

and from paulion switches to the boson GF using Wick's theorem for Bose operators. Thus, the following relations are obtained:

$$\Gamma_{\vec{a}\vec{b}}(t) \approx (1 - 4\langle B^+ B\rangle) G_{\vec{a}\vec{b}}(t); \qquad \left\langle\left\langle P^+_{\vec{a}}(t) P_{\vec{b}}(t) P_{\vec{c}}(t) \mid P^+_{\vec{d}}(0)\right\rangle\right\rangle$$
$$\approx \left\langle\left\langle B^+_{\vec{a}}(t) B_{\vec{b}}(t) B_{\vec{c}}(t) | B^+_{\vec{d}}(0)\right\rangle\right\rangle = \langle B^+_{\vec{a}} B_{\vec{b}}\rangle G_{\vec{c}\vec{d}}(t) + \langle B^+_{\vec{a}} B_{\vec{c}}\rangle G_{\vec{b}\vec{d}}(t) \tag{7.101}$$

If it is still taken $\langle P^+ P = \langle B^+ B\rangle$ (which is valid in the zero approximation), after substitution (3.65) in (3.55) and the corresponding Fourier transforms, we obtain

$$G_{\vec{k}}(\omega) = \frac{i}{2\pi} \frac{1 + 2\langle B^+_{\vec{a}} B_{\vec{b}}\rangle}{\omega - \omega^{(1)}_{\vec{k}}} \tag{7.102}$$

wherefrom

$$\langle B^+ B\rangle = (1 + 2\langle B^+_{\vec{a}} B_{\vec{b}}\rangle) \frac{1}{N} \sum_{\vec{k}} \left(e^{E^{(1)}_{\vec{k}}/\theta} - 1\right)^{-1} \tag{7.103}$$

The quantity $\langle P^+ P\rangle$ can be calculated in the following approximation:

$$\langle P^+ P\rangle = \langle B^+ B\rangle - \langle B^{+2} B^2\rangle \approx \langle B^+ B\rangle - 2\langle B^+ B\rangle^2. \tag{7.104}$$

It follows

$$\langle P^+ P\rangle = \frac{1}{N} \sum_{\vec{k}} \frac{1}{e^{E^{(1)}_{\vec{k}}/\theta} - 1} + 2\langle B^+ B\rangle \frac{1}{N} \sum_{\vec{k}} \frac{1}{e^{E^{(1)}_{\vec{k}}/\theta} - 1} - 2\langle B^+ B\rangle^2$$
$$= \frac{1}{N} \sum_{\vec{k}} \frac{1}{e^{E^{(1)}_{\vec{k}}/\theta} - 1} + O(\tau^{9/2}). \tag{7.105}$$

As can be seen, the term $\langle P^+ P\rangle$ obtained for the boson representation differs from the same expression (3.63) obtained in the paulion representation. The magnetization whose mean value $\langle P^+ P\rangle$ is given by (3.69) has the form

$$\sigma = 1 - 2\zeta_{3/2}\tau^{3/2} - \frac{3\pi}{2}\zeta_{5/2}\tau^{5/2} - \frac{33\pi^2}{16}\zeta_{7/2}\tau^{7/2} - 6\pi\zeta_{3/2}\zeta_{5/2}\tau^4 + O(\tau^{9/2}) \tag{7.106}$$

This is the correct expression for the magnetization that Dyson received in 1956, but with far more complicated procedure.

Magnetization with Accuracy up to $(k_B T/I)^{5/2}$

We start with the general magnetization expression:

$$\sigma = 1 - \frac{2}{N}\sum_{\vec{k}} \frac{1}{\mathrm{e}^{E^{(1)}_{\vec{k}}/\theta} - 1} \tag{7.107}$$

and from the sum we turn to integral as a rule:

$$\frac{1}{N}\sum_{\vec{k}} \to \frac{1}{N}\frac{V}{(2\pi)^3}\int d^3\vec{k} = \frac{a^3}{(2\pi)^3}\int d^3\vec{k}; \qquad V = Na^3$$

where a^3 is the volume of the elemental cell. For the ferromagnet of the free cubic structure and the spin $S = 1/2$ energy of spin waves with accuracy to the fourth degree of the wave vector and in the approximation of the nearest neighbors, it is calculated in the following way:

$$\begin{aligned} E_{\vec{k}} &= \frac{J_0 - J_{\vec{k}}}{2} - \frac{1}{2}[6I - (\cos ak_x + \cos ak_y + \cos ak_y)] \\ &\approx \left[6I - 6I + Ia^2k^2 - \frac{1}{12}Ia^4k^4(\cos^4\varphi\sin^4\theta + \sin^4\varphi\sin^4\theta + \cos^4\theta)\right] \\ &= \frac{1}{2}Ia^2k^2 - \frac{1}{24}Ia^4k^4\Lambda(\varphi,\theta); \end{aligned}$$

$$\Lambda(\varphi,\theta) = \cos^4\varphi\sin^4\theta + \sin^4\varphi\sin^4\theta + \cos^4\theta \tag{7.108}$$

By substituting this in (3.71), after replacing $ak = x$, the magnetization is obtained:

$$\sigma = 1 - \frac{2}{(2\pi)^3}\int_0^{2\pi} d\varphi \int_0^{\pi} d\theta\,\sin\theta \int_0^{k_{\max}} \frac{dx\,x^2}{\mathrm{e}^{I\,x^2[1 - x^2\Lambda(\varphi,\theta)/12]/(2\theta)} - 1}. \tag{7.109}$$

Based on the developed expression $\frac{1}{\mathrm{e}^{\alpha}-1} = \frac{\mathrm{e}^{-\alpha}}{1-\mathrm{e}^{-\alpha}} = \sum_{m=0}^{\infty}\mathrm{e}^{-(m+1)\,\alpha} = \sum_{n=1}^{\infty}\mathrm{e}^{-n\,\alpha}$, the previous equation can be written as

$$\sigma = 1 - \frac{2}{(2\pi)^3}\sum_{n=1}^{\infty}\int_0^{2\pi} d\varphi \int_0^{\pi} d\theta\,\sin\theta \int_0^{k_{\max}} dx\,x^2\mathrm{e}^{n\,Ix^2[1 - x^2\Lambda(\varphi,\theta)/12]/(2\theta)} \tag{7.110}$$

Now, the change is introduced, $nIx^2/\theta = y$, well, since it is a low temperature, the upper limit, $nI(ak_{\max})^2/\theta$, can be roughly taken to be equal to infinity. After this, the magnetization formula becomes

$$\sigma = 1 - \frac{1}{(2\pi)^3}\left(\frac{2\theta}{I}\right)^{3/2}\sum_{n=1}^{\infty} n^{-3/2}\int_0^{2\pi} d\varphi \int_0^{\pi} d\theta \sin\theta$$

$$\times \int_0^{\infty} dy\, y^{1/2} e^{-y[1-\Lambda\theta y/(6nI)]}. \tag{7.111}$$

Next, an approximate development is carried out:

$$e^{-y\left(1-y\Lambda\frac{\theta}{6nI}\right)} \approx e^{-y}\left(1+\frac{\theta}{6nI}y^2\Lambda\right)$$

when the term for magnetization finally becomes

$$\sigma = 1 - \frac{2^{1/2}}{\pi^2}\left(\frac{\theta}{I}\right)^{3/2}\sum_{n=1}^{\infty} n^{-3/2}\left(\int_0^{\infty} dy\, y^{1/2}e^{-y} + \frac{\theta}{10\,I}n^{-1}\int_0^{\infty} dy\, y^{5/2}e^{-y}\right)$$

$$= 1 - \frac{2}{(2\pi)^{3/2}}\left(\frac{\theta}{I}\right)^{3/2}\sum_{n=1}^{\infty} n^{-3/2}\left(1+\frac{3}{8}\frac{\theta}{I}n^{-1}\right) \tag{7.112}$$

$$= 1 - 2\,\varsigma_{3/2}\tau^{3/2} - \frac{3\pi}{2}\varsigma_{5/2}\tau^{5/2}.$$

Kinematic Excitation in the Optical Excitation System

Optical excitations arise in crystals under the influence of incident light [138–143]. A photon switches an electron from the base to a triggered state and thus creates a pair of electron cavities. Such pairs in semiconductors usually have a large radius (the cavity is in the valence band, and the electron in the conductive), while in molecular crystals they have a small radius, because they remain on the same molecule. Optical excitations in semiconductors are called the Wannier–Mott excitons, while the excitations in molecular crystals are Frenkel's excitons.

The Frenkel's exciton system will be analyzed here. In order for the calculations to be relaxed, the case where the quanta of incident light transpose an electron into only one type of excited state (the so-called double-circuit scheme) will be considered. Then the operators of creation and annihilation of the exciton — Pauli operators — satisfy the already advanced date relational relation (3.52):

$$\left[P_{\vec{n}},\, P_{\vec{m}}{}^{+}\right] = \left(1 - 2P_{\vec{n}}^{+} P_{\vec{n}},\right) \delta_{\vec{n}\vec{m}}; \qquad \langle P_{\vec{n}}{}^{+} P_{\vec{n}} \rangle \simeq 0 \,\text{or}\, 1;$$

$$[P_{\vec{n}}, P_{\vec{m}}] = [P_{\vec{n}}{}^{+}, P_{\vec{m}}{}^{+}] = 0; \qquad P_{\vec{n}}{}^{2} = P_{\vec{n}}{}^{+2} = 0$$

The emergence of excitons as collective crystalline excitations, essentially and physically, is represented as follows: the quantum of light initiates one molecule and this excitation is "expanded" in the form of a wave through the crystal due to the interaction between the molecules. Intermolecular interactions are dipole–dipole type and in order of magnitude are about 0.01 eV. The excitation molecule's excitation energy is substantially higher and amounts to several eV, depending on the energy that the molecule's quantum of light that stimulates the molecule.

Concentrations of exits produced by classical light sources were of the order 10^{-10}–10^{-7}. After the appearance of the laser, the concentrations of optical stimulation reached values up to 10^{-3}, and then the interaction between these excitations could not be ignored. One type of interaction is a dynamic type and represents the scattering of optical motions on the intermolecular potential. The second type of interaction is of a kinematic type and is due to the difference in the commutation rules between the Pauli and the Bose operators. It is that, in the case of low concentrations, Pauli operators were simply replaced by the Bose operators, and this approximation was completely satisfactory. On it is "built" the entire harmonic theory of optical excitations [5, 143–145]. At higher concentrations, it was necessary to take into account the fact that, in kinematic sense, Pauli operators are not Bose operators.

Pauli operators have two behaviors [89, 90, 143–145]: both as Fermi, but also as a Bose operators. When the subject of the investigation is individual (quasi) particles, then the fermion properties are used, but when investigating the collective behavior of elementary excitations in some structure, the consequences of which are certain macroscopic characteristics of this structure, then the boson properties come to expression. In this part, a boson representation will be analyzed.

Pauli operators are represented by the help of the Bose operator in different ways. There is Holstein–Primakoff, as well as the Dyson–Maleev boson representation of the Pauli operators [5, 89, 90, 138–140]. Both these representations are approximate and therefore the exact boson representation, developed by Agranovich and Toshich [90], which is also already mentioned in (3.53), will be used here:

$$P = \left[\sum_{\nu=0}^{\infty} \frac{(-2)^{\nu}}{(1+\nu)!} B^{+\nu} B^{\nu}\right]^{1/2} B; \quad P^{+} = B^{+} \left[\sum_{\nu=0}^{\infty} \frac{(-2)^{\nu}}{(1+\nu)!} B^{+\nu} B^{\nu}\right]^{1/2};$$

$$P^{+} P = \left[\sum_{\nu=0}^{\infty} \frac{(-2)^{\nu}}{(1+\nu)!} B^{+(\nu+1)} B^{\nu+1}\right]^{1/2}$$

As can be seen, Pauli operators are here presented to infinite boson summations, which mean that processes between Pauli excitations can be described by multi-boson processes. Here, a case of the fusion of two exits into one and its corresponding decay will be considered. In the second part of the process, there are new excitations in relation to the original "normal" exits. These initiatives are short lived in the literature [144–150] (and in our further text), which will be called kinematical excitations.

The Hamiltonian of such an exciton system has the following form:

$$H = \Delta \sum_{\vec{n}} P_{\vec{n}}^{+} P_{\vec{n}} + \sum_{\vec{n},\vec{m}} T_{\vec{n}\vec{m}} P_{\vec{n}}^{+} P_{\vec{m}} + \sum_{\vec{n},\vec{m}} D_{\vec{n}\vec{m}} P_{\vec{n}}^{+} P_{\vec{n}} P_{\vec{m}}^{+} P_{\vec{m}} \tag{7.113}$$

where Δ is the excitation molecule energy of the isolated molecule and is 3–5 eV, $T_{\vec{n}\vec{m}}$ is characterized by the transfer of the exit from the node $\vec{n}$ to the node $\vec{m}$ of the crystal lattice, and $D_{\vec{n}\vec{m}}$ describes their scattering. Both of these sizes are of the order of 0.1 eV and less.

The system will be analyzed via a commutative two-time Paulian Green function, defined by following expression (3.56):

$$\Gamma_{\vec{a}\vec{b}}(t) = \Theta(t)\, \langle [P_{\vec{a}}(t),\, P_{\vec{b}}(0)] \rangle \equiv \langle\langle P_{\vec{a}}(t) | P_{\vec{b}}(0) \rangle\rangle$$

By differentiating this equation by time using the Heisenberg equation of motion, we arrive at the following relation:

$$\begin{aligned} i\hbar \frac{d}{dt} \langle\langle P_{\vec{a}}(t) | P_{\vec{b}}(0) \rangle\rangle &= i\hbar\delta(t) \langle [P_{\vec{a}}(0), P_{\vec{b}}(0)] \rangle + \Delta \langle [P_{\vec{a}}(t), P_{\vec{b}}(0)] \rangle \\ &+ \sum_{\vec{m}} T_{\vec{a}\vec{m}} \langle\langle P_{\vec{a}}(t) | P_{\vec{b}}(0) \rangle\rangle - 2 \sum_{\vec{m}} T_{\vec{a}\vec{m}} \langle\langle P_{\vec{a}}^{+}(t) P_{\vec{a}}(t) P_{\vec{m}}(t) | P_{\vec{b}}^{+}(0) \rangle\rangle \\ &+ 2 \sum_{\vec{m}} D_{\vec{n}\vec{m}} \langle\langle P_{\vec{m}}^{+}(t) P_{\vec{m}}(t) P_{\vec{a}}(t) | P_{\vec{b}}^{+}(0) \rangle\rangle \end{aligned} \tag{7.114}$$

Now the paulion GF is replaced by the boson GF, where the exact boson representation (3.53) is used in the following approximation:

$$P = B - B^{+}BB; \qquad P^{+} = B^{+} - B^{+}B^{+}B$$

same as in (3.64). After replacing this in all paulion GF forms (3.77), the mean values of the Bose operator are kept up to six operators — finishing. The middle products of the Bose operator are broken up into the sums of the mean products of the pairs of operators by Wick's theorem [5, 143–145]. Subsequently, in Eq. (3.78), there appear the higher-order boson GFs containing products of four and six Bose operators. These more boson GFs are "decoupled" also by using Wick's theorem for the Bose operators, whereby the Bose operators act at the same time, and also the Bose operators acting at different time moments. In addition to the retarded commutative boson GFs, the described decompression process in the budget also includes pseudo-causal boson GFs of the following forms:

$$\begin{aligned} g^{pr}_{\vec{a}\vec{b}}(t) &= \Theta(t)\langle[B_{\vec{a}}(t), B^{+}_{\vec{b}}(0)]\rangle; & g^{pa}_{\vec{a}\vec{b}}(t) &= \Theta(t)\langle[B^{+}_{\vec{a}}(t), B_{\vec{b}}(0)]\rangle; \\ g^{pr}_{\vec{a}\vec{b}}(t) &= \Theta(t)\langle[B^{+}_{\vec{b}}(0), B_{\vec{a}}(t)]\rangle; & g^{pa}_{\vec{a}\vec{b}}(t) &= \Theta(t)\,\langle[B_{\vec{b}}(0), B^{+}_{\vec{a}}(t)]\rangle \end{aligned} \tag{7.115}$$

These functions are calculated in the harmonic approximation with the "square" Hamiltonian:

$$H - \Delta \sum_{\vec{n}} B^{+}_{\vec{n}} B_{\vec{n}} \mid \sum_{\vec{n},\vec{m}} T_{\vec{n}\vec{m}} B^{+}_{\vec{n}} B_{\vec{m}} \tag{7.116}$$

The Fourier components of the functions in (3.79) are defined, as in (3.59), by the following standard transformation [5, 143–145]:

$$\psi_{\vec{a}\vec{b}}(t) = \frac{1}{N}\sum_{\vec{k}} \mathrm{e}^{i\vec{k}(\vec{a}-\vec{b})} \int_{-\infty}^{+\infty} d\omega \mathrm{e}^{-i\omega t}\, \psi_{\vec{k}}(\omega)$$

where N is the number of molecules in the crystal, which have the following form:

$$\begin{aligned} g^{pr}_{\vec{k}}(\omega + i\delta) &= g^{pa}_{\vec{k}}(\omega + i\delta) = \frac{i}{2\pi}\frac{1+\bar{N}}{\omega + \Omega_{\vec{k}} + i\delta}; \\ g^{pa}_{\vec{k}}(\omega - i\delta) &= g^{pr}_{\vec{k}}(\omega - i\delta) = \frac{i}{2\pi}\frac{\bar{N}}{\omega + \Omega_{\vec{k}} - i\delta} \end{aligned} \tag{7.117}$$

where they figure as $\Omega_{\vec{k}} = \Omega_{\Delta} + \Omega_{T_{\vec{k}}}; \Omega_{\Delta} = \Delta/\hbar;\ \Omega_{T_{\vec{k}}} = T_{\vec{k}}/\hbar;\ T_{\vec{k}} = \sum_{\vec{m}} T_{\vec{n}\vec{m}} \mathrm{e}^{i\vec{k}\vec{m}}$ and $\bar{N} = \langle B^{+}_{\vec{k}} B_{\vec{k}}\rangle$ Paulion GF, which is expressed in the way described above by the boson, is replaced by the expression (3.78). The obtained equation, which now contains the boson GFs only, translates into algebraic equations according to their Fourier images — by transformations of the type (3.59). After that, there is an explicit solution found in the retarded commutator boson GF. On this occasion, all contributions to the proportional square or the higher degree of boson concentration are neglected. In this way, the following result is obtained:

$$G_{\vec{k}}(\omega) = \frac{i}{2\pi} \frac{1 + 2\bar{N}}{\omega - \Omega_{\vec{k}} + M_{\vec{k}} Q_{\vec{k}}(\omega)} \tag{7.118}$$

where they figure as $M_{\vec{k}} = \frac{2}{N} \sum_{\vec{q}} (\Omega_{\vec{k}} + \Omega_{\vec{q}} - \omega_{\vec{k}} - \bar{N}\omega_{\vec{k}-\vec{q}})$, with $\omega_{\vec{k}} = D_{\vec{k}}/\hbar$ and $D_{\vec{k}} = \sum_{\vec{m}} D_{\vec{n}\vec{m}}\, e^{i\vec{k}\vec{m}}$ represents the correction of the harmonic exciton frequency, which arises from nonlinear effects. Based on this, the frequencies of normal exciton levels are given by the expression

$$\omega_n(\vec{k}) = \Omega_{\vec{k}} - M_{\vec{k}} \tag{7.119}$$

As follows from the GF poles (3.81), the function $Q_{\vec{k}}(\omega)$ is given by

$$\begin{aligned}
Q_{\vec{k}}(\omega) &= 1 + \frac{3}{2N^2} \sum_{\vec{q}_1 \vec{q}_2} [\omega - \Omega_{\vec{k}} - \Phi_{\vec{k},\vec{q}_1\vec{q}_2}] \Lambda_{\vec{k},\vec{q}_1\vec{q}_2}(\omega)\, \bar{N}_{\vec{q}_1};\\
\Phi_{\vec{k},\vec{q}_1\vec{q}_2} &= \Omega_{\vec{k}-\vec{q}_1+\vec{q}_2} + \Omega_{\vec{q}_2} - \omega_{\vec{q}_1-\vec{q}_2} - \Omega_{\vec{k}-\vec{q}_2};\\
\Lambda_{\vec{k},\vec{q}_1\vec{q}_2}(\omega) &= \frac{1}{\omega + \Omega_{\vec{q}_1} - \Omega_{\vec{q}_2} - \Omega_{\vec{k}-\vec{q}_1+\vec{q}_2} + \delta/3}\\
&= \frac{1}{\omega + \Omega_{\vec{q}_1} - \Omega_{\vec{q}_2} - \Omega_{\vec{k}-\vec{q}_1+\vec{q}_2}} - \frac{i\pi}{3}\delta(\omega + \Omega_{\vec{q}_1} - \Omega_{\vec{q}_2} - \Omega_{\vec{k}-\vec{q}_1+\vec{q}_2})
\end{aligned} \tag{7.120}$$

The following equation

$$Q_{\vec{k}}(\omega) = 0, \tag{7.121}$$

gives additional poles of GF $G_{\vec{k}}(\omega)$ in relation to pole $\omega_n(\vec{k})$. The solutions of this equation represent the frequencies of the kinematic excitations that arise in three-step boson processes.

In general, the frequencies satisfying Eq. (3.84) are complex sizes. This means that kinematical excitations are muted and have the final lifetime. The last equation will be solved approximate, because analytic cannot be solved precisely, but only numerically. If the spatial dispersion is neglected, i.e., $\Omega_{\vec{k}} \to \Omega_0, \omega_{\vec{k}} \to \omega_0$ and assumes that $\omega \neq \Omega_0$, Eq. (3.84) is reduced to the following form:

$$1 + \frac{3}{2}\bar{N} - \frac{3(1-\eta)\bar{N}\Omega_0}{\omega - \omega_0} = 0$$

from which, after neglecting the squares of exciton concentrations, we obtain

$$\omega_0 = \Omega_\Delta + [1 + 3(1-\eta)\bar{N}]\Omega_0; \qquad \eta \equiv \omega_0/\Omega_0 \tag{7.122}$$

If the same approximation is used, and in (3.82) then the energies of the normal exciton levels in the form are obtained:

$$\omega_n = \Omega_\Delta + [1 - 2(1-\eta)\bar{N}]\Omega_0 \tag{7.123}$$

and the boson GF (3.81) is reduced to the following form:

$$G_0 = \frac{i}{2\pi} - \frac{(1+\bar{N}/2)(\omega - \Omega_\Delta - \Omega_0)}{(\omega-\omega_n)(\omega-\omega_0)} \tag{7.124}$$

Comparing (3.85) with (3.86), one can easily conclude that the kinematic and normal exciton levels are energetically very close to one another due to the fact that the maximum exciton concentrations are 10^{-3}–10^{-2}. It should be emphasized that, in the context of the approximation used, the kinematic level is suppressed.

In the end, it must be noted that the use of kinematical excitations explains Urbach's rule [133, 151–155] for the absorption of light by which the absorption decree does not depend on the crystal structure and is approximately equal to the unit.

7.4. Application of Green's Functions and Operator Solving of Difference Equations[5]

Over the last decade, we have witnessed a rapid development of nanotechnology, i.e., production technologies of structures with nanodimensions [6, 91, 157]. The trend of rapid development of nanotechnologies is not accidental since nanostructures possess physical characteristics that are substantially different from the physical characteristics of the structures which are larger in size. Nanostructures are characterized by a series of qualitatively new effects. Superconductive, thermal insulating, acoustic and other features that characterize nanomaterials are better than those in bulk structures. It is also expected that the use of nanostructured elements may increase the sensitivity of measuring instruments which, in turn,

[5]In this section, parts of the text from three cited papers were used with permission [85] *Reviews in Theoretical Science* **4**/4, 353–401 (2016), [86] *Quantum Matter* **6**/1, 28–30 (2017) and [156] *Journal of Computational and Theoretical Nanoscience* **2**, 1–6 (2005); Copyright@ASP.

leads to new experimental findings. All of the above-mentioned lead to the conclusion that nanomaterials can be a good basis for further development of physics. In a way, nanostructure effects are connected to the effects in bulk structures, similarly to the relationship between quantum and classical effects. Therefore, further and even faster development of the science of nanostructure materials and nanotechnology is to be expected.

Science is nowadays very interested in low-dimensional systems, the dimensions of order of even few nanometers, which, in practical application, demonstrate exceptional characteristics in various fields. The need to minimize dimensions imposed by a number of interrelated and mutually dependent requests of modern civilization, probably key to its further survival and sustainable development, which can generally be categorized as belonging to the fields of energy, health and ecology. In that sense, the subject of the research in this work will include nanomaterials in the field of biopharmaceutical technology for biomedical application [81, 93, 158]. This includes the very precise encapsulated drug delivery, to exactly defined place in the human tissue or organ and disintegration of capsule–drug carrier, so that the medicament can start producing its effect.

The development of biomedical science is directly related to the development of natural and mathematical sciences, especially materials science. Enormous resources have been invested in the research that would lead to the solutions for many deadly diseases and in the end it is not possible not to ask a completely logical question: whether the results of modern research justify the invested resources [81, 158]?.

Let us consider in what ways we can make contributions to scientific research. One approach is experimental, while the other is theoretical. There are big differences in the popularity of these approaches, yet both are said to have advantages and disadvantages.

There are two aspects of theoretical approach. One uses computer resources, while the other uses analytics (literally speaking paper and pencil). The aim of our research group is that within materials science we understand the essential mechanisms of transport processes in low-dimensional systems, so we chose an analytical approach with the ambition that in the very near future, we will study with the help of computer resources, too. Our method of choice is Green's function which we use in the field of molecular crystals.

The precise structuring of materials to the dimensions of the order of nanometer is of great importance in electronics, optoelectronics, high temperature superconductivity, biology, medicine, environmental protection and many other scientific and technological disciplines. Theoretical and experimental studies of properties of low-dimensional systems (thin films, superlattices, quantum wires and quantum points) have become very intense in recent decades, so it could be concluded that they represent one of the key directions of research in modern condensed matter physics [159, 160].

The development of sophisticated techniques for the growth control of very thin layers caused intensive research of the nanocrystalline structures. One of the basic principles for their production is based on a process of self-construction or self-assembly [102, 103, 107], which is one of the "bottom-up" techniques, in which atoms or molecules arrange themselves into low-dimensional structures on the basis of mutual physical–chemical interactions. Although the self-assembly has been known for many years, its use in industry is still at the beginning. Improved knowledge of thermodynamics and kinetics of the nanoscale processes, with the advancement of characterization techniques and computer modeling, should lead to the development of complex systems.

Nanoscience continues to advance at a dramatic pace and is making revolutionary contributions in diverse fields, including electronics, optoelectronics, quantum electronics, materials science, chemistry, and biology. The technologies needed to fabricate nanoscale structures and devices are advancing rapidly. These technologies have facilitated the design and study of a vast array of novel devices, structures and systems confined dimensionally on the scale of 10 nm or less in one or more dimensions [161–163]. Moreover, nanotechnology is continuing to mature rapidly and will, no doubt, lead to further revolutionary breakthroughs like those exemplified by quantum-dot semiconductor lasers operating at room temperature, intersubband multiple quantum-well and quantum wire lasers, double-barrier quantum-well diodes operating in the terahertz frequency range, single-electron transistors and memories.

Here we will illustrate the operator method for solving difference equations and the method of Green's functions in the phonon, exciton and electron subsystems in different nanostructured models.

7.4.1. *Phonons in ultrathin films*

Application of nanostructures requires knowledge of their fundamental physical (mechanical, electromagnetic, optical, etc.) characteristics [92, 164]. Thermodynamic properties [165–167] associated with phonon displacements through the nanosamples are particularly interesting. Independent of the type of lattices, the thermodynamics of their subsystems (electrons, excitons, spin waves, etc.) is determined when the subsystem is in thermodynamic equilibrium with phonons [168–170]. Phonons are collective mechanical oscillations of molecules or atoms and represent the most important system of excitations. The reason is simple: phonons are present in all systems and their influence, more or less, changes the behavior of all other objects or excitations of the system. The mentioned influence is reflected primarily in the fact that they play the role of a thermostat of the system and determine its thermodynamics. Besides, the acoustical characteristics, as well as conductive and superconductive properties, etc., could not be realistically explained without phonons [170, 172]. All quoted is well known and all applications of phonons in bulk structures have been intensively exploited for more than a century.

The fact which must be specifically mentioned is that the role of phonons in nanostructures is much more impressive than in bulk structures. The main fact concerning phonon properties in nanostructures is the absence of the so-called acoustical phonons, i.e., phonons whose energy tends to zero when phonon momentum tends to zero. An activation energy different from zero is necessary for exciting phonons in nanostructures. Such unexpected characteristics require revision of all conclusions obtained regarding bulk theories of phonons. Therefore, the contribution of phonon subsystems to thermodynamic and energy transfering analysis is the first step in the research of nanostructure properties.

This part describes a major aspect of the effort to understand nanostructures, namely the study of phonons and phonon-mediated effects in structures with nanoscale-dimensional confinement in one or more spatial dimensions [173–175]. During the last two decades there has been a steady effort to understand the optical and acoustic phonons in nanostructures such as the superlattice, quantum wires, nanotubes and quantum dots. The central theme of this paper is the description of the acoustic phonons of the optical type in these nanostructures. As a preliminary to describing the dispersion relations and mode structures for phonons in nanostructures, phonon amplitudes are quantized in terms of the harmonic oscillator approximation, and anharmonic effects leading to phonon decay are described in terms of the dominant phonon decay channels. These elastic and discontinued models are applied to describe the deformation potential and interactions in a variety of nanostructures including quantum wells, nanowires and quantum dots. Finally, this paper describes how the

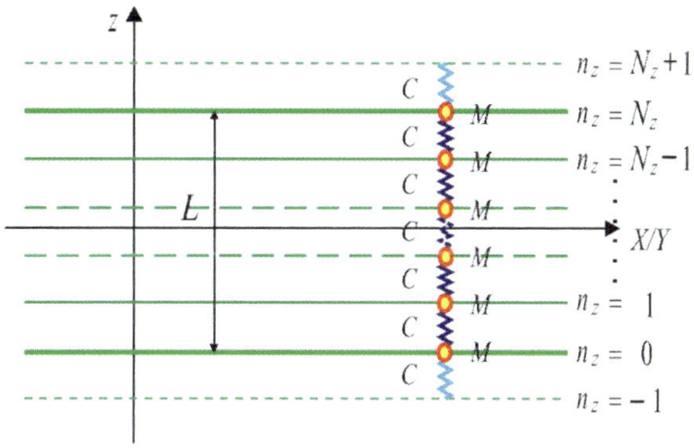

Fig. 7.1: Cross section of the thin crystalline film model (from [85]).

dimensional confinement of phonons in nanostructures leads to modifications in the electronic, optical, acoustic, superconductive and thermodynamic properties of the quantum.

7.4.2. *Model of phonons in thin films*

Our model of an ultrathin film represents a crystalline structure with broken space-translational symmetry. Here, we will consider a nanothin film, which is infinite in XY planes having finite (very small) thickness in z-direction. It means that symmetry break appears in z-direction (Fig. 7.1).

The main consequence of symmetry break is the spatial dependence of physical properties of the system. In the considered case, physical properties will depend on the index $n_z = 0, 1, \ldots, N_z$; $2 \leq N_z \leq 10$, labeling layers of the film in z-direction. Spatial dependence is determined by boundary conditions. The simplest boundary conditions consisting in the absence of layers labeled with $n_z = -1$ and $n_z = N_z + 1$ [60, 176–178] will be taken here. The analysis of the film-properties will be carried out with the help of Green's functions [131–133, 170, 179, 180] of displacement as well as momentum types and their spectral intensities will be used for determination of some relevant kinetic and thermodynamic characteristics of the film.

Green's Function of the Displacement Type

As in [85], the calculation procedure will be done in two steps. Firstly, the equation defining Green's function of the ideal system will be found. Thereafter, this equation will be translated into the system of equations in accordance to boundary conditions.

The Hamiltonian of mechanical oscillations in the system, in which the torsion effects are neglected, taken in the nearest-neighbor approximation is of the form [1–5, 167–170]:

$$H = \frac{1}{2M}\sum_{\vec{n}} p_{\vec{n}}^2 + \frac{C}{2}\sum_{\vec{n}} [(u_{\vec{n}} - u_{\vec{n}-\vec{a}_x})^2 + (u_{\vec{n}} - u_{\vec{n}-\vec{a}_y})^2 + (u_{\vec{n}} - u_{\vec{n}-\vec{a}_z})^2] \tag{7.125}$$

In this formula, M is the mass of the atom, C is Hook's constant, u are the atom displacements and p are the corresponding momenta ($p = M\,\dot{u}_n$). The symbol $\vec{n} \pm \vec{a}_x$ has to be read as follows: $u_{\vec{n}\pm\vec{a}_x} \equiv u_{n_x\pm 1,n_y,n_z}$ and the other symbols of the quoted type are similar.

Green's function of the displacement–displacement type is given by

$$\Psi_{\vec{n},\vec{m}}(t) \equiv \Psi_{n_x,n_y,n_z;m_x,m_y,m_z}(t) \equiv \langle\langle u_{\vec{n}}(t)|u_{\vec{m}}(0)\rangle\rangle = \Theta(t)\langle[u_{\vec{n}}(t), u_{\vec{m}}(0)]\rangle \tag{7.126}$$

where $\Theta(t)$ is the Heaviside step function. Differentiating (7.126) with respect to time and using the equations of motions, we obtain

$$\frac{d}{dt}\Psi_{\vec{n},\vec{m}}(t) = \frac{1}{M}\Theta(t)\langle[p_{\vec{n}}(t), u_{\vec{m}}(0)]\rangle = \frac{1}{M}\langle\langle p_{\vec{n}}(t)|u_{\vec{m}}(0)\rangle\rangle \tag{7.127}$$

After differentiating Green's function $\frac{d}{dt}\langle\langle p_{\vec{n}}(t)|u_{\vec{m}}(0)\rangle\rangle$ according to time and the use of equations of motions, we obtain

$$\begin{aligned}\frac{d}{dt}\langle\langle p_{\vec{n}}(t)|u_{\vec{m}}(0)\rangle\rangle &= -i\hbar\delta_{\vec{n},\vec{m}}\delta(t) + C[\Psi_{\vec{n}+\vec{a}_x,\vec{m}}(t) + \Psi_{\vec{n}-\vec{a}_x,\vec{m}}(t)\\ &+ \Psi_{\vec{n}+\vec{a}_y,\vec{m}}(t) + \Psi_{\vec{n}-\vec{a}_y,\vec{m}}(t) + \Psi_{\vec{n}+\vec{a}_z,\vec{m}}(t) + \Psi_{\vec{n}-\vec{a}_z,\vec{m}}(t) - 6\,\Psi_{\vec{n},\vec{m}}(t)]\end{aligned} \tag{7.128}$$

Carrying out the Fourier time–frequency transformations

$$\Psi_{\vec{n},\vec{m}}(t) = \int_{-\infty}^{+\infty} d\omega e^{-i\omega t}\Psi_{\vec{n},\vec{m}}(\omega); \quad \delta(t) = \int_{-\infty}^{+\infty} d\omega e^{-i\omega t} \tag{7.129}$$

in (7.127) and (7.128) and by combining these transformed equations, we obtain

$$\begin{aligned}\omega^2\Psi_{\vec{n},\vec{m}}(\omega) &= \frac{i\hbar}{2\pi}\frac{1}{M}\delta_{\vec{n},\vec{m}} - \frac{C}{M}[\Psi_{\vec{n}+\vec{a}_x,\vec{m}}(\omega) + \Psi_{\vec{n}-\vec{a}_x,\vec{m}}(\omega) + \Psi_{\vec{n}+\vec{a}_y,\vec{m}}(\omega)\\ &+ \Psi_{\vec{n}-\vec{a}_y,\vec{m}}(\omega) + \Psi_{\vec{n}+\vec{a}_z,\vec{m}}(\omega) + \Psi_{\vec{n}-\vec{a}_z,\vec{m}}(\omega) - 6\Psi_{\vec{n},\vec{m}}(\omega)]\end{aligned} \tag{7.130}$$

Keeping in mind that translational invariance in x and y directions is conserved, we introduce the following space-momentum Fourier's transformations:

$$\Psi_{\vec{n},\vec{m}}(\omega) = \frac{1}{N_x N_y}\sum_{k_x,k_y} \mathrm{e}^{ik_x a_x(n_x-m_x)+ik_y a_y(n_y-m_y)} A_{n_z,m_z}(k_x,k_y,\omega);$$
$$\delta_{\vec{n},\vec{m}} \equiv \delta_{n_x,m_x}\delta_{n_y,m_y}\delta_{n_z,m_z} = \frac{\delta_{n_z,m_z}}{N_x N_y}\sum_{k_x,k_y} \mathrm{e}^{ik_x a_x(n_x-m_x)+ik_y a_y(n_y-m_y)} \tag{7.131}$$

After the application of the last two formulas, Eq. (7.130) is reduced to

$$A_{n_z+1,m_z} + A_{n_z-1,m_z} + \rho A_{n_z,m_z} = F_{n_z,m_z} \tag{7.132}$$

where $A_{n_z,m_z} \equiv A_{n_z,m_z}(k_x,k_y,\omega)$, $F_{n_z,m_z} = \frac{i\hbar}{2\pi}\frac{1}{C}\delta_{n_z,m_z}$ and

$$\rho = \frac{M}{C}\omega^2 - 4\sin^2\frac{a_x k_x}{2} - 4\sin^2\frac{a_y k_y}{2} - 2 \tag{7.133}$$

At this stage of calculation, we shall take into account the boundary conditions, consisting in the absence of the layers $n_z = -1$ and $n_z = N_z + 1$. These boundary conditions can be expressed as $u_{n_x,n_y,-1} = u_{n_x,n_y,N_z+1} = 0$. Due to these conditions, Eq. (7.132) goes over to the system of difference equations:

$$-\text{for } n_z = 0: \quad A_{1,m_z} + \rho\, A_{0,m_z} = F_{0,m_z}, \tag{7.134}$$

$$-\text{for } n_z = N_z: \quad A_{N_z-1,m_z} + \rho\, A_{N_z,m_z} = F_{N_z,m_z} \tag{7.135}$$

$$-\text{for } 1 \le n_z \le N_z: \quad A_{n_z+1,m_z} + A_{n_z-1,m_z} + \rho\, A_{n_z,m_z} = F_{n_z,m_z} \tag{7.136}$$

The system of Eqs. (7.134)–(7.136) is the system of N_z+1 nonhomogenous difference equations of the second order with nonlinear coefficients [6, 112, 128, 157, 181–183]. The solution of (7.136) will be looked for in the following form:

$$A_{n_z,m_z} = \sum_{\mu=\mu_0}^{N_z'} a_{\mu,m_z}\sin(n_z+1)\phi_\mu \tag{7.137}$$

and after its substitution in (7.136), we obtain

$$\sum_{\mu=\mu_0}^{N_z'} (2\cos\phi_\mu + \rho)\, a_{\mu,m_z}\sin(n_z+1)\phi_\mu = \frac{i\hbar}{2\pi}\frac{1}{C}\delta_{n_z,m_z} \tag{7.138}$$

The substitution of (7.139) into (7.134) leads to the following relation:

$$\sum_{\mu=\mu_0}^{N_z'} (2\cos\phi_\mu + \rho) a_{\mu,m_z} \sin\phi_\mu = \frac{i\hbar}{2\pi}\frac{1}{C}\delta_{0,m_z} \tag{7.139}$$

where it is seen that it represents (7.140) for $n_z = 0$. Now, introducing (7.137) into (7.135), it becomes

$$\sum_{\mu=\mu_0}^{N_z'} (2\cos\phi_\mu + \rho) a_{\mu,m_z} [\sin N_z\phi_\mu + \rho\sin(N_z+1)\phi_\mu] = \frac{i\hbar}{2\pi}\frac{1}{C}\delta_{N_z,m_z} \tag{7.140}$$

The simple trigonometric transformations give $\sin N_z\phi_\mu + \rho\sin(N_z + 1)\phi_\mu = (2\cos\phi_\mu + \rho) \cdot \sin(N_z+1)\phi_\mu - \sin(N_z+2)\phi_\mu$, and determining the parameter ϕ_μ under the condition

$$\sin(N_z+2)\,\phi_\mu = 0 \quad \Rightarrow \quad \phi_\mu = \frac{\pi\,\mu}{N_z+2}; \quad \mu = 1, 2, \ldots, N_z+1 \tag{7.141}$$

It should be pointed out that index μ may not take the value $\mu = 0$ since it leads to a trivial solution (7.137), i.e., $A_{n_z,m_z} = 0$. Besides, this index must take the same set of values as n_z and therefore the maximal value of μ is $N_z + 1$. Due to the condition (7.141), the relation (7.138) becomes

$$\sum_{\mu=1}^{N_z+1} (2\cos\phi_\mu + \rho)\, a_{\mu,m_z} \sin(n_z+1)\frac{\pi\mu}{N_z+2} = \frac{i\hbar}{2\pi}\frac{1}{C}\delta_{n_z,m_z} \tag{7.142}$$

and the obtained relation corresponds to (7.140) for $n_z = N_z$. Since the set of values of index μ is determined in the formula (7.141) we finally conclude: the solution

$$A_{n_z,m_z} = \sum_{\mu=\mu_0}^{N_z'} a_{\mu,m_z} \sin(n_z+1)\frac{\pi\,\mu}{N_z+2} \tag{7.143}$$

fulfills all of $N_z + 1$ equations of the system of Eqs. (7.134) to (7.136) and reduces them to a unique one:

$$\sum_{\mu=1}^{N_z+1} a_{\mu,m'} \left(2\cos\frac{\pi\,\mu}{N_z+2} + \rho\right) a_{\mu,m_z} \sin(n_z+1)\frac{\pi\,\mu}{N_z+2} = \frac{i\hbar}{2\pi}\frac{1}{C}\delta_{n_z,m_z};$$

$$n_z = 0, 1, 2, \ldots, N_z \tag{7.144}$$

Now, we shall introduce the standing waves representation of Kronecker's symbol. Taking into account that $\sum_{\mu=1}^{N_z+1} \sin(n_z+1)\frac{\pi\,\mu}{N_z+2}\sin(m_z+1)\frac{\pi\,\mu}{N_z+2} = \frac{N_z+2}{2}\delta_{n_z,m_z}$, we obtain the defining expression:

$$\delta_{n_z,m_z} = \frac{2}{N_z+2}\sum_{\mu=1}^{N_z+1} \sin(n_z+1)\frac{\pi\,\mu}{N_z+2}\sin(m_z+1)\frac{\pi\,\mu}{N_z+2} \tag{7.145}$$

Taking undetermined coefficient $a_{\mu,m'}$ in the form:

$$a_{\mu,m_z} = \frac{2}{N_z+2}\,\alpha(k_x,k_y,\mu,\omega)\sin(m_z+1)\frac{\pi\,\mu}{N_z+2} \tag{7.146}$$

and substituting (7.145) and (7.146) in (7.144), we obtain: $\alpha(k_x,k_y,\mu,\omega) = \frac{i\hbar}{2\pi}\frac{1}{C}(\rho+2\cos\frac{\pi\,\mu}{N_z+2})^{-1}$, wherefrom, after the substitution of the parameter ρ from (7.133), it follows

$$\alpha(k_x,k_y,\mu,\omega) = \frac{i\hbar}{2\pi}\frac{1}{2\omega_{k_x,k_y,\mu}}\frac{1}{M}\left(\frac{1}{\omega-\omega_{k_x,k_y,\mu}} - \frac{1}{\omega+\omega_{k_x,k_y,\mu}}\right) \tag{7.147}$$

where

$$\omega_{k_x,k_y,\mu} = 2\,\Omega\sqrt{\sin^2\frac{a_xk_x}{2} + \sin^2\frac{a_yk_y}{2} + \sin^2\frac{\pi\mu}{2(N_z+2)}} \tag{7.148}$$

and $\Omega = \sqrt{C/M}$. Subsequently, combining (7.146) and (7.149), and substituting the result of combining into (7.137), we obtain

$$A_{n_z,m_z}(k_x,k_y,\omega) = \frac{i\hbar}{2\pi}\frac{1}{M}\frac{1}{N_xN_y}\frac{1}{N_z+2}\sum_{k_x,k_y}\sum_{\mu=1}^{N_z+1}\frac{1}{\omega_{k_x,k_y,\mu}} \times\left(\frac{1}{\omega-\omega_{k_x,k_y,\mu}} - \frac{1}{\omega+\omega_{k_x,k_y,\mu}}\right)\cdot\sin(n_z+1)\frac{\pi\mu}{N_z+2}\sin(m_z+1)\frac{\pi\mu}{N_z+2} \tag{7.149}$$

Consequently, the Green function given by the formula (7.131) is the following:

$$\Psi_{\vec{n},\vec{m}}(\omega) = \frac{i\hbar}{2\pi}\frac{1}{M}\frac{1}{N_xN_y}\frac{1}{N_z+2}\sum_{k_x,k_y}\sum_{\mu=1}^{N_z+1}\frac{e^{ik_xa_x(n_x-m_x)+ik_ya_y(n_y-m_y)}}{\omega_{k_x,k_y,\mu}} \times\left(\frac{1}{\omega-\omega_{k_x,k_y,\mu}} - \frac{1}{\omega+\omega_{k_x,k_y,\mu}}\right)\cdot\sin(n_z+1)\frac{\pi\mu}{N_z+2}\sin(m_z+1)\frac{\pi\mu}{N_z+2} \tag{7.150}$$

The spectral intensity of the function $\Psi_{\vec{m},\vec{n}}(\omega)$ is defined [1–5, 92, 184–186] by $I_\Psi = \frac{\Psi_{\vec{n},\vec{m}}(\omega+i\delta)-\Psi_{\vec{n},\vec{m}}(\omega-i\delta)}{e^{\hbar\omega/\theta}-1}$; $\delta\to 0^+$; $\theta = k_\mathrm{B}T$. After the substitution of (7.150) into this relation, we finally obtain the expression for spectral intensity of the function Ψ in the following form:

$$I_\Psi = \frac{\hbar}{M}\frac{1}{N_xN_y}\frac{1}{N_z+2}\sum_{k_x,k_y}\sum_{\mu=1}^{N_z+1} e^{ik_xa_x(n_x-m_x)+ik_ya_y(n_y-m_y)}$$
$$\times \frac{\delta(\omega-\omega_{k_x,k_y,\mu})-\delta(\omega+\omega_{k_x,k_y,\mu})}{e^{\hbar\omega/\theta}-1}\sin(n_z+1)\frac{\pi\mu}{N_z+2} \quad (7.151)$$
$$\times \sin(m_z+1)\frac{\pi\mu}{N_z+2}\frac{1}{\omega_{k_x,k_y,\mu}}$$

By the definition of corresponding correlation function $\langle u_{\vec{m}}(0)\, u_{\vec{n}}(t)\rangle$ [96, 132, 170, 187–189]

$$\langle u_{\vec{m}}(0)u_{\vec{n}}(t)\rangle = \int_{-\infty}^{+\infty} d\omega e^{-i\omega t}I_\psi(\omega) \quad (7.152)$$

we obtain

$$\langle u_{\vec{m}}(0)\, u_{\vec{n}}(t)\rangle = \frac{\hbar}{M}\frac{1}{N_xN_y}\frac{1}{N_z+2}\sum_{k_x,k_y}$$
$$\times\sum_{\mu=1}^{N_z+1}\frac{e^{ik_xa_x(n_x-m_x)+ik_ya_y(n_y-m_y)-it\omega_{k_x,k_y,\mu}}}{\omega_{k_x,k_y,\mu}} \quad (7.153)$$
$$\times\left(\frac{e^{-i\omega_{k_x,k_y,\mu}t}}{e^{\hbar\omega_{k_x,k_y,\mu}/\theta}-1}-\frac{e^{i\omega_{k_x,k_y,\mu}t}}{e^{-\hbar\omega_{k_x,k_y,\mu}/\theta}-1}\right)$$
$$\times\sin(n_z+1)\frac{\pi\mu}{N_z+2}\sin(m_z+1)\frac{\pi\mu}{N_z+2}$$

From the last formula, we find the expression for the mean value of the square of atom displacements. Putting in (7.153) $\vec{m}=\vec{n}$ and $t=0$, we have

$$\langle u_{\vec{n}}^2\rangle = \frac{\hbar}{M}\frac{1}{N_xN_y}\frac{1}{N_z+2}\sum_{k_x,k_y}\sum_{\mu=1}^{N_z+1}\frac{1}{\omega_{k_x,k_y,\mu}}\sin^2(n_z+1)\frac{\pi\mu}{N_z+2}$$
$$\times\coth\frac{\hbar\omega_{k_x,k_y,\mu}}{2\theta} \quad (7.154)$$

This formula is very important since it is entered into the expressions for potential energy, crystal density and the other physical characteristics of the system.

Green's Function of Momentum Type

The momentum–momentum Green's function defined as [1–5, 85, 167–170]:

$$\Phi_{\vec{n},\vec{m}}(t) \equiv \Phi_{n_x,n_y,n_z;m_x,m_y,m_z}(t) = \langle\langle p_{\vec{n}}(t)|p_{\vec{m}}(0)\rangle\rangle = \Theta(t)\langle[p_{\vec{n}}(t), p_{\vec{m}}(0)]\rangle \tag{7.155}$$

is also important for defining physical characteristics of the film. It can be found by a procedure which is similar to that which was used for displacement–displacement Green's function in the previous paragraph. Defining: $\Pi_{\vec{n},\vec{m}}(t) \equiv \langle\langle \vec{u}_{\vec{n}}(t)|p_{\vec{m}}(0)\rangle\rangle$ and differentiating (7.155) with respect to time and using the equations of motion, we find:

$$\begin{aligned}\tfrac{d}{dt}\Phi_{\vec{n},\vec{m}}(t) &= C[\Pi_{\vec{n}+\vec{a}_x,\vec{m}}(t) + \Pi_{\vec{n}-\vec{a}_x,\vec{m}}(t) + \Pi_{\vec{n}+\vec{a}_y,\vec{m}}(t)\\ &+ \Pi_{\vec{n}-\vec{a}_y,\vec{m}}(t) + \Pi_{\vec{n}+\vec{a}_z,\vec{m}}(t) + \Pi_{\vec{n}-\vec{a}_z,\vec{m}}(t) - 6\Pi_{\vec{n},\vec{m}}(t)]\end{aligned} \tag{7.156}$$

Differentiating the function Π with respect to time and using the equation of motion, we have

$$\frac{d}{dt}\Pi_{\vec{n},\vec{m}}(t) = i\hbar\delta_{\vec{n},\vec{m}}\delta(t) + \frac{1}{M}\Phi_{\vec{n},\vec{m}}(t) \tag{7.157}$$

After Fourier's transformations of time–frequency

$$\begin{aligned}\Phi_{\vec{n},\vec{m}}(t) &= \int_{-\infty}^{+\infty} d\omega \mathrm{e}^{-i\omega t}\Phi_{\vec{n},\vec{m}}(\omega);\\ \Pi_{\vec{n},\vec{m}}(t) &= \int_{-\infty}^{+\infty} d\omega \mathrm{e}^{-i\omega t}\Pi_{\vec{n},\vec{m}}(\omega); \quad \delta(t) = \frac{1}{2\pi}\int_{-\infty}^{+\infty} d\omega \mathrm{e}^{-i\omega t}\end{aligned} \tag{7.158}$$

in (7.157) and (7.156), the last equation goes over to

$$\begin{aligned}-i\omega\Phi_{\vec{n},\vec{m}}(\omega) &= -\frac{C\hbar}{2\pi\omega}(\delta_{\vec{n}+\vec{a}_x,\vec{m}} + \delta_{\vec{n}-\vec{a}_x,\vec{m}} + \delta_{\vec{n}+\vec{a}_y,\vec{m}} + \delta_{\vec{n}-\vec{a}_y,\vec{m}}\\ &+\delta_{\vec{n}+\vec{a}_z,\vec{m}} + \delta_{\vec{n}-\vec{a}_z,\vec{m}} - 6\delta_{\vec{n},\vec{m}}) - \frac{C}{iM\omega}[\Phi_{\vec{n}+\vec{a}_x,\vec{m}}(\omega) + \Phi_{\vec{n}-\vec{a}_x,\vec{m}}(\omega)\\ &+\Phi_{\vec{n}+\vec{a}_y,\vec{m}}(\omega) + \Phi_{\vec{n}-\vec{a}_y,\vec{m}}(\omega) + \Phi_{\vec{n}+\vec{a}_z,\vec{m}}(\omega) + \Phi_{\vec{n}-\vec{a}_z,\vec{m}}(\omega) - 6\,\Phi_{\vec{n},\vec{m}}(\omega)]\end{aligned} \tag{7.159}$$

Due to the fact that translational symmetry is conserved in x and y directions we shall represent the function $\Phi_{n,m}(\omega)$ in the following way:

$$\Phi_{\vec{n},\vec{m}}(\omega) = \frac{1}{N_x N_y}\sum_{k_x,k_y} \mathrm{e}^{ik_x a_x(n_x-m_x)+ik_y a_y(n_y-m_y)} B_{n_z,m_z}(k_x,k_y,\omega) \tag{7.160}$$

Besides, taking into account

$$\delta_{n_x,m_x} = \frac{1}{N_x}\sum_{k_x} \mathrm{e}^{ik_x a_x(n_x-m_x)}; \quad \delta_{n_y,m_y} = \frac{1}{N_y}\sum_{k_y} \mathrm{e}^{ik_y a_y(n_y-m_y)} \tag{7.161}$$

with expression (4.21), and introducing the notation $B_{n_z,m_z} \equiv B_{n_z,m_z}(k_x, k_y, \omega)$, we reduce (7.160) to

$$\begin{aligned} -i\omega B_{n_z,m_z} = &-\frac{C}{iM\omega}\Bigg[\left(4\sin^2\frac{a_x k_x}{2} + 4\sin^2\frac{a_y k_y}{2}\right) \\ &\times B_{n_z,m_z} + B_{n_z+1,m_z} + B_{n_z-1,m_z} - 2B_{n_z,m_z}\Bigg] \\ &-\frac{C\hbar}{\pi\omega}\frac{4}{N_z+2}\sum_{\mu=1}^{N_z+1}\left[\sin^2\frac{a_x k_x}{2} + sin^2\frac{a_y k_y}{2} + \sin^2\frac{\pi\mu}{2(N_z+2)}\right] \\ &\times \sin(n_z+1)\frac{\pi\mu}{N_z+2}\sin(m_z+1)\frac{\pi\mu}{N_z+2} \end{aligned} \tag{7.162}$$

In the next step, we use the formulae (7.141) and (7.148) which enable us to translate (7.162) into the following form:

$$\begin{aligned} B_{n_z+1,m_z} + B_{n_z-1,m_z} + \rho B_{n_z,m_z} = \frac{i\hbar}{2\pi}\frac{M^2}{C}\frac{2}{N_z+2} \\ \times \sum_{\mu=1}^{N_z+1}\omega^2_{k_x,k_y,\omega}\sin(n_z+1)\frac{\pi\mu}{N_z+2}\sin(m_z+1)\frac{\pi\mu}{N_z+2} \end{aligned} \tag{7.163}$$

where the parameter ρ is given by the formula (7.133).

At this stage, the boundary conditions can be applied, consisting in the absence of the layers $n_z = -1$ and $n_z = N_z + 1$. These conditions can be expressed as $p_{n_x,n_y,-1} = p_{n_x,n_y,N_z+1} = 0$. Due to these conditions, the system (7.163) goes over to three subsystems:

$$-\text{for } n_z = 0: \quad B_{1,m_z} + \rho\, B_{0,m_z} = J_{0,m_z}; \tag{7.164}$$

$$-\text{for } n_z = N_z: \quad B_{N_z,m_z} + \rho\, B_{N_z,m_z} = J_{N_z,m_z} \tag{7.165}$$

$$-\text{for } 1 \le n_z \le N_z: \quad B_{n_z+1,m_z} + B_{n_z-1,m_z} + \rho\, B_{n_z,m_z} = J_{n_z,m_z} \tag{7.166}$$

where we introduce the following notation: $J_{n_z,m_z} = \frac{i\hbar}{2\pi}\frac{M^2}{C}\frac{2}{N_z+2}\sum_{\mu=1}^{N_z+1} \omega^2_{k_x,k_y,\omega}\sin(n_z+1)\frac{\pi\mu}{N_z+2}\sin(m_z+1)\frac{\pi\mu}{N_z+2}$ The system of Eqs. (7.164)–(7.166) has the same structure as the system of Eqs. (7.134)–(7.136). For this reason, it is not necessary to repeat the solution procedure, which was exposed in the previous section, we will quote only the results of the solution. Taking

$$B_{n_z,m_z} = \sum_{\mu=1}^{N_z+1} b_{\mu,m_z} \sin(n_z+1)\frac{\pi\,\mu}{N_z+2}\,;$$
$$b_{\mu,m_z} = \frac{2}{N_z+2}\beta(k_x,k_y,\mu,\omega)\sin(m_z+1)\frac{\pi\,\mu}{N_z+2} \tag{7.167}$$

we obtain that Eqs. (7.164)–(7.166) are satisfied if equality is valid:

$$\beta(k_x,k_y,\mu,\omega) = \frac{i\hbar}{2\pi}\frac{M^2\omega^2_{k_x,k_y,\omega}}{C}\left(\rho+2\cos\frac{\pi\,\mu}{N_z+2}\right)^{-1}$$
$$= \frac{i\hbar}{2\pi}\frac{M\omega_{k_x,k_y,\mu}}{2}\left(\frac{1}{\omega-\omega_{k_x,k_y,\mu}}-\frac{1}{\omega+\omega_{k_x,k_y,\mu}}\right) \tag{7.168}$$

Combining this and (7.167), we obtain

$$B_{n_z,m_z}(k_x,k_y,\omega) = \frac{i\hbar}{2\pi}\frac{M}{N_z+2}\sum_{\mu=1}^{N_z+1}\omega_{k_x,k_y,\mu}$$
$$\times\left(\frac{1}{\omega-\omega_{k_x,k_y,\mu}}-\frac{1}{\omega+\omega_{k_x,k_y,\mu}}\right) \tag{7.169}$$
$$\times\sin(n_z+1)\frac{\pi\mu}{N_z+2}\cdot\sin(m_z+1)\frac{\pi\mu}{N_z+2}$$

Now, by introducing this result into (7.160), we get

$$\Phi_{\vec{n},\vec{m}}(\omega) = \frac{i\hbar}{2\pi}M\frac{1}{N_xN_y}\frac{1}{N_z+2}\sum_{k_x,k_y}\sum_{\mu=1}^{N_z+1}\mathrm{e}^{ik_xa_x(n_x-m_x)+ik_ya_y(n_y-m_y)}\omega_{k_x,k_y,\mu}$$
$$\times\left(\frac{1}{\omega-\omega_{k_x,k_y,\mu}}-\frac{1}{\omega+\omega_{k_x,k_y,\mu}}\right) \tag{7.170}$$
$$\times\sin(n_z+1)\frac{\pi\mu}{N_z+2}\cdot\sin(m_z+1)\frac{\pi\mu}{N_z+2}$$

By the definition of the spectral intensity of momentum–momentum Green's function $\Phi_{n,m}(\omega)$ [68–73]: $I_\Phi = (\mathrm{e}^{\hbar\omega/\theta}-1)^{-1}[\Phi_{\vec{n},\vec{m}}(\omega+i\delta)-$

$\Phi_{\vec{n},\vec{m}}(\omega - i\delta)]|_{\delta\to 0^+}$, we give:

$$I_\Phi(\omega) = \hbar M \frac{1}{N_xN_y}\frac{1}{N_z+2}\sum_{k_x,k_y}\sum_{\mu=1}^{N_z+1} \mathrm{e}^{ik_xa_x(n_x-m_x)+ik_ya_y(n_y-m_y)}$$
$$\times\, \omega_{k_x,k_y,\mu}\sin(n_z+1)\frac{\pi\mu}{N_z+2}\times\sin(m_z+1) \tag{7.171}$$
$$\times\frac{\pi\mu}{N_z+2}(\mathrm{e}^{\hbar\omega/\theta}-1)^{-1}[\delta(\omega-\omega_{k_x,k_y,\mu})-\delta(\omega+\omega_{k_x,k_y,\mu})]$$

From the definition of the correlation function [71–73]: $\langle p_{\vec{m}}(0)p_{\vec{n}}(t)\rangle = \int_{-\infty}^{+\infty} d\omega\, \mathrm{e}^{-i\omega t} I_\Phi(\omega)$, after the substitution (7.171), we obtain

$$\langle p_{\vec{m}}(0)\, p_{\vec{n}}(t)\rangle = \frac{\hbar M}{N_xN_y(N_z+2)}\sum_{k_x,k_y}\sum_{\mu=1}^{N_z+1}$$
$$\times\, \omega_{k_x,k_y,\mu}\, \mathrm{e}^{ik_xa_x(n_x-m_x)+ik_ya_y(n_y-m_y)-it\omega_{k_x,k_y,\mu}} \tag{7.172}$$
$$\times\left(\frac{\mathrm{e}^{-i\omega_{k_x,k_y,\mu}\,t}}{\mathrm{e}^{\hbar\omega_{k_x,k_y,\mu}/\theta}-1}-\frac{\mathrm{e}^{i\omega_{k_x,k_y,\mu}\,t}}{\mathrm{e}^{-\,\hbar\omega_{k_x,k_y,\mu}/\theta}-1}\right)$$
$$\times\sin(n_z+1)\frac{\pi\mu}{N_z+2}\sin(m_z+1)\frac{\pi\mu}{N_z+2}$$

Finally, putting: $\vec{m}=\vec{n}$ and $t=0$ in (7.172), we obtain the expression of the square of the momentum of oscillations in the following form:

$$\langle p_{\vec{n}}^2\rangle = \frac{\hbar M}{N_xN_y(N_z+2)}\sum_{k_x,k_y}\sum_{\mu=1}^{N_z+1}\omega_{k_x,k_y,\mu}\sin^2(n_z+1)\frac{\pi\mu}{N_z+2}\coth\frac{\hbar\omega_{k_x,k_y,\mu}}{2\,\theta} \tag{7.173}$$

The square of momentum enables us to find diffusion tensor [1–5, 183–186], kinetic energy and other characteristics of the film connected with them.

Dispersion Law of Phonons in Ultrathin Films

For better understanding of thin films physics, it is necessary to emphasize their main feature: physical characteristics of thin films are spatially dependent. This spatial dependence is the consequence of non-equivalency spatial distribution of their crystallographic points. As far as we know, this obvious fact was not theoretically reproduced. The formal reason for the need for the existence of theory of broken symmetry structures is the lack of procedures for calculating Green's functions which depend on two spatial coordinates.

In the theory of broken symmetry structures, up to now, Green's functions have been used depending on the difference of two spatial coordinates [131–133] (this approach is correct only in the case of ideal structures) or Green's functions with diagonal spatial coordinates [179, 180]. In this work, the calculation techniques for Green's functions depending on two spatial coordinates are developed separately. These unique correct techniques of calculation immediately reproduced the expected physical result, i.e., the spatial dependence of the broken symmetry structure properties. In this section, this will be demonstrated on a set of examples. Apart from the confirmation of spatially dependent physical properties, some very interesting effects of structure deformation are ascertained, as well.

We shall start from the dispersion law of thin film phonons. It is obtained by solving algebraic equations, polynomial identity (Chebyshev) of the second type and order $N_z + 1$, which represents the developed form of the determinant system of difference equations (7.134)–(7.146) and/or (7.164)–(7.166) and zero [47–49]. The setting of this identity is the result of seeking singularity of Green's functions (real parts of singularity pose potential energy of elementary excitations [131–133]). The roots of the determinants of the mentioned systems of equations have solutions of the form: $\rho \equiv \rho_{kxky,\mu}$, $\mu = 1, 2, 3, \ldots, N_z + 1$, where, by means of the relation (7.133), phonon energies can be expressed as $E \equiv E_{kx,ky,\mu}$. Consequently, the energy of thin film phonons is given by

$$E_{k_x,k_y,\mu} = 2\hbar\Omega\sqrt{\sin^2\frac{a_x k_x}{2} + \sin^2\frac{a_y k_y}{2} + \sin^2\frac{\pi\mu}{2(N_z+2)}} \tag{7.174}$$

These energy spectra can be represented graphically (Fig. 7.2), for example, as in [190], where it is marked:

$$\varepsilon_\mu \equiv \frac{E_{k_x,k_y,\mu}}{2\hbar\,\Omega}; \quad R_{xy} \equiv \sin^2\frac{a_x k_x}{2} + \sin^2\frac{a_y k_y}{2} \in [0,\, 2],$$

since we observe the first Brillouin zone with $a_x k_x \wedge a_y k_y \in [-\pi/a, +\pi/a]$.

The physically most interesting energetic states are those with the lowest energy. Since minimal values of k_x and k_y, as well as the function R, are equal to zero (centre of the first 2D Brillouin zone), while the minimal value of μ is equal to 1, the minimal energy of thin film phonons is

$$\min E_{k_x,k_y,\mu} = E_{0,0,1} = 2\hbar\,\Omega\sin^2\frac{\pi}{2(N_z+2)} \tag{7.175}$$

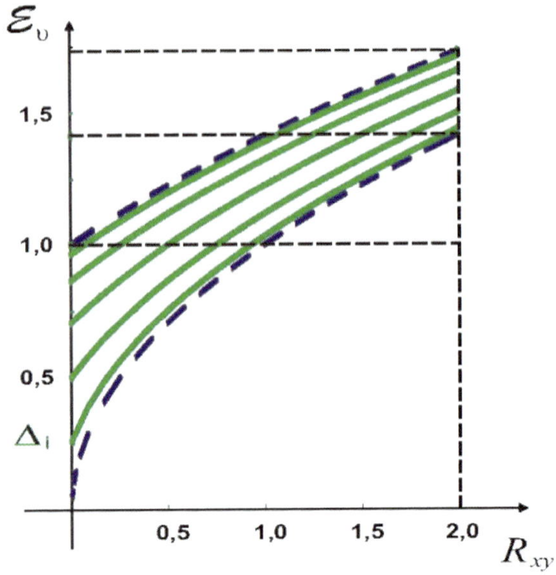

Fig. 7.2: Energy spectra of phonons in the thin crystalline film model.

Table 7.1: Values of the energy gaps of phonos in thin films.

N_z	$E_{\min}[k_B T]$	$E_{\min}[\mathrm{eV}]$
2	153.0730	0.01320260
5	89.0084	0.00767697
10	52.2105	0.00450315

These minimal values are given in Table 7.1 for films containing 3, 6 and 11 layers and $\hbar\Omega = 200\,k_{\mathrm{B}}$.

It is seen from this table that the excitation of collective mechanical oscillations requires heat quanta corresponding to the temperatures $T_3 = 153.073\,\mathrm{K}$, $T_6 = 89.008\,\mathrm{K}$ and $T_{11} = 52.210\,\mathrm{K}$, respectively. It means that up to the quoted temperatures, the collective mechanical waves (acoustical phonons) cannot appear. Up to these temperatures, atom oscillations are mutually independent. Consequently, for a film with three layers, the temperature of 153 K practically represents the absolute zero. From the point of view of superconductive behavior of crystals, this means that the three-layered film has the critical superconductive temperature equal to

153 K. This conclusion points out to extraordinary superconductive properties of ultrathin films [190–195].

The most practically important result is the presence of the energy gap in phonon spectra of ultrathin films. In accordance with formula (7.175), the minimal phonon energy is given by $E_{\min} = 2\hbar\,\Omega \sin\frac{\pi}{2(N_z+2)}$. This formula is valid for simple cubic structure films with $N_z + 1$ layer. In accordance with this formula, for three-layered films, the energy gap is about 25% of maximal phonon energy. Taking $\Theta_D = 300\,k_{\mathbf{B}}$, we conclude that the phonon gap of three-layered simple cubic film is about $60\,k_B$.

The experiments with MgB_2 films [196–199] have shown that superconductive critical temperature of this nanostructure is about 37 K. This corresponds to the energy gap of $37\,k_{\rm B}$, which is two times smaller than the value which follows from our theory. Taking into account that MgB_2 has hexagonal structure without a central atom, it is understandable that it possesses a smaller energy gap than a simple cubic film. For the three-layered MgB_2 film, $N_z = 12$ [200], wherefrom it follows $E_{\min}(MgB_2) = 2\hbar\,\Omega\sin\frac{\pi}{6(N_z+2)}$. This value is about three times smaller than $E_{\min}$ of simple cubic structure. Independently, this comparison leads to the belief that the phonon gap is the main mechanism of superconductivity of films.

7.4.3. *Excitons in nanolayered structures*

As we mentioned in [86], theoretical research of low-dimensional crystalline systems (nanostructures: ultrathin films, nanowires, quantum dots, etc.) has been intensified recently in order to obtain fundamental information relating to extremely different physical and chemical properties of materials, and also due to their wide practical application (technical and technological) in nano-, opto- and bioelectronics [93, 164, 201, 202]. One specificity of these "tiny" structures is that the presence of near boundary planes causes highly changed general properties of these materials and occurrence of nonspecific phenomena (as a consequence of confinement effects) in comparison to the properties of corresponding "large" samples [125].

The interest in the study of exciton subsystem appeared because excitons are responsible for dielectric, optic (absorption, dispersion, luminescence), photoelectric and other properties of crystals [203, 204].

This part 47ehaviou the influence of presence of film-structure boundaries on excitons, and special attention was paid to the occurrence of localized states. In addition, optical properties were researched (their dielectric permeability was determined). Analyses were carried out using the innovated method of Green's functions, adapted for the research of quantum structures [205, 206].

Exciton Model

Optic and dielectric properties of molecular crystals are dictated by the 47ehaviour of excitons of small concentration in Frenkel's exciton subsystem [1–5, 43, 128, 138–145]. This model mirrors the standard exciton Hamiltonian in the configuration space:

$$H_{\text{ex}} = H_0 + \sum_{\vec{n}} \Delta_{\vec{n}} P_{\vec{n}}^{+} P_{\vec{n}} + \sum_{\vec{n}} X_{\vec{n}\vec{m}} P_{\vec{n}}^{+} P_{\vec{m}} + \sum_{\vec{n}} Y_{\vec{n}\vec{m}} P_{\vec{n}}^{+} P_{\vec{n}} P_{\vec{m}}^{+} P_{\vec{m}} \quad (7.176)$$

where $P_{\vec{n}}^{+}$ and $P_{\vec{n}}^{\square}$ are creation and annihilation operators of excitons on node $\vec{n}$ in a crystal lattice. $\Delta_{\vec{n}}$ is the energy of exciton localized at node $\vec{n}$, while $X_{\vec{n}\vec{m}}$ and $Y_{\vec{n}\vec{m}}$ are matrix elements of exciton transfer from node $\vec{n}$ to node $\vec{m}$.

The main difficulty in the exciton theory is the fact that Pauli operators P^{+} and $P_{\square}^{\square}$ are neither Bose nor Fermi operators, but a certain hybrid of both [125, 144], and their kinematics is fermion for one node and boson for different nodes. For a precise analysis of exciton systems, by encompassing effects of inter-exciton interaction, it is not enough to simply replace Pauli operators by Bose operators [90]: P, $P_{\square}^{+} \longrightarrow B$, B^{+}.

Theoretical analysis will be performed using commutator Green's function [132, 133, 179, 180]:

$$G_{\vec{n}\vec{m}} \equiv \langle\langle B_{\vec{n}}(t) | B_{\vec{n}}^{+}(0) \rangle\rangle = \Theta(t) \langle [B_{\vec{n}}(t), B_{\vec{n}}^{+}(0)] \rangle$$

which satisfies the equation of motion:

$$\begin{aligned}
\hbar\omega\, G_{n_x n_y n_z;\, m_x m_y m_z}(\omega) &= \frac{i\hbar}{2\pi} \delta_{n_x n_y n_z;\, m_x m_y m_z} \\
&+ \Delta_{n_x n_y n_z} G_{n_x n_y n_z;\, m_x m_y m_z}(\omega) \\
&+ [X_{n_x n_y n_z;\, n_x+1, n_y n_z}(\omega) G_{n_x+1, n_y n_z;\, m_x m_y m_z}(\omega) \\
&+ X_{n_x n_y n_z;\, n_x-1, n_y n_z}(\omega) G_{n_x-1, n_y n_z;\, m_x m_y m_z}(\omega) \\
&+ \big[X_{n_x n_y n_z;\, n_x+1, n_y n_z}(\omega) G_{n_x+1, n_y n_z;\, m_x m_y m_z}(\omega) \\
&+ X_{n_x n_y n_z;\, n_x-1, n_y n_z}(\omega) G_{n_x-1, n_y n_z;\, m_x m_y m_z}(\omega) \qquad (7.177) \\
&+ X_{n_x n_y n_z;\, n_x n_y+1, n_z}(\omega) G_{n_x n_y+1, n_z;\, m_x m_y m_z}(\omega) \\
&+ X_{n_x n_y n_z;\, n_x n_y-1, n_z}(\omega) G_{n_x n_y-1, n_z;\, m_x m_y m_z}(\omega) \\
&+ X_{n_x n_y n_z;\, n_x n_y n_z+1,}(\omega) G_{n_x n_y n_z+1;\, m_x m_y m_z}(\omega) \\
&+ X_{n_x n_y n_z;\, n_x n_y n_z-1}(\omega) G_{n_x n_y n_z-1;\, m_x m_y m_z}(\omega)].
\end{aligned}$$

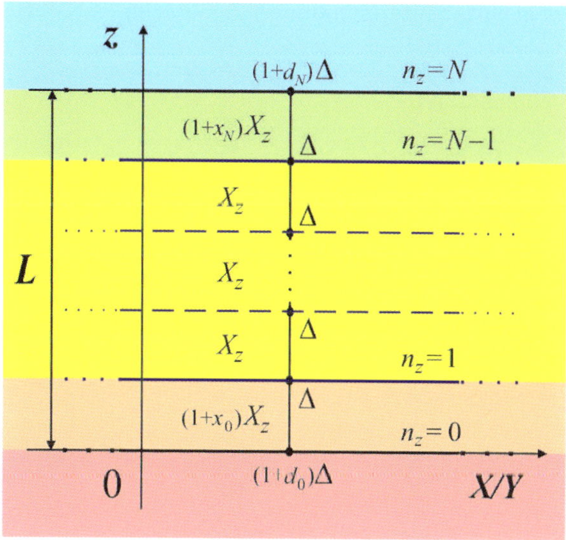

Fig. 7.3: Film cross-section in $X(Y)Z$ plane (from [86]).

Ultrathin Exciton Film

Here, we will observe a very thin (several nanometers) film, made on substrate by some technological process (steaming, reaching, spraying, etc.) [125]. The film's dimensions are such that they are infinite in XY planes, while along z-axis has final thickness (L). Therefore, the film has two infinite border areas parallel to XY-plane, for $z = 0$ and $z = L \equiv Na$ (Fig. 7.3). As a consequence of changed boundary conditions, we take that exciton energies in boundary ($n_z = 0$, $n_z = N$), and in adjacent planes ($n_z = 1$, $n_z = N$–1) are perturbed:

$$\Delta_{\vec{n}} \equiv \Delta(1 + d_0\delta_{n_z,0} + d_N\delta_{n_z,N});$$

$$n_z = 0, 1, 2, \ldots, N; \quad N \in [2, 10];$$

$$X_{\vec{n},\vec{n}+\vec{\lambda}_z} \equiv X_z(1 + x_0\delta_{n_z,0} + x_N\delta_{n_z,N-1});$$

$$X_{\vec{n},\vec{n}-\vec{\lambda}_z} \equiv X_z(1 + x_0\delta_{n_z,1} + x_N\delta_{n_z,N})$$

Due to the spatial limitation of the film along z-direction, only a partial spatial Fourier transformation may be applied (only along x and y axes) to

Eq. (7.177). It is suitable to introduce the following denotations:

$$G_{n_z,m_z} \equiv G_{n_z,m_z}(k_x, k_y, \omega), \quad \mathcal{K} \equiv \frac{i\hbar}{2\pi\,|X|},$$
$$X_x = X_y = X_z \equiv -|X|, \quad a_x = a_y = a_z \equiv a,$$
$$\rho \equiv \frac{\hbar\omega - \Delta}{|X|} - \mathcal{F}_{xy}, \quad \mathcal{F}_{xy} \equiv 2(\cos\, ak_x + \cos ak_y)$$

Then instead of (7.201), we obtain

$$G_{n_z+1,m_z}(1 + x_0\delta_{n_z,0} + x_N\delta_{n_z,N-1}) + G_{n_z-1,m_z}(1 + x_0\delta_{n_z,1} + x_N\delta_{n_z,N})$$
$$+G_{n_z,m_z}[\rho - \frac{\Delta}{|X|}(d\delta_{n_z,0} + d_N\delta_{n_z,N})] = \mathcal{K}. \tag{7.178}$$

This equation is in fact a system of $N+1$ nonhomogeneous algebraic-differential equations with initial-boundary conditions: $G_{n_z,m_z} \equiv G_{n_z} = 0$, for $n_z < 0$ and $n_z > N+1$.

Films Permittivity

To determine the crystalline film dynamic permittivity, we will use formula of Dzyaloshinski–Pitaevski [207], with a note that permittivity here depends on the position of film plane (n_z). Substituting the expression for Green's functions [125, 204, 205], we obtain

$$\varepsilon_{n_z}(\omega) \equiv \left\{1 - \frac{\pi S|X|}{i\hbar}[G_{n_z}(\omega) + G_{n_z}(-\omega)]\right\}^{-1} \tag{7.179}$$
$$= \left[1 + S\sum_{\nu=1}^{N+1} g_{n_z}^{\nu} \frac{\rho_\nu - |p| + F_{xy}}{|f|^2 - (\rho_\nu - |p| + F_{xy})^2}\right]^{-1}$$

where $|p| \equiv \frac{\Delta}{|X|}$, $|f| \equiv \frac{\hbar\omega}{|X|}$ and $\rho_\nu\,(\nu = 1, 2, 3, \ldots, N+1)$ are roots of determinant of system (7.178).

Based on the last formula, it is clear that relative permittivity depends on position, e.g., number of crystallographic planes ($n_z = 0, 1, 2, \ldots, N$) along the direction of broken translational symmetry by existing of border areas (perpendicular to them), of course and on non-dimensional energy/frequency of external electromagnetic field. These functional dependencies are mathematically very complicated and can be determined only numerically. Besides that, in this case we have 5 (e.g., 6) parameters: $N = 4\,(n_z = 0, 1, 2, 3, 4)$, $d_{0/N} \in \{-0.3, +0.3\}$ and $x_{0/N} \in \{-0.8, +2.0\}$, which means that we have 6,000 different combinations of their possible values that are performed with the help of adequate software and graphically arranged.

Figure 7.4 shows the dependence of dynamic permittivity (ε) on reduced relative energy of external electromagnetic field (f), for four-layered or five-planed molecular film ($N = 4$). Dependence was calculated for the center of the plane (2D) Brillouin's zone ($k_x = k_y = 0$), but individually per plane (parallel boundary planes) of crystalline film, therefore for $n_z = 0$ and/or $n_z = 4$, then $n_z = 1$ and/or $n_z = 3$ and finally for $n_z = 2$.

Based on earlier results [208–210] here briefly outlined, the following can be concluded. In the molecular film-structures, resonant peaks exist at precisely determined discrete energies. The number of these peaks depends on the position of planes in the film for which permittivity is calculated, but also on the size of perturbation parameters.

(1) By changing energy values for molecules at the boundary areas, absorption zone widens, and dominating resonant peaks are visible only at those boundary regions.
(2) By increasing vault energy of excitons in the border layer of the film, symmetrical widening of absorption zone occurs, and also quenching of certain inner resonant peaks.

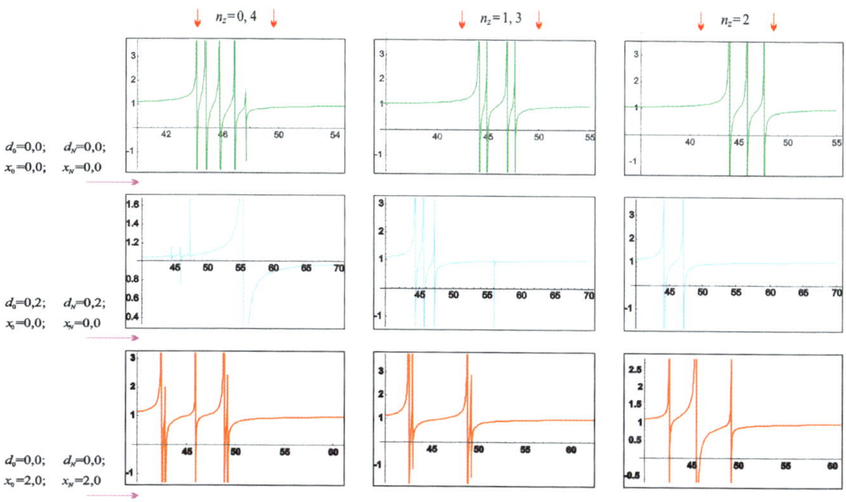

Fig. 7.4: Dynamic permittivity of molecular film (from [86]).

7.4.4. *Electrons in quantum dots*

The research of electron behaviour in quantum tubes (quantum wires as well as quantum dots) is very popular research theme. Practically, all aspects of electron theory in tubes are analyzed [211–215].

The vacant point of all these investigations is the formulation of Green's function method suitable for analyses of the systems with broken symmetry. It should be pointed out that this is not a simple problem since in structures with broken symmetry, Green's function depends on two coordinates, separately. In the ideal structure, Green's functions depend on coordinate difference and this practically means that in ideal structure, Green's function depends on only one configurational characteristic. The attempts of Green's functions analyses of broken symmetry systems have been up to now reduced to the finding of Green's function with equal configrational indices [216, 217].

It is quite clear that physical characteristics of a broken symmetry structure must depend on coordinates. It will be proved here for electrons. Hall's constant, mobility of electrons, relaxation time and other characteristics will be found and compared to corresponding results of approaches of other authors [1–5, 218].

Electron Green's Function

We shall analyze cylindrical quantum dot. The electron Hamiltonian of simplified Hubbard model [219–221] is the following:

$$\begin{aligned} H = \frac{1}{2}\sum_{m,n}&(W_{m,n;m+1,n} + W_{l,m,n;l,m-1,n} + W_{m,n;m,n+1} \\ &+ W_{m,n;m,n-1})\, a^{+}_{mn} a_{mn} \\ &- \frac{1}{2}\sum_{n,m} a^{+}_{m,n}(W_{m,n;m+1,n} a_{m+1,n} \\ &+ W_{m,n;m-1,n} a_{m-1,n} W_{m,n;m,n+1} a_{m,n+1} + W_{m,n;m,n-1} a_{m,n-1}) \end{aligned} \tag{7.180}$$

In this formula, index m changes along z-axis and determines the position of discs. Index n denotes atoms in the disc. In discs, cyclical symmetry takes place [222]. The mentioned indices take the following values: $m = 0, 1, 2, \ldots, M; n = 0, 1, 2, \ldots, N$, where M and N are of the order 10.

The Hamiltonian (4.57) will be taken in the nearest neighbour's approximation [223]. Then it goes over to

$$H = \sum_{m,n} \left[(P+Q)\, a^{+}_{m,n} a_{m,n} - \frac{1}{2} Q\, a^{+}_{m,n}(a_{m+1,n} + a_{m-1,n}) \right.$$
$$\left. - \frac{1}{2} R a^{+}_{m,n}(a_{m,n+1} + a_{m,n-1}) \right] \tag{7.181}$$

where the notations — $W_{l,m,n;l,m\pm1,n} \equiv Q$ and $W_{l,m,n;l,m,n\pm1} \equiv R$ — are used. Operators a^{+} and a are Fermi operators creating and annihilating electrons. As it is well known [1–5] , Fermi operators satisfy the following commutator rules (the large brackets denote anticommutator):

$$\{a_{m,n}, a^{+}_{m'n'}\} = \delta_{m,m'}\, \delta_{n,n'}; \quad \{a_{m,n}, a_{m',n'}\} = \{a^{+}_{m,n}, a^{+}_{m'n'}\} = 0\,;$$
$$a^{2}_{m,n} = (a^{+}_{m,n})^{2} = 0$$

Using these relations, we find the commutator of operator $a_{m,n}$ with the Hamiltonian (7.181):

$$[a_{m,n}, H] = (Q+R)a_{m,n} - \frac{1}{2}Q(a_{m+1,n} + a_{m+1,n}) - \frac{1}{2}R(a_{m,n+1} + a_{m,n-1}) \tag{7.182}$$

The system will be analyzed with the help of anticommutator Green's function

$$\Gamma_{m,n;m',n'}(t) = \langle\langle a_{m,n}(t) | a^{+}_{m',n'}(t) \rangle\rangle = \Theta(t) \langle \{a_{m,n}(t), a^{+}_{m',n'}(0)\} \rangle \tag{7.183}$$

where $\Theta(t)$ is the Heaviside step function. Differentiating (7.183) with respect to time, we obtain

$$\frac{d}{dt}\langle\langle a_{m,n}(t) | a^{+}_{m',n'}(0) \rangle\rangle = \delta(t)\delta_{m,m'}\delta_{n,n'} + \Theta(t)\langle \{\dot{a}_{m,n}(t), a^{+}_{m',n'}(0)\} \rangle \tag{7.184}$$

Using the equation of motion for operator $a_{m,n}$ in the form $\dot{a}_{m,n}(t) = \frac{1}{i\hbar}[a_{m,n}, H]$, we reduce (7.184) to

$$\begin{aligned} i\hbar \frac{d}{dt} & \langle\langle a_{m,n}(t) | a^{+}_{m',n'}(0) \rangle\rangle \\ &= i\hbar \delta_{m,m'}\delta_{n,n'}\delta(t) + (Q+R)\langle\langle a_{m,n}(t) | a^{+}_{m',n'}(0) \rangle\rangle \\ &\quad - \frac{1}{2} Q \left[\langle\langle a_{m+1,n}(t) | a^{+}_{m',n'}(0) \rangle\rangle \langle\langle a_{m-1,n}(t) | a^{+}_{m',n'}(0) \rangle\rangle \right] \\ &\quad - \frac{1}{2} R \left[\langle\langle a_{m,n+1}(t) | a^{+}_{m',n'}(0) \rangle\rangle \langle\langle a_{m,n-1}(t) | a^{+}_{m',n'}(0) \rangle\rangle \right] \end{aligned} \tag{7.185}$$

In (7.185), we will use time–frequency Fourier transformations $\Gamma_{m,n;m',n'}(t) = \int_{-\infty}^{+\infty} d\omega e^{-i\omega t}\Gamma_{m,n;m',n'}(\omega)$; $\delta(t) = \frac{1}{2\pi}\int_{-\infty}^{+\infty} d\omega e^{-i\omega t}$, after which we obtain

$$\frac{\Gamma_{m+1,n;m',n'}(\omega) + \Gamma_{m-1,n;m',n'}(\omega)}{2}Q + \frac{\Gamma_{m,n+1;m',n'}(\omega) + \Gamma_{m,n-1;m',n'}(\omega)}{2}R$$

$$+ (E - Q - R)\,\Gamma_{m,n;m',n'}(\omega) = \frac{i\hbar}{2\pi}\delta_{m,m'}\,\delta_{n,n'} \tag{7.186}$$

Electrons in discs obey the cyclicity rule (see [224]). Consequently, the Green function Γ can be expressed in terms of cosine function:

$$\Gamma_{m,n;m',n'}(\omega) = \sum_{\nu=0}^{N} G_{m,m'}(\omega,\nu)\cos\frac{2\pi n}{N+1}\nu\cos\frac{2\pi n'}{N+1}\nu \tag{7.187}$$

It can be shown [218, 224] that

$$\delta_{n,n'} = \begin{cases} \dfrac{1}{N+1}\displaystyle\sum_{\nu=0}^{\infty}\cos\frac{2\pi n}{N+1}\nu\,\cos\frac{2\pi n'}{N+1}\nu; & \dfrac{2\nu}{N+1} - \text{integer}, \\ \dfrac{2}{N+1}\displaystyle\sum_{\nu=0}^{\infty}\cos\frac{2\pi n}{N+1}\nu\,\cos\frac{2\pi n'}{N+1}\nu; & \dfrac{2\nu}{N+1} - \text{not integer} \end{cases} \tag{7.188}$$

After this, Eq. (7.186) becomes

$$\frac{G_{m+1,m'}(\omega,\nu) + G_{m-1,m'}(\omega,\nu)}{2}Q + \rho\,G_{m,m'}(\omega,\nu) = F_{m,m'} \tag{7.189}$$

where we introduced the following notations:

$$\rho = E - R\left(1 - \cos\frac{2\pi\nu}{N+1}\right) - Q; \tag{7.190}$$

$$F_{m,m'} = \frac{i\hbar}{2\pi}\delta_{m,m'}\cdot\begin{cases} \dfrac{1}{N+1}, & \dfrac{2\nu}{N+1} - \text{integer}; \\ \dfrac{2}{N+1} & \dfrac{2\nu}{N+1} - \text{not integer}. \end{cases}$$

Now, we shall look for $G_{m,m'}$ in the form

$$G_{m,m'} = \sum_{\mu}\alpha_{\mu,m'}\left[A_{\mu}\sin(m+1)\xi_{\mu} + B_{\mu}\sin m\xi_{\mu}\right] \tag{7.191}$$

where $\alpha_{m'}(\mu, \nu, \omega) \equiv \alpha_{\mu,m'}$ are unknown coefficients. It is obvious that the following is valid:

$$G_{m+1,m'} + G_{m-1,m'} = \sum_{\mu} 2\, \alpha_{\mu,m'}[A_\mu \sin(m+1)\xi_\mu + B_\mu \sin m\xi_\mu]\cos\xi_\mu \tag{7.192}$$

Taking into account that the discs with indices $m = -1$ and $m = M + 1$ are absent, the subsystem of difference equation (7.189) is valid for $m = 1, 2, 3, \ldots, M-1$, with two boundary difference equations:

$$m = 0: \quad \frac{Q}{2} G_{1,m'} + \rho G_{0,m'} = F_{0,m'} \tag{7.193}$$

$$m = M: \quad \frac{Q}{2} G_{M-1,m'} + \rho G_{M,m'} = F_{M,m'} \tag{7.194}$$

Electron Spectra in Quantum Dots

Introducing (7.191) into (7.189), we obtain the following expression:

$$\sum_{\mu} \alpha_{\mu,m'}(Q\cos\zeta_\mu + \rho)[A_\mu \sin(m+1)\xi_\mu + B_\mu \sin m\, \xi_\mu] = F_{m,m'} \tag{7.195}$$

The functions $G_{0,m'}$ and $G_{1,m'}$ obtained from (7.191) will be substituted in (7.193) and the obtained equation will be written in the form which is analogous to (7.195). So, we obtain

$$B_\mu = 0 \tag{7.196}$$

and the analogous form becomes

$$\sum_{\mu} \alpha_{\mu,m'}(Q\cos\xi_\mu + \rho)A_\mu \sin\xi_\mu = F_{0,m'} \tag{7.197}$$

In the boundary subsystem (7.194) will be taken $B_\mu = 0$, and consequently

$$G_{M-1,m'} = \sum_{\mu} \alpha_{\mu,m'} A_\mu \sin M\xi_\mu; \quad G_{M,m'} = \sum_{\mu} \alpha_{\mu,m'} A_\mu \sin(M+1)\xi_\mu \tag{7.198}$$

Introducing (7.198) into (7.194), we obtain

$$\sum_{\mu} \alpha_{\mu,m'} \left[\frac{Q}{2} \frac{A_\mu \sin M\xi_\mu}{A_\mu \sin(M+1)\xi_\mu} + \rho \right] A_\mu \sin\xi_\mu = F_{m,m'} \tag{7.199}$$

This expression becomes analogous with (7.195) and (7.197) if the following condition is fulfilled:

$$\frac{Q}{2}\frac{A_\mu \sin M\xi_\mu}{A_\mu \sin(M+1)\xi_\mu} = Q\cos\xi_\mu \quad \Rightarrow \quad \sin(M+2)\xi_\mu = 0 \tag{7.200}$$

wherefrom follow the values of the parameter ξ_μ

$$\xi_\mu = \frac{\pi\,\mu}{M+2}; \quad \mu = 1, 2, 3, \ldots, M+1 \tag{7.201}$$

Using Eq. (7.197), we calculate the quantity ρ, and by (7.190) and the allowed energy of electrons in disks:

In the formula, we substitute ρ from (7.190) and we obtain

$$\begin{aligned} E_{\mu,\nu} &= Q\left(1 - \cos\frac{\pi\mu}{M+2}\right) + R\left(1 - \cos\frac{2\pi\nu}{N+1}\right) \\ &= 2Q\sin^2\frac{\pi\mu}{2(M+2)} + 2R\sin^2\frac{\pi\nu}{N+1} \end{aligned} \tag{7.202}$$

The results obtained in this section can be summarized as follows:

(1) In quantum dots, electron states are standing waves; consequently, the electron current is equal to zero.
(2) Maximal electron energy in the observed quantum dot is $E_{\max} = 2Q + 2R$. This has no serious physical significance.
(3) Minimal electron energy in quantum dot is $E_{\min} = 2Q\sin^2\frac{\pi\mu}{2(M+2)}$. It means that, for three layers ($M = 2$), this energy is equal to $Q/3$, while for 11 discs ($M = 10$) this minimal value is: $Q/32$. These minimal values change (decrease) chemical potential of electrons, and this has major repercussions on the change in all of the physical characteristics of this nanosample compared to bulk ones.

However, this is not the goal here, but only the presentation of a methodological approach to overcoming problems in the theoretical research of nanostructures.

7.5. Concluding Remarks

Research on the nanostructures presented in this chapter is more of a starting point for nanostructure studies than a complete theory. Since the study of nanotechnology is still a relatively new area which is still explored, especially on a methodological basis, the main goal here was to formulate the correct methodology of nanostructures of different geometries and in various physical systems and thus create a basis for further research of nanostructures. As shown by the key applications of the operator solving of difference equations and the modified method of Green's functions, the authors hope that this foundation is sufficiently wide and solid. Taking into account that nanoscience has a more rapid development in technological and experimental field than in the theoretical one, we have paid more attention to theoretical problems connected with analyses of nanostructures. The reader can get acquainted with technological and experimental advantages of nanoscience through numerous scientific papers and monographs, as well as with the research of nanostructures with simulation techniques, which can be found in magazine articles. Our aim here is to make the theoretical approaches more analytical and as systematic as possible. The goal of this review of approaches in nanostructures theory was to accelerate the theory of nanophenomena and to follow the rapid development of nanotechnology. The authors' intention was to find methods which could give analytical solutions to nanoscience problems. It was the reason for dividing nanostructures into planned and random ones.

This work contains two relatively different parts. The first part is rather of the methodological mathematical character, demonstrating the method of operator solving of the basic solid state difference equation with variable coefficient. Since the exposed general operator method gives the compact solution if the variable coefficient is of exponential type, it can be applied to the problem of finding molecular displacements in crystalline chain of finite length. It is shown that the atom displacement represents the specific superposition of plane and standing waves. In this part, we have demonstrated that the characteristic difference equation in translatory non-invariant crystalline structures theory could be of importance in some practical problems.

The second part of this chapter is dedicated to the methodology of theoretical research of structures with broken translational symmetry and nanostructures. In particular, this concerns the use of Green's function techniques, which are mainly connected to the research of properties of structures of different crystallographic and spatial symmetries. The analyses of nanostructures (with the presence of confinement effects) require a different Green's technique, which has also been elaborated at length. Therefore, we have presented here an adaptation of the well-known and recognized theoretical method from the quantum static physics of solid states — the method of two-time temperature-dependent Green's functions. The formulated method of calculation of Green's functions in broken symmetry structures has shown that these Green's functions depend on two separated spatial coordinates. The analyses with the help of this Green's function method reproduced a physically sound fact: the properties of a broken symmetry structure are spatially dependent.

The technique introduced in this part is applied to some important problems concerned with order-disorder nanostructures. That problem is low temperature magnetization of nanostructures. The similar analyses can be extended to other systems (electron, phonon, excitons, etc.). Concerning methodology — its further development is absolutely necessary — it has been shown that in the molecular film-structures, resonant peaks exist at precisely determined energies. The number of these peaks depends on the position of planes in the film for which permittivity is calculated, but also from the size of perturbation parameters. The results of the analyses of quantum dots electrons could be summarized as follows: the concentration of quantum dots charge carriers depends on spatial coordinates. It increases proportionally to the square of disc indices.

Appendix: Statistical and Dynamical Equivalence of Different Elementary Cells[6]

In the last two to three decades, theoretical investigations of nanostructures became very intensive. Nanotechnology has sharp progress in last time and produces different rectangular and cylindrical structures [121, 156, 167, 216, 224–227]. The main problem of theoretical investigations is correct including of boundary conditions into physical behaviour of nanostructures. The geometrical form of produced nanostructures is rectangular (thin films, quantum rods and quantum parallelepiped) or cylindrical (cylinders with nanocross-section and macroscopic height and cylinders having nanoheight and nanocross-section). These geometrical forms enable correct inclusion of boundary conditions into evaluations [121, 224–227]. Concerning the lower symmetry structure (monoclinic, triclinic, tetragonal and others), the correct inclusion of boundary conditions is very complicated and in some cases is impossible.

Taking this into account, we shall try, in the first part of this work, to achieve statistical equivalence between rectangular cell and lower symmetry cell. The statistical equivalence means equating of momentum volumes the

---

[6]Reprinted with permission from Ref. [116] *J. Comput. Theor. Nanosci.* **4**, 619–626 (2007) and [10] *Encyclopedia of Nanoscience and Nanotechnology* **50**, 281–350 (2011); Copyright @ASP.

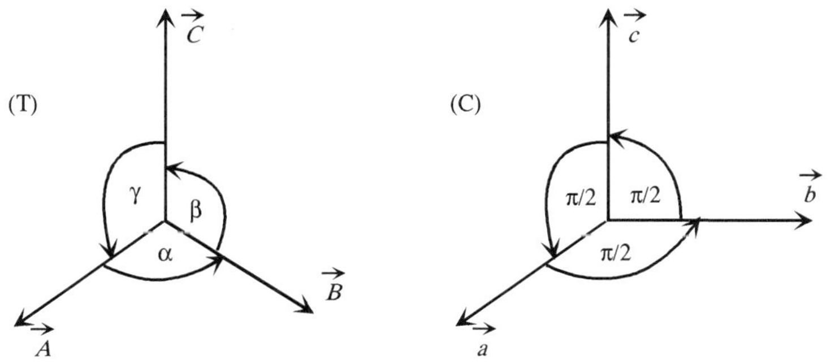

Fig. A.1: Elementary cell of triclinic (left) and rectangular (right) structure.

mentioned cells since statistical averages are usually calculated in momentum space.

In the second part, the rectangular cell will be reduced to simple cubic one which has the same momentum volume as the lower symmetry one. After this, we can try to achieve the dynamical equivalence, too. In the nearest neighbour approximation, the dynamics of simple cubic cell is defined by six nearest neighbours. For lower symmetry cell, the dynamics is defined by six or more neighbours. The last is dependent on angles between vectors of lower symmetry cell. If some of the angles are less than 90°, then the interactions between ends of corresponding vectors have to be included into total interactions of atoms of lower symmetry cells. We are of the opinion that the achievement of described equivalence will noticeably accelerate theoretical investigations of nanostructures.

We shall consider the elementary cell of triclinic crystal structure [2, 3, 189, 228] whose lattice constants are A, B and C. The lengths of these lattice constants are different and the angles between vectors $\vec{A}$, $\vec{B}$ and $\vec{C}$ are different: $\angle(\vec{A},\vec{B}) = \alpha$, $\angle(\vec{B},\vec{C}) = \beta$ and $\angle(\vec{C},\vec{A}) = \gamma$. The notations used are as follows: $|\vec{A}| = A$, $|\vec{B}| = B$ and $|\vec{C}| = C$. The lattice constant of rectangular structure will be denoted with a, b and c. Their lengths are different in the general case. The angles between vectors $\vec{a}$, $\vec{b}$ and $\vec{c}$ are $90°$. The notations used are $|\vec{a}| = a$, $|\vec{b}| = b$ and $|\vec{c}| = c$. The elementary cells introduced are presented in Fig. A.1.

In order to find statistically equivalent rectangular structure for the given triclinic one, we shall start from the fact that phase volume of cells of crystals with simple lattice is equal to h^3 [189], where h is Planck's constant. The phase volume Φ is the product of configuration volume V

and momentum volume W:

$$\Phi = VW \tag{A.1}$$

If the configuration volume of triclinic cell is denoted with V_T and the momentum cell volume with W_T, we have

$$V_T W_T = \Phi_T = h^3 \tag{A.2}$$

For rectangular structure, we have the relation

$$V_C W_C = \Phi_C = h^3 \tag{A.3}$$

where V_C is configuration volume and W_C is momentum cell volume. If we achieve equality

$$V_C = V_T \tag{A.4}$$

then automatically follows $W_C = W_T$. The last means that statistical equivalence is achieved.

In order to satisfy (A.4) we shall put rectangular vector $\vec{c}$ in the direction of vector $\vec{C}$ of three-clinic cell and both of them put into direction of z-axis:

$$\vec{C} = C\vec{k}; \quad \vec{c} = c\vec{k} \tag{A.5}$$

Vectors $\vec{a}$ and $\vec{b}$ of rectangular cell will be put in directions of x-axis and y-axis, respectively. In this way for rectangular structure, the following can be written:

$$\vec{a} = a\vec{i}; \quad \vec{b} = b\vec{j}; \quad \vec{c} = c\vec{k} \tag{A.6}$$

where $\vec{i}$, $\vec{j}$ and $\vec{k}$ are orts of Descartes coordinate system. Taking this into account, we can find the expression for configuration volume of rectangular structure:

$$V_C = \vec{a} \circ (\vec{b} \times \vec{c}) = \begin{vmatrix} a & 0 & 0 \\ 0 & b & 0 \\ 0 & 0 & c \end{vmatrix} = abc \tag{A.7}$$

On the basis of (A.5), the configuration volume of triclinic structure is given by

$$V_T = \vec{A} \circ (\vec{B} \times \vec{C}) = \begin{vmatrix} A_x & A_y & A_z \\ B_x & B_y & B_z \\ 0 & 0 & C \end{vmatrix} = C(A_x B_y - A_y B_x) \tag{A.8}$$

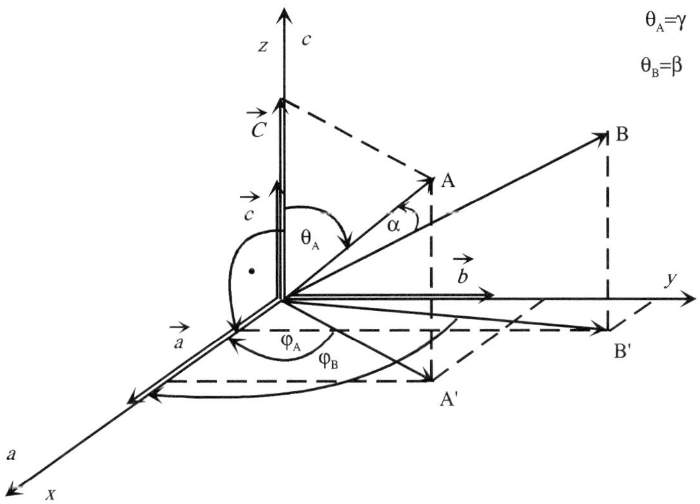

Fig. A.2: Triclinic rectangular lattice vectors in spherical coordinates.

Considering this, we obtain

$$abc = (A_x B_y - A_y B_x)C \tag{A.9}$$

Further analysis will be continued in spherical coordinates (see Fig. A.2).

Since vector $\vec{C}$ lies in z-axis direction and the angle between vectors A and $\vec{C}$ is γ, the angle γ is azimuthal angle of vector $\vec{A}$. Projection of vector $\vec{A}$ in xOy plane is equal to $A\sin\gamma$. If the angle between vector $\vec{A}$ projection and axis x are denoted with $\varphi_{A,}$ then we can write

$$A_x = A\sin\gamma\cos\varphi_A; \quad A_y = A\sin\gamma\sin\varphi_A; \quad A_z = A\cos\gamma \tag{A.10}$$

The angle β between vectors $\vec{B}$ and $\vec{C}$ is azimuthal one for vector $\vec{B}$. Denoting with φ_B the angle between $\vec{B}$ and x-axis, we get the following:

$$B_x = B\sin\beta\cos\varphi_B; \quad B_y = B\sin\beta\sin\varphi_B; \quad B_z = B\cos\beta \tag{A.11}$$

Combining last two equations, we obtain

$$A_x B_y - A_y B_x = AB\sin\beta\sin\gamma\sin(\varphi_B - \varphi_A) \tag{A.12}$$

Since

$$\vec{A}\circ\vec{B} = A_x B_x + A_y B_y + A_z B_z$$

the substitution of (A.10) and (A.11) in last one leads to

$$AB\cos\alpha = AB\sin\beta\sin\gamma\cos(\varphi_B - \varphi_A) + AB\cos\beta\cos\gamma$$

Combining this and (A.12), we obtain

$$\cos(\varphi_B - \varphi_A) = \frac{\cos\alpha - \cos\beta\cos\gamma}{\sin\beta\sin\gamma} \tag{A.13}$$

wherefrom it follows

$$\sin(\varphi_B - \varphi_A) = \sqrt{1 - \left(\frac{\cos\alpha - \cos\beta\cos\gamma}{\sin\beta\sin\gamma}\right)^2} \tag{A.14}$$

Substituting (A.11) and (A.14) into (A.9), we finally obtain the connection between configuration volumes of triclinic and rectangular cell.

$$abc = XABC \tag{A.15}$$

where

$$X = 1 + 2\cos\alpha\cos\beta\cos\gamma - \cos^2\alpha - \cos^2\beta - \cos^2\gamma \tag{A.16}$$

The formula (A.15) can be applied only if $X > 0$. We wish to point out that the condition $X > 0$ must not be satisfied always. In such cases, the equivalence cannot be achieved. As an illustration of this statement, we can quote the following. If $\alpha = 90^\circ$, $\beta = 60^\circ$ and $\gamma = 30^\circ$, in accordance with (A.16), we obtain that $X = 0$. For $\alpha = 150^\circ$, $\beta = 90^\circ$ and $\gamma = 30^\circ$ it follows that $X = -0.5$. In both cases, the equivalence cannot be achieved. The cases when the exposed procedure gives satisfactory result are, for example, $\alpha = \beta = \gamma = 1^\circ$, with $X = 6.9 \cdot 10^{-8}$, for $\alpha = 70^\circ$, $\beta = 50^\circ$ and $\gamma = 30^\circ$, with $X = -0.312$, but for $\alpha = 110^\circ$, $\beta = 100^\circ$ and $\gamma = 80^\circ$, with $X = -0.843$, etc. For TGS [229], adequate data are $\alpha = 125.9^\circ$, $\beta = 115.3^\circ$ and $\gamma = 115.1^\circ$, and the corresponding one is $X = -0.081$, i.e., TGS has equivalent cubic cell.

It should be pointed out that transition to statistically equivalent cell is possible for all monoclinic structures [159]. In these structures, two angles are equal 90° and in accordance with (A.16) it follows: $X = 1 - \cos^2\beta$, where $\beta \neq 90^\circ$, i.e., we always have that $X > 0$.

It is clear, also, that the equivalence conditions are not uniquely defined. We shall quote some variants which naturally impose. One of the variants is

$$a = X^{\frac{1}{6}}A; \quad b = X^{\frac{1}{6}}B; \quad c = X^{\frac{1}{6}}C \tag{A.17}$$

The second possibility is

$$\begin{aligned} a &= X^{\frac{1}{4}}A; \quad b = X^{\frac{1}{4}}B; \quad c = C \\ a &= X^{\frac{1}{4}}A; \quad b = X^{\frac{1}{4}}B; \quad c = C \\ a &= X^{\frac{1}{4}}A; \quad c = X^{\frac{1}{4}}C; \quad b = B \end{aligned} \tag{A.18}$$

The variants

$$
\begin{aligned}
a &= X^{\frac{1}{2}}A; \quad b = B; \quad c = C \\
a &= A; \quad b = X^{\frac{1}{2}}B; \quad c = C \\
a &= A; \quad b = B; \quad c = X^{\frac{1}{2}}C
\end{aligned}
\tag{A.19}
$$

can be considered as most natural, since only one of triclinic lattice constant changes the length.

At the end of this section, we shall determine the momentum volume of the equivalent cell. Momentum cell volume is defined [230] as product of reciprocal configuration cell and factor $\left(\frac{h}{2\pi}\right)^3$. The vectors: $\vec{k}_a$, $\vec{k}_b$ and $\vec{k}_c$ are defined as follows:

$$
\begin{aligned}
&\vec{k}_a = 2\pi\frac{\vec{a}\times\vec{c}}{\vec{a}\cdot(\vec{b}\times\vec{c})} = \frac{2\pi}{a}\vec{i}; \quad \vec{k}_b = 2\pi\frac{\vec{c}\times\vec{a}}{\vec{b}\cdot(\vec{c}\times\vec{a})} = \frac{2\pi}{b}\vec{j}; \\
&\vec{k}_c = 2\pi\frac{\vec{a}\times\vec{b}}{\vec{c}\cdot(\vec{a}\times\vec{b})} = \frac{2\pi}{c}\vec{k}
\end{aligned}
\tag{A.20}
$$

Taking into account formula (7.48), we easily conclude on the basis of (7.63) that the volume of reciprocal cell is

$$
R = \frac{(2\pi)^3}{abc} \tag{A.21}
$$

and multiplying this with $\hbar^3$ we obtain the volume of the momentum cell for rectangular structure

$$
W_C = \frac{h^3}{abc} = W_T \tag{A.22}
$$

The last equality sign follows from the fact that equality of configuration cells V_C and V_T has been achieved.

Illustrative example 1

In this stage of analysis, we shall not demonstrate how statistical averages of triclinic structures can be calculated using statistical equivalence with rectangular cell. Instead of that, we shall reduce the equivalent rectangular cell to simple cubic equivalent one.

We shall start from the formula (A.15), i.e., $abc = XABC$, and we shall take

$$
a = X^pA; \quad b = X^qB; \quad c = X^rC \tag{A.23}
$$

where p, q and k are arbitrary constants. Substituting this into definition (A.15) gives us an interesting connection between arbitrary parameters:

$$p + q + r = \frac{1}{2} \tag{A.24}$$

In order to obtain an equivalent simple cubic cell, we must take

$$\frac{a}{b} = X^{p-q}\frac{A}{B} = 1; \quad \frac{a}{c} = X^{p-r}\frac{A}{C} = 1 \tag{A.25}$$

Combining the last three relations, we find that

$$p = \frac{1}{6} - \frac{2\ln A - \ln B - \ln C}{3\ln X} \tag{A.26}$$

and taking $\frac{b}{a} = X^{q-p}\frac{B}{A}$ and $\frac{b}{c} = X^{q-r}\frac{B}{C}$, we have

$$q = \frac{1}{6} - \frac{2\ln B - \ln A - \ln C}{3\ln X} \tag{A.27}$$

Finally, taking that: $\frac{c}{a} = X^{r-p}\frac{C}{A}$ and $\frac{c}{b} = X^{r-q}\frac{C}{B}$, we get

$$r = \frac{1}{6} - \frac{2\ln C - \ln A - \ln B}{3\ln X} \tag{A.28}$$

Substituting last three expressions (7.71), (7.72) and (7.73) into (A.15), we obtain the constant d of simple cubic lattice which is statistically equivalent to triclinic one with lattice constants A, B and C in the following form:

$$d = X^p A = X^q B = X^r C \tag{A.29}$$

For TGS, the angles are: $\alpha = 125.9^\circ$, $\beta = 115.3^\circ$ and $\gamma = 115.1^\circ$, while the lattice constants are: $A = 0.942\,\text{nm}$, $B = 1.264\,\text{nm}$ and $C = 0.573\,\text{nm}$, the value of d is calculated by means of (A.25–A.27) is: $d = 0.839\,\text{nm}$. Now, we can look for dynamical equivalence of triclinic cell with constants A, B and C, and simple cubic cell with constant d.

The achievement of this equivalence requires knowing of interactions between atoms (molecules) of triclinic cell. We shall use Landau's estimate [189] of intermolecular interactions, i.e., we shall assume that interaction between two atoms is of the form

$$I(l) = \begin{cases} \dfrac{\phi}{l^\eta}; & l > 1\text{nm}; \\ \dfrac{\phi}{l^{-\eta}}; & l < 1\,\text{nm}, \end{cases} \quad , \eta > 0 \tag{A.30}$$

where $\eta > 0$, and φ is the interaction constant expressed in $J \times (10^{-9}\text{m})^{\pm\eta}$, while l is the distance between nearest neighbours expressed in 10^{-9} m.

The total interaction energy of triclinic cell is given as

$$J_T = 2\phi\left[\frac{1}{A^{-\eta}} + \frac{1}{B^{-\eta}} + \frac{1}{C^{-\eta}} + \frac{1}{(AB)^{-\eta}} + \frac{1}{(AC)^{-\eta}} + \frac{1}{(BC)^{-\eta}}\right] \quad \text{(A.31)}$$

since all lengths are less than 1 nm. The formula is valid if $A \geq B, C, AB, AC, BC$. If some of the lengths AB, AC and BC are higher than A, this term (or terms) has to be omitted from (A.32).

The total interaction energy of the equivalent simple cubic cell is given by

$$J_{SC} = \frac{6\Phi}{d^{-\eta}}, \quad d < 1\text{nm} \quad \text{(A.32)}$$

where Φ is the interaction constant of equivalent simple cubic cell. It is obvious that the structures will be dynamically equivalent if

$$\Phi = \frac{\phi}{3}d^{\eta}\left[\frac{1}{A^{-\eta}} + \frac{1}{B^{-\eta}} + \frac{1}{C^{-\eta}} + \frac{1}{(AB)^{-\eta}} + \frac{1}{(AC)^{-\eta}} + \frac{1}{(BC)^{-\eta}}\right] \quad \text{(A.33)}$$

For calculation of some physical characteristics of simple cubic lattice, most often interaction between two neighbours is used. This interaction is $j_{CE} = \Phi/d^{-\eta}$, i.e., in accordance with (A.32):

$$j_{CE} = \frac{\phi}{3}\left[\frac{1}{A^{-\eta}} + \frac{1}{B^{-\eta}} + \frac{1}{C^{-\eta}} + \frac{1}{(AB)^{-\eta}} + \frac{1}{(AC)^{-\eta}} + \frac{1}{(BC)^{-\eta}}\right] \quad \text{(A.34)}$$

Illustrative example 2

As an illustrative example, we shall first consider the ideal Heisenberg ferromagnetic with spin $S = \frac{1}{2}$ in Bloch's approximation [5, 134]. Before analysis, we wish to note that exchange interactions are, in accordance with general opinion, of exponential type [230]. It means that in formula (A.33), every term has to be prescribed in the following way: $\frac{1}{L^{-\eta}} = \mathrm{e}^{\ln \frac{1}{L^{-\eta}}} = \mathrm{e}^{\eta \ln L}$ $(L < 1)$. The Hamiltonian of this system in nearest neighbours' approximation is

$$\begin{aligned} H = {} & 3j_{CE}\sum_{n_x,n_y,n_z} B^{+}_{n_x,n_y,n_z}B_{n_x,n_y,n_z} \\ & -\frac{1}{2}j_{CE}\sum_{n_x,n_y,n_z} B^{+}_{n_x,n_y,n_z}\big(B_{n_x+1,n_y,n_z} + B_{n_x-1,n_y,n_z} \\ & +B_{n_x,n_y+1,n_z} + B_{n_x,n_y-1,n_z} + B_{n_x,n_y,n_z+1} + B_{n_x,n_y,n_z-1}\big) \end{aligned} \quad \text{(A.35)}$$

wherefrom follows the dispersion law of spin waves

$$E_{k_x,k_y,k_z} = (3 - \cos k_x d - \cos k_y d - \cos k_z d) j_{CE} \approx \frac{1}{2} j_{CE} k^2 d^2 \tag{A.36}$$

The ordering parameter in Bloch's approximation is given by

$$\sigma = 1 - 2 < B^+ B >= 1 - 2\zeta_{3/2}\tau^{3/2} \tag{A.37}$$

where

$$\zeta_{3/2} = \sum_{n=1}^{\infty} \frac{1}{n^{3/2}} \tag{A.38}$$

is Riemann's function, while

$$\tau = \frac{k_B T}{2\pi} \frac{1}{j_{CE}} \tag{A.39}$$

As it is seen, the value j_{CE} is expressed in parameters of triclinic structure (see formula (A.33)) and, consequently, the result (A.34) with τ expressed by (A.36) represents ordering parameter of ferromagnetic with triclinic lattice which is evaluated by means of equivalent simple cubic lattice.

The given example demonstrates the advantages of translation of the lower symmetry structure into equivalent simple cubic one. The statistical averages can be evaluated without mathematical complication and by means of standard and well-developed approaches.

Illustrative example 3

The more complicated problem is determining of ordering parameter of thin film cut off from triclinic crystal. In this case, we immediately go over to thin film of statistically and dynamically equivalent simple cubic structure. The interactions j_{CE} are given by formula (A.33). The triclinic film and the equivalent cubic one are presented in Fig. A.3.

The Hamiltonian of equivalent film in nearest neighbour's approximation is given by

$$\begin{aligned} H = \frac{1}{2} \sum_{n_x,n_y,n_z} \big(& j_{n_x+1,n_y,n_z;n_x,n_y,n_z} + j_{n_x-1,n_y,n_z;n_x,n_y,n_z} \\ & + j_{n_x,n_y+1,n_z;n_x,n_y,n_z} + j_{n_x,n_y-1,n_z;n_x,n_y,n_z} + j_{n_z,n_y,n_z+1;n_x,n_y,n_z} \\ & + j_{n_x,n_y,n_z-1;n_x,n_y,n_z} \big) B^+_{n_x,n_y,n_z} B_{n_x,n_y,n_z} \\ & - \frac{1}{2} \sum_{n_x,n_y,n_z} B^+_{n_x,n_y,n_z} \big(j_{n_x+1,n_y,n_z;n_x,n_y,n_z} B^+_{n_x+1,n_y,n_z} \\ & + j_{n_x-1,n_y,n_z;n_x,n_y,n_z} B^+_{n_x-1,n_y,n_z} + j_{n_x,n_y+1,n_z;n_x,n_y,n_z} B^+_{n_x,n_y+1,n_z} \\ & + j_{n_x,n_y,n_z+1;n_x,n_y,n_z} B^+_{n_x,n_y,n_z+1} + j_{n_x,n_y,n_z-1;n_x,n_y,n_z} B^+_{n_x,n_y,n_z-1} \big) \end{aligned} \tag{A.40}$$

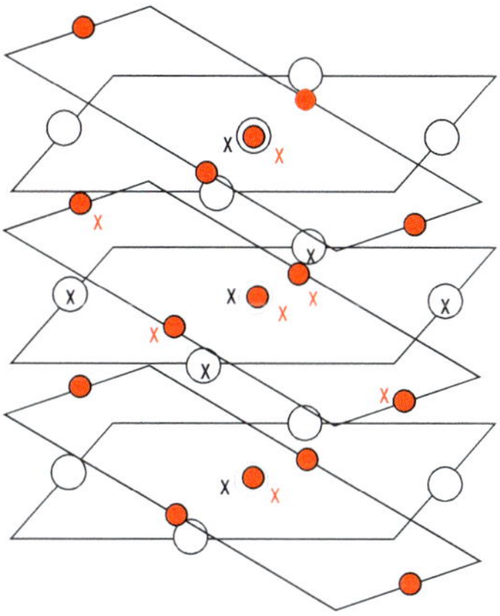

Fig. A.3: Full circles are atoms of triclinic film, while the empty circles are atoms of equivalent cubic cell.

For functions j, the following is valid:

$$j_{n_x\pm1,n_y,n_z;n_x,n_y,n_z} = j_{n_x,n_y\pm,n_z+1;n_x,n_y,n_z} = j_{n_x,n_y,n_z\pm1;n_x,n_y,n_z} = j_{CE} \tag{A.41}$$

The cut off in z direction leads to the following boundary conditions:

$$j_{n_x,n_y-1;n_x,n_y,0} = j_{n_x,n_y,N_z+1;n_x,n_y,N_z} = 0 \tag{A.42}$$

The film will be analyzed by means of Green's function

$$\begin{aligned} G_{n_x,n_y,n_z;m_x,m_y,m_z}(t) &= \ll B_{n_x,n_y,n_z}(t)|B^+_{n_x,n_y,n_z}(0) \gg \\ &= \Theta(t) < [B_{n_x,n_y,nz}(t), B^+_{n_x,n_y,n_z}(0)] > \end{aligned} \tag{A.43}$$

where the step-function is defined as $\Theta(t) = \begin{cases} 1, & t > 0; \\ 0, & t < 0 \end{cases}$

Steps of further calculation are [5.134] as follows:

- Formula (A.40) is differentiated with respect to time;
- Equations of motion for operator $B_{n_x,n_y,n_z}(t)$ are used;
- Fourier transformations:

$$G_{n_x,n_y,n_z;m_x,m_y,m_z}(t) = \int_{-\infty}^{+\infty} d\omega e^{-i\omega t} G_{n_x,n_y,n_z;m_x,m_y,m_z}(\omega);$$

$$\delta(t) = \frac{1}{2\pi} \int_{-\infty}^{+\infty} d\omega e^{-i\omega t} \tag{A.44}$$

are made;

– In order to use translational invariance in x, y planes, the following transformation is used [224]:

$$G_{n_x,n_y,n_z;m_x,m_y,m_z}(\omega) = \frac{1}{N_x N_y} \sum_{k_x,k_y} \mathrm{e}^{idk_x(n_x-m_x)+idk_y(n_y-m_y)} \Gamma(k_x,k_y,\omega) \tag{A.45}$$

– The boundary conditions (A.39) are applied.

The quoted procedure gives the system of three difference equations

$$1 \le n_z \le N_z - 1 \quad \frac{1}{2} j_{CE}(\Gamma_{n_z+1,m_z} + \Gamma_{n_z-1,m_z}) + \rho \Gamma_{n_z,m_z} = \frac{i\hbar}{2\pi} \delta_{n_z,m_z} \tag{A.46}$$

$$n_z = 0 \quad \frac{1}{2} j_{CE} \Gamma_{1,m_z} + \left(\rho + \frac{1}{2} j_{CE}\right) \Gamma_{0,m_z} = \frac{i\hbar}{2\pi} \delta_{0,m_z} \tag{A.47}$$

$$n_z = N_z \quad \frac{1}{2} j_{CE} \Gamma_{N_z-1,m_z} + \left(\rho + \frac{1}{2} j_{CE}\right) \Gamma_{N_z,m_z} = \frac{i\hbar}{2\pi} \delta_{N_z,m_z} \tag{A.48}$$

where

$$\rho = E - 3j_{CE} + j_{CE}(\cos dk_x + \cos dk_y) \tag{A.49}$$

It can be easily shown that by the substitution

$$\Gamma_{n_z;m_z}(\omega) = \sum_{\lambda=1}^{N_z} g(k_x,k_y,\lambda;\omega) \Omega_\lambda(m_z) F_\lambda(n_z) \tag{A.50}$$

where Ω are undetermined functions and F is given by

$$F_\lambda(n_z) = \sin(n_z+1)\frac{\pi\lambda}{N_z+1} - \sin n_z \frac{\pi\lambda}{N_z+1}; \quad \lambda = 1,2,3,\ldots,N_z \tag{A.51}$$

the system of Eqs. (A.43) to (A.45) reduces into one equation of the following form:

$$\sum_{\lambda=1}^{N_z} g(k_x,k_y,\lambda;\omega) \left(j_{CE} \cos \frac{\pi\lambda}{N_z+1} + \rho \right) \Omega_\lambda(m_z) F_\lambda(n_z) = \frac{i\hbar}{2\pi} \delta_{n_z,m_z} \tag{A.52}$$

for all $n_z = 0,1,2,\ldots,N_z$. Kronecker's symbol will be taken in the form $\delta_{n_z,m_z} = \sum_{\lambda=1}^{N_z} \Omega_\lambda(m_z) F_\lambda(n_z) = \begin{cases} 1, & n_z = m_z; \\ 0, & n_z \neq m_z, \end{cases}$ and we obtain the system of n_z^2 algebraic equations determining unknown functions Ω, so that the relation (A.49) gives

$$g(k_x, k_y, \lambda; \omega) = \frac{i\hbar}{2\pi} \frac{1}{j_{CE}\cos\frac{\pi\lambda}{N_z+1} + \rho} = \frac{i\hbar}{2\pi} \frac{1}{E - E_{k_x,k_y}} \tag{A.53}$$

where

$$E_{k_x,k_y,\lambda} = 3j_{CE} - j_{CE}\left(\cos k_x d + \cos k_y d + \cos\frac{\pi\lambda}{N_z+1}\right) \tag{A.54}$$

Substituting (A.50) into (A.47), i.e., in (A.42), we get final result for Green's function expression:

$$G_{n_x,n_y,n_z;m_x,m_y,m_z}(\omega) = \frac{i}{2\pi} \frac{1}{N_x N_y} \sum_{k_x,k_y} \mathrm{e}^{idk_x(n_x-m_x)+idk_y(n_y-m_y)}$$

$$\times \sum_{\lambda=1}^{N_z} \frac{\Omega_\lambda(m_z) F_\lambda(n_z)}{\omega - \omega_{k_x,k_y,\lambda}} \tag{A.55}$$

where

$$\omega = \frac{E}{\hbar}, \quad \omega_{k_x,k_y,\lambda} = \frac{E_{k_x,k_y,\lambda}}{\hbar} \tag{A.56}$$

By means of spectral intensity of the function G [133, 134]

$$I_G(\omega) = \frac{G(\omega + i\delta) - G(\omega - i\delta)}{\mathrm{e}^{\hbar\omega/\theta} - 1}$$

where $\vartheta = k_B T$, we obtain correlation function in the form

$$\langle B^+_{n_x,n_y,n_z}(0) B_{n_x,n_y,n_z}(t)\rangle = \int_{-\infty}^{+\infty} d\omega \mathrm{e}^{-it\omega} I_G(\omega)$$

The concentration of spin waves follows from this expression for $t = 0, n_i = m_i$, $i = (x, y, z)$:

$$\langle B^+_{n_x,n_y,n_z}(0) B_{n_x,n_y,n_z}(t)\rangle = \frac{1}{N_x N_y} \sum_{k_x,k_y} \sum_{\lambda=1}^{N_z} \frac{\Omega_\lambda(m_z) F_\lambda(n_z)}{\omega - \omega_{k_x,k_y,\lambda}} \tag{A.57}$$

It is very important to note that the transformation from configuration space to momentum space is not isomorphic one. It is transition of the type $n \to n - 1$ [227]. It is seen that for $N_z + 1$ values of index n_z the momentum index λ takes N_z values. It practically means that in one of the layers of the film there are no spin waves. Further, we shall consider three layers of film (see Fig. A.4). Due to the mentioned reduction, it goes over to three subfilms containing two layers.

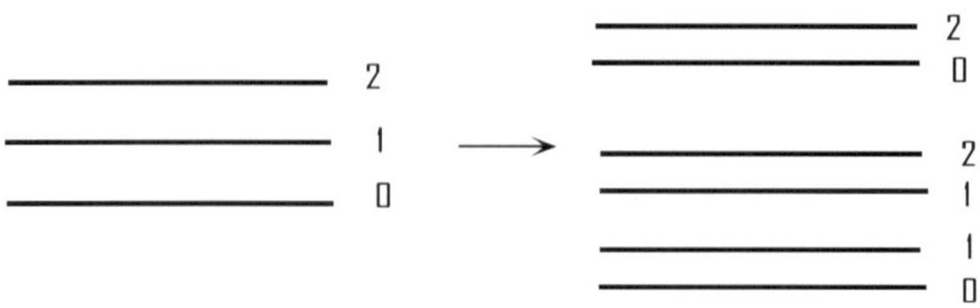

Fig. A.4: Autoreduction in three-layered film.

Now, we can find spin wave concentrations in subfilms. For subfilm 0–1, where it is $\delta_{m_z,n_z} = \sum_{\lambda=1}^{2} \Omega_\lambda(m_z)F_\lambda(m_z)$; $n_z, m_z \in (0,1)$, the solutions of the equation are

$$\Omega_1(0) = 2\frac{\sqrt{3}}{3}; \quad \Omega_2(0) = 0; \quad \Omega_1(1) = \frac{\sqrt{3}}{3}; \quad \Omega_2(1) = -\frac{\sqrt{3}}{3}. \tag{A.58}$$

Substituting this into (A.54), we obtain the requested concentrations:

$$\begin{aligned} \langle B^+B\rangle_{n_z=0} &= \frac{1}{4\sqrt{\pi}}\frac{\theta}{\jmath_{CE}}\sum_{n=1}^{\infty}\frac{\mathrm{e}^{-\frac{\jmath_{CE}}{2\theta}n}}{n}; \\ \langle B^+B\rangle_{n_z=1} &= \frac{1}{4\sqrt{\pi}}\frac{\theta}{\jmath_{CE}}\sum_{n=1}^{\infty}\frac{\mathrm{e}^{-\frac{3\jmath_{CE}}{2\theta}n}}{n} \end{aligned} \tag{A.59}$$

It is seen that concentration of the layers 0 and 1 differs. It is the consequence of broken symmetry. The corresponding ordering parameters are

$$\sigma_{n_z=0} = 1 - \frac{1}{2\sqrt{\pi}}\frac{\theta}{\jmath_{CE}}\sum_{n=1}^{\infty}\frac{\mathrm{e}^{-\frac{\jmath_{CE}}{2\theta}n}}{n}; \quad \sigma_{n_z=1} = 1 - \frac{1}{2\sqrt{\pi}}\frac{\theta}{\jmath_{CE}}\sum_{n=1}^{\infty}\frac{\mathrm{e}^{-\frac{3\jmath_{CE}}{2\theta}n}}{n} \tag{A.60}$$

The last result shows that the layer $n_z = 1$ is better ordered. It is physically less than stable since to spin in layer $n_z = 1$, exchange forces act from two sides, while to spin in layer $n_z = 0$, exchange forces act from one side, only.

In the same way, for subfilm 1–2, we have

$$\langle B^+B\rangle_{n_z=2} = \frac{1}{4\sqrt{\pi}}\frac{\theta}{j_{CE}}\sum_{n=1}^{\infty}\frac{e^{-\frac{j_{CE}}{2\theta}n}}{n};$$

$$\langle B^+B\rangle_{n_z=1} = \frac{1}{4\sqrt{\pi}}\frac{\theta}{j_{CE}}\sum_{n=1}^{\infty}\frac{e^{-\frac{3j_{CE}}{2\theta}n}}{n};$$

$$\sigma_{n_z=2} = 1 - \frac{1}{2\sqrt{\pi}}\frac{\theta}{j_{CE}}\sum_{n=1}^{\infty}\frac{e^{-\frac{j_{CE}}{2\theta}n}}{n};$$

$$\sigma_{n_z=1} = 1 - \frac{1}{2\sqrt{\pi}}\frac{\theta}{j_{CE}}\sum_{n=1}^{\infty}\frac{e^{-\frac{3j_{CE}}{2\theta}n}}{n} \tag{A.61}$$

Finally, for subfilm 0–2, we get

$$\langle B^+B\rangle_{n_z=0} = \langle B^+B\rangle_{n_z=2} = \frac{1}{8\sqrt{\pi}}\frac{\theta}{j_{CE}}\left(\sum_{n=1}^{\infty}\frac{e^{-\frac{j_{CE}}{2\theta}n}}{n} + \sum_{n=1}^{\infty}\frac{e^{-\frac{3j_{CE}}{2\theta}n}}{n}\right)$$

$$\sigma_{n_z=0} = \sigma_{n_z=2} = 1 - \frac{1}{4\sqrt{\pi}}\frac{\theta}{j_{CE}}\left(\sum_{n=1}^{\infty}\frac{e^{-\frac{j_{CE}}{2\theta}n}}{n} + \sum_{n=1}^{\infty}\frac{e^{-\frac{3j_{CE}}{2\theta}n}}{n}\right) \tag{A.62}$$

As it is seen, the results obtained are in full accordance with symmetry of structure.

The analysis of this section, in our opinion, represents the importance of transition from lower symmetry structure to higher symmetry one, since this transition reduces mathematical problems to minimum.

The recovery of equivalence between rectangular structures and the structures of lower symmetry (triclinic, monoclinic, etc.) is one of the most important necessities in the theory of nanostructures. The theoretical analysis of nanostructures is developed for rectangular and cylindrical ones. Concerning nanostructures of lower symmetry, up to now, there does not exist any systematically approach. Taking this into account, we attempted to solve this problem by introducing the concept of statistical and dynamical equivalence between structures of higher and lower symmetry.

The proposed method is not universal, unfortunately. As it was seen for some triclinic structures, the equivalent rectangular one does not exist. Nevertheless, in many actual cases, the method of equivalence can be applied. Concerning the monoclinic structure, this method is always successful.

Acknowledgements

This work was partially supported by the Ministry for Scientific and Technological Development, Higher Education and Information Society, Government of Republic of Srpska, Bosnia and Herzegovina (Grants: 19/6-020/961-21/18 and 19/6-020/961-35/18, as well as Grants: 19.032/961-42/19 and 19.032/961-36/19).

References

1. Ch. Kittel, *Introduction to Solid State Physics*, 8 ed. (J. Wiley & Sons, New York, **2005**).
2. J. M. Ziman, *Principles of the Theory of Solids* (University Press, Cambridge, **1972**).
3. L. D. Landau and E. M. Lifshitz, *Statistical Physics*, 22 edn. (Pergamon Press, Oxford, **2013**).
4. R. Kubo, Statistical Mechanics, *An Advanced Course with Problems and Solutions* (North-Holland Physics Publishing, Amsterdam, **1965**).
5. B. S. Tošić, *Statistical Physics* (Institute of Physics, Novi Sad **1978**) (in Serbian).
6. V. D. Sajfert and B. S. Tošić, The research of nanoscience progress, *J. Comput. Theor. Nanosci.* 7/1, 15–84 (**2010**); doi: 10.1166/jctn.2010.1333.
7. M. L. Bellac, *Quantum Physics* (Cambridge University Press, Cambridge, **2006**).
8. V. D. Sajfert, S. K. Jaćimovski, D. Popov, and B. S. Tošić, Statistical and dynamical equivalence of different elementary cells, *J. Comput. Theor. Nanosci.* 4, 619–626 (**2007**); doi: 10.1166/jctn.2007.025.
9. V. D. Sajfert, M. Garić, and B. S. Tošić, Nanostructure rectangular rod, *J. Comput. Theor. Nanosci.* 3, 431–439 (**2006**); doi: 10.1166/jctn.2006.3026.
10. V. D. Sajfert and B. S. Tošić, Order–disorder excitations in nanostructures. In *Series: Encyclopedia of Nanoscience and Nanotechnology*, H. S. Nalwa (Ed.) 50, 281–350 (**2011**); (American Scientific Publishers, Valencia, California, **2004–2011**).
11. A. Nabok, *Organic and Inorganic Nanostructures*, (Artech House MEMS Series, Boston **2005**).
12. I. Langmuir, Forces near the surfaces of molecules, *Chem. Rev.* 6, 451–479 (**1930**); doi: 10.1021/cr60024a002.
13. K. B. Blodgett, Monomolecular films of fatty acids on glass, *J. Am. Chem. Soc.* 56, 495–495 (**1934**); doi: 10.1021/ja01317a513.
14. H. Kühn, Energy transfer in monomolecular layers, *Naturwissenschaften.* 54, 429 (**1967**) (in German).
15. B. Mann and H. Kühn, Tunneling through fatty acid salt monolayers, *J. Appl. Phys.* 42, 4398–4405 (**1971**); doi: 10.1063/1.1659785.

16. B. W. Rossiter and J. F. Hamilton (Eds.), *Physical Methods of Chemistry* Vol. 3, Part A, Determination of chemical composition and molecular structure, 2 edn. (Wiley, New York, **1987**).
17. A. Aviram and M. Ratner, Molecular rectifiers, *Chem. Phys. Lett.* **29**, 277–283 (**1974**); doi: 10.1016/0009-2614(56)85031-1.
18. F. L. Carter, The molecular device computer: Point of departure for large scale cellular automata, *Physica D* 10, 175–194 (**1984**); doi: 10.1016/0167-2789(66)90260-4.
19. L. J. Blum, *Bio- and Chemi-Luminescent Sensors,* (World Scientific, London, **1997**).
20. T. E. Edmonds (Ed.), *Chemical Sensors* (Springer Science, New York, **1988**).
21. Proceedings of the 17th International Meeting on Chemical Sensors, Viena, Austria, July 15–19, **2018**; *AMA* 2018; https://www.ama-science.org/proceedings/listing/2835.
22. S. Y. Yurish (Ed.), Chemical sensors and biosensors. In *Book Series: Advances in Sensors: Reviews* Vol. 6, (IFSA, S.I., **2018**).
23. N. V. Kirianaki, N. V. Kirianaki, S. Y. Yurish, N. O. Shpak, and V. P. Deynega, *Data Acquisition and Signal Processing for Smart Sensors* (John Wiley & Sons, New York, **2002**).
24. J. R. Carstens, *Electrical Sensors and Transducers* (Prentice-Hall, New York, **1993**).
25. D. A. Krohn, *Fiber Optic Sensors: Fundamentals and Applications* 3rd edn. (ISA, North Carolina, **2000**).
26. A. Ekimov, Growth and optical properties of semiconductor nanocrystals in a glass matrix, *J. Lumin.* 70, 1–20 (**1996**); doi: 10.1016/0022-2313(96)00040-3.
27. Y. Ohfuti and K. Cho, Size- and shape-dependent optical properties of assembled semiconductor spheres, *J. Lumin.* 70, 203–211 (**1996**); doi: https://doi.org/10.1016/0022-2313(96)00055-5.
28. C. C. Yang and S. Li, Size dependence of optical properties in semiconductor nanocrystals, *Key Eng. Mater.* **444**, 133–162 (**2010**); doi: 10.4028/www.scientific.net/KEM.444.133.
29. T. Takagahara, Electron-phonon interactions in semiconductor quantum dots. In *Semiconductor Quantum Dots, NanoScience and Technology* Y. Masumoto and T. Takagahara (eds.) (Springer, Berlin, **2002**).
30. M. A. Stroscio and M. Dutta, Advances in quantum-dot research and technology: The path to application in biology, *Int. J. High Speed Electron. Syst.* 12/4, 1039–1056 (**2002**); doi:10.1142/S0129156402001915.
31. W. Zhoua and J. J. Coleman, Semiconductor quantum dots, *Curr. Opin. Solid State Mater. Sci.* 20/6, 352–360 (**2016**); doi: 10.1016/j.cossms.2016.06.006.
32. U. Woggon, *Optical Properties of Semiconductor Quantum Dots* (Springer, Berlin, **1997**).
33. V. Vasilevskiy, A. G. Rolo, and M. J. M. Gomes, One-phonon raman scattering from arrays of semiconductor nano-crystals, *Solid State Commun.* 104/7, 381–386 (**1997**); doi: 10.1016/S0038-1098(74)00356-6.

34. M. Dutta and M. A. Stroscio (Eds.), *Selected Topics in Electronics and Systems* Vol. 14: Quantum-based electronic devices and systems (World Scientific, Singapore, **1998**).
35. Z. W. Yan, S. L. Ban, and X. X. Liang, Polaron properties in ternary group-iii nitride mixed crystals, *AIP Conf. Proc.* 850/1, 1546 (**2006**); doi: 10.1063/1.2355294.
36. X.-Q. Li and Y. Arakawa, Optical linewidths in an individual quantum dot, *Phys. Rev. B* 60, 1915–1920 (**1999**); doi: 10.1103/PhysRevB.60.1915.
37. C. Wood and D. Jena (Eds.), Polarization effects in semiconductors: From *Ab Initio* theory to device application (Springer Science, Berlin, **2007**).
38. X.-Q. Li, H. Nakayama, and Y. Arakawa, Phonon Bottleneck in Quantum Dots: Role of lifetime of the confined optical phonons, *Phys. Rev. B* 59, 5069–5073 (**1999**); doi: 10.1103/PhysRevB.59.5069.
39. G. A. Narvaez, G. Bester, and A. Zunger, Carrier relaxation mechanisms in self-assembled (In, Ga)A/GaAs quantum dots: Efficient P $\rightarrow$ S auger relaxation of electrons, *Phys. Rev. B* 74, 075403 (**2006**); doi: https://doi.org/10.1103/PhysRevB.74.075403.
40. T. Takagahara, Theory of exciton dephasing in semiconductor quantum dots, *Phys. Rev. B* 60, 2638–2652 (**1999**); doi: https://doi.org/10.1103/PhysRevB.60.2638.
41. L. Li, X. Luo, X. Yang, M. Wang, X. Lü, and Y. Wu, An analog of photon-assisted tunneling in a periodically modulated waveguide array, *Sci. Rep.* 6, 35744 (**2016**); doi: 10.1038/srep35744.
42. H. Gotoh, H. Ando, and T. Takagahara, Radiative recombination lifetime of excitons in thin quantum boxes, *J. Appl. Phys.* 81, 1785–1789 (**1997**); doi: 10.1063/1.364034.
43. S.-W. Feng, Radiative recombination lifetime and exciton dimension in zinc blende $Al_xGa_{1-x}N$/AlN quantum boxes, *Jpn. J. Appl. Phys.* 48/5R, 051001 (**2009**); doi: 10.1143/JJAP.48.051001.
44. M. C. Löbl, I. Söllner, A. Javadi, T. Pregnolato, R. Schott, L. Midolo, A. V. Kuhlmann, S. Stobbe, A. D. Wieck, P. Lodahl, A. Ludwig, and R. J. Warburton, Narrow optical linewidths and spin pumping on charge-tunable close-to-surface self-assembled quantum dots in an ultrathin diode, *Phys. Rev. B* 96, 165440 (**2017**); doi: 10.1103/PhysRevB.96.165440.
45. M. Grundmann, J. Christen, N. N. Ledentsov, J. Bohrer, D. Bimberg, S. S. Ruvimov, P. Werner, U. Richter, U. Gosele, J. Heydenreich, V. M. Ustinov, A. E. Zhukov, P. S. Kop′ev, and Zh. I. Alferov, Ultranarrow luminescence lines from single quantum dots, *Phys. Rev. Lett.* 74, 4043–4046 (**1995**); doi: 10.1103/PhysRevLett.74.4043.
46. M. El-Said, Study of the energy level-crossings in $GaAsAl_xGa_{1-x}As$ quantum dots, *Solid State Commun.* 97, 971–974 (**1996**); doi: 10.1016/0038-1098(73)00684-2.
47. J. J. H. A. van Hest, G. A. Blab, H. C. Gerritsen, C.de M. Donega, and A. Meijerink, The role of a phonon bottleneck in relaxation processes for LnDoped $NaYF_4$ nanocrystals, *J. Phys. Chem. C* 122/7, 3985–3993 (**2018**); doi: 10.1021/acs.jpcc.7b11171.

48. S. Komiyama, O. Astafiev, V. Antonov, T. Kutsuwa, and H. Hirai, A single-photon detector in the far-infrared range, *Nature* 403, 405–407 (**2000**); doi: 10.1038/35000166.
49. V. N. Selvakumar and T. Takagahara, Weakly correlated exciton pair states in large quantum dots, *Phys. Rev. B* 53, R10516–R10519 (**1996**); doi: 10.1 103/PhysRevB.53.R10516.
50. J. Groenen, C. Priester, and R. Carles, Strain distribution and optical phonons in InAs/InP self-assembled quantum dots, *Phys. Rev. B* 60, 16013 (**1999**); doi: 10.1103/PhysRevB.60.16013.
51. U. W. Pohl, A. Strittmatter, Control of self-organized In(Ga)As/GaAs quantum dot growth. In *Semiconductor Nanostructures. NanoScience and Technology* D. Bimberg (Ed.) (Springer, Berlin, **2008**); doi: 10.1007/978-3 -540-77899-8_2.
52. P. Boucaud, K. S. Gill, J. B. Williams, M. S. Sherwin, W. V. Schoenfeld, and P. M. Petroff, Saturation of THz-frequency intraband absorption in InAs/GaAs quantum dot molecules, *Appl. Phys. Lett.* 77, 510–512 (**2000**); doi: 10.1063/1.127027.
53. S. Sauvage, P. Boucaud, F. H. Julien, J.-M. Gerard, and V. Thierry-Mieg, Intraband absorption in N-doped InAs/GaAs quantum dots, *Appl. Phys. Lett.* 71, 2785–2787 (**1997**); doi: 10.1063/1.120133.
54. H. Zhou, A. V. Nurmikko, C.-C. Chu, J. Han, R. L. Gunshor, and T. Takagahara, Observation of antibound states for exciton pairs by four-wave-mixing experiments in a single ZnSe quantum well, *Phys. Rev. B* 58, R10131–10134 (**1998**); doi: 10.1103/PhysRevB.58.R10131.
55. B. Legrand, S. Grandidier, J. P. Nyes, D. Stievenard, J. M. Gerard, and V. Thierry-Mieg, Scanning tunneling microscopy and scanning tunneling spectroscopy of self-assembled InAs quantum dots, *Appl. Phys. Lett.* 73, 96–98 (**1998**); doi: 10.1063/1.121792.
56. Z. M. Wang, *Self-Assembled Quantum Dots* (Springer Science, Berlin **2010**).
57. M. Souaf, M, Baira, O. Nasr, M. H. H. Alouane, H. Maaref, L. Sfaxi, and B. Ilahi, Investigation of the InAs/GaAs quantum dots' size: Dependence on the strain reducing layer's position, *Mater. (Basel)* 8(8), 4699–4709 (**2015**); doi: 10.3390/ma8084699.
58. C. Y. Jina, H. Y. Liu, S. Y. Zhang, Q. Jiang, S. L. Liew, and M. Hopkinson, Optical transitions in type-II InAs/GaAs quantum dots covered by a GaAsSb strain-reducing layer, *Appl. Phys. Lett.* 91, 021102 (**2007**); doi: 10 .1063/1.2752778
59. J.-Z. Zhang and I. Galbraith, Intraband absorption for InAs/GaAs quantum dot infrared photodetectors, *Appl. Phys. Lett.* 84, 1934 (**2004**); doi: 10.106 3/1.1687459.
60. T. Bruches, P. Boucaud, S. Sauvage, F. Glotin, R. Prazeres, J.-M. Ortega, A. Lemaitre, and J.-M. Gerard, Midinfrared second-harmonic generation in p-type InAs/GaAs self-assembled quantum dots, *Appl. Phys. Lett.* 75(6), 835–837 (**1999**); doi: 10.1063/1.124529.
61. S. Sauvage, P. Boucaud, T. Brunhes, V. Immer, E. Finkman, and J.-M. Gerard, Midinfrared absorption and photocurrent spectroscopy of InAs/GaAs

self-assembled quantum dots, *Appl. Phys. Lett.* 78, 2327–2329 (**2001**); doi: 10.1063/1.1365411.
62. V. V. Mitin, V. I. Pipa, A. V. Sergeev, M. Dutta, and M. Stroscio, High-gain quantum-dot infrared photodetector, *Infrared Phys. Techn.* 42(3–5), 467–472 (**2001**); doi: 10.1016/S1350-4495(01)00107-4.
63. V. Ryzhii, I. Khmyrova, V. Mitin, M. Stroscio, and M. Willander, On the detectivity of quantum-dot infrared photodetectors, *Appl. Phys. Lett.* 78(15), 3523–3525 (**2001**); doi: 10.1063/1.1376435.
64. J. Phillips, P. Bhattacharya, S. W. Kennedy, D. W. Beekman, and M. Dutta, Self-assembled InAs-GaAs quantum-dot intersubband detector, *IEEE J. Quantum Elec.* 35(6), 936–943 (**1999**); doi: 10.1109/3.766837.
65. G. Yusa and H. Sakaki, Coulomb blockade gap and coulomb staircase of a new type in various tunnel-junction arrays with two branches, *Superlattices and Microst.* 25(1)–2, 247–250 (**1999**); doi: 10.1006/spmi.1998.0688.
66. R. L. Sellin, Ch. Ribbat, M. Grundmann, N. N. Ledentsov, and D. Bimberg, Close-to-ideal device characteristics of high-power InGaAs/GaAs quantum dot lasers, *Appl. Phys. Lett.* 78(9), 1207–1209 (**2001**); doi: 10.1063/1.13505 96.
67. N. R. Jana, L. Gearhearth, and C. Murphy, Wet chemical synthesis of silver nanorods and nanowires of controllable aspect ratio, *Chem. Commun.* 7, 617–618 (**2001**); doi: 10.1039/b100521i.
68. V. M. Cepak and C. R. Martin, Preparation and stability of template-synthesized metal nanorod sols in organic solvents, *J. Phys. Chem.* 102(49), 9985–9990 (**1998**); doi: 10.1021/jp982882i.
69. Z. L. Wang, R. P. Gao, B. Nikoobakht, and M. A. El Sayed, Surface reconstruction of the unstable {110} surface in gold nanorods, *J. Phys. Chem.* 104(23), 5417–5420 (**2000**); doi: 10.1021/jp000800w.
70. M. Bianca, I. Van der Zande, M. R. Böhmer, G. Lambertus, J. Fokkink, and C. Schönenberger, Aqueous gold sols of rod-shaped particles, *J. Phys. Chem.* 101(6), 852–854 (**1997**); doi: 10.1021/jp963348i.
71. M. A. Stroscio and M. Dutta (Eds.), *Biological Nanostructures and Applications of Nanostructures in Biology: Electrical, Mechanical, and Optical Properties* (Springer Science, Berlin, **2004**).
72. N. R. Jana, L. Gearhearth, and C. J. Murphy, Wet chemical synthesis of high aspect ratio cylindrical gold nanorods, *J. Phys. Chem.* 105(19), 4065–4067 (**2001**); doi: 10.1021/jp0107964.
73. E. Yasuda, M. Inagaki, K. Kaneko, M. Endo, A. Oya, and Y. Tanabe, *Carbon Alloys: Novel Concepts to Develop Carbon Science and Technology* (*Elsevier*, **2003**).
74. A. Henglein and J. Giersig, Reduction of Pt(II) by H2: Effects of Citrate and NaOH and reaction mechanism, *Phys. Chem.* 104(29), 6767–6772 (**2000**); doi: 10.1021/jp000801o.
75. M. E. Toimil Molares, V. Buschmann, D. Dobrev, R. Neumann, R. Scholz, I. U. Schuchert, and J. Vetter, Single-crystalline copper nanowires produced by electrochemical deposition in polymeric ion track membranes, *Adv. Mater.* 13(1), 62–65 (**2001**); doi: 10.1002/1521-4095(200101)13:1⟨62::AID-ADMA 62⟩3.0.CO;2-7.

76. S. Wang and S. Yang, Preparation and characterization of oriented PbS crystalline nanorods in polymer films, *Langmuir* 16(2), 389–397 (**2000**); doi: 10.1021/la990780t.
77. C. C. Chen, C. Y. Chao, and Z. H. Lang, Simple solution-phase synthesis of soluble CdS and CdSe nanorods, *Chem. Mater.* 12(6), 1516–1518 (**2000**); doi: 10.1021/cm9907920.
78. J. G. Korvink and A. Greiner, *Semiconductors for Micro- and Nanotechnology* (Wiley-VCH, Weinheim, **2002**).
79. A. J. Šetrajčić–Tomić, J. K. Popović, M. Vojnović, Lj. D. Džambas, and J. P. Šetrajčić, Review of core-multishell nanostructured models for nanobiomedical and nano-biopharmaceutical application, *Bio-Med. Mater. Eng.* 29(4), 451–471 (**2018**); doi: 10.3233/BME-181002.
80. K. A. Phillips, The intersection of biotechnology and pharmacogenomics: Health policy implications, *Health Aff.* 25(5), 1271–1280 (**2006**); doi: 10.13 77/hlthaff.25.5.1271.
81. K. J. Morrow, R. Bawa, and C. Wei, Recent advances in basic and clinical nanomedicine, *Med. Clin. N. Am.* 91(5), 805–843 (**2007**); doi: 10.1016/j.m cna.2007.05.009.
82. K. A. Wani and R. Kothari, Nano technology in medicine and future implications: A mini review, *J. Nanomed. Nanotechnol.* 9(2), 497 (**2018**); doi: 1 0.4172/2157-7439.1000497.
83. B. Sarmento and J. das Neves (Eds.), *Biomedical Applications of Functionalized Nanomaterials Concepts, Development and Clinical Translation, A volume in Micro and Nano Technologies* Eds. (Elsevier, **2018**).
84. D. Popov, V. D. Sajfert, J. P. Šetrajčić, and N. Pop, Coherent states formalism applied to the quantum well model, *Quantum Matter* 3(4), 388–393 (**2014**); doi: 10.1166/qm.2014.1137.
85. J. P. Šetrajčić, V. D. Sajfert, and S. K. Jaćimovski, Fundamental preferences of the phonon engineering for nanostructural samples, *Rev. Theor. Sci.* **4**(4), 353–401 (**2016**); doi: 10.1166/rits.2016.1067.
86. V. D. Sajfert, J. P. Šetrajčić, S. K. Jaćimovski, and D. Popov, Theoretical basis for phonon engineering of nanostructures, *Quantum Matter* 6/1, 18–27 (**2017**); doi: 10.1166/qm.2017.1391.
87. A. J. Šetrajčić-Tomić, D. Rodić, I. J. Šetrajčić, V. D. Sajfert, and J. P. Šetrajčić, Basics of optical engineering – analysis of environmental and quantum size effects on the optical characteristics of molecular crystalline nanofilms, *Photon. Nanostruct.* 31, 115–128 (**2018**). doi: 10.1016/j.photoni cs.2018.05.011.
88. V. E. Demidov, O. Dzyapko, S. O. Demokritov, G. A. Melkov, and A. N. Slavin, Observation of spontaneous coherence in Bose-Einstein condensate of magnons, *Phys. Rev. Lett.* 100, 047205 (**2008**); doi: 10.1103/PhysRevLe tt.100.047205.
89. V. M. Agranovich, Dispersion of Electromagnetic Waves in Crystals, *Zh. Eksper. Teoret. Fiz.* 37, 430 (1959); *JETP* 10(2), 307 (**1960**); http://www. jetp.ac.ru/cgi-bin/dn/e_010_02_0307.pdf.

V. M. Agranovich, L. N. Ovander, and B. S. Toshich, Contribution to the theory of nonlinear polarizability of crystals, *Zh. Eksper. Teoret. Fiz.* 50(5), 1332–1342 (**1965**); *JETP* 23, 885–892 (1968); http://www.jetp.ac.ru/cgi-bin/dn/e_023_05_0885.pdf.

90. V. M. Agranovich and B. S. Toshich, Collective properties of frenkel excitons, *Zh. Eksper. Teoret. Fiz.* 53, 149 (1967); *JETP* 26, 104 (**1968**); http://www.jetp.ac.ru/cgi-bin/dn/e_026_01_0104.pdf.
91. G. Cao, Nanostructures & Nanomaterials, Synthesis, Properties & Applications (Imperial College Press, London, **2004**); *J. Am. Chem. Soc.* 126(29), 14679 (**2004**); doi: 10.1021/ja0409457.
92. C. Delerue and M. Lannoo, *Nanostructures – Theory and Modelling* (Springer, Berlin, **2009**).
93. H. E. Schaefer, *Nanoscience — The Science of the Small in Physics, Engineering, Chemistry, Biology and Medicine* (Springer, Berlin, **2010**).
94. J. Jeevanandam, A. Barhoum, Y. S. Chan, A. Dufresne, and M. K. Danqua, Review on nanoparticles and nanostructured materials: History, sources, toxicity and regulations, *Beilstein J. Nanotechnol.* 9, 1050–1074 (**2018**); doi: 10.3762/bjnano.9.98.
95. G. Zhang and H. Zhang, Thermal conduction and rectification in few-layer graphene Y junctions, *Nanoscale* 3(11), 4604–4607 (**2011**); doi: 10.1039/c1nr10945f.
96. D. L. Nika, A. I. Cocemasov, and A. A. Balandin, Specific heat of twisted bilayer graphene: Engineering phonons by atomic plane rotations, *Appl. Phys. Lett.* 105(3), 031904 (**2014**); doi: 10.1063/1.4890622.
97. L. Lindsay, D. A. Broido, and N. Mingo, Flexural phonons and thermal transport in multilayer graphene and graphite, *Phys. Rev. B* 83, 235428 (**2011**); doi: 10.1103/PhysRevB.83.235428.
98. J.-W. Jiang, B.-Sh. Wang, and T. Rabczuk, Acoustic and breathing phonon modes in bilayer graphene with Moiré patterns, *Appl. Phys. Lett.* 101, 023113 (**2012**); doi: 10.1063/1.4735246.
99. J. P. Šetrajčić, S. K. Jaćimovski, D. I. Raković, and D. I. Ilić, Phonon spectra in crystalline nanostructures. In *Electrical and Computer Engineering Series: Advances in Simulation, Systems Theory and Systems Engineering*, N. E. Mastorakis, V. V. Kluev, Dj. Koruga (Eds.), (WSEAS Press, Athens, **2003**) 146–151.
100. M. A. Stroscio and M. Dutta, *Phonons in Nanostructures* (University Press, Cambridge, **2004**).
101. A. A. Balandin and D. L. Nika, Phononics in low-dimensional materials, *Mater. Today* **15**(6), 266 (**2012**); doi: 10.1016/S1369-7021(12)70117-7.
102. W. T. S. Huck (Ed.), *Nanoscale Assembly — Chemical Techniques* (Springer, New York, **2005**).
103. Self-Organized Nanoscale Materials, Eds. M. Adachi, D. J. Lockwood (Springer, New York, **2006**).
104. C. Delerue and M. Lannoo, Nanostructures — Theory and Modelling (Springer, Berlin, **2009**).
105. H. E. Schaefer, Nanoscience — The Science of the Small in Physics, Engineering, Chemistry, Biology and Medicine (Springer, Berlin, **2010**).

106. J. P. Šetrajčić, S. K. Jaćimovski, and V. D. Sajfert, Phonon Engineering Theory of Crystalline Layered Nanostructure (Lambert Academic Publication, Saarbrcken, **2015**).
107. J. P. Šetrajčić, S. K. Jaćimovski, and V. D. Sajfert, Phonon contribution to heat capacitance of nanolayered crystalline structures, *Mod. Phys. Lett. B* 29(4), 1550008 (**2015**); doi: 10.1142/S0217984915500086.
108. J. P. Šetrajčić, S. K. Jaćimovski, and S. M. Vučenović, Diffusion of phonons through (along and across) the ultrathin crystalline films, *Physica A* 486, 839–848 (**2017**), doi: 10.1016/j.physa.2017.06.003.
109. J. P. Šetrajčić, D. I. Ilić, and S. K. Jaćimovski, The influence of the surface parameter changes onto the phonon states in ultrathin crystalline films, *Physica A* 496, 434–445 (**2018**), doi: 10.1016/j.physa.2017.12.138.
110. S. K. Jaćimovski, B. S. Tošić, J. P. Šetrajčić, and V. D. Sajfert, Charge tunneling in doped quantum rod, *Int. J. Mod. Phys. B* 24(27), 5345–5358 (**2010**); doi: 10.1142/S0217979210054725.
111. B. S. Tošić, J. P. Šetrajčić, and S. K. Jaćimovski, Methods of Theoretical Physics, Ch. 7, *University of Criminal Investigation and Police Studies* (Belgrade, **2018**) (in Serbian).
112. S. Elaydi, *An Introduction to Difference Equations* (Springer-Verlag, New York, **2005**).
113. W. G. Kelley, A. C. *Peterson, Difference Equations, An Introduction with Applications* (Academic Press, San Diego, **2001**).
114. L. C. Maximon, *Differential and Difference Equations — A Comparison of Methods of Solution* (Springer, Washington DC, **2016**).
115. S. S. Cheng, *Partial Difference Equations, Advances in Discrete Mathematics and Applications* (CRC Press, Boca Raton, **2003**).
116. P. Cull, M. Flahive, and R. Robson, *Difference Equations, From Rabbits to Chaos* (Springer, New York, **2005**).
117. D. L. Jagerman, *Difference Equations with Applications to Queues* (Marcel Dekker, New York, **2000**).
118. V. Lakshmikantham and D. Trigiante, *Theory of Difference Equations, Numerical Methods and Applications* (Marcel Dekker Inc, New York, **2002**).
119. J. A. Schwarz, C. I. Contescu, and K. Putyera, *Dekker Encyclopedia of Nanoscience and Nanotechnology* Vol. 1 (CRC Press, London, **2004**).
120. R. E. Mickens, *Difference Equations, Theory, Applications and Advanced Topics* (CRC Press, London, **2015**).
121. B. S. Tošić, V. D. Sajfert, D. Popov, J. P. Šetrajčić, and D. Ćirić, Application of Difference Calculus in Analysis of Nanostructures (Vojvodina Academy of Sciences and Arts, Novi Sad, **2005**) (in Serbian).
122. Biological Nanostructures, Materials, and Applications, Eds. M. C. Demirel, H. De Long, J. L. Moran Lopez, J. Rusling (ECS Transactions, 3(19), **2007**).
123. S. N. Elaydi, S. Axler, F. W. Gehring, and K. A. Ribet, An Introduction to Difference Equations (Springer, New York, **2000**).
124. E. R. Benton, Ordinary Differential and Difference Equations. In *Handbook of Applied Mathematics*, Ed. C. E. Pearson (Springer, Boston, **1990**).

125. S. G. Davison and M. Steslicka, Basic Theory of Surface States (Clarendon, Oxford, **1996**).
126. Nanostructured Materials, Ed. G. Wilde, Vol. 1; Atomic and Molecular Manipulation, Eds. A. J. Mayne, G. Dujardin, Vol. 2; In Series: Frontiers of Nanoscience (Elsevier Ltd., **2009**, **2011**).
127. S. M. Stojković, D. Lj. Mirjanić, J. P. Šetrajčić, D. D. Šijačić, and I. K. Junger, Spectra and states of electrons in surface perturbed quantum wires, *Surf. Sci.* 477, 235–242 (**2001**); doi: 10.1016/S0039-6028(01)00887-1.
128. V. D. Sajfert, J. P. Šetrajčić, D. Popov, and B. S. Tošić, Difference equations in condensed matter physics and their applications to the exciton system in thin molecular film, *Physica A* 353, 217–234 (**2005**); doi: 10.1016/j.physa.2005.01.022.
129. S. K. Jaćimovski, B. S. Tošić, J. P. Šetrajčić, V. M. Zorić, V. D. Sajfert, and D. I. Ilić, Thermodynamics of mechanical oscillations in crystalline superlattices, *Int. J. Mod. Phys. B* 21(6), 917–930 (**2007**); doi: 10.1142/S0217979207036783.
130. J. P. Šetrajčić, S. K. Jaćimovski, and S. M. Vučenović, *Phonon Engineering Theory of Crystalline Layered Nanostructures* (Academy of Criminalistic and Police Studies, Zemun – Belgrade, **2017**).
131. B. S. Tošić, J. P. Šetrajčić, and S. K. Jaćimovski, Methods of theoretical physics, Ch. 12, *University of Criminal Investigation and Police Studies* (Belgrade, **2018**) (In Serbian).
132. G. Rickayzen, *Green's Functions and Condensed Matter* (Academic Press, London, **1980**).
133. G. D. Mahan, *Many Particle Physics* (Kluwer Academic, New York, **2000**).
134. S. V. Tyablikov, *Methods in the Quantum Theory in Magnetism* (Plenum Press, New York, **1967**).
135. A. Aharoni, *Introduction to the Theory of Ferromagnetism* (University Press, Oxford, **2000**).
136. S. Chikazumi, *Physics of Ferromagnetism* (University Press, Oxford, **2009**).
137. E. Pavarini, Magnetism: Models and mechanisms. In E. Pavarini, E. Koch, and U. Schollwöck (Eds.) *Emergent Phenomena in Correlated Matter* (Jülich, **2013**).
138. R. S. Knox, *Theory of Excitons* (Academic Press, Waltham, **1963**).
139. D. C. Reynolds, *Excitons* (Academic Press, Waltham, **1981**).
140. A. S. Davydov, *Theory of Molecular Excitons* (*Springer*, Berlin, **2013**).
141. V. M. Agranovich, *Theory of Excitons* (Nauka, Moskwa, **1968**) (in Russian).
142. V. M. Agranovich, *Excitations in Organic Solids* (University Press, Oxford, **2014**).
143. V. M. Agranovich, *Crystal Optics with Spatial Dispersion, and Excitons* (Springer, Berlin, **2014**).
144. F. J. García de Abajo, Optical excitations in electron microscopy, *Rev. Mod. Phys.* 82(1), 209–276 (**2010**); doi: 10.1103/RevModPhys.82.209.
145. J. Singh and R. T. Williams (Eds.), *Excitonic and Photonic Processes in Materials* (Springer, Berlin, **2016**).

146. S. S. Pelemiš, J. P. Šetrajčić, B. Markoski, N. V. Delić, and S. M. Vučenović, Selective absorption in twolayered optic films, *J. Comput. Theor. Nanosci.* 6(7), 1474–1477 (**2009**); doi: 10.1166/jctn.2009.1195.
147. B. Škipina, D. Lj. Mirjanić, S. M. Vučenović, J. P. Šetrajčić, I. J. Šetrajčić, A. J. Šetrajčić-Tomić, S. S. Pelemiš, and B. Markoski, Selective IR absorption in molecular nanofilms, *Opt. Mater.* 33, 1578–1584 (**2011**); doi: 10.10 16/j.optmat.2011.04.008.
148. J. P. Šetrajčić, S. K. Jaćimovski, V. D. Sajfert, and D. Rodić, Some optical exceptions of molecular nanofilm-structures, *Quantum Matter* 3(4), 394–399 (**2014**); doi: 10.1166/qm.2014.1138.
149. J. P. Šetrajčić, D. Rodić, V. D. Sajfert, N. Pop, and M. D. Popov, Confinement consequences on optics of molecular crystalline nanofilms, *Quantum Matter* 6(1), 28–30 (**2017**); doi: 10.1166/qm.2017.1392.
150. A. J. Šetrajčić-Tomić, D. Rodić, I. J. Šetrajčić, V. D. Sajfert, and J. P. Šetrajčić, Basics of optical engineering — analysis of environmental and quantum size, effects on the optical characteristics of molecular crystalline nanofilms, *Photon. Nanostruct.* 31, 115–128 (**2018**); doi: 10.1016/j.photoni cs.2018.05.011.
151. Y. Toyozawa, A proposed model for the explanation of the *Urbach Rule, Prog. Theor. Phys.* 22(3), 455–457 (**1959**); doi: 10.1143/PTP.22.455.
152. T. H. Keil, Theory of the Urbach Rule, *Phys. Rev.* 144, 582 (**1966**); doi: 1 0.1103/PhysRev.144.582.
153. R. K. Waring and W. Y. Hsu, Urbach rule behavior in strongly absorbing fine particle solids, *J. App. Phys.* 54, 4093 (**1983**); doi: 10.1063/1.332541.
154. L. Bai, Ch. Xu, P. G. Schunemann, K. Nagashio, R. S. Feigelson, and N. C. Giles, Urbach rule used to explain the variation of the absorption edge in $CdGeAs_2$ crystals, *J. Phys.* 17(3), 549 (**2005**); doi: 10.1088/0953-8984/17/ 3/013.
155. I. Studenyak, M. Kranjčec, and M. Kurik, Urbach rule in solid state physics, *Int. J. Optics App.* 4(3), 76–83 (**2014**); doi: 10.5923/j.optics.20140403.02.
156. V. D. Sajfert and B. S. Tošić, Conductance properties of cylindrical quantum nano dots, *J. Comput. Theor. Nanosci.* 2, 148–153 (**2005**); doi:10.1166/jct n.2005.013.
157. R. P. Agarwal, *Difference Equations and Inequalities: Theory, Methods and Applications*, (Marcel Dekker, New York, **2000**).
158. Y. Pathak and D. Thassu, *Drug Delivery Nanoparticles — Formulation and Characterization* (Informa, New York, **2009**).
159. P. M. Chaikin and T. C. Lubensky, *Principles of Condensed Matter Physics* (University Press, Cambridge, **1995**).
160. J. D. Martin, What's in a Name Change? Solid state physics, condensed matter physics, and materials science, *Physics in Perspective.* 17(1), 3–32 (**2015**); doi: 10.1007/s00016-014-0151-7.
161. A. N. Cleland, *Foundations of Nanomechanics* (*Springer*, Berlin, **2003**).
162. C. Dupas, P. Houdy, and M. Lahmani (Eds.) *Nanoscience — Nanotechnologies and Nanophysics* (Springer, Berlin, **2007**).

163. O. Manasreh, *Semiconductor Heterojunctions and Nanostructures* (McGraw-Hill, New York, **2005**).
164. C. Politis and W. Schommers (Eds.), *Series on the Foundations of Natural Science and Technology* Eds., ISSN (print): 2010–1961 (World Scientific, Singapore, **2006**).
165. M. C. Tringides, M. Jałochovski, and E. Bauer, Quantum size effect in metallic nanostructures, *Physics Today* 60(4), 50–54 (**2007**); doi: 10.1063/1.2731973.
166. M. Pruton, *Introduction to Surface Physics* (*Clarendon*, Oxford, **1995**).
167. V. D. Sajfert, J. P. Šetrajčić, S. K. Jaćimovski, and B. S. Tošić, Thermodynamic and kinetic properties of phonons in cylindrical quantum dots, *Physica E* 353, 479–491 (**2005**); doi: 10.1016/j.physe.2004.07.008.
168. D. N. Zubarjev, V. Morozov, and G. Ropke, *Statistical Mechanics of Nonequilibrium Processes*, Vol. 1: Basic concepts, kinetic theory; Vol. 2: Relaxation and hydrodynamic processes (Wiley-VCH, Berlin, **1996**, **1997**) (in Russian).
169. Ch. Kittel, *Quantum Theory of Solids* (Wiley, New York, **1987**).
170. G. D. Mahan, *Condensed Matter in a Nutshell* (University Press, Princeton, **2011**).
171. F. Marsiglio and J. P. Carbotte, Electron-Phonon Superconductivity, http://arxiv.org/find/cond-mat/, No: 0106143 (**2001**).
172. A. K. Saxena, *High-Temperature Superconductors* (Springer, Berlin, **2010**).
173. J. M. Ziman, *Electrons and Phonons: The Theory of Transport Phenomena in Solids* (Oxford University Press, New York, **2001**).
174. A. A. Balandin and K. L. Wang, Significant decrease of the lattice thermal conductivity due to phonon confinement in a free-standing semiconductor quantum well, *Phys. Rev. B* 58(3), 1544 (**1998**); doi: 10.1103/PhysRevB.58.1544.
175. A. A. Balandin and D. L. Nika, Phononics in low-dimensional materials, *Materials Today* 15(6), 266 (**2012**); doi: 10.1016/S1369-7021(12)70117-7.
176. B. S. Tošić, J. P. Šetrajčić, R. P. Đajić, and D. Lj. Mirjanić, Phonons in broken-symmetry structures, *Phys. Rev. B* 36, 9094–9097 (**1987**); doi: 10.1103/PhysRevB.36.9094.
177. B. S. Tošić, J. P. Šetrajčić, D. Lj. Mirjanić, and Z. V. Bundalo, Low temperature properties of thin films, *Physica A* 184, 354–366 (**1992**); doi: 10.1016/0378-4371(71)90311-D.
178. D. Popov, V. D. Sajfert, and B. S. Tošić, Thermodynamic and kinetic properties of mechanical oscillations in thin films, *Int. J. Mod. Phys. B* 20(24), 3507–3532 (**2006**); doi: 10.1142/S0217979206035564.
179. E. N. Economou, *Green's Functions in Quantum Physics* (*Springer*, Berlin, **2006**).
180. S. Doniach, E. H. Sondheimer, *Green's Functions for Solid State Physicists* (Imperial College Press, London, **1998**).
181. J. P. Šetrajčić, V. D. Sajfert, D. Lj. Mirjanić, and S. M. Vučenović, Operator method for solving the difference equations, *Proceedings 5th WSEAS International Conference on Applied Mathematics*, 488(116), 1–4 (**2004**).

182. J. P. Šetrajčić, S. K. Jaćimovski, V. D. Sajfert, and I. J. Šetrajčić, Specific quantum mechanical solution of difference equation of hyperbolic type, *Commun. Nonlinear Sci. Numer. Simulat.* 19(5), 1313 (**2014**); doi: 10.1016/j.cnsns.2013.08.026.
183. R. E. Mickens, *Difference Equations: Theory, Applications and Advanced Topics* 3rd edn. (CRC Press, New York, **2015**).
184. D. L. Nika, A. I. Cocemasov, and A. A. Balandin, Specific heat of twisted bilayer graphene: Engineering phonons by atomic plane rotations, *Appl. Phys. Lett.* 105, 031904 (**2014**); doi: 10.1063/1.4890622.
185. G. Strobl, *Condensed Matter Physics: Crystals, Liquid Crystals, and Polymers* (Springer-Verlag, Berlin, **2004**).
186. W. Harrison, *Solid State Theory*, (Dover, New York, **1979**).
187. N. N. Bogolyubov, *Problems of the Dynamic Theory in Statistical Physics* (*Nauka*, Moscow, **1946**) (in Russian)
188. I. Prigogine, *Non-Equilibrium Statistical Mechanics* (Dover, New York, **2017**).
189. D. N. Zubarev, *Nonequilibrium Statistical Thermodynamics* (Consultants Bureau, New York, **1974**); D. Zubarev, *Statistical Mechanics of Nonequilibrium Processes, Basic Concepts, Kinetic Theory* (Wiley, New York, **1996**).
190. S. Lazarev, M. R. Pantić, S. M. Stojković, B. S. ToŠić, and J. P. Šetrajčić, Thermodynamical properties of ultrathin layered structures, *J. Phys. Chem. Sol.* 61, 931 (**2000**); doi: 10.1016/S0022-3697(76)00392-3.
191. D. Popov, S. K. Jaćimovski, B. S. Tošić, and J. P. Šetrajčić, Kinetics of thin films mechanical oscillations, *Physica A* 317, 129–139 (**2003**); doi: 10.1016/S0378-4371(02)01328-6.
192. S. K. Jaćimovski, J. P. Šetrajčić, B. S. Tošić, and V. D. Sajfert, Thermodynamics of mechanical oscillations in superlattices, *Mater. Sci. Forum* 453–454, 33–36 (**2004**); doi: 10.4028/www.scientific.net/MSF.453-454.33.
193. J. P. Šetrajčić, V. M. Zorić, N. V. Delić, D. Lj. Mirjanić, and S. K. Jaćimovski, Phonon participation in thermodynamics and superconductive properties of thin ceramic films, Ch. 15, 317–348. In *Thermodynamics*, M.Tadashi (Ed.) (*InTech*, Vienna, **2011**).
194. J. P. Šetrajčić, D. Lj. Mirjanić, S. M. Vučenović, D. I. Ilić, B. Markoski, S. K. Jaćimovski, V. D. Sajfert, and V. M. Zorić, Phonon contribution in thermodynamics of nano-crystalline films and wires, *Acta Phys. Pol. A* 115/4, 778 (**2009**); doi: 10.12693/APhysPolA.115.778.
195. S. K. Jaćimovski, J. P. Šetrajčić, M. S. Jaćimovski, and V. V. Stojanović, Phonon contribution in thermodynamics and transport properties of ultrathin ceramic films, *Acta Phys. Pol. A* 126/3, 811 (**2014**); doi: 10.12693/APhysPolA.126.811.
196. D. I. Ilić, S. M. Vučenović, S. K. Jaćimovski, V. M. Zorić, and J. P. Šetrajčić, Phonon spectra and thermodynamics of crystalline nanowires, *Proceedings Materials Research Society Spring'07 Conference; Symposium DD: Low-Dimensional Materials — Synthesis, Assembly, Property Scaling, and Modeling* Vol.1017, 76–81 (**2007**).

197. B. Sahoo, V. Kuncser, S. Kirsch, and W. Kleune, Superconductivity of MgB_2 thin films prepared by coevaporation and magnetism of MgB_2/Fe multilayers studied by Mössbauer spectroscopy, *Phase Transitions* 76/4–5, 423–435 (**2003**); doi: 10.1080/0141159031000102985.
198. H.-J. Kim, W. N. Kang, E.-Mi Choi, M.-S. Kim, K. H. P. Kim, and S.-Ik. Lee, High current-carrying capability in *c*-axis-oriented superconducting MgB_2 thin films, *Phys. Rev. Lett.* 87, 087002 (**2001**); doi: 10.1103/PhysRevLett.87.087002.
199. H. J. Choi, D. Hinks, H. Sun, M. L. Cohen, and S. G. Louie, The origin of the anomalous superconducting properties of MgB_2, *Nature* 418, 758–760 (**2002**).
200. Shu-H. Cheng, Y. Zhang, H.-Zh. Wang, Yu.-L. Li, C. Yang, and Y. Wanga, Fabrication and characterization of superconducting MgB_2 thin film on graphene, *AIP Advances* 8, 075015 (**2018**); doi: 10.1063/1.5023658.
201. P. M. Chaikin and T. C. Lubensky, *Principles of Condensed Matter Physics*, (University Press, Cambridge, **1995**).
202. M. Rieth and W. Schommers (Eds.), *Handbook of Theoretical and Computational Nanotechnology* (10-Volume Set) (World Scientific, Singapore, **2006**).
203. A. A. Maradudin (Ed.), Light scattering and nanoscale surface roughness, In *Nanostructure Science and Technology*, (Springer, New York **2007**).
204. J. P. Šetrajčić, D. I. Ilić, B. Markoski, A. J. Šetrajčić, S. M. Vučenović, D. Lj. Mirjanić, B. Škipina, and S. S. Pelemiš, Adapting and application of the Green's functions method onto research of the molecular ultrathin film optical properties, *Physica Scripta T* 135, 014043, 1–4 (**2009**); doi: 10.1088/0031-8949/2009/T135/014043.
205. J. P. Šetrajčić, Adequate determination of micro and macro properties of optical nano-crystals, *Opto-Electron. Rev.* 25/4, 303–310 (**2017**); doi: 10.1016/j.opelre.2017.08.003.
206. V. D. Sajfert, S. K. Jaćimovski, J. P. Šetrajčić, Lj. Mašković, N. Bednar, N. Pop, and B. S. Tošić, Optical properties of nanostructures, *J. Comput. Theor. Nanosci.* 8(11), 2285–2290 (**2011**); doi: 10.1166/jctn.2011.1957.
207. I. E. Dzyaloshinskii and L. P. Pitaevskii, Van der Waals forces in an inhomogeneous dielectric, *Zh. Eksp. Teor. Fiz.* 36, 1797 (**1959**) [*Sov. Phys. JETP* 9, 1282 (**1959**)]; http://www.jetp.ac.ru/cgi-bin/e/index/e/9/6/p1282?a=list.
208. I. D. Vragovich, J. P. Šetrajčić, and R. Scholz, Quantum size effects in the optical properties of organic superlattices containing 3,4,9,10 perylene tetracarboxylic dianhydride (PTCDA), *Eur. Phys. J. B* 66, 185 (**2008**); doi: 10.1140%2Fepjb%2Fe2008-00409-1.
209. J. P. Šetrajčić and S. K. Jaćimovski, Different regions of exciton localized states in ultrathin dielectric films, *J. Phys. Chem. Sol.* 105, 1–8 (**2017**); doi: 10.1016/j.jpcs.2017.02.001.
210. M. Vojnović, A. J. Šetrajčić-Tomić, S. M. Vučenović, and J. P. Šetrajčić, Discrete and selective absorption in crystalline molecular nanofilms, *Opt. Quant. Electron.* 50/4, 198 [1–13] (**2018**); doi: 10.1007/s11082-018-1443-y.

211. F. Banhart, N. Grobert, M. Terrones, J. C. Charlier, and P. M. Ajayan, Metal atoms in carbon nanotubes and related nanoparticles, *Int. J. Mod. Phys. B* 15/31, 4037–4069 (**2001**); doi: 10.1142/S0217979201007944.
212. A. Šiber, Phonons and specific heat of linear dense phases of atoms Physisorbed in the grooves of carbon nanotube bundles, *Phys. Rev. B*, 66, 235414 (**2002**); doi: 10.1103/PhysRevB.66.235414.
213. M. A. Shah, M. A. Bhat, and J. P. Davim (Eds.), Nanotechnology applications for improvements in energy efficiency and environmental management, (IGI Global, Hershey, **2014**).
214. G. Moos, R. Fasel, and T. Hertel, Temperature dependence of electron to lattice energy-transfer in single-wall carbon nanotube bundles, *J. Nanosci. Nanotech.* 3, 145 (**2002**); doi: 10.1166/jnn.2003.163.
215. G.-M. Zhao, Identification of the bulk pairing symmetry in high-temperature superconductors: Evidence for an extended S-save with eight line nodes, *Phys. Rev. B* 64, 024503 (**2001**); doi: 10.1103/PhysRevB.64.024503.
216. J. M. Wesselinova, On the theory of thin ferroelectric films, *Phys. Stat. Sol.* (b) 223, 737 (**2001**); doi: 10.1002/1521-3951(200102)223:3⟨737::AID-PSSB737⟩3.0.CO;2-8; Electric field dependence of phase transitions in ferroelectric thin films, *Phys. Stat. Sol.* (b) 229, 1329 (**2002**); doi: 10.1002/1521-3951(200202)229:3⟨1329::AID-PSSB1329⟩3.0.CO;2-V.
217. N. Settera and D. Damjanovic, Ferroelectric thin films: Review of materials, properties, and applications, *J. App. Phys.* 100, 051606 (**2006**); doi: 10.1063/1.2336999.
218. H. Ibach and H. Lüth, *Solid State Physics — An Introduction to Principles of Materials Science*, 3rd edn. (Springer-Verlag, Berlin, **2003**).
219. M. A. Baranov, A. M. M. Pruisken, and B. Škorić, The problem of Coulomb interactions in the theory of the quantum hall effect, *Uspekhi Fizicheskikh Nauk Suppl.* 44–48 (**2000**); doi: 10.1070/1063-7869/44/10S/S08.
220. T. Baier, E. Bick, and C. Weterich, Antiferromagnetic gap in the Hubbard model, *Phys. Lett. B* 605/1–2, 144–150 (**2005**); doi: 10.1016/j.physletb.2004.11.022.
221. R. T. Scalettar, An Introduction to the Hubbard hamiltonian. In *Quantum Materials: Experiments and Theory Modeling and Simulation* Vol. 6, E. Pavarini, E. Koch, J. van den Brink and G. Sawatzky (Eds)., *Forschungszentrum Jülich*, Jülich (**2016**).
222. J. P. Šetrajčić, S. Armaković, I. J. Šetrajčić, and Lj. D. Džambas, Surface localization of electrons in ultrathin crystalline structures, *Mod. Phys. Lett. B* 28(4), 1450023 (**2014**); doi: 10.1142/S0217984914500237.
223. V. Sajfert, B. S. Tošić, M. M. Marinković, and U. F. Kozmidis-Luburić, Surface deformation in films and exciton concentration, *Physica A* 166, 430–446 (**1990**); doi: https://doi.org/10.1016/0378-4371(70)90066-2.
224. V. Sajfert, R. Đajić, M. Ćetković, and B. Tošić, Cylindrical quantum dots with hydrogen-bonded materials, *Nanotechnology* 14, 358–365 (**2003**); doi: 10.1088/0957-4484/14/3/302; V. D. Sajfert, N. Bucalović, Lj. D. Mašković and B. S. Tošić, Electrons in thin films, *Czech J. Phys.* 56/3, 253–266 (**2006**); doi: 10.1007/s10582-006-0086-6.

225. D. B. Hayrapetyan, T. V. Kotanjyan, and H. Kh. Tevosyan, Modeling of confinement potential for cylindrical quantum dot, *J. Contemp. Phys.* 49/6, 272–276 (**2014**); doi: 10.3103/S106833721406005X.
226. V. D. Sajfert, J. P. Šetrajčić, B. S. Tošić, and R. P. Đajić, Excitonic diffusion in thin molecular films, *Czech. J. Phys.* 54/9, 975–988 (**2004**); doi: 10.1023/B:CJOP.0000042649.90593.dd.
227. D. Backović, V. D. Sajfert, J. P. Šetrajčić, and B. S. Tošić, Magnetic nanorod in the transition temperature vicinity, *J. Comput. Theor. Nanosci.* 2/3, 448–455 (**2005**); doi: 10.1166/jctn.2005.216.
228. V. D. Sajfert, R. P. Đajić, and B. S. Tošić, Hydrogen-bonded nanotubes as a model for DNA transcription, *J. Nanosci. Nanotechnol.* 4, 886–890 (**2004**); doi: 10.1166/jnn.2004.102.
229. H. Huang and J. F. Scott (Eds.), Ferroelectric Materials for Energy Applications. (Wiley-VCH Verlag, Weinheim, **2018**).
230. J. S. Blakemore, *Solid State Physics*, 2nd edn. (Graduate Center, Oregon, **1985**).

Chapter 8

Heisenberg's Uncertainty Principle and the Quantum Capacitance in Quantum Wire Field Effect Transistors

J. Pal* and K. P. Ghatak†,‡

**Department of Physics,*
Meghnad Saha Institute of Technology,
Nazirabad, P.O. Uchepota, Anandapur,
Kolkata 700150, India
†Department of Basic Science and Humanities,
Institute of Engineering and Management, D1,
Management House, Salt Lake,
Sector V, Kolkata 700091, India
‡kamakhyaghatakcu@gmail.com

8.1. Introduction

It is well known that in quantum wires (QWs), the restriction of the motion of the carriers along two directions may be viewed as carrier confinement by two infinitely deep 1D rectangular potential wells along any two orthogonal directions, leading to quantization of the wave vectors along the said directions, allowing 1D carrier transport [1]. With the help of modern fabrication techniques, such one-dimensional quantized structures have been experimentally realized and enjoy an enormous range of important applications in the realm of nanoscience in quantum regime. They have generated much interest in the analysis of nanostructured devices for investigating their electronic, optical and allied properties [2–8]. Examples of such new applications are based on the different transport properties of ballistic charge carriers, which include quantum resistors [9–11], resonant tunneling diodes and band filters [12, 13], quantum switches [14], quantum sensors [15, 16], quantum logic gates [17, 18], quantum transistors and sub-tuners [19, 20], heterojunction FETs [21], high-speed digital networks [22], high-frequency microwave circuits [23], optical modulators [24], optical switching systems [25], and other nanoscale devices.

In this chapter, we study the Quantum Capacitance (QC) in QWFETs of non-parabolic compounds having different band structures. At first, we shall investigate the QC of nonlinear optical compounds which are being used in nonlinear optics and light-emitting diodes [26]. The quasi-cubic model can be used to investigate the symmetric properties of both the bands at the zone center of wave vector space of the same compound. Kildal [27] formulated the electron dispersion law under the assumptions of isotropic momentum matrix and the isotropic spin–orbit splitting constant, respectively, although the anisotropies in the two aforementioned band constants are the significant physical features of the said materials [28]. In Section 8.2.1, the QC in QWFETs of nonlinear optical materials has been investigated by considering the combined influence of the anisotropies of the said energy band constants together with the inclusion of the crystal field splitting, respectively, within the framework of **k.p** formalism. The III–V compounds find applications in infrared detectors [29], quantum dot light-emitting diodes [30], quantum cascade lasers [31], quantum well wires [32], optoelectronic sensors [33], high-electron-mobility transistors [34], etc. The electron energy spectrum of III–V compounds can be described by the three- and two-band models of Kane [35], together with the models of Stillman *et al.* [36], Newson and Kurobe [37] and Palik *et al.* [38], respectively. In this context, it may be noted that the ternary and quaternary compounds enjoy the singular position in the entire spectrum of optoelectronic materials. The ternary alloy $Hg_{1-x}Cd_xTe$ is a classic narrow gap compound. The band gap of this ternary alloy can be varied to cover the spectral range from 0.8 to over 30 μm [39] by adjusting the alloy composition. $Hg_{1-x}Cd_xTe$ finds extensive applications in infrared detector materials and photovoltaic detector arrays in the 8–12 μm wave bands [40]. The above uses have generated the $Hg_{1-x}Cd_xTe$ technology for the experimental realization of high-mobility single crystal with specially prepared surfaces. The same compound has emerged to be the optimum choice for illuminating the narrow sub-band physics because the relevant material constants can easily be experimentally measured [41]. Besides, the quaternary alloy $In_{1-x}Ga_xAs_yP_{1-y}$ lattice matched to InP also finds wide use in the fabrication of avalanche photo-detectors [42], heterojunction lasers [43], light-emitting diodes [44] and avalanche photodiodes [45], field effect transistors, detectors, switches, modulators, solar cells, filters and new types of integrated optical devices are made from the quaternary systems [46]. It may be noted that all types of band models as discussed for III–V compounds are also applicable for ternary and quaternary compounds. In Section 8.2.2, the QC in QWFETs of III–V, ternary and quaternary materials has been studied in accordance with the said band models and the simplified results for wide gap materials having parabolic energy bands under certain limiting conditions have further been demonstrated as a special case, thus confirming the compatibility test.

The II–VI compounds are being used in nanoribbons, blue green diode lasers, photosensitive thin films, infrared detectors, ultra-high-speed bipolar transistors, fiber optic communications, microwave devices, solar cells, semiconductor gamma-ray detector arrays, semiconductor detector gamma camera and allow for a greater density of data storage on optically addressed compact discs [47–49]. The carrier energy spectra in II–VI compounds are defined by the Hopfield model [50] where the splitting of the two-spin states by the spin–orbit coupling and the crystalline field has been taken into account. Section 8.2.3 contains the investigation of the QC in QWFETs of II–VI compounds.

Lead chalcogenides (PbTe, PbSe, and PbS) are IV–VI non-parabolic compounds whose studies over several decades have been motivated by their importance in infrared IR detectors, lasers, light-emitting devices, photovoltaics and high-temperature thermoelectrics [51]. PbTe, in particular, is the end compound of several ternary and quaternary high-performance high-temperature thermoelectric materials [52]. It has been used not only as bulk but also as films [53], quantum wells [54], superlattices [55], nanowires [56], and colloidal and embedded nanocrystals [57] and PbTe films doped with various impurities have also been investigated [58]. These studies revealed some of the interesting features that had been seen in bulk PbTe, such as Fermi level pinning and, in the case of superconductivity [59]. In Section 8.2.4, the QC in QWFETs of IV–VI compounds has been studied taking PbTe, PbSe, and PbS as examples.

The stressed compounds are being investigated for strained silicon transistors, quantum cascade lasers, strain gages, thermal detectors and strained-layer structures [60]. The QC in QWFETs of stressed compounds (taking stressed n-InSb as an example) has been investigated in Section 8.2.5. The vacuum-deposited Tellurium (Te) has been used as the semiconductor layer in thin-film transistors (TFT) [61] which is being used in CO_2 laser detectors [62], electronic imaging, strain sensitive devices [63] and multichannel Bragg cell [64]. Section 8.2.6 contains the investigation of QC in QWFETs of Tellurium. The n-Gallium Phosphide (n-GaP) is being used in quantum dot light-emitting diode [65], high-efficiency yellow solid state lamps, light sources, high-peak current pulse for high gain tubes. The green and yellow light-emitting diodes made of nitrogen-doped n-GaP possess a longer device life at high drive currents [66]. In Section 8.2.7, the QC in QWFETs of n-GaP has been studied. The Platinum Antimonide ($PtSb_2$) finds application in device miniaturization, colloidal nanoparticle synthesis, sensors and detector materials and thermo-photovoltaic devices [67]. Section 8.2.8 explores the QC in QWFETs of $PtSb_2$. Bismuth telluride (Bi_2Te_3) was first identified as a material for thermoelectric refrigeration in 1954 [68] and its physical properties were later improved by the addition of bismuth selenide and antimony telluride to form solid solutions [69]. The alloys of Bi_2Te_3 are useful compounds for the thermoelectric industry and have been investigated in the literature [69]. In Section 8.2.9, the QC in QWFETs of Bi_2Te_3 has been considered.

The usefulness of elemental semiconductor Germanium is already well known since the inception of transistor technology, and it is also being used in memory circuits, single photon detectors, single photon avalanche diode, ultrafast optical switch, THz lasers and THz spectrometers [70]. In Section 8.2.10, the QC in QWFETs of Ge has been studied. Gallium Antimonide (GaSb) finds applications in the fiber optic transmission window, heterojunctions and quantum wells. A complementary heterojunction field effect transistor in which the channels for the p-FET device and the n-FET device forming the complementary FET are formed from GaSb is considered. The band gap energy of GaSb makes it suitable for low-power operation [71]. In Section 8.2.11, the QC in QWFETs of GaSb has been studied. The II–V compounds are being used in photovoltaic cells constructed of single crystal materials in contact with electrolyte solutions. Cadmium selenide shows an open-circuit voltage of 0.8 V and power conservation coefficients near 6 percent for 720-nm light [72]. They are also used in ultrasonic amplification [73]. The development of an evaporated thin-film transistor using cadmium selenide as the semiconductor has also been reported [74, 75]. In Section 8.2.12, we shall study the QC in QWFETs of II–V materials. Section 8.3 contains the results and discussion.

8.2. Theoretical Background

8.2.1. *The quantum capacitance QWFET of nonlinear optical materials*

In QWFET, the QC (hereafter referred to as C_g) can be expressed by the following equation:

$$C_g = e\bar{D}n_s \tag{8.1}$$

where e is the electron charge, $\bar{D} = \frac{\partial}{\partial V_g}$ and n_s is the electron statistics per unit length.

For two-dimensional quantizations along x and y directions, the 1D dispersion relation in this case assumes the form

$$k_z^2 = A_{11}(E, n_x, n_y) \tag{8.2a}$$

where $A_{11}(E, n_x, n_y) = [f_2(E)]^{-1}[\gamma(E) - \phi_1(n_x, n_y)f_1(E)]$, $\phi_1(n_x, n_y) = \left(\frac{n_x\pi}{d_x}\right)^2 + \left(\frac{n_y\pi}{d_y}\right)^2$ $n_x = (1, 2, 3, \ldots)$, $n_y = (1, 2, 3, \ldots)$ are the size quantum numbers along x and y directions, respectively, d_x and d_y are the nanothickness along x and y directions, respectively, and the other notations have been defined elsewhere [41].

The quantized sub-band energy (E_{11}) is given by

$$\gamma(E_{11}) = f_1(E_{11})\phi_1(n_x, n_y) \tag{8.2b}$$

The application of Heisenberg's Uncertainty Principle leads to the expression of electron statistics per unit length under the condition of extreme degeneracy as

$$n_s = \frac{2g_v}{\pi} S[B_{11}(E'_F, n_x, n_y)] \tag{8.2c}$$

where $S = \sum_{n_x=1}^{n_{z\max}} \sum_{n_y=1}^{n_{y\max}}$, $E'_F = [e\bar{V}_g - E_{fb} - \frac{e^2 n_s}{C_{ins}}]$, E_{fb} is the Fermi energy of the bulk material, $C_{ins} = (2\pi\varepsilon_{ins})\left[\ln\left(\frac{2t_{ins}+t_{wire}}{t_{wire}}\right)\right]^{-1}$, ε_{ins} is the permittivity of the insulator layer, t_{ins} is the thickness of the insulator layer and t_{wire} is the diameter of the wire and the other notations are given in [41].

Using (8.1) and (8.2c), the QC can be expressed as

$$C_g = \left[\frac{2g_v e^2}{\pi} S(\theta_1)\right]\left[1 + \frac{2g_v e^2}{\pi C_{ins}} S(\theta_1)\right]^{-1} \tag{8.3}$$

where

$$\theta_1 = \frac{\partial}{\partial E'_F}(B_{11}(E'_F, n_x, n_y))$$

8.2.2. *The quantum capacitance in QWFET of III–V, ternary and quaternary materials*

(i) The three-band model of Kane

The application of Heisenberg's Uncertainty Principle leads to the expression of electron statistics per unit length under the condition of extreme degeneracy in this case as [41]

$$n_s = \frac{2g_v}{\pi} S[B_{13}(E'_F, n_x, n_y)] \tag{8.4}$$

Using (8.1) and (8.3), the QC can be expressed as

$$\boldsymbol{C_g = \left[\frac{2g_v e^2}{\pi} S(\theta_2)\right]\left[1 + \frac{2g_v e^2}{\pi C_{ins}} S(\theta_2)\right]^{-1}}$$
$$\text{where} \quad \boldsymbol{\theta_2 = \frac{\partial}{\partial E'_F}(B_{13}(E'_F, n_x, n_y))} \tag{8.5}$$

(ii) The two-band model of Kane

The application of Heisenberg's Uncertainty Principle leads to the expression of electron statistics per unit length under the condition of extreme degeneracy as [41]

$$n_s = \frac{2g_v}{\pi} S[B_{15}(E'_F, n_x, n_y)] \tag{8.6}$$

Using (8.1) and (8.6), the QC can be expressed as

$$C_g = \left[\frac{2g_v e^2}{\pi} S(\theta_3)\right]\left[1 + \frac{2g_v e^2}{\pi C_{ins}} S(\theta_3)\right]^{-1}$$
$$\text{where} \quad \theta_3 = \frac{\partial}{\partial E'_F}(B_{15}(E'_F, n_x, n_y)) \tag{8.7}$$

(iii) The model of Stillman *et al.*

The application of Heisenberg's Uncertainty Principle leads to the expression of electron statistics per unit length under the condition of extreme degeneracy as [41]

$$n_s = \frac{2g_v}{\pi} S[B_{17}(E'_F, n_x, n_y)] \tag{8.8}$$

Using (8.1) and (8.8), the quantum capacitance can be expressed as

$$\begin{aligned} C_g &= \left[\frac{2g_v e^2}{\pi} S(\theta_4)\right]\left[1 + \frac{2g_v e^2}{\pi C_{ins}} S(\theta_4)\right]^{-1} \\ \text{where} \quad \theta_4 &= \frac{\partial}{\partial E'_F}(B_{17}(E'_F, n_x, n_y)) \end{aligned} \tag{8.9}$$

(iv) The model of Newson and Kurobe

The application of Heisenberg's Uncertainty Principle leads to the expression of electron statistics per unit length under the condition of extreme degeneracy as [41]

$$n_s = \frac{2g_v}{\pi} S[B_{19}(E'_F, n_x, n_y)] \tag{8.10}$$

Using (8.1) and (8.10), the quantum capacitance can be expressed as

$$\begin{aligned} C_g &= \left[\frac{2g_v e^2}{\pi} S(\theta_5)\right]\left[1 + \frac{2g_v e^2}{\pi C_{ins}} S(\theta_5)\right]^{-1} \\ \text{where} \quad \theta_5 &= \frac{\partial}{\partial E'_F}(B_{19}(E'_F, n_x, n_y)) \end{aligned} \tag{8.11}$$

(v) The model of Palik *et al.*

The application of Heisenberg's Uncertainty Principle leads to the expression of electron statistics per unit length under the condition of extreme degeneracy as [41]

$$n_s = \frac{2g_v}{\pi} S[B_{21}(E'_F, n_x, n_y)] \tag{8.12}$$

Using (8.1) and (8.12), the quantum capacitance can be expressed as

$$\begin{aligned} C_g &= \left[\frac{2g_v e^2}{\pi} S(\theta_6)\right]\left[1 + \frac{2g_v e^2}{\pi C_{ins}} S(\theta_6)\right]^{-1} \\ \text{where} \quad \theta_6 &= \frac{\partial}{\partial E'_F}(B_{21}(E'_F, n_x, n_y)) \end{aligned} \tag{8.13}$$

8.2.3. *The quantum capacitance in QWFET of II–VI*

The application of Heisenberg's Uncertainty Principle leads to the expression of electron statistics per unit length under the condition of extreme degeneracy as [41]

$$n_s = \frac{g_v}{\pi\sqrt{B_0}} S[t_7(E'_F, n_x, n_y)] \tag{8.14}$$

Using (8.1) and (8.14), the quantum capacitance can be expressed as

$$C_g = \left[\frac{g_v e^2}{\pi\sqrt{B_0}} S(\theta_7)\right]\left[1 + \frac{g_v e^2}{C_{ins}\pi\sqrt{B_0}} S(\theta_7)\right]^{-1}$$
$$\text{where} \quad \theta_7 = \frac{\partial}{\partial E'_F}(t_7(E'_F, n_x, n_y)) \tag{8.15}$$

8.2.4. *The quantum capacitance in QWFET of IV–VI*

The application of Heisenberg's Uncertainty Principle leads to the expression of electron statistics per unit length under the condition of extreme degeneracy as [41]

$$n_s = \frac{2g_v}{\pi} S[B_{32}(E'_F, n_x, n_y)] \tag{8.16}$$

Using (8.1) and (8.16), the quantum capacitance can be expressed as

$$C_g = \left[\frac{2g_v e^2}{\pi} S(\theta_8)\right]\left[1 + \frac{2g_v e^2}{\pi C_{ins}} S(\theta_8)\right]^{-1}$$
$$\text{where} \quad \theta_8 = \frac{\partial}{\partial E'_F}(B_{32}(E'_F, n_x, n_y)) \tag{8.17}$$

8.2.5. *The quantum capacitance in QWFET of stressed Kane-type materials*

The application of Heisenberg's Uncertainty Principle leads to the expression of electron statistics per unit length under the condition of extreme degeneracy as [41]

$$n_s = \frac{2g_v}{\pi} S[B_{34}(E'_F, n_x, n_y)] \tag{8.18a}$$

Using (8.1) and (8.18a), the quantum capacitance can be expressed as

$$C_g = \left[\frac{2g_v e^2}{\pi} S(\theta_9)\right]\left[1 + \frac{2g_v e^2}{\pi C_{ins}} S(\theta_9)\right]^{-1}$$
$$\text{where} \quad \theta_9 = \frac{\partial}{\partial E'_F}(B_{34}(E'_F, n_x, n_y)) \tag{8.18b}$$

8.2.6. *The quantum capacitance in QWFET of stressed Kane-type materials*

The application of Heisenberg's Uncertainty Principle leads to the expression of electron statistics per unit length under the condition of extreme degeneracy as [41]

$$n_s = \frac{g_v}{\pi} S[B_{36,\pm}(E'_F, n_x, n_y)] \tag{8.19a}$$

Using (8.1) and (8.19a), the quantum capacitance can be expressed as

$$\begin{aligned} C_g &= \left[\frac{g_v e^2}{\pi} S(\bar{\theta}_9)\right]\left[1 + \frac{g_v e^2}{\pi C_{ins}} S(\bar{\theta}_9)\right]^{-1} \\ \text{where} \quad \bar{\theta}_9 &= \frac{\partial}{\partial E'_F}(B_{36,\pm}(E'_F, n_x, n_y)) \end{aligned} \tag{8.19b}$$

8.2.7. *The quantum capacitance in QWFET of GaP*

The application of Heisenberg's Uncertainty Principle leads to the expression of electron statistics per unit length under the condition of extreme degeneracy as [41]

$$n_s = \frac{2g_v}{\pi} S[B_{38}(E'_F, n_x, n_y)] \tag{8.20}$$

Using (8.1) and (8.20), the quantum capacitance can be expressed as

$$\begin{aligned} C_g &= \left[\frac{2g_v e^2}{\pi} S(\theta_{10})\right]\left[1 + \frac{2g_v e^2}{\pi C_{ins}} S(\theta_{10})\right]^{-1} \\ \text{where} \quad \theta_{10} &= \frac{\partial}{\partial E'_F}(B_{38}(E'_F, n_x, n_y)) \end{aligned} \tag{8.21}$$

8.2.8. *The quantum capacitance in QWFET of $PtSb_2$*

The application of Heisenberg's Uncertainty Principle leads to the expression of electron statistics per unit length under the condition of extreme degeneracy as [41]

$$n_s = \frac{2g_v}{\pi} S[B_{40}(E'_F, n_x, n_y)] \tag{8.22}$$

Using (8.1) and (8.23), the quantum capacitance can be expressed as

$$\begin{aligned} C_g &= \left[\frac{2g_v e^2}{\pi} S(\theta_{11})\right]\left[1 + \frac{2g_v e^2}{\pi C_{ins}} S(\theta_{11})\right]^{-1} \\ \text{where} \quad \theta_{11} &= \frac{\partial}{\partial E'_F}(B_{40}(E'_F, n_x, n_y)) \end{aligned} \tag{8.23}$$

8.2.9. *The quantum capacitance in QWFET of Bi_2Te_3*

The application of Heisenberg's Uncertainty Principle leads to the expression of electron statistics per unit length under the condition of extreme degeneracy as [41]

$$n_s = \frac{2g_v}{\pi} S[B_{42}(E'_F, n_x, n_y)] \tag{8.24}$$

Using (8.1) and (8.24), the quantum capacitance can be expressed as

$$C_g = \left[\frac{2g_v e^2}{\pi} S(\theta_{12})\right]\left[1 + \frac{2g_v e^2}{\pi C_{ins}} S(\theta_{12})\right]^{-1}$$
$$\text{where} \quad \theta_{12} = \frac{\partial}{\partial E'_F}(B_{42}(E'_F, n_x, n_y)) \tag{8.25}$$

8.2.10. *The quantum capacitance in QWFET of Ge*

(i) Model of Cardona *et al.*

The application of Heisenberg's Uncertainty Principle leads to the expression of electron statistics per unit length under the condition of extreme degeneracy as [41]

$$n_s = \frac{2g_v}{\pi} S[B_{44}(E'_F, n_x, n_y)] \tag{8.26}$$

Using (8.1) and (8.26), the quantum capacitance can be expressed as

$$C_g = \left[\frac{2g_v e^2}{\pi} S(\theta_{13})\right]\left[1 + \frac{2g_v e^2}{\pi C_{ins}} S(\theta_{13})\right]^{-1}$$
$$\text{where} \quad \theta_{13} = \frac{\partial}{\partial E'_F}(B_{44}(E'_F, n_x, n_y)) \tag{8.27}$$

(ii) Model of Wang and Ressler

The application of Heisenberg's Uncertainty Principle leads to the expression of electron statistics per unit length under the condition of extreme degeneracy as [41]

$$n_s = \frac{2g_v}{\pi} S[B_{46}(E'_F, n_x, n_y)] \tag{8.28}$$

Using (8.1) and (8.28), the quantum capacitance can be expressed as

$$C_g = \left[\frac{2g_v e^2}{\pi} S(\theta_{14})\right] \left[1 + \frac{2g_v e^2}{\pi C_{ins}} S(\theta_{14})\right]^{-1}$$
$$\text{where} \quad \theta_{14} = \frac{\partial}{\partial E'_F} (B_{46}(E'_F, n_x, n_y)) \tag{8.29}$$

8.2.11. *The quantum capacitance in QWFET of GaSb*

The application of Heisenberg's Uncertainty Principle leads to the expression of electron statistics per unit length under the condition of extreme degeneracy as [41]

$$n_s = \frac{2g_v}{\pi} S[B_{48}(E'_F, n_x, n_y)] \tag{8.30}$$

Using (8.1) and (8.31), the quantum capacitance can be expressed as

$$C_g = \left[\frac{2g_v e^2}{\pi} S(\theta_{15})\right] \left[1 + \frac{2g_v e^2}{\pi C_{ins}} S(\theta_{15})\right]^{-1}$$
$$\text{where} \quad \theta_{15} = \frac{\partial}{\partial E'_F} (B_{48}(E'_F, n_x, n_y)) \tag{8.31}$$

8.2.12. *The quantum capacitance in QWFET of II–V*

The application of Heisenberg's Uncertainty Principle leads to the expression of electron statistics per unit length under the condition of extreme degeneracy as [41]

$$n_s = \frac{2g_v}{\pi} S[B_{49}(E'_F, n_x, n_y)] \tag{8.32}$$

Using (8.1) and (8.33), the quantum capacitance can be expressed as

$$C_g = \left[\frac{2g_v e^2}{\pi} S(\theta_{16})\right] \left[1 + \frac{2g_v e^2}{\pi C_{ins}} S(\theta_{16})\right]^{-1}$$
$$\text{where} \quad \theta_{16} = \frac{\partial}{\partial E'_F} (B_{49}(E'_F, n_x, n_y)). \tag{8.33}$$

Under certain limiting conditions, all the results for all the models get simplified as

$$n_s = \frac{2g_v}{\pi} S[B_{50}(E'_F, n_x, n_y)] \tag{8.34}$$

$$C_g = \left[\frac{2g_v e^2}{\pi} S(\theta_{50})\right] \left[1 + \frac{2g_v e^2}{\pi C_{ins}} S(\theta_{50})\right]^{-1}$$

$$\text{where} \quad \theta_{50} = \frac{\partial}{\partial E'_F}(B_{50}(E'_F, n_x, n_y)),$$

$$B_{50}(E'_F, n_x, n_y) = \sqrt{a_{50}(E'_F, n_x, n_y)} \tag{8.35}$$

$$\text{and} \quad a_{50}\left(E'_F, n_x, n_y\right) = \left[\frac{2m_c E'_F}{\hbar^2} \phi_1(n_x, n_y)\right],$$

which are valid for QWFETs made of wide band gap materials.

8.3. Results and Discussion

In Fig. 8.1, the normalized C_g in QWFETs made of $CdGeAs_2$ with film thickness as the independent variable has been premeditated using (a) generalized band model, (b) generalized band model with $\delta = 0$, (c) three-band model of Kane, (d) two-band model of Kane, and (e) parabolic band model. From Fig. 8.1, it appears that the normalized C_g enhances in quantized saw-tooth manner with increasing film thickness. In Fig. 8.2, the normalized C_g in QWFETs made of $InAs$ with film thickness as the independent variable has been premeditated using (a) three-band model of Kane, (b) two-band model of Kane, and (c) parabolic band model. From Fig. 8.2, it appears that the normalized C_g enhances in quantized distorted saw-tooth manner with increasing film thickness. In Fig. 8.3, the normalized C_g in QWFETs made of $InSb$ with film thickness as the independent variable has been premeditated using (a) three-band model of Kane, (b) two-band model of Kane, and (c) parabolic band model. From Fig. 8.3, it appears that the normalized C_g in this case enhances in quantized saw-tooth manner with increasing film thickness. In Fig. 8.4, the normalized C_g in QWFETs made of $Hg_{1-x}Cd_xTe$ with film thickness as the independent variable has been premeditated using (a) three-band model of Kane, (b) two-band model of Kane, and (c) parabolic band model. From Fig. 8.4, it appears that the normalized C_g in this case enhances in quantized saw-tooth manner with increasing film thickness. In Fig. 8.5, the normalized C_g in QWFETs made of $Ga_{1-x}Al_xAs$ with film thickness as the independent variable has been

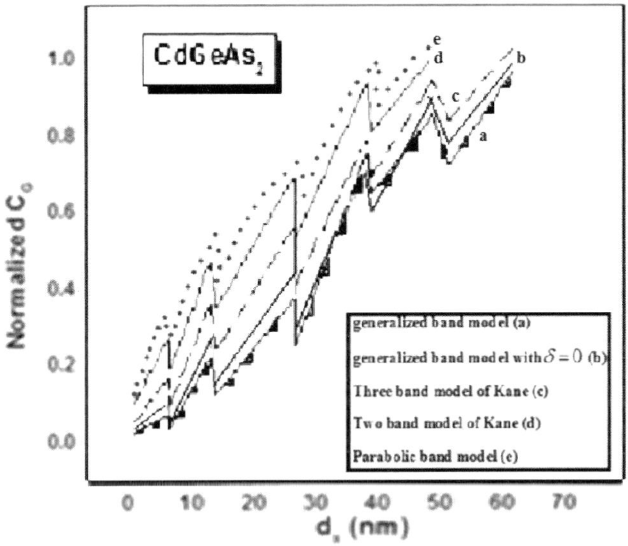

Fig. 8.1: The normalized C_g in QWFETs made of $CdGeAs_2$ with film thickness as the independent variable has been premeditated using (a) generalized band model, (b) generalized band model with $\delta = 0$, (c) three-band model of Kane, (d) two-band model of Kane, and (e) parabolic band model.

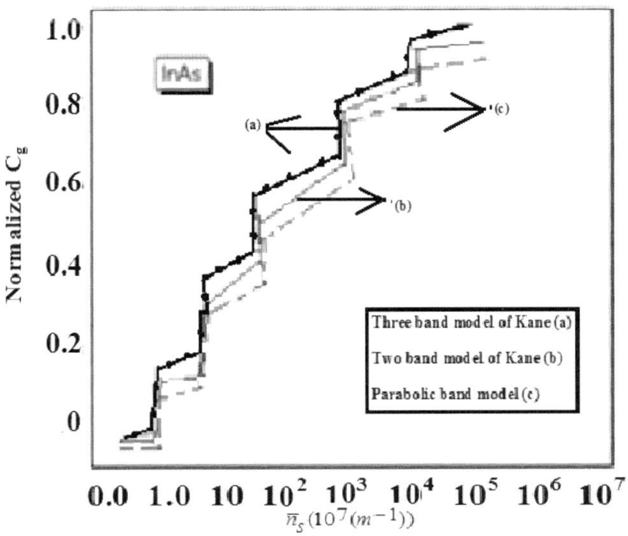

Fig. 8.2: The normalized C_g in QWFETs made of $InAs$ with film thickness as the independent variable has been premeditated using (a) three-band model of Kane, (b) two-band model of Kane, and (c) parabolic band model.

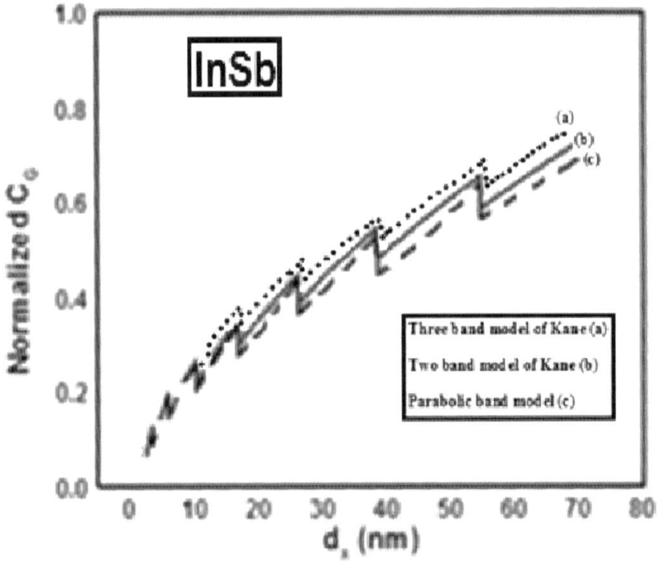

Fig. 8.3: The normalized C_g in QWFETs made of $InSb$ with film thickness as the independent variable has been premeditated using (a) three-band model of Kane, (b) two-band model of Kane, and (c) parabolic band model.

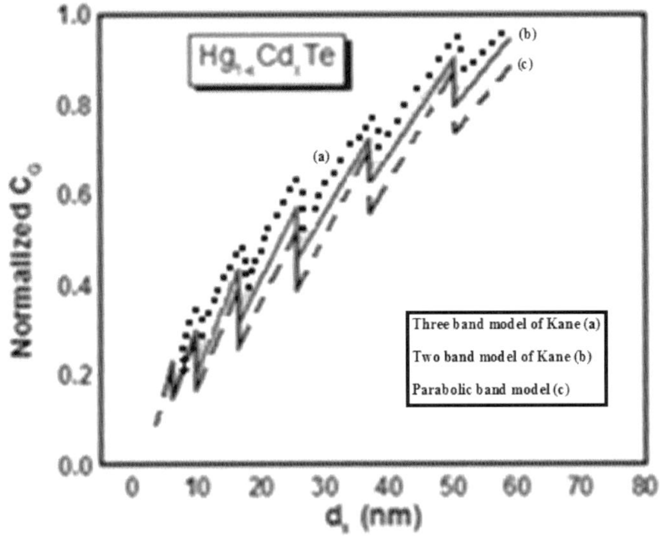

Fig. 8.4: The normalized C_g in QWFETs made of $Hg_{1-x}Cd_xTe$ with film thickness as the independent variable has been premeditated using (a) three-band model of Kane, (b) two-band model of Kane, and (c) parabolic band model.

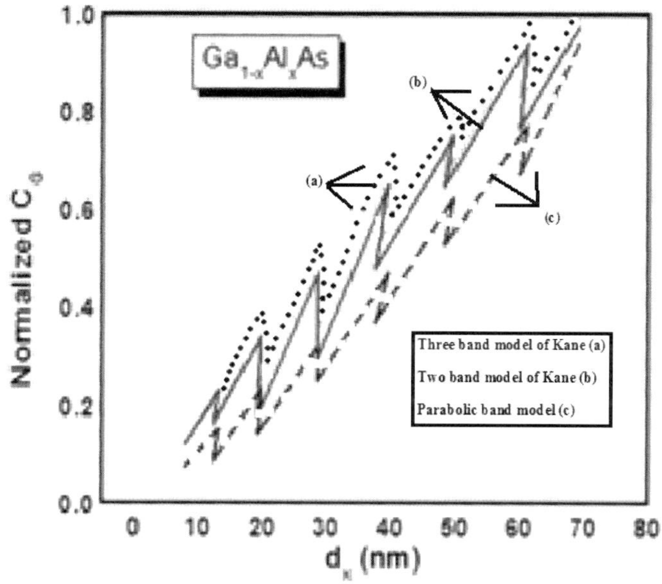

Fig. 8.5: The normalized C_g in QWFETs made of $Ga_{1-x}Al_xAs$ with film thickness as the independent variable has been premeditated using (a) three-band model of Kane, (b) two-band model of Kane, and (c) parabolic band model.

premeditated using (a) three-band model of Kane, (b) two-band model of Kane, and (c) parabolic band model. From Fig. 8.5, it appears that the normalized C_g in this case enhances in quantized saw-tooth manner with increasing film thickness. In Fig. 8.6, the normalized C_g in QWFETs made of $In_{1-x}Ga_xAs_yP_{1-y}$ with film thickness as the independent variable has been premeditated using (a) three-band model of Kane, (b) two-band model of Kane, and (c) parabolic band model.

From Fig. 8.6, it appears that the normalized C_g enhances in quantized saw-tooth manner with increasing film thickness. In Fig. 8.7, the normalized C_g in QWFETs made of Cd_3As_2 with 1D electron statistics as the independent variable has been premeditated using (a) generalized band model, (b) generalized band model with $\delta = 0$, (c) three-band model of Kane, (d) two-band model of Kane, and (e) parabolic band model. From Fig. 8.7, it appears that the normalized C_g enhances in quantized step-like manner with increasing 1D electron statistics. In Fig. 8.8, the normalized C_g in QWFETs made of $CdGeAs_2$ with 1D electron statistics as the independent variable has been premeditated using (a) generalized band model,

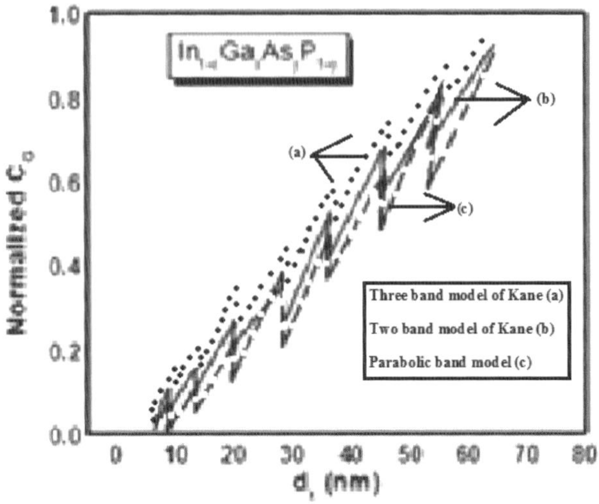

Fig. 8.6: The normalized C_g in QWFETs made of $In_{1-x}Ga_xAs_yP_{1-y}$ with film thickness as the independent variable has been premeditated using (a) three-band model of Kane, (b) two-band model of Kane, and (c) parabolic band model.

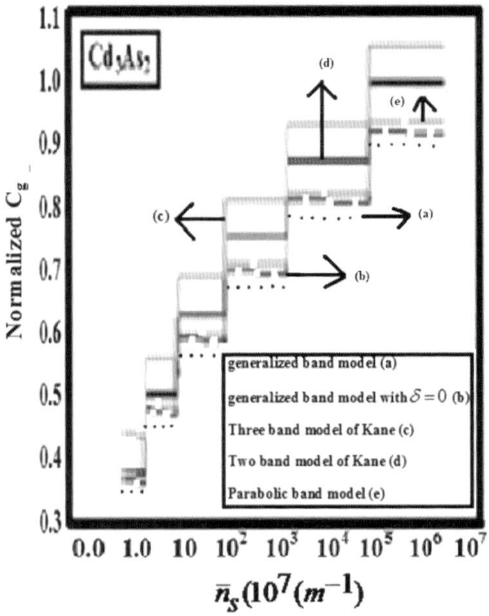

Fig. 8.7: The normalized C_g in QWFETs made of Cd_3As_2 with 1D electron statistics as the independent variable has been premeditated using (a) generalized band model, (b) generalized band model with $\delta = 0$, (c) three-band model of Kane, (d) two-band model of Kane, and (e) parabolic band model.

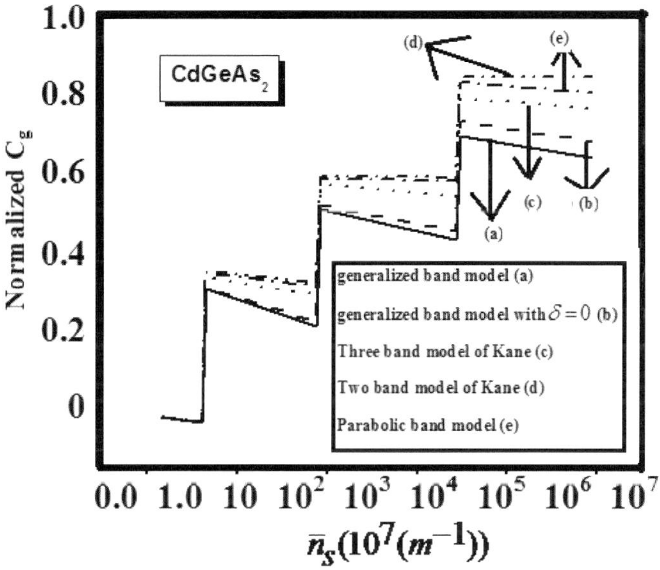

Fig. 8.8: The normalized C_g in QWFETs made of $CdGeAs_2$ with 1D electron statistics as the independent variable has been premeditated using (a) generalized band model, (b) generalized band model with $\delta = 0$, (c) three-band model of Kane, (d) two-band model of Kane, and (e) parabolic band model.

(b) generalized band model with $\delta = 0$, (c) three-band model of Kane, (d) two-band model of Kane, and (e) parabolic band model. From Fig. 8.8, it appears that the normalized C_g in this case enhances in quantized step-like manner with increasing 1D electron statistics. In Fig. 8.9, the normalized C_g in QWFETs made of $InAs$ with 1D electron statistics as the independent variable has been premeditated using (a) three-band model of Kane, (b) two-band model of Kane, and (c) parabolic band model. From Fig. 8.9, it appears that the normalized C_g enhances in quantized step-like manner with increasing 1D electron statistics.

In Fig. 8.10, the normalized C_g in QWFETs made of $InSb$ with 1D electron statistics as the independent variable has been premeditated using (a) three-band model of Kane, (b) two-band model of Kane, and (c) parabolic band model. From Fig. 8.10, it appears that the normalized C_g enhances in quantized step-like manner with increasing 1D electron statistics. In Fig. 8.11, the normalized C_g in QWFETs made of $Hg_{1-x}Cd_xTe$ with 1D electron statistics as the independent variable has been premeditated using (a) three-band model of Kane, (b) two-band model of Kane, and

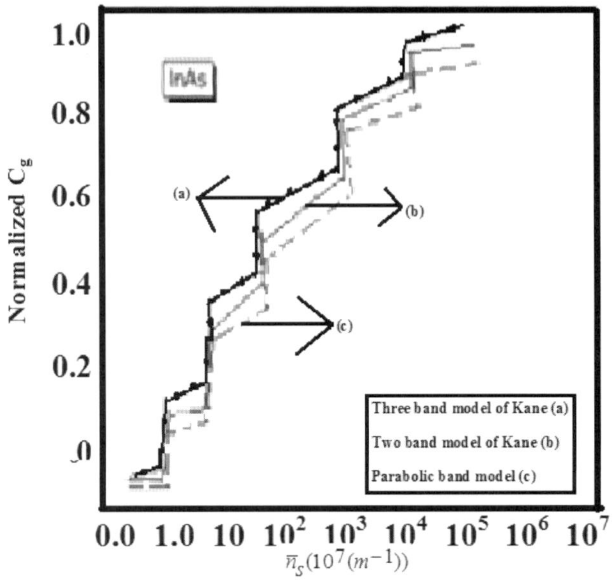

Fig. 8.9: The normalized C_g in QWFETs made of $InAs$ with 1D electron statistics as the independent variable has been premeditated using (a) three-band model of Kane, (b) two-band model of Kane, and (c) parabolic band model.

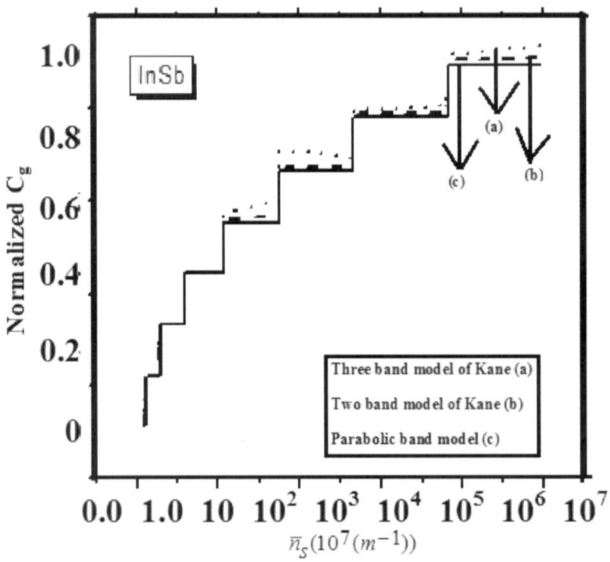

Fig. 8.10: The normalized C_g in QWFETs made of $InSb$ with 1D electron statistics as the independent variable has been premeditated using (a) three-band model of Kane, (b) two-band model of Kane, and (c) parabolic band model.

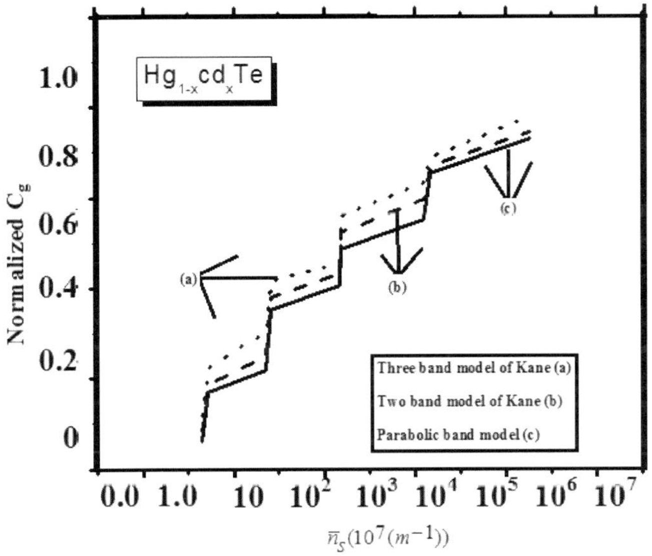

Fig. 8.11: The normalized C_g in QWFETs made of $Hg_{1-x}Cd_xTe$ with 1D electron statistics as the independent variable has been premeditated using (a) three-band model of Kane, (b) two-band model of Kane, and (c) parabolic band model.

(c) parabolic band model. From Fig. 8.11, it appears that the normalized C_g enhances in quantized step-like manner with increasing 1D electron statistics. In Fig. 8.12, the normalized C_g in QWFETs made of $Ga_{1-x}Al_xAs$ with 1D electron statistics as the independent variable has been premeditated using (a) three-band model of Kane, (b) two-band model of Kane, and (c) parabolic band model. From Fig. 8.12, it appears that the normalized C_g enhances in quantized step-like manner with increasing 1D electron statistics. In Fig. 8.13, the normalized C_g in QWFETs made of $In_{1-x}Ga_xAs_yP_{1-y}$ *in* with 1D electron statistics as the independent variable has been premeditated using (a) three-band model of Kane, (b) two-band model of Kane, and (c) parabolic band model. From Fig. 8.13, it appears that the normalized C_g enhances in quantized step-like manner with increasing 1D electron statistics. In Fig. 8.14, the normalized C_g in QWFETs made of Cd_3As_2 with V_g as the independent variable has been premeditated using (a) generalized band model, (b) generalized band model with $\delta = 0$, (c) three-band model of Kane, (d) two-band model of Kane, and (e) parabolic band model. From Fig. 8.14, it appears that the normalized C_g decreases with increasing gate voltage in quantized oscillatory fashion for all cases, as considered in this case.

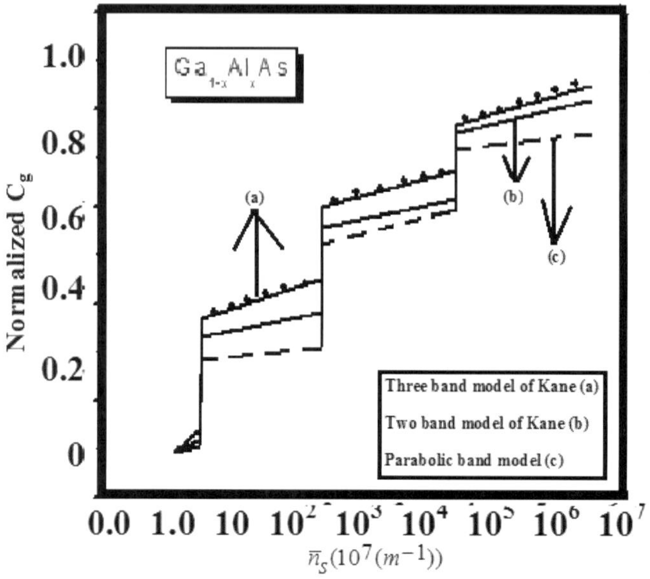

Fig. 8.12: The normalized C_g in QWFETs made of $Ga_{1-x}Al_xAs$ with 1D electron statistics as the independent variable has been premeditated using (a) three-band model of Kane, (b) two-band model of Kane, and (c) parabolic band model.

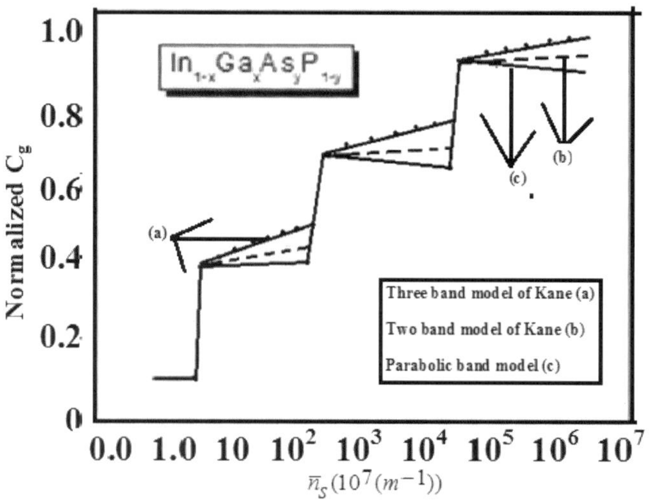

Fig. 8.13: The normalized C_g in QWFETs made of $In_{1-x}Ga_xAs_yP_{1-y}$ *in* with 1D electron statistics as the independent variable has been premeditated using (a) three-band model of Kane, (b) two-band model of Kane, and (c) parabolic band model.

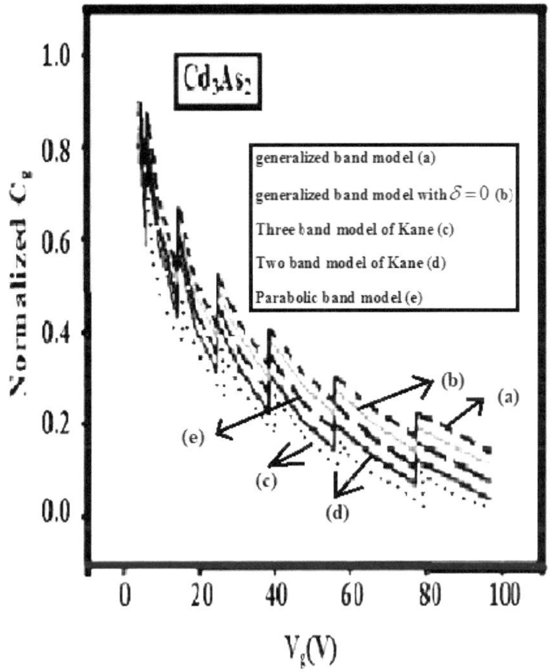

Fig. 8.14: The normalized C_g in QWFETs made of $Cd3As_2$ with V_g as the independent variable has been premeditated using (a) generalized band model, (b) generalized band model with $\delta = 0$, (c) three-band model of Kane, (d) two-band model of Kane, and (e) parabolic band model.

The rate of oscillatory decrease changes from one type of band model to another type. In Fig. 8.15, the normalized C_g in QWFETs made of $CdGeAs_2$ with V_g as the independent variable has been premeditated using (a) generalized band model, (b) generalized band model with $\delta = 0$, (c) three-band model of Kane, (d) two-band model of Kane, and (e) parabolic band model. From Fig. 8.15, it appears that the normalized C_g decreases with increasing gate voltage in quantized oscillatory fashion for all cases, as considered in this case. The rate of oscillatory decrease changes from one type of band model to another type. In Fig. 8.16, the normalized C_g in QWFETs made of $InAs$ with V_g as the independent variable has been premeditated using (a) three-band model of Kane, (b) two-band model of Kane, and (c) parabolic band model. From Fig. 8.16, it appears that the normalized C_g decreases with increasing gate voltage in quantized oscillatory fashion for all cases, as considered in this case. The rate of oscillatory

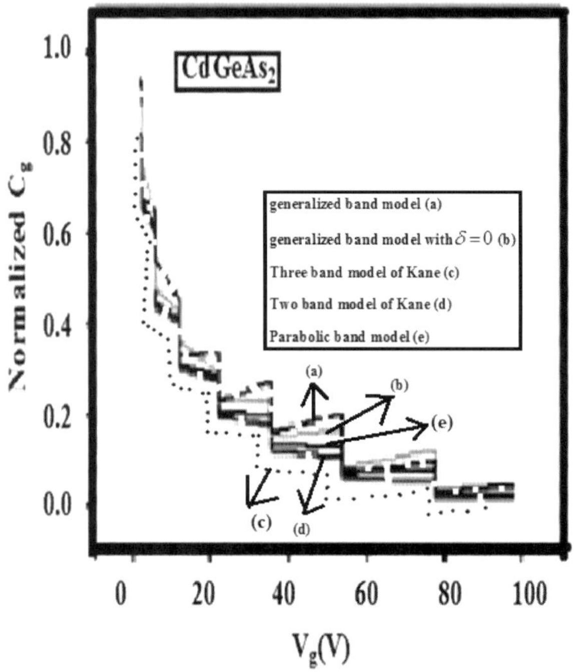

Fig. 8.15: The normalized C_g in QWFETs made of $CdGeAs_2$ with V_g as the independent variable has been premeditated using (a) generalized band model, (b) generalized band model with $\delta = 0$, (c) three-band model of Kane, (d) two-band model of Kane, and (e) parabolic band model.

decrease changes from one type of band model to another type. In Fig. 8.17, the normalized C_g in QWFETs made of $InSb$ with V_g as the independent variable has been premeditated using (a) three-band model of Kane, (b) two-band model of Kane, and (c) parabolic band model. From Fig. 8.17, it appears that the normalized C_g decreases with increasing gate voltage in quantized oscillatory fashion for all cases, as considered in this case. The rate of oscillatory decrease changes from one type of band model to another type. In Fig. 8.18, the normalized C_g in QWFETs made of $Hg_{1-x}Cd_xTe$ with V_g as the independent variable has been premeditated using (a) three-band model of Kane, (b) two-band model of Kane, and (c) parabolic band model. From Fig. 8.18, it appears that the normalized C_g decreases with increasing gate voltage in quantized oscillatory fashion for all cases, as considered in this case. The rate of oscillatory decrease changes from one type of band model to another type.

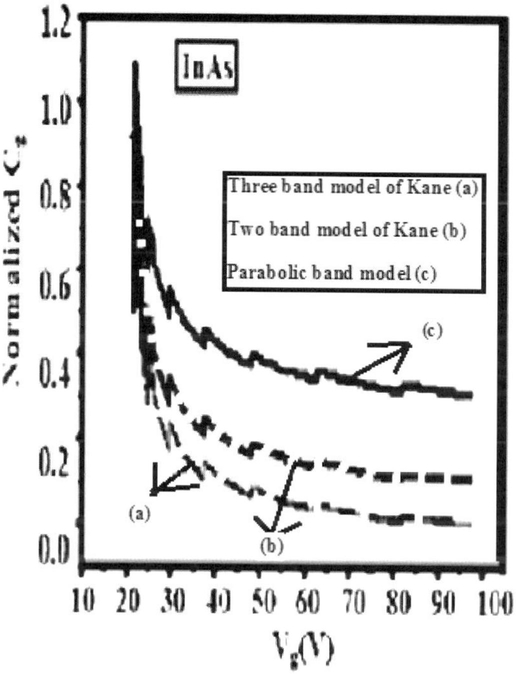

Fig. 8.16: The normalized C_g in QWFETs made of $InAs$ with V_g as the independent variable has been premeditated using (a) three-band model of Kane, (b) two-band model of Kane, and (c) parabolic band model.

In Fig. 8.19, the normalized C_g in QWFETs made of $Ga_{1-x}Al_xAs$ with V_g as the independent variable has been premeditated using (a) three-band model of Kane, (b) two-band model of Kane, and (c) parabolic band model. From Fig. 8.19, it appears that the normalized C_g decreases with increasing gate voltage in quantized oscillatory fashion for all cases, as considered in this case. The rate of oscillatory decrease changes from one type of band model to another type. In Fig. 8.20, the normalized C_g in QWFETs made of $In_{1-x}Ga_xAs_yP_{1-y}$ with V_g as the independent variable has been premeditated using (a) three-band model of Kane, (b) two-band model of Kane, and (c) parabolic band model. From Fig. 8.20, it appears that the normalized C_g decreases with increasing gate voltage in quantized oscillatory fashion for all cases, as considered in this case. The rate of oscillatory decrease changes from one type of band model to another type.

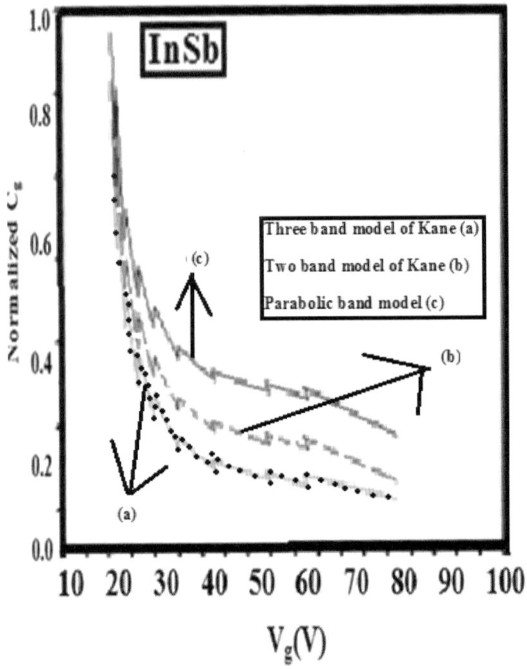

Fig. 8.17: The normalized C_g in QWFETs made of $InSb$ with V_g as the independent variable has been premeditated using (a) three-band model of Kane, (b) two-band model of Kane, and (c) parabolic band model.

With varying electron degeneracy, a change is reflected in the QC of QWFETs through the redistribution of the electrons among the size-quantized levels. It may be noted that at the transition zone from one sub-band to another, the height of the peaks between any two sub-bands decreases with the increase in the degree of quantum confinement and is clearly shown in all the curves. It should be noted that although the QC changes in various manners with all the variables as evident from all the figures, the rates of variations are totally band structure-dependent.

It is imperative to state that the present investigation excludes the many-body, hot-electron, broadening and the allied effects in the simplified theoretical formalism due to the absence of proper analytical techniques for including them for generalized systems, as considered here. Our simplified approach will be appropriate for the purpose of comparisons when the methods of tackling the formidable problems after inclusion of the said effects for the generalized systems emerge. The results of this simplified

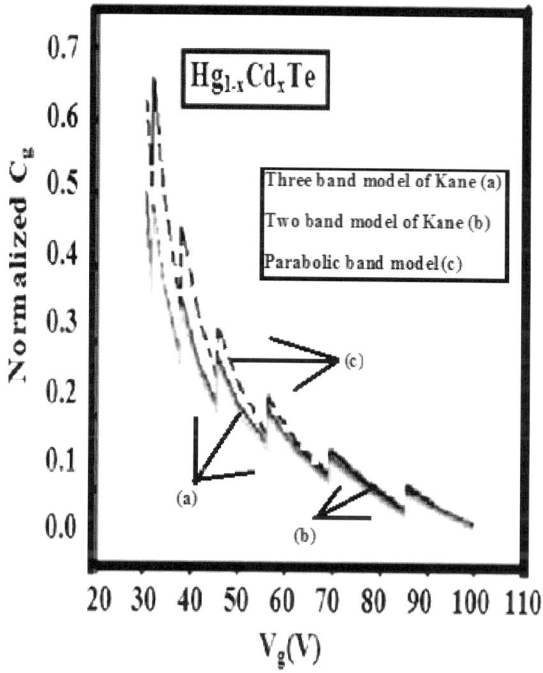

Fig. 8.18: The normalized C_g in QWFETs made of $Hg_{1-x}Cd_xTe$ with V_g as the independent variable has been premeditated using (a) three-band model of Kane, (b) two-band model of Kane, and (c) parabolic band model.

approach get transformed to the well-known formulation of the QC for wide gap materials having parabolic energy bands. This indirect test not only exhibits the mathematical compatibility of the formulation but also shows the fact that this simple analysis is a more generalized one, since one can obtain the corresponding results for materials having parabolic energy bands under certain limiting conditions from the present derivation. For the purpose of computer simulations for obtaining the plots of QC versus various external variables, we have taken very low temperatures since the quantization effects are basically low-temperature phenomena together with the fact that the temperature dependence of all the energy band constants of all the compounds and their nanostructures as considered in this chapter are not available in the literature. The experimental results for the verification of theoretical formulations of QC are still not available in the literature. It is worth noting that the nature of the curves of field emitted current with various physical variables based on our simplified formulations

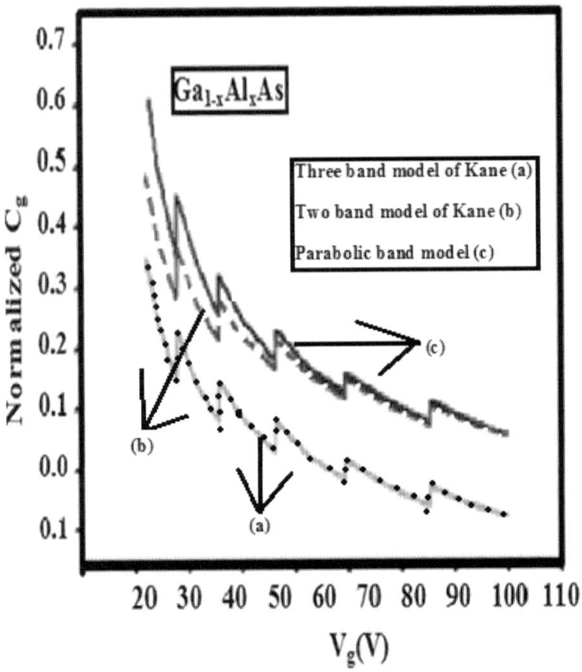

Fig. 8.19: The normalized C_g in QWFETs made of $Ga_{1-x}Al_xAs$ with V_g as the independent variable has been premeditated using (a) three-band model of Kane, (b) two-band model of Kane, and (c) parabolic band model.

as presented here would be useful to analyze the experimental results when they materialize. The inclusion of the said effects would certainly increase the accuracy of the results although the qualitative features of QC would not change in the presence of the aforementioned effects.

It can be noted that on the basis of the dispersion relations of the various quantized structures as discussed above, the effective electron mass, the Debye screening length, the plasma frequency, the activity coefficient, the carrier contribution to the elastic constants, the diffusion coefficient of minority carriers, the third-order nonlinear optical susceptibility, the heat capacity, the diamagnetic and paramagnetic susceptibilities and the various important dc/ac transport coefficients can be probed for all types of QWFETs, as considered here. Thus, our theoretical formulation comprises the dispersion relation-dependent properties of various technologically important quantum-confined semiconductors having different band structures. We have not considered other types of compounds in order to

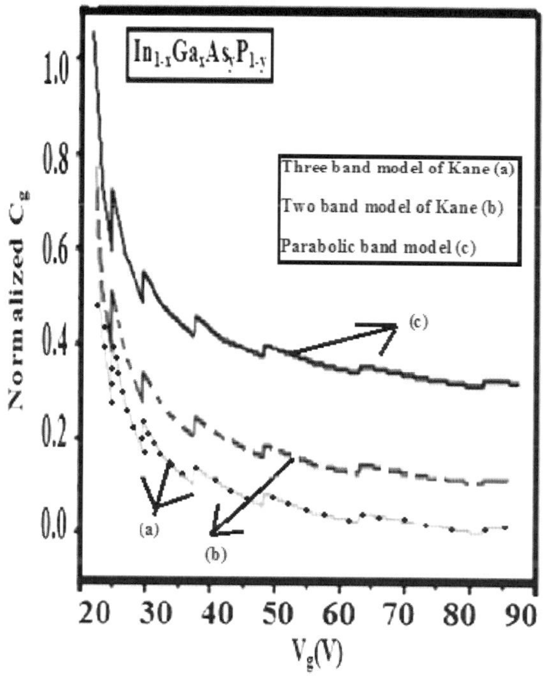

Fig. 8.20: The normalized C_g in QWFETs made of $In_{1-x}Ga_xAs_yP_{1-y}$ with V_g as the independent variable has been premeditated using (a) three-band model of Kane, (b) two-band model of Kane, and (c) parabolic band model.

keep the presentation concise and succinct. With different sets of energy band parameters, we shall get different numerical values of the QCs. The nature of variations of the QCs as shown here would be similar for the other types of materials and the simplified analysis of this chapter exhibits the basic qualitative features of the QC. It may be noted that the basic aim of this chapter is not solely to demonstrate the influence of quantum confinement on the QC for a wide class of QWFETs but also to formulate the appropriate electron statistics per unit length in the most generalized form, since the transport and other phenomena in QWFETs having different band structures and the derivation of the expressions of many important carrier properties are based on the carrier statistics in such systems.

We note that the impact of investigation in this context is rather large, since the QC in QWFET is a very important quantity for the purpose of technical application and experimental studies of physical properties of quantum-confined transistors. Our simplified approach shows the fundamental qualitative behavior of the QC in QWFETs of different technologically important small gap materials. These compounds obey the model of Kane, as our study predicts by extending $\vec{k}.\vec{p}$ theory. Finally, we observe that the Heisenberg's Uncertainty Principle of modern physics not only generates a simplified expression of the QC in QWFETs in a compact and consistent manner but also exhibits the fact the origin of the QC in QWFETs.

References

1. P. Harrison, Quantum wells, *Wires and Dots*, John Wiley and Sons, Ltd, (**2002**); B. K. Ridley, *Electrons and Phonons in Semiconductors Multilayers*, Cambridge University Press, Cambridge (**1997**); V. V. Martin, A. A. Kochelap, and M. A. Stroscio, *Quantum Heterostructures*, Cambridge University Press, Cambridge (**1999**).
2. C. S. Lent and D. J. Kirkner, *J. Appl. Phys.* 67, 6353 (**1990**); F. Sols, M. Macucci, U. Ravaioli, and K. Hess, *Appl. Phys. Lett.* 54, 350 (1980).
3. C. S. Kim, A. M. Satanin, Y. S. Joe, and R. M. Cosby, *Phys. Rev. B* 60, 10962 (**1999**).
4. S. Midgley and J. B. Wang, *Phys. Rev. B* 64, 153304 (**2001**).
5. T. Sugaya, J. P. Bird, M. Ogura, Y. Sugiyama, D. K. Ferry, and K. Y. Jang, *Appl. Phys. Lett.* 80, 434 (**2002**).
6. B. Kane, G. Facer, A. Dzurak, N. Lumpkin, R. Clark, L. PfeiKer, and K. West, *Appl. Phys. Lett.* 72, 3506 (**1998**).
7. C. Dekker, *Physics Today* 52, 22 (**1999**).
8. A. Yacoby, H. L. Stormer, N. S. Wingreen, L. N. Pfeiffer, K. W. Baldwin, and K. W. West, *Phys. Rev. Lett.* 77, 4612 (**1996**).
9. Y. Hayamizu, M. Yoshita, S. Watanabe, H. A. L. PfeiKer, and K. West, *Appl. Phys. Lett.* 81, 4937 (**2002**).
10. S. Frank, P. Poncharal, Z. L. Wang, and W. A. Heer, *Science* 280, 1744 (**1998**).
11. I. Kamiya, I. I. Tanaka, K. Tanaka, F. Yamada, Y. Shinozuka, and H. Sakaki, *Physica E* 13, 131 (**2002**).
12. A. K. Geim, P. C. Main, N. LaScala, L. Eaves, T. J. Foster, P. H. Beton, J. W. Sakai, F. W. Sheard, M. Henini, G. Hill *et al.*, *Phys. Rev. Lett.* 72, 2061 (**1994**).
13. A. S. Melinkov and V. M. Vinokur, *Nature* 415, 60 (**2002**).
14. K. Schwab, E. A. Henriksen, J. M. Worlock, and M. L. Roukes, *Nature* 404, 974 (**2000**).
15. L. Kouwenhoven, *Nature* 403, 374 (**2000**).
16. S. Komiyama, O. Astafiev, V. Antonov, and H. Hirai, *Nature* 403, 405 (**2000**).
17. E. Paspalakis, Z. Kis, E. Voutsinas, and A. F. Terziz, *Phys. Rev. B* 69, 155316 (**2004**).
18. J. H. Jefferson, M. Fearn, D. L. J. Tipton, and T. P. Spiller, *Phys. Rev. A* 66, 042328 (**2002**).

19. J. Appenzeller, C. Schroer, T. Schapers, A. Hart, A. Froster, B. Lengeler, and H. Luth, *Phys. Rev. B* 53, 9959 (**1996**).
20. J. Appenzeller and C. Schroer, *J. Appl. Phys.* 87, 31659 (**2002**).
21. P. Debray, O. E. Raichev, M. Rahman, R. Akis, and W. C. Mitchel, *Appl. Phys. Lett.* 74, 768 (**1999**).
22. D. Kasemet, C. S. Hong, N. B. Patel, and P. D. Dapkus, *Appl. Phys. Lett.* 41, 912 (**1982**); K. Woodbridge, P. Blood, E. D. Pletcher, and P. J. Hulyer, *Appl. Phys. Lett.* 45, 16 (**1984**); S. Tarucha and H. O. Okamoto, *Appl. Phys. Lett.* 45, 16 (**1984**).
23. M. I Nathan, *Appl. Phys. Lett.* 47, 1105 (**1985**).
24. Suemune, L. A. Coldren, IEEE, *J. Quant. Electronic.* 24, 1178 (**1988**).
25. J. L. Shay and J. W. Wernick, *Ternary Chalcopyrite Semiconductors — Growth, Electronic Properties and Applications*, (Pergamon Press, London **1975**).
26. J. W. Rowe and J. L. Shay, *Phys. Rev. B* 3, 451 (**1973**).
27. H. Kildal, *Phys. Rev. B* 10, 5082 (**1974**).
28. J. Bodnar, in Proc. Int. Conf. of the Physics of Narrow-Gap Semiconductors (Polish Science Publishers, Warsaw, 1978); G. P. Chuiko, N. N. Chuiko, *Sov. Phys. Semicond.* 15, 739 (**1981**); K. P. Ghatak and S. N. Biswas, *Proc. of SPIE*, 1484, 149 (**1991**).
29. A. Rogalski, *J. Alloys Comp.* 371, 53, (**2004**).
30. A. Baumgartner, A. Chaggar, A. Patanè, L. Eaves, and M. Henini, *Appl. Phys. Lett.* 92, 091121 (**2008**).
31. J. Devenson, R. Teissier, O. Cathabard, and A. N. Baranov, *Proc. SPIE* 6909, 69090U (**2008**).
32. B. S. Passmore, J. Wu, M. O. Manasreh, and G. J. Salamo, *Appl. Phys. Lett.* 91, 233508 (**2007**).
33. M. Mikhailova, N. Stoyanov, I. Andreev, B. Zhurtanov, S. Kizhaev, E. Kunitsyna, K. Salikhov, and Y. Yakovlev, *Proc. SPIE* 6585, 658526 (**2007**).
34. W. Kruppa, J. B. Boos, B. R. Bennett, N. A. Papanicolaou, D. Park, and R. Bass, *Electron. Lett.* 42, 688 (**2006**).
35. B. R. Nag, *Electron Transport in Compound Semiconductors* (Springer Verlag, **1980**); E. O. Kane, in *Semiconductors and Semimetals*, Vol. 1, Ed. By R. K. Willardson, A. C. Beer (Academic Press, New York **1966**) p. 75.
36. G. E. Stillman, C. M. Wolfe, and J. O. Dimmock, in *Semiconductors and Semimetals*, Ed. R. K. Willardon and A. C. Beer 12, p. 169 (**1977**) Academic Press, IV, V, USA.
37. D. J. Newson and A. Karobe, *Semicond. Sci. Tech.* 3, 786 (**1988**).
38. E. D. Palik, G. S. Picus, S. Teither, and R. E. Wallis, *Phys. Rev.* 122, 475 (**1961**).
39. P. Y. Lu, C. H. Wung, C. M. Williams, S. N. G. Chu, and C. M. Stiles, *Appl. Phys. Lett.* 49, 1372 (**1986**).
40. N. R. Taskar, I. B. Bhat, K. K. Prat, D. Terry, H. Ehasani, and S. K. Ghandhi, *J. Vac. Sci. Tech.* 7A (**1989**) 281.
41. S. Bhattacharya and K. P. Ghatak, *Springer Series in Solid State Sciences* (Vol. 170, Springer-Verlag, Germany, **2012**).

42. L. R. Tomasetta, H. D. Law, R. C. Eden, I. Reyhimy, and K. Nakano, *IEEE, J. Quant. Electron.* 14, 800 (**1978**).
43. T. Yamato, K. Sakai, S. Akiba, and Y. Suematsu, *IEEE J. Quant. Electron.* 14, 95 (**1978**).
44. T. P. Pearsall, B. I. Miller, and R. J. Capik, *Appl. Phys. Lett.* 28 (**1976**) 499.
45. M. A. Washington, R. E. Nahory, M. A. Pollack, and E. D. Beeke, *Appl. Phys. Lett.* 33, 854 (**1978**).
46. M. I. Timmons, S. M. Bedair, R. J. Markunas, and J. A. Hutchby, *Proceedings of the 16th IEEE Photovoltaic Specialist Conference* (IEEE, San Diego, California 666 (**1982**)).
47. J. A. Zapien, Y. K. Liu, Y. Y. Shan, H. Tang, C. S. Lee, and S. T. Lee, *Appl. Phys. Lett.* 90, 213114 (**2007**).
48. M. Park, *Proc. SPIE* 2524, 142 (**1995**).
49. S.-G. Hur, E. T. -Kim, J. H. Lee, G. H. Kim, and S. G. Yoon, *Electrochem. Solid-State Lett.* 11, H176 (**2008**); H. Kroemer, *Rev. Mod. Phys.* 73, 783, (**2001**); T. Nguyen Duy, J. Meslage, and G. Pichard, *J. Crys. Growth* 72, 490 (**1985**); T. Aramoto, F. Adurodija, Y. Nishiyama, T. Arita, A. Hanafusa, K. Omura, and A. Morita, *Solar Energy Mat. & Solar Cells* 75, 211, (**2003**); H. B. Barber, *J. Elect. Mat.* 25, 1232 (**1996**); S. Taniguchi, T. Hino, S. Itoh, K. Nakano, N. Nakayama, A. Ishibashi, and M. Ikeda, *Elect. Lett.* 32, 552 (**1996**).
50. J. J. Hopfield, *J. Appl. Phys.* 32, 2277 (**1961**).
51. G. P. Agrawal and N. K. Dutta, *Semiconductor Lasers*, Van Nostrand Reinhold, New York, (**1993**), p. 547; S. Chatterjee, U. Pal, *Opt. Eng., Bellingham*, 32, 2923 (**1993**); T. K. Chaudhuri, *Int. J. Energy Res.* 16, 481 (**1992**); J. H. Dughaish, *Physica B* 322, 205 (**2002**); C. Wood, *Rep. Prog. Phys.* 51, 459 (**1988**).
52. K.-F. Hsu, S. Loo, F. Guo, W. Chen, J. S. Dyck, C. Uher, T. Hogan, E. K. Polychroniadis, and M. G. Kanatzidis, *Science* 303, 818 (**2004**); J. Androulakis, K. F. Hsu, R. Pcionek, H. Kong, C. Uher, J. J. D'Angelo, A. Downey, T. Hogan, and M. G. Kanatzidis, *Adv. Mater. Weinheim Ger.* 18, 1170 (**2006**); P. F. P. Poudeu, J. D'Angelo, A. D. Downey, J. L. Short, T. P. Hogan, M. G. Kanatzidis, *Angew. Chem. Int. Ed.* 45, 3835 (**2006**); P. F. Poudeu, J. D'Angelo, H. Kong, A. Downey, J. L. Short, R. Pcionek, T. P. Hogan, C. Uher, and M. G. Kanatzidis, *J. Am. Chem. Soc.* 128, 14347 (**2006**); J. R. Sootsman, R. J. Pcionek, H. Kong, C. Uher, and M. G. Kanatzidis, *Chem. Mater.* 18, 4993 (**2006**).
53. A. J. Mountvala and G. Abowitz, *J. Am. Ceram. Soc.* 48, 651 (**1965**); E. I. Rogacheva, I. M. Krivulkin, O. N. Nashchekina, A. Yu. Sipatov, V. A. Volobuev, and M. S. Dresselhaus, *Appl. Phys. Lett.* 78, 3238 (**2001**); H. S. Lee, B. Cheong, T. S. Lee, K. S. Lee, W. M. Kim, J. W. Lee, S. H. Cho, and J. Y. Huh, *Appl. Phys. Lett.* 85, 2782 (2004); K. Kishimoto, M. Tsukamoto, and T. Koyanagi, *J. Appl. Phys.* 92, 5331 (**2002**).
54. E. I. Rogacheva, O. N. Nashchekina, S. N. Grigorov, M. A. Us, M. S. Dresselhaus, and S. B. Cronin, *Nanotechnology* 14, 53 (**2003**).

55. E. I. Rogacheva, O. N. Nashchekina, A. V. Meriuts, S. G. Lyubchenko, M. S. Dresselhaus, and G. Dresselhaus, *Appl. Phys. Lett.* 86, 063103 (**2005**); E. I. Rogacheva, S. N. Grigorov, O. N. Nashchekina, T. V. Tavrina, S. G. Lyubchenko, A. Yu. Sipatov, V. V. Volobuev, A. G. Fedorov, and M. S. Dresselhaus, *Thin Solid Films* 493, 41 (**2005**).
56. X. Qiu, Y. Lou, A. C. S. Samia, A. Devadoss, J. D. Burgess, S. Dayal, and C. Burda, *Angew. Chem. Int. Ed.* 44, 5855 (**2005**).
57. C. Wang, G. Zhang, S. Fan, and Y. Li, *J. Phys. Chem. Solids* 62, 1957 (**2001**); B. Poudel, W. Z. Wang, D. Z. Wang, J. Y. Huang, and Z. F. Ren, *J. Nanosci. Nanotechnol.* 6, 1050 (**2006**); B. Zhang, J. He, and T. M. Tritt, *Appl. Phys. Lett.* 88, 043119 (**2006**); W. Heiss, H. Groiss, E. Kaufmann, G. Hesser, M. Böberl, G. Springholz, F. Schäffler, K. Koike, H. Harada, and M. Yano, *Appl. Phys. Lett.* 88, 192109 (**2006**).
58. B. A. Akimov, V. A. Bogoyavlenskiy, L. I. Ryabova, and V. N. Vasil'kov, *Phys. Rev. B* 61, 16045 (**2000**); Ya. A. Ugai, A. M. Samoilov, M. K. Sharov, O. B. Yatsenko, and B. A. Akimov, *Inorg. Mater.* 38, 12 (**2002**); Ya. A. Ugai, A. M. Samoilov, S. A. Buchnev, Yu. V. Synorov, and M. K. Sharov, *Inorg. Mater.* 38, 450 (**2002**); A. M. Samoilov, S. A. Buchnev, Yu. V. Synorov, B. L. Agapov, and A. M. Khoviv, *Inorg. Mater.* 39, 1132 (**2003**); A. M. Samoilov, S. A. Buchnev, E. A. Dolgopolova, Yu. V. Synorov, and A. M. Khoviv, *Inorg. Mater.* 40, 349 (2004); H. Murakami, W. Hattori, and R. Aoki, *Physica C* 269, 83 (**1996**); H. Murakami, W. Hattori, Y. Mizomata, and R. Aoki, *Physica C* 273, 41 (**1996**); H. Murakami, R. Aoki, and K. Sakai, *Thin Solid Films*, 27, 343 (**1999**).
59. B. A. Volkov, L. I. Ryabova, and D. R. Khokhlov, *Phys. Usp.* 45, 819 (**2002**), and references therein.
60. F. Hüe, M. Hÿtch, H. Bender, F. Houdellier, and A. Claverie, *Phys. Rev. Lett.* 100, 156602 (**2008**); S. Banerjee, K. A. Shore, C. J. Mitchell, J. L. Sly, and M. Missous, *IEE Proc., Circuits Devices Syst.* 152, 497 (**2005**); M. Razeghi, A. Evans, S. Slivken, J. S. Yu, J. G. Zheng, and V. P. Dravid, *Proc. SPIE* 5840, 54 (**2005**); R. A. Stradling, *Semicond. Sci. Technol.*, 6, *C52* (**1991**).
61. P. K. Weimer, *Proc. IEEE* 52, 608 (**1964**).
62. G. Ribakovs and A. A. Gundjian, *IEEE, J. Quant. Electron.* QE-14, 42, (**1978**).
63. S. K. Dey, *J. Vac. Sc. Tech.* 10, 227 (**1973**); S. J. Lynch, *Thin Solid Films* 102, 47, (**1983**).
64. V. V. Kudzin, V. S. Kulakov, D. R. Pape', S. V. Kulakov, and V. V. Molotok, *IEEE, Ultrasonics Symposium* 1, 749, (**1997**).
65. F. Hatami, V. Lordi, J. S. Harris, H. Kostial, and W. T. Masselink, *J. Appl. Phys.* 97, 096106 (**2005**).
66. B. W. Wessels, *J. Electrochem. Soc.* 122, 402 (**1975**); D. W. L. Tolfree, *J. Sci. Instrum.* 41, 788, 1964; P. B. Hart, *Proc. the IEEE* 61, 880, (**1973**).
67. M. A. Hines and G. D. Scholes, *Adv. Mater.* 15, 1844 (**2003**); C. A. Wang, R. K. Huang, D. A. Shiau, M. K. Connors, P. G. Murphy, P. W. O'Brien, A. C. Anderson, D. M. DePoy, G. Nichols, and M. N. Palmisiano, *Appl. Phys. Lett.* 83, 1286 (**2003**); C. W. Hitchcock, R. J. Gutmann, J. M. Borrego,

I. B. Bhat, and G. W. Charache, *IEEE Trans. Electron. Devices* 46, 2154 (**1999**).
68. H. J. Goldsmid and R. W. Douglas, *Brit. J. Appl. Phys.* 5, 386 (**1954**).
69. F. D. Rosi, B. Abeles, and R. V. Jensen, *J. Phys. Chem. Sol.* 10, 191 (**1959**); T. M. Tritt (ed.), *Semiconductors and Semimetals, Vols. 69, 70 and 71: Recent Trends in Thermoelectric Materials Research I, II and III* (Academic Press, USA, **2000**); D. M. Rowe (Ed.), *CRC Handbook of Thermoelectrics* (CRC Press, USA, **1995**); D. M. Rowe, C. M. Bhandari, *Modern Thermoelectrics* (Reston Publishing Company, Virginia, **1983**); D. M. Rowe (ed.), *Thermoelectrics Handbook: Macro to Nano* (CRC Press, USA, **2006**).
70. H. Choi, M. Chang, M. Jo, S. J. Jung, and H. Hwang, *Electrochem. Solid-State Lett.* 11, H154 (**2008**); S. Cova, M. Ghioni, A. Lacaita, C. Samori, and F. Zappa, *Appl. Opt.* 35, 1956, (**1996**); H. W. H. Lee, B. R. Taylor, and S. M. Kauzlarich, *Nonlinear Optics: Materials, Fundamentals, and Applications* (Technical Digest, 12, **2000**); E. Brundermann, U. Heugen, A. Bergner, R. Schiwon, G. W. Schwaab, S. Ebbinghaus, D. R. Chamberlin, E. E. Haller, and M. Havenith, *29th International Conference on Infrared and Millimeter Waves and 12th International Conference on Terahertz Electronics*, 283, (**2004**).
71. A. N. Baranov, T. I. Voronina, N. S. Zimogorova, L. M. Kauskaya, and Y. P. Yakoviev, *Sov. Phys. Semicond.* 19, 1676 (**1985**); M. Yano, Y. Suzuki, T. Ishii, Y. Matsushima, and M. Kimata, *Jpn. J. Appl. Phys.* 17, 2091 (**1978**); F. S. Yuang, Y.K. Su, and N.Y. Li, *Jpn. J. Appl. Phys.* 30, 207 (**1991**); F. S. Yuang, Y. K. Su, N. Y. Li, and K. J. Gan, *J. Appl. Phys.* 68, 6383 (**1990**); Y. K. Su and S. M. Chen, *J. Appl. Phys.* 73, 8349 (**1993**); S. K. Haywood, A. B. Henriques, N. J. Mason, R. J. Nicholas, and P. J. Walker, *Semicond. Sci. Technol.* 3, 315 (**1988**).
72. C. Young, W. W. Anderson, and L. B. Anderson, *Trans. Electron Dev. IEEE* 24, 492 (**1977**).
73. R. L. Gordon, V. I. Neeley, and H. R. Curtin, *Proc. IEEE* 54, 2014 (**1966**).
74. P. K. Weimer, *Proc. IRE* 50, 1462 (**1962**).
75. M. J. Lee, S. W. Wright, C. P. Judge, and P. Y. Cheung, Display Research Conference, International Conference Record, 211 (**1991**).

Index

activation energy, 320, 360
autoreduction, 313, 396

ballistic transport theory, 245
basic equations, 69, 70, 75
Born's probability interpretation, 68, 74, 75
Born, Max, 68, 74, 75
boundary scattering, 179–184, 186, 188
broken symmetry problem, 312
broken symmetry structure, 312, 320–322, 326, 370, 371, 378, 383, 384
broken translational symmetry, 321, 376, 383
Bunge, Mario, 82, 91

Carbon allotropes, 181, 187, 188, 204, 208
carbon nanotubes, 179, 187, 189, 209, 213, 221, 317
carrier confinement, 413
carrier transport, 413
causal green's functions, 346
ceramics, 140, 141, 156, 159, 160
chemical sensors, 181
chlorophyll molecules, 2
classical approximation, 83, 99, 105, 106, 133
classical size effect, 180, 181
classical time, 82
clock time, 39–42, 44, 46, 48, 49, 51, 54–59, 62–66, 68, 72, 73, 76–78, 81, 87–91
coherent conductor, 251–253
collective crystalline excitation, 353
complete information, 52, 54, 56, 57, 59, 62, 63
container model, 50–53, 83, 92
continuum mechanics, 177
correlational-interaction, 80
correlations, 60, 74, 75, 80, 96, 100, 216
coupling, 31, 32, 43–45, 49, 55, 101, 107–112, 217, 266, 278, 279, 282, 286, 294–296, 415
creep, 140, 141
crystalline diamond stones, 142
crystalline structure, 29, 31, 188, 312, 321, 361, 383

Debye screening length, 438
density functional perturbation theory, 20
density of states, 28, 215, 283
deposition process, 144
description element, 40, 55
DFT, 8, 28, 29
diamond coatings, 162
diamond formation process, 142
diamond growth, 142, 143, 146, 149, 150, 163, 167
diamond microtechnology, 161

Dirac spin, 96, 101, 102, 112–115, 117, 121, 124, 129
dispersion law, 370, 371, 392, 414
DNA, 292, 294–302, 304–306
double-barrier quantum-well, 359

effective-phonon approximation, 316
elastic continuum models, 177
electrical conductance, 243–247, 249, 287, 288, 295, 306
electron transport, 182, 184, 225
elementary excitations, 315, 316, 320, 331, 336, 338, 342, 344, 345, 347, 353, 371
energy conversion, 9, 178, 181, 203, 204, 206, 241, 247, 284, 295, 305, 306
energy conversion devices, 181
energy-representation, 83, 84
engineering materials, 140, 187
exciton, 9, 15, 19, 24, 26, 27, 29, 31, 32, 96, 97, 108, 319, 352–354, 356, 357, 359, 360, 373–375, 377, 384
exciton model, 374
extensions, 42, 91
external time, 39, 40, 43–50, 55, 57, 58, 64, 70, 85, 91

Fermi level pinning, 415
ferromagnet, 351
field-related effects, 315
fluctuating quantum entity, 88
fluctuations, 67, 73–75, 78–80, 87, 88, 192
Fraser, Simon, 46
Frenkel exciton system, 352
Fuchs-Sondheimer solution, 184

general nonlinear problem, 325
grain boundaries, 140, 146, 148, 153, 154, 167, 180, 187
grain boundary engineering, 141
grain boundary sliding, 140, 141
graphene layers, 209, 213, 214
graphene sheets, 189
graphite, 184, 185, 222

Green's function, v, 218, 250, 256–269, 271–274, 281, 284, 311, 312, 328, 330–332, 338, 340–342, 344, 346, 357–359, 361, 362, 367, 369–371, 374, 376, 378, 379, 383, 384, 393, 395

Hamiltonian, 40, 82, 97, 100, 101, 107, 109, 111, 257, 260, 263, 266, 267, 269, 272, 280–282, 285–287, 292–294, 330, 332, 334, 342, 343, 348, 354, 355, 362, 374, 378, 379, 391, 392
Hamiltonians of low-dimensional quantum systems, 266
heat transport, v, 150, 177, 181, 182
Heisenberg's equations of motion, 341, 348
high-frequency microwave circuits, 414
holography, 15, 17, 18
HOMO, 11, 12, 19, 22, 23, 29
Hopfield model, 415
hot-filament CVD, 142, 144

impurity-scattering, 179, 182
incoherent conductor, 254
inelastic scattering effects, 245
infrared photodetectors, 316
inner picture, 52
interface-controlled nanostructures, 140
intermetallics, 140
internal and surface chemistry, 139
internal time, 48, 49, 55
isotropic momentum matrix, 414

Jung, C.G., 51, 52

kinematic excitation, 352, 354, 356, 357

lead chalcogenides, 415
low-dimensional structures, 203, 208, 320, 359
LUMO, 11, 12, 19, 22, 23

macrocycle molecule, 3
macroscopic objects, 44
magnesium, 1, 2, 6
magnesium tetraphenylporphyrin, 1
magnetic devices, 181
magnetization at low temperatures, 347
magnetoconductivity measurements, 96, 110
materials informatics, 215, 216, 225
MD simulations, 8
membranes, 7, 198–200, 202, 203
Metal magnesium, 2
metalloporphyrins, 2, 11, 13, 24
metaphysical, 45–47, 64
microelectronic devices, 183, 208, 223
micromechanical systems, 139
model of Wang and Ressler, 422
molecular junctions, 245, 246, 284, 285, 291, 295, 300, 306
molecular sub-rings, 285
molecule-to-electrode junction, 289, 290
morphology control, 9

naphthalene molecule, 284–288, 306
Newton's mechanics, 45, 46, 50
Newton, Isaac, 44–46, 50, 66
non-equilibrium condition, 247, 249
nonequilibrium quantum transport equations, 97, 99, 111
nonlinear optics, 8, 414

OLED device, 13, 19
one-dimensional nanostructure, 179, 181, 203
one-to-one-correspondence, 43, 52
operator complexes, 331–333, 340–342, 345
operator geometry progressions, 323
optoelectronic materials, 414

Peltier coefficient, 242
phenomenological characteristics, 329, 331
phonon group velocity, 178, 189
phonon transport, 180, 182, 190, 217, 223
phonon-phonon scattering, 179, 182
phononic crystals, 223
phonons in crystalline chain, 327
phonons in ultrathin films, 359, 370
photolithography, 161, 169, 192
physical laws, vii, viii, 43–46, 48, 49
polarons, 316
polycrystalline diamond, 141, 142, 144, 146, 148, 156, 158
porphyrin nanostructures, 8, 9
porphyrins, 1–3, 5–11, 18, 19, 22, 24, 28, 29, 32
protein molecule, 295–299, 301, 302, 304–306
pseudo-classical particles, 178
pulsating heat pipe model, 209, 210
pyrrolic hydrogen, 11, 13

quantum capacitance, v, 413, 417–423
quantum confinement, 180, 245, 436, 439
quantum dots, 245, 315, 316, 320, 360, 373, 378, 381, 382, 384
quantum information processing, 26
quantum logic gates, 414
quantum sensors, 414
quantum size effect, 180, 181
quantum spin Hall effect, 110
quantum switches, 414
quantum time, 39, 85, 89
quantum transistors, 414
quantum transport, v, 95–97, 99–102, 104, 107, 109, 111, 112, 116
quantum treatment of time, 39
quantum wells, 99, 360, 415, 416
quantum wires, 245, 315, 359, 360, 378, 413

Raman scattering, 316
random symmetry, 312
raw information, 56, 62
Rayleigh scattering, 179, 186
reference time, 40, 48, 49, 54, 58, 91
resonance tunneling devices, 315

San Francisco cable car, 45–48
Schrödinger, Erwin, 40
Seebeck effect, 241, 242, 248
self-assembly, 6–10, 26, 32, 359
self-construction, 359
self-organizing processes, viii, 39, 85–90
semiconducting nanoparticles, 245
separated objects, 78
single-electron devices, 315
single-photon detection, 316
singularity, 43, 45, 48, 49, 336, 371
space impression, 40, 41, 52, 53
space-structure, 51
space-time structure, 42, 43, 59
spintronic devices, 102
stationary states, 85
structure designing, 215, 217
sub-Hamiltonians, 286
superfluid properties, 316
superlattices, 359, 415
surface chemistry, 139
surface roughness, 148, 149, 157–159, 165, 166
system-specific time, 47–50, 53–55, 58, 63, 64, 76, 77, 81, 83, 85, 87, 88, 90, 91

tetraphenylporphyrin, 1, 8, 13, 32
Theory of Relativity, 44, 45
thermal bridge method, 198
thermal conductance, 180, 200–202, 219, 243, 245–247, 249, 250, 288, 291, 295, 297, 298, 300, 306
thermal interface material, 210–214
thermal microactuators, 141
thermal transport, 177–184, 190, 199, 203, 214–216, 220, 225
thermal transport properties, 180, 203, 214, 215, 225
thermoelectric materials, 187, 205–208, 216, 220, 243–245, 415
thermoelectric phenomena, v, 241, 246, 284, 292, 306
thermoelectric power generator, 243
thermoelectric refrigerator, 244
thermoelectricity, 244, 245
thermoelectrics, 204, 207, 208, 223, 415
thermopower measurements, 246
thermoreflectance measurements, 196
thin film thermal conductivity, 184
thin-film transistors, 416
three omega method, 185, 192
three-band model, 418, 424–439
time conception, 46
time experience, 40
time feeling, 40, 42–44, 91
time operator, 40, 84
time representation, 83, 84
time-machine, 43–51, 53–55, 58
topological insulators, v, 95, 97, 106, 108, 111
traditional view, 43, 50, 58, 64, 67, 78, 85, 88, 89, 92
transmission probability, 245, 248–251, 253, 255, 256, 279, 280
transport mechanism, 244
two-band model, 414, 418, 424–439
Two-dimensional nanostructure, 179, 183
two-dimensional quantizations, 417

ultra-nanocrystalline diamond, 147, 148, 153, 157, 166, 167, 169
uncertainty relations, 81, 83, 91

valley spin, 96, 118, 121, 126, 129
varying structures, 42, 47, 49

Wannier-Mott excitons, 352
wear behavior, 157, 159
Wick's theorem, 329, 348, 350, 355

Young's modulus, 141, 153–156

Series on the Foundations of Natural Science and Technology

(Continued from page ii)

Vol. 5 *Grasping Reality: An Interpretation-Realistic Epistemology*
by H. Lenk

Vol. 4 *What is Life? Scientific Approaches and Philosophical Positions*
by H.-P. Dürr, F.-A. Popp and W. Schommers

Vol. 3 *The Visible and the Invisible: Matter and Mind in Physics*
by W. Schommers

Vol. 2 *Symbols, Pictures and Quantum Reality: On the Theoretical Foundations of the Physical Universe*
by W. Schommers

Vol. 1 *Space and Time, Matter and Mind: The Relationship between Reality and Space-Time*
by W. Schommers

Forthcoming:

Vol. 10 *Space-Time Reality: Mach's Principle, Relativity and Novel Aspects*
by W. Schommers

Vol. 11 *The Origin of Mind: Interplay between Matter and Mind*
by W. Schommers